Springer Series in

OPTICAL SCIENCES 82

founded by H.K.V. Lotsch

Springer
Berlin
Heidelberg
New York
Barcelona
Hong Kong
London
Milan
Paris
Tokyo

Springer Series in
OPTICAL SCIENCES

The Springer Series in Optical Sciences, under the leadership of Editor-in-Chief *William T. Rhodes*, Georgia Institute of Technology, USA, and Georgia Tech Lorraine, France, provides an expanding selection of research monographs in all major areas of optics: lasers and quantum optics, ultrafast phenomena, optical spectroscopy techniques, optoelectronics, information optics, applied laser technology, industrial applications, and other topics of contemporary interest.
With this broad coverage of topics, the series is of use to all research scientists and engineers who need up-to-date reference books.

The editors encourage prospective authors to correspond with them in advance of submitting a manuscript. Submission of manuscripts should be made to the Editor-in-Chief or one of the Editors. See also http://www.springer.de/phys/books/optical_science

Stefano Trillo William Torruellas (Eds.)

Spatial Solitons

With 194 Figures and 7 Tables

Springer

Professor Stefano Trillo
University of Ferrara
Dipartimento di Ingeneria
Via Saragat 1
44100 Ferrara
Italy
E-mail: strillo@ing.unife.it

Dr. William Torruellas
Corvis Corporation
7015 Albert Einstein Drive
P.O. Box 9400
Columbia, MD 21046-9400
USA
E-mail: wtorruellas@corvis.com

ISSN 0342-4111

ISBN 3-540-41653-6 Springer-Verlag Berlin Heidelberg New York

Library of Congress Cataloging-in-Publication Data

Spatial solitons / Stefano Trillo, William E. Torruellas (eds.).
p.cm. – (Springer series in optical siences ; v. 82)
Includes bibliographical references and index.
ISBN 3540416536 (alk. paper)
1. Solitons. I. Trillo, Stefano, 1957- II. Torruellas, William E., 1961- III . Series.
QC174.26.W28 S68 2001   530.12'4–dc21   2001034457

Springer-Verlag Berlin Heidelberg New York
a member of BertelsmannSpringer Science+Business Media GmbH

http://www.springer.de

Printed in Germany

Data conversion by Steingraeber Satztechnik GmbH, Heidelberg
Cover concept by eStudio Calamar Steinen using a background picture from The Optics Project. Courtesy of John T. Foley, Professor, Department of Physics and Astronomy, Mississippi State University, USA.
Cover production: *design & production* GmbH, Heidelberg

Printed on acid-free paper   SPIN 10726365   56/3141/di   5 4 3 2 1 0

Preface

Though not often perceived in our everyday life, nonlinearity permeates our physical world. The evidence for nonlinear behavior appears in so many aspects of physics, chemistry, biology, economics, etc., that it may be nearly impossible to mention them all. Among the most striking and aesthetically appealing manifestations of nonlinearity is the propagation of solitons or, more generally, solitary waves. Strictly speaking, solitons differ from solitary waves because of the remarkable property of integrability of the governing models and its physical consequences. Such a property is, however, scarcely relevant to the physics discussed in this book. Therefore we shall make use of the term "soliton" in its broader sense, to mean a wavepacket held together by the interplay of mutually counterbalancing linear and nonlinear effects.

Historically, the first documented observation of a soliton (in a shallow-water canal in Scotland) is attributed to J.S. Russell in 1834. Since then, solitons in physics have received increasing growing interest. In the twentieth century, the soliton became one of the most fruitful concepts of nonlinear physics and a key to understanding how nonlinearity acts in nature. The main reason of this success lies in the particle-like behavior of solitons and their intrinsic nature as "modes" of nonlinear systems. Solitons have become a widespread interdisciplinary field of theoretical and experimental studies, in branches ranging from fluid dynamics, oceanography, plasma physics, mechanical vibrations, electronics, solid-state physics, cosmology, superconductivity, and acoustics, to electrodynamics. In optics, the advent of laser sources has allowed easy experimental access to the excitation of a nonlinear response in dielectric materials. It took only a few years to develop, in the early 1960s (see below in the introductory chapter by Chiao), the idea of compensating the diffractive spreading of a beam (i.e. a bell-shaped spatial wavepacket) with the self-induced lensing effect. The latter effect is associated with a graded, intensity-dependent change of refractive index, known as the optical Kerr effect, which entails a cubic (in the field) material response. This balance gives rise to self-trapping into a *spatial soliton*, i.e. a shape-invariant *self-guided beam of light*, which can be viewed as a *self-induced waveguide*. Diffraction is overcome because the different plane-wave components do not suffer phase delays but, rather, are forced to travel in phase, with a common propagation constant determined by the nonlinear response. Apart from its genuine self-induced nonlinear origin, the underlying guiding mechanism behind spatial solitons is identical to that which allows virtually unlimited propagation in a graded optical fiber or waveguide, that is, a properly tailored gradient of the refractive index, leading to repeated total reflections. In retrospect, it is worth noting that the idea of self-trapping came even earlier than the proposal to employ silica fibers for transmitting information (K.C. Kao, G.A. Hockham, Proc. IEE **133**, 1151 (1966); A. Werts, L'Onde Électrique **46**, 967 (1966)), which was the very origin of the modern revolution in optical communications.

Nowadays, the world of optical solitons has grown enormously and includes temporal solitons, such as those observed in dispersive optical fibers, stimulated scattering, and self-induced transparency. In particular, dispersion-compensating fiber solitons, although proposed more recently (A. Hasegawa, F. Tappert, Appl. Phys. Lett. **23**, 142 (1973)), have overshadowed spatial solitons, owing to their enormous applicative importance as natural information carriers which can propagate thousands of kilometers in fiber links. In comparison, spatial solitons exploit the nonlinearity of bulk materials (or slab waveguides) and hence usually propagate only few centimeters. Moreover, diffraction is usually stronger than fiber dispersion, thereby requiring, in principle, a larger nonlinear response (higher intensities). In spite of this, the physics of spatial solitons is much richer, basically because, unlike the case for temporal solitons, trapping occurs in both one and two transverse dimensions and involves different types of nonlinear material response. Fascinating phenomena that are specifically linked to the higher number of dimensions, which have no counterpart for temporal solitons, can take place, for example self-focusing and transverse instabilities, spiraling and hybrido-dimensional soliton–soliton interaction, the existence of optical vortices, and the formation of complex patterns and localized structures in cavities. These aspects have stimulated a renewed interest in the field, as witnessed by several important achievements obtained during the last ten years. For instance, the common beliefs that spatial solitons need extremely high intensities, a self-induced refractive index-change (Kerr effect), and a coherent excitation were demolished by the observation of solitons in photorefractive materials, by second-harmonic generation, and by means of incoherent light, respectively. The recent outstanding results have motivated us to put together the material contained in this book, which is specifically dedicated to spatial solitons. Although several books on optical solitons are available, none of them give a proper description of spatial solitons. The ultimate aim of the book is twofold: (i) to give an updated overview of the field, which will serve as a reference and guide for PhD students and scientists working in the field; and (ii) to attract scientists from other areas of science, specifically in nonlinear dynamics, convincing them that optics offers unique opportunities to investigate soliton physics owing to the vast number of physical situations where solitons are observed, and the ease and reliability of experiments. This said, we shall be glad if new, unexpected results and directions make the book outdated. In particular, we hope to see spatial solitons find their definitive way into applications, such as reconfigurable waveguides, all-optical devices, beam-quality control, and many others.

The book includes invited contributions from the majority of world leading experts, often pioneers, in the field. Not all the groups who have contributed substantially to our current understanding of spatial solitons are adequately represented, for a number of reasons. We regret that it was not possible to find room for all the respected colleagues and friends who took part in the development of this exciting topical area.

The content of the book covers a wide spectrum of issues concerning the propagation of spatial solitons, ranging from the most challenging theoretical problems to the most exciting experimental achievements to date. The general topic of spatial solitons is introduced by one of its harbingers, Prof. R. Chiao, who reviews the field, sharing with us his unique historical perspective. Then, a first group of chapters deals with the most recent and important experimental results. The different materials and mechanisms commonly employed in the recent self-trapping experiments are introduced by Luther-Davies and Stegeman, who describe and compare their features. The discussion of experimental results which exploit these specific materials is deepened in the following four chapters. Silberberg and Stegeman review the activity on Kerr-like materials, with special emphasis on observation of solitons in semiconductor bulk and waveguide arrays. Then Del Re et al. focus on solitons in photorefractive materials. As far as photorefractives are concerned, their slow nonlinear response has permitted researchers to open up a fascinating new avenue, namely incoherent solitons, or self-trapping with incoherent light. This area is reviewed by Segev and Christodoulides. Multicolor trapping in frequency conversion processes due to quadratic $\left(\chi^{(2)}\right)$ nonlinearities is then described by Torruellas et al. In the following chapter, de Sterke et al. give an overview of trapping in Bragg gratings, which occurs in the form of gap solitons. These are the only objects described in this book that are not related to diffraction-compensating trapping in the transverse plane; rather, they are confined along the longitudinal direction of the grating. The reason to consider them as spatial solitons, though of a different kind, stems from the fact that they also exist in the strict stationary limit as zero-velocity envelopes.

A second group of chapters shares a more theoretical viewpoint and reveals the problems faced by the mathematical description of spatial solitons. The crucial problem of stability is tackled in the general chapter by Kivshar and Sukhorukov, while the aspects related to collapse (catastrophic self-focusing) are specifically addressed by Bergé. Lederer et al. acquaint us with the physics and application of discrete solitons in waveguide arrays, for which the coupling between adjacent waveguides introduces a diffraction-like spreading mechanism. The principles of optical vortex solitons, self-trapped beams with a screw dislocation in their phase, are discussed by Swartzlander.

In a third group of chapters, we have grouped together contributions dealing with spatial solitons in dissipative systems. Dissipation is first introduced by Akhmediev and Ankiewicz, who discuss the rich world of solitons of the Ginzburg–Landau equation and its application in nonlinear optics. Then, Firth and Harkness focus more specifically on cavity solitons, giving both a historical perspective and a tutorial introduction to their description. Trillo and Haelterman discuss the opportunity offered by the interplay of parametric nonlinearity and passive feedback to confine the light either in passive cavities (parametric oscillators) or in gratings. The chapter by Weiss and collaborators summarizes the main recent experimental achievements in this field.

The last two chapters show how the importance of self-trapping is widened by new areas of applications: Boardman and Xie deal with propagation in magneto-optical (anisotropic) materials, whereas Saffman and Skryabin introduce us to the basic physics of joint trapping of radiation and matter waves, which is likely to be more than a dream in view of the progress in atom cooling and Bose–Einstein condensation.

Let us conclude with a dedication. Since both of us have fully respected the tradition that requires physicists to meet very understanding companions and generate mainly daughters, let us dedicate this book to the six beautiful women in our lifes (three plus three), from whom we have subtracted the time needed to complete this book.

Rome,
Columbia,
January 2001

Stefano Trillo
William E. Torruellas

Contents

List of Contributors

Nail Akhmediev
Australian Photonics CRC
Optical Sciences Centre
Research School of Physical Sciences and Engineering
Institute of Advanced Studies
Australian National University
Canberra, ACT 0200, Australia
nna124@rsphysse.anu.edu.au

Adrian Ankiewicz
Australian Photonics CRC
Optical Sciences Centre
Research School of Physical Sciences and Engineering
Institute of Advanced Studies
Australian National University
Canberra, ACT 0200, Australia
ana124@rsphysse.anu.edu.au

Luc Bergé
Commisariat à l'Energie Atomique
CEA/Bruyères-le-Chatel, B.P. 12
91680 Bruyères-le-Chatel, France
luc.berge@cea.fr

Allan D. Boardman
Joule Physics Laboratory
School of Sciences
University of Salford
Salford, M5 4WT, UK
a.d.boardman@physics.salford.ac.uk

Raymond Y. Chiao
Department of Physics
University of California
Berkeley, CA 94720, USA
chiao@physics.berkeley.edu

Demetrios N. Christodoulides
Department of Electrical Engineering
and Computer Science
Lehigh University
Bethlehem, PA 18015, USA
dnc0@lehigh.edu

Bruno Crosignani
Dipartimento di Fisica
Università dell'Aquila
67010 L'Aquila, Italy
crosignani@aquila.infn.it

Sergey Darmanyan
Friedrich-Schiller-Universitaet Jena
Institute of Solid-State Theory
and Theoretical Optics
Max-Wien-Platz 1
07743 Jena, Germany
sergej@pinet.uni-jena.de

Eugenio DelRe
Fondazione Ugo Bordoni
Via B. Castiglione 59
00142 Rome, Italy
edelre@fub.it

C. Martijn de Sterke
School of Physics
University of Sydney
Sydney 2006, Australia;
also with the Australian Photonics CRC
desterke@physics.usyd.edu.au

Paolo Di Porto
Dipartimento di Fisica
Università dell'Aquila
67010 L'Aquila, Italy
diporto@aquila.infn.it

Benjamin J. Eggleton
Optical Fiber Research Dept.
Bell Labs., Lucent Technologies
700 Mountain Ave
Murray Hill, NJ 07974, USA
egg@lucent.com

William J. Firth
Department of Physics
and Applied Physics
University of Strathclyde
Glasgow G4 0NG, Scotland
willie@phys.strath.ac.uk

Marc Haelterman
Service d'Optique et Acoustique
Universite Libre de Bruxelles
50, av. Roosevelt, CP 194/5
1050 Brussels, Belgium
mhaelter@pop.ulb.ac.be

Graeme K. Harkness
Department of Physics
and Applied Physics
University of Strathclyde
Glasgow G4 0NG, Scotland
G.Harkness@phys.strath.ac.uk

Yuri S. Kivshar
Australian Photonics CRC
Nonlinear Physics Group, RSPhysSE
Australian National University
Canberra ACT 0200, Australia
ysk124@rsphy1.anu.edu.au

Andrey Kobyakov
Friedrich-Schiller-Universitaet Jena
Institute of Solid-State Theory
and Theoretical Optics
Max-Wien-Platz 1
07743 Jena, Germany
pfl@uni-jena.de

Robert Kuszelewicz
Centre National d'Etudes
de Telecommunications (CNET)
92220 Bagneux, France
robert.kuszelewicz
@cnet.francetelecom.fr

Falk Lederer
Friedrich-Schiller-Universitaet Jena
Institute of Solid-State Theory
and Theoretical Optics
Max-Wien-Platz 1
07743 Jena, Germany
2pfl@uni-jena.de

Barry Luther-Davies
Laser Physics Centre, RSPhysSE
Australian National University
P.O. Box 4
Canberra ACT 2601, Australia
bld111@anu.edu.au

Mark Saffman
Department of Physics
University of Wisconsin
1150 University Avenue
Madison, WI 53706, USA
msaffman@facstaff.wisc.edu

Mordechai Segev
Department of Physics
and Solid State Institute
Technion, Institute of Technology
32000 Haifa, Israel
and
Electrical Engineering Department
and Center for Photonics and
Opto-Electronic Materials (POEM)
Princeton University
Princeton, NJ 08544, USA
msegev@techunix.technion.ac.il

Yaron Silberberg
Department of Physics
of Complex Systems
The Weizmann Institute of Science
Rehovot 76100, Israel
feyaron@wis.weizmann.ac.il

John E. Sipe
Department of Physics
University of Toronto
Toronto, Canada M5S 1A7
sipe@physics.utoronto.ca

Dmitry Skryabin
Department of Physics
and Applied Physics
University of Strathclyde
Glasgow G4 0NG, Scotland
dmitry@phys.strath.ac.uk

Gintas Slekys
Physikalisch-Technische
Bundesanstalt
38116 Braunschweig, Germany
gintas.slekys
@cnet.francetelecom.fr

Kestutis Staliunas
Physikalisch-Technische
Bundesanstalt
38116 Braunschweig, Germany
kestutis.staliunas@ptb.de

George I. Stegeman
School of Optics and CREOL
University of Central Florida
4000 Central Florida Blvd.
P.O. Box 162700
Orlando, FL 32816-2700, USA
george@mail.creol.ucf.edu

Andrey A. Sukhorukov
Australian Photonics CRC
Nonlinear Physics Group, RSPhysSE
Australian National University
Canberra ACT 0200, Australia
ans124@rsphysse.anu.edu.au

Grover A. Swartzlander, Jr.
University of Arizona
Optical Sciences Center
1630 E. University Blvd
Tucson, AZ 85721, USA
gswartzlander
@optics.arizona.edu

Victor B. Taranenko
Physikalisch-Technische
Bundesanstalt
38116 Braunschweig, Germany
victor.taranenko@ptb.de

William Torruellas
Corvis Corporation
7015 Albert Einstein Drive
P.O. Box 9400
Columbia, MD 21046-9400, USA
wtorruellas@corvis.com

Stefano Trillo
Dipartimento di Ingegneria
Università di Ferrara
Via Saragat 1, 44100 Ferrara, Italy
strillo@ing.unife.it

Carl O. Weiss
Physikalisch-Technische
Bundesanstalt
38116 Braunschweig, Germany
carl.weiss@ptb.de

Ming Xie
Joule Physics Laboratory
School of Sciences
University of Salford
Salford, M5 4WT, UK
m.xie@pgr.salford.ac.uk

Introduction to Spatial Solitons

Raymond Y. Chiao

Summary. A historical review of spatial solitons in nonlinear optics will be given, followed by a discussion of the outlook concerning possible future developments.

1 Introduction

A collimated beam of light with a diameter d will usually spread with an angle of λ/d owing to diffraction. It would seem that this kind of spreading is inevitable and therefore irreducible, since it originates at a fundamental level from the uncertainty principle of quantum mechanics. However, in 1964, Chiao, Garmire, and Townes [1,2][1] showed that the spreading of an optical beam could in principle be avoided in a *nonlinear* optical medium, and therefore that the expansion of a beam of light due to diffraction was neither inevitable nor irreducible *in such a medium.* We considered the case of a medium which possesses an intensity-dependent index of refraction, in which the index *increases* with the intensity of the light. Then the beam could form a dielectric waveguide for itself, in which the refractive index would be greater at the center of the beam than in its wings. Since the light in this self-formed dielectric waveguide could then propagate without spreading, the phenomenon of "self-trapping" of an optical beam was thereby predicted to occur. One could also understand this phenomenon as a dynamic balancing of two opposing tendencies, namely, the tendency for the beam to expand due to diffraction, and the tendency for the beam to contract due to self-focusing. [3] This kind of dynamic balancing is stable in one transverse spatial dimension, as was shown in the work of Zakharov and Shabat [4]. The inverse scattering method indicated that the hyberbolic secant solution found in [1,2] for one transverse spatial dimension was in fact a spatial *bright soliton.* The 1D classical field theory corresponding to the self-trapping problem is fully integrable , since it possesses an infinite number of conserved quantities. In particular, when two solitons were directed against each other in a scattering configuration, it was predicted that the solitons would survive the resulting collision intact.

The number of transverse dimensions D in this nonlinear optical problem plays an important role in determining the stability or instability of the so-

[1] See also the contemporaneous Soviet work [1a] and the earlier work [1b].

lutions. For one transverse dimension, it follows from the inverse scattering method [4] that the self-trapping solution is stable against all perturbations, linear or nonlinear, and therefore forms a genuine spatial soliton. However, for a cylindrically symmetric beam with two transverse dimensions, a self-focusing instability occurs for all positive Kerr nonlinearities, leading in general to unstable solutions. In the early history of this subject, it was proposed that the small filaments of light [5][2] which were observed to form inside self-focusing Kerr media were merely the loci of a moving self-focus singularity: as the self-focus moved with time in accordance with the time dependence of the laser pulse intensity, it generated the observed filaments. Therefore it was claimed that no true stationary-state self-trapping of optical beams with a cylindrical symmetry actually existed [6]. However, this claim could not be completely correct, since an optical filament containing a *stationary* self-trapped optical beam with a cylindrical symmetry was in fact observed in a *continuous-wave* experiment in a sodium vapor cell [7]. Such cylindrical solitary-wave solutions stabilize themselves on account of the saturation of the Kerr nonlinearity in two-level atoms. Therefore, not only the lowest-order Kerr effect, which depends linearly on the intensity of the light, but also the higher-order Kerr effects which depend on the *square* of the light intensity, etc., are important for stabilizing these cylindrical solitary-wave solutions. Solitary-wave solutions have also been predicted to occur when there are parametric-wave or second-harmonic interactions in nonlinear optics [8], and have more recently been observed to form in phase-matched quadratic media [9].

The two-dimensional instability is closely connected with Derrick's theorem, [10] which states that the only nonsingular, time-independent solutions of finite energy for $D \geq 2$ are the ground states, for any classical field theory with a Lagrangian density of the form

$$\mathcal{L} = \frac{1}{2}\partial_\mu \partial^\mu \phi - U(\phi) \;, \tag{1}$$

where the spacetime index μ is summed up to D spatial dimensions, ϕ represents any set of scalar fields, and $U(\phi)$ is any nonnegative potential-energy functional which is equal to zero for the ground states of the theory. In particular, for $D = 2$, no nonsingular, nonzero soliton solution with finite energy can exist.

2 Solitons in one Transverse Dimension

In contrast to the situation for two transverse spatial dimensions, spatial solitons can indeed form in one transverse dimension in a medium with an optical Kerr nonlinearity

$$n(E) = n_0 + n_2\overline{E^2} + \ldots , \tag{2}$$

[2] For a more recent discussion of the formation of multiple light filemants, see [5a].

where n_0 is the linear refraction index, n_2 is the optical Kerr coefficient, E is the instantaneous electric field of the optical beam, and the overbar denotes the time-averaging appropriate for a typical optical nonlinearity, which cannot respond at optical frequencies. Empirically, one finds that for most optical media, the Kerr coefficient is a positive quantity ($n_2 > 0$). The Taylor series expansion of the index of refraction in powers of the electric field given in (2) is the most general possible for a centrosymmetric medium, in which odd powers of the electric field are forbidden to appear by inversion symmetry.

A "dielectric-waveguide" model for the spatial soliton starts with a slab-shaped geometry (see Fig. 1), in which we make the approximation that the the light is propagating inside a slab waveguide whose index of refraction is *rectangular* in profile, i.e. it is constant inside the waveguide and abruptly drops to a smaller constant value outside. The change in the refractive index is proportional to the intensity of the light itself, and is given by

$$\Delta n = n_2 \overline{E^2} > 0 \,, \tag{3}$$

where E is the typical value of the optical electric field inside the waveguide. The critical angle for total internal reflection of light rays inside the waveguide is given by

$$\cos\theta_{\mathrm{crit}} = \frac{n_0}{n_0 + \Delta n} = \frac{n_0}{n_0 + n_2\overline{E^2}} \,. \tag{4}$$

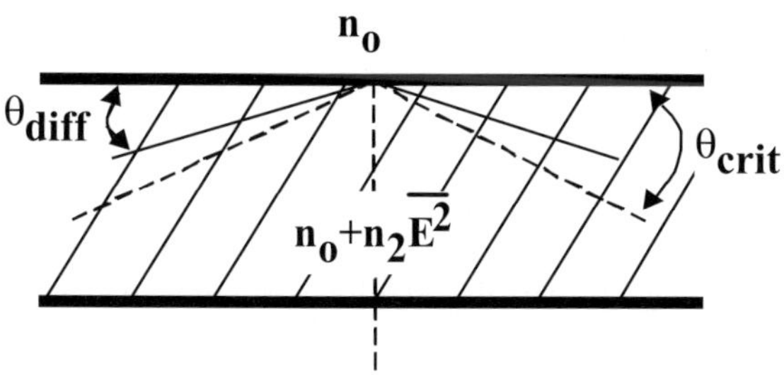

Fig. 1. A slab-shaped di- electric-waveguide model for the spatial soliton in 1D

We expect under most circumstances that both the critical angle θ_{crit} and the index change Δn are small quantities. Taking the Taylor series expansions of both sides of (4), we obtain

$$1 - \frac{1}{2}{\theta_{\mathrm{crit}}}^2 + \ldots = 1 - \frac{n_2\overline{E^2}}{n_0} + \ldots \,, \tag{5}$$

and therefore that for small angles, the critical angle is given by the relation

$$\theta_{\text{crit}} = \left(\frac{2n_2\overline{E^2}}{n_0} \right)^{1/2} . \tag{6}$$

Now for self-trapping of optical beams, and hence spatial solitons, to occur, we require that the diffraction angle must be less than the critical angle, so that the diffracting rays from the interior of the waveguide, and hence from the interior of the self-trapped beam of light, will be totally internally reflected back into the waveguide. Thus the beam could self-consistently propagate without an expansion due to diffraction. Hence we require that

$$\theta_{\text{crit}} = \left(\frac{2n_2\overline{E^2}}{n_0} \right)^{1/2} \geq \theta_{\text{diff}} = \frac{\lambda}{d} . \tag{7}$$

This implies a relationship between the typical size of the soliton d and the typical optical field strength $\sqrt{\overline{E^2}}$, given by

$$d \approx \left(\frac{n_0}{2n_2} \right)^{1/2} \frac{1}{\sqrt{\overline{E^2}}} . \tag{8}$$

Now we shall review how we first arrived at the hyperbolic-secant spatial-soliton solution, starting from Maxwell's equations in a nonlinear medium [2]. Instead of the nonlinear index expansion, here it is more convenient to start with the Taylor series expansion of the nonlinear dielectric function

$$\epsilon = \epsilon_0 + \epsilon_2\overline{E^2} + \dots , \tag{9}$$

where, for weak nonlinearities and because $n = \sqrt{\epsilon}$, the nonlinear dielectric coefficient ϵ_2 is related to the nonlinear-index Kerr coefficient n_2 by

$$\epsilon_2 = 2n_0n_2 . \tag{10}$$

One then obtains the following nonlinear wave equation for a soliton with a profile in only one transverse dimension, the x direction, propagating along the z axis:

$$\frac{\partial^2 E}{\partial x^2} + \frac{\partial^2 E}{\partial z^2} - \frac{\epsilon_0}{c^2}\frac{\partial^2 E}{\partial t^2} - \frac{\epsilon_2}{c^2}\overline{E^2}\frac{\partial^2 E}{\partial t^2} = 0 . \tag{11}$$

The soliton ansatz, which describes a steady-state propagation solution along the z axis with no change in the transverse profile $\mathcal{E}(x)$ during this propagation, is

$$E(x, z, t) = \mathcal{E}(x)\cos(k_z z - \omega t) . \tag{12}$$

The unknown propagation wavenumber k_z is to be found in a self-consistent manner along with the unknown transverse field profile $\mathcal{E}(x)$, and corresponds to some unique peak soliton amplitude $\mathcal{E}(0)$. Thus this soliton ansatz represents a *nonlinear* eigenmode of propagation. Upon substitution of this ansatz into the nonlinear wave equation (11), we obtain the ordinary differential equation

$$\frac{d^2}{dx^2}\mathcal{E}(x) - \Gamma^2\mathcal{E}(x) + \frac{1}{2}\epsilon_2 k_0^2 \mathcal{E}(x)^3 = 0 , \tag{13}$$

where $\Gamma = \sqrt{k_z^2 - k^2}$ and $k = n_0 k_0 = n_0 \omega / c$.

Since $\mathcal{E}(x)$ represents a slab-shaped beam confined in the x direction, we shall require the boundary conditions that $\mathcal{E}(x) \to 0$ as $x \to \pm\infty$. Since it is also natural to expect that the lowest order solution is one which is even in x, we shall also require the boundary condition that $d\mathcal{E}(x)/dx = 0$ at $x = 0$. Since the formation of a dielectric waveguide means that there must exist evanescent tails in $\mathcal{E}(x)$ as $x \to \pm\infty$, this implies that $\Gamma^2 > 0$. These evanescent-wave solutions are also what one would also expect on the basis of the above dielectric-waveguide model.

A mechanical problem with equations identical to (13) is that of a classical particle moving in one dimension along the x axis in a double-well quartic potential (see Fig. 2)

$$V(x) = -\frac{1}{2}x^2 + \frac{1}{4}bx^4 , \tag{14}$$

where $b = \epsilon_2 k_0^2 / 2\Gamma^2$. This leads to the classical, Newtonian equation of motion for the particle

$$\ddot{x} - x + bx^3 = 0 , \tag{15}$$

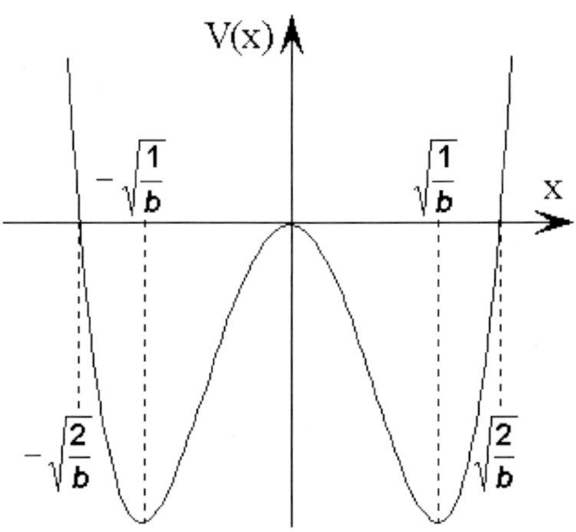

Fig. 2. Potential-energy curve for a particle moving in the x dimension: a mechanical analogue used in obtaining the optical spatial soliton solution in 1D

where the correspondences between the mechanical and the electromagnetic quantities are as follows:

$$x(t) \Leftrightarrow \mathcal{E}(x) \ , \tag{16}$$

$$t \Leftrightarrow \Gamma x \ . \tag{17}$$

The corresponding boundary conditions become

$$x(t) \to 0 \text{ as } t \to \pm\infty \Leftrightarrow \mathcal{E}(x) \to 0 \text{ as } x \to \pm\infty \ , \tag{18}$$

$$\dot{x}(0) = 0 \Leftrightarrow \frac{\mathrm{d}\mathcal{E}(x)}{\mathrm{d}x} = 0 \text{ at } x = 0 \ . \tag{19}$$

By inspection of Fig. 2, we see that the solution to the mechanical problem can be obtained from the first integral of the equation of motion (15), which represents the conservation of the total mechanical energy E of the system, viz.,

$$T(\dot{x}) + V(x) = E \ , \tag{20}$$

where $T(\dot{x}) = (1/2)\dot{x}^2$ represents the kinetic energy of a particle with a unit mass, and $V(x) = -(1/2)x^2 + (1/4)bx^4$ represents its potential energy.

We seek solutions where $x \to 0$ as $t \to \infty$. Clearly, by inspection of Fig. 2, the only possible solution satisfying this final condition corresponds to placing the particle initially at $\pm\sqrt{2/b}$ with zero initial velocity, which will then also automatically satisfy the initial condition $\dot{x}(0) = 0$. This corresponds to a total energy for the particle of $E = 0$. The particle will then fall into the well and reach $x = 0$ after an infinite amount of time. This leads to the implicit solution

$$\int_{\sqrt{2/b}}^{x} \frac{\mathrm{d}x}{\sqrt{x^2 - (1/2)bx^4}} = t \ , \tag{21}$$

and hence to the explicit solution

$$x(t) = \sqrt{\frac{b}{2}} \frac{1}{\cosh t} \ . \tag{22}$$

Using the correspondences given in (17) to translate this mechanical solution back to the original electromagnetic problem, we obtain the soliton solution

$$\mathcal{E}(x) = \frac{\mathcal{E}(0)}{\cosh \Gamma x} = \mathcal{E}(0) \ \mathrm{sech}\left((k_z^2 - k^2)^{1/2} \, x \right) \ , \tag{23}$$

where the initial condition for the mechanical problem uniquely determines the following relationship between the peak soliton amplitude $\mathcal{E}(0)$ and the soliton wavenumber k_z:

$$\mathcal{E}(0) = \sqrt{\frac{2}{b}} = 2\frac{\sqrt{k_z^2 - k^2}}{k_0} \frac{1}{\sqrt{\epsilon_2}} = \left(\frac{k_z^2}{k^2} - 1 \right)^{1/2} \left(\frac{2n_0}{n_2} \right)^{1/2} \ . \tag{24}$$

3 Experiments

Early experiments to observe the sech spatial soliton were performed with beams with an elliptical cross section passing through samples of carbon disulphide (CS_2) by Barthélemy et al. [11]. The idea of "soliton waveguiding", in which the soliton is viewed as a reconfigurable index profile which produces a waveguide for another light beam, was introduced by this French group [12], and contemporaneously by an Australian group [13], who introduced the same idea in the context of the self-*de*focusing sign of the nonlinearity, where a black (vortex) self-guided beam forms a single-mode waveguide for a weak signal beam at a higher frequency. However, the kind of beam geometry used by the French group still possessed two transverse dimensions in the degrees of freedom of the field for the self-focusing sign of the nonlinearity, and it was doubtful if the observed self-trapped beams were truly stable solitons.

In 1990, a self-trapped beam of light was clearly observed to form inside a thin, surface, linear-dielectric waveguide on top of a Schott B270 glass sample in an experiment performed at Bellcore [14]. Mode-locked dye-laser pulses of 75 fs duration and of peak power around half a megawatt were injected into this surface waveguide. A self-trapped beam of light, which did not diffract in the single remaining unconfined transverse dimension parallel to the glass surface (the x direction in Fig. 3), was observed to possess the expected hyperbolic secant shape. The relationship between the soliton width and the peak soliton amplitude was also observed to be consistent with the above theory.

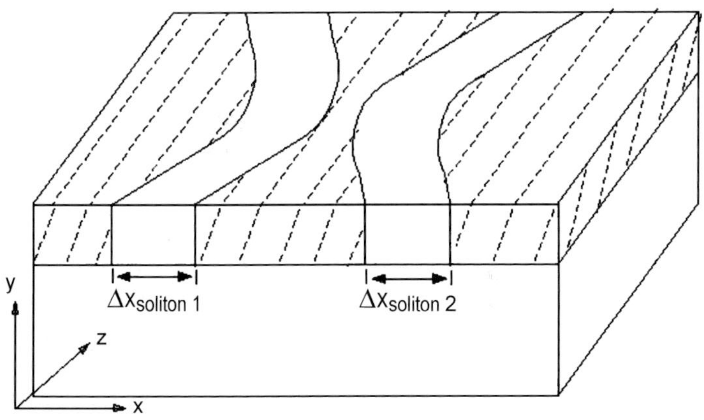

Fig. 3. Soliton–soliton collision experiment involving two incident sech solitons propagating inside a surface glass waveguide (adapted from [15]). The solitons formed inside the *shaded region* on the surface of the glass sample, which was doped with ions so that it formed a thin, linear surface waveguide with an index increase of 0.007 above the bulk glass index of 1.53

The most convincing evidence that the self-trapped beams of light inside the surface waveguide of the Bellcore experiment were indeed spatial solitons was obtained in a subsequent soliton–soliton collision experiment [15] In this experiment, two solitons were injected at an angle with respect to each other, such that they collided inside the middle of the top surface of the glass sample (see Fig. 3). These two solitons were observed to survive the collision intact. Furthermore, if the relative optical phase between the two solitons was adjusted, their behavior during the collision could be adjusted continuously from that of a mutual attraction to that of a mutual repulsion. These kinds of spatial solitons have also been observed in a self-focusing semiconductor gain medium [16].

Now that we have discussed the case of spatial solitons in one transverse dimension, it is natural to consider next the case of two transverse dimensions, since most optical beams possess cylindrical symmetry. As mentioned above, for the self-focusing sign of the nonlinearity ($n_2 > 0$), no true soliton solutions should exist because of Derrick's theorem, but cylindrically symmetric, stationary, solitary-wave-type solutions have been observed in sodium vapor. However, for the self-defocusing sign of the nonlinearity ($n_2 < 0$), genuine spatial soliton solutions in the form of optical vortex solitons were predicted by Snyder et al. [13] and were observed by Swartzlander and Law [17]. Vortex solitons, which were generated by means of a computer-generated holographic mask, have also been observed in rubidium vapor [18]. Collision between these kinds of spatial solitons, and also of 2D bright solitary waves, has been observed [19]. Vortex streets, i.e, periodic vortex shedding, have also been observed in walking parametric wave mixing [20]. These optical vortices are important examples of *topological* solitons, in that they possess nonzero "topological charges", i.e. quantized vorticities which correspond to certain topological winding numbers, corresponding to phase shifts of multiples of 2π around the vortex cores. Furthermore, they exert a Coulomb-like, long-range, "gauge field" force on each other. They are similar to the quantized vortices seen in superfluid helium.

Spatial solitons have also been observed to form inside optical cavities. [21] Recently, we have observed a "nonlinear mode", i.e. a nonlinear spatial pattern whose fields oscillate everywhere at the same frequency. The "nonlinear mode" is a generalization of the idea of the spatial soliton, and has been observed to form inside cavities with one transverse degree of freedom [22]. Discrete, left–right symmetry breaking in the formation of such nonlinear modes has been predicted to occur in such 1D cavities, which should be associated with the occurrence of optical tristability [23].

4 New Directions

Now we shall begin the discussion of possible new directions for research in spatial-soliton physics. Such a discussion will necessarily be somewhat subjective, and I shall take the liberty of choosing some of my own favorite topics. In my opinion, there are two main directions which this research will take. One direction is towards *applications* of such solitons, and the other direction is towards *basic research* in soliton physics. Of course, the distinction between these two directions is never completely a sharp one. Concerning the first direction, one of the main ideas for applications is that of all-optical switching of light beams by means of spatial solitons, i.e. the control of light by light, [24] which can occur at both the classical and quantum levels. The control of one light beam by another could be useful for optical communications and optical information processing. The impressive recent developments in this direction use large photorefractive nonlinearities to accomplish the control of light by light [25]. In one version of a quantum nondemolition, experiment [26] it is possible to use one quantum soliton to impart a phase shift to another quantum soliton during a soliton–soliton collision, so that a measurement of photon number can be made in which the photon number is not significantly disturbed. In a perhaps speculative application, a soliton–soliton collision could in principle also be used as the heart of a "controlled NOT gate", in which one quantum soliton imparts a phase shift on another quantum soliton, but only when the second quantum soliton is simultaneously present with the first. This could be useful in principle for constructing a quantum computer.

Concerning the second main direction, which is towards basic physics research, we start with the remark that almost all the extant examples of solitons are solutions of *classical* field theories. A natural extension in terms of basic physics would be towards the direction of *quantum* solitons, in which one investigates the soliton-like solutions of the underlying *quantum* field theories. One should then recover, in the correspondence-principle limit of large quantum numbers, the known classical soliton solutions. There are many interesting avenues of research along these lines, some of which have already been initiated for 1D *temporal* solitons in optical fibers by Lai and Haus [27], and which have been extended to 1D *spatial* solitons by Chiao et al. [28]. One consequence of the formation of spatial solitons is that the light within them should no longer remain a coherent state, but should become a squeezed state, [29], in analogy with the formation of squeezed states observed in temporal solitons in optical fibers [30]. The fact that the quantum field problem in 1D has exact solutions, which can be found through the Bethe ansatz method [31], provides a powerful tool for the study of quantum soliton physics in 1D. In particular, we shall discuss below the correspondence-principle limit of the 1D two-particle correlation function, in which the classical sech soliton behavior is recovered from the quantum soliton solutions, with the "diphoton", a two-photon soliton, as an example of a quantum soliton in the deep quantum limit [28].

The quantum theory for the analogue of a sech soliton, along the x direction with ω_0 as its fast optical frequency, could start from the Hamiltonian

$$H = H_{\text{free}} + H_{\text{interaction}} ,$$

where

$$H_{\text{free}} = \hbar\omega_0 \int \psi^\dagger \psi \, \mathrm{d}x + \frac{\hbar\omega_0}{2c^2} \int \frac{\partial \psi^\dagger}{\partial x} \frac{\partial \psi}{\partial x} \, \mathrm{d}x . \tag{25}$$

Here ψ is an operator corresponding to a slowly varying envelope of the soliton electromagnetic field; it satisfies the equal-time Bose commutation relation,

$$[\psi(x,t), \psi^\dagger(x',t)] = \delta(x - x') . \tag{26}$$

The interaction Hamiltonian is chosen so that the Heisenberg equations of motion go over to the nonlinear Schrödinger equation in the classical-field-theory limit, viz.,

$$H_{\text{interaction}} = -\frac{1}{2} G \int \psi^\dagger(x_1)\psi^\dagger(x_2)\delta(x_1 - x_2)\psi(x_1)\psi(x_2) \, \mathrm{d}x , \tag{27}$$

where G is a coupling constant which is proportional to the Kerr coefficient n_2. Thus the problem reduces to that of the weakly interacting nonrelativistic Bose gas, where the effective mass of the photon is $m_{\text{eff}} = \hbar\omega_0/c^2$, and where the photons interact with each other via pairwise delta-function potentials whose sign and strength depend on n_2. This 1D quantum field theory corresponds to a quantum N-body problem which has been solved exactly by means of the Bethe ansatz method [31]. In the two-body sector, for the case of the self-focusing sign, the problem reduces to that of two particles which attract each other via an attractive two-body delta-function potential. There exists one and only one bound state for the two-body problem, which has a wavefunction u_{diphoton} in the relative coordinate $x_1 - x_2$ with soliton size d and bound-state energy E_{diphoton}. These are

$$u_{\text{diphoton}}(x_1 - x_2) = (2d)^{1/2} \exp\left(-\frac{|x_1 - x_2|}{2d}\right) , \tag{28}$$

$$d = \frac{\hbar^2}{m_{\text{eff}} G} , \tag{29}$$

$$E_{\text{diphoton}} = \frac{m_{\text{eff}} G^2}{4\hbar^2} , \tag{30}$$

a solution which we have dubbed the "diphoton". It represents the smallest number of photons which can form a bound state, and hence the lowest-order quantum solution which corresponds to the classical sech soliton. A plot of the two-point correlation function is shown in Fig. 4 as a function of the relative coordinate $x - x'$ of two detectors at x and x', which could

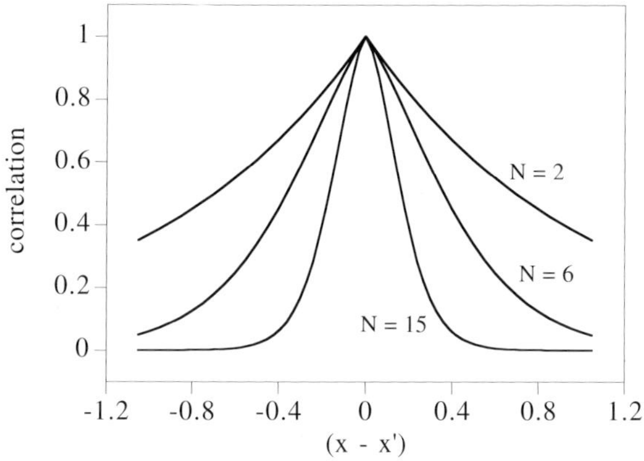

Fig. 4. Two-point correlation function for the N-body ground state solutions, starting with $N = 2$, the "diphoton" given by (28). As $N \to \infty$, the shape of this correlation function approaches the hyperbolic-secant shape of the classical sech soliton (from [32])

be measured by coincidence detection using two detectors placed in the far field outside the nonlinear medium as a function of their separation. This two-point correlation function is also plotted for various larger values of N, where N is the number of photons in the lowest Bethe ansatz N-body bound state [32]. As N becomes very large, we recover the hyperbolic-secant shape of the classical sech soliton. The picture which emerges from these results is that the self-trapping of an optical beam occurs because of the formation of bound states of the photons inside this beam, due to their mutual attraction. This is the reason why a self-trapped beam of light does not spread owing to diffraction.

However, one should also go on to ask the following important question: are there any *qualitatively* new features of soliton physics at the quantum level which clearly display fundamentally new features of quantum physics per se, for which there exist no classical, correspondence-principle analogies? I shall suggest below that one such feature might be the nonstandard quantum statistics obeyed by the quantum spatial solitons in 1D and 2D. In particular, I shall argue that the optical 2D vortex soliton and the 1D sech and tanh solitons might be "anyons".

When the classical soliton solutions of the classical field theory are quantized, the resulting quantum solitons or "quasiparticles" possess discrete quantum numbers, such as the quantized vorticity of the vortex soliton in 2D. When all of these quantum numbers are identical for two quantum solitons, they become truly identical, i.e. *indistinguishable*. The question naturally arises: what kind of quantum statistics do such indistinguishable quantum solitons obey? The first guess would be that they would obey either Bose–Einstein or Fermi–Dirac statistics, as do particles or quasiparticles in three spatial dimensions. For example, we know that the sound waves in 3D condensed-matter systems, when quantized, become phonons, which are ordinary bosons.

However, in systems of one and two spatial dimensions, there exist "reduced-dimensionality" effects in condensed-matter systems, which, owing to the nontrivial topology of their many-body configuration spaces, might lead to "anyonic", or "intermediate", quantum statistics, i.e. quantum statistics which interpolate between the usual Bose and Fermi quantum statistics [33]. Upon interchange of the positions of two identical particles or quasiparticles at positions $\boldsymbol{x}_1$ and $\boldsymbol{x}_2$ in one or two spatial dimensions, the wavefunction for the system picks up a phase factor $\mathrm{e}^{-\mathrm{i}\theta}$; thus,

$$\psi(\boldsymbol{x}_1, \boldsymbol{x}_2) \rightarrow \psi(\boldsymbol{x}_2, \boldsymbol{x}_1) = \mathrm{e}^{-\mathrm{i}\theta} \psi(\boldsymbol{x}_1, \boldsymbol{x}_2) , \tag{31}$$

where the quantum-statistical phase θ need not be, in the case of one or two spatial dimensions, the usual Bose–Einstein phase ($\theta = 0$) or Fermi–Dirac phase ($\theta = \pi$), but instead could be "anything" (hence "anyonic" [34]) in the range $0 \leq \theta \leq \pi$. An example of an anyon in a 2D condensed-matter system may be the $(1/3)e$ charged quasiparticle discovered in the 2D quantum Hall system [35], i.e. a system of electrons moving on a two-dimensional plane subjected to a strong magnetic field. Owing to the fractional charge $(1/3)e$ of this vortex-like quasiparticle, and the unit flux quantum hc/e frozen onto it during its motions, the Aharonov–Bohm phase picked up by the charge of one quasiparticle encircling the flux tube of the other particle, upon the interchange of the positions of two such identical quasiparticles, must be a fraction of 2π, specifically, in this case, $\theta = \pi/3$, which is clearly neither bosonic nor fermionic (see Fig. 5). Since this quasiparticle is vortex-like, this suggests that we look generally at other 2D vortex solitons to see if they might also exhibit anyonic statistics [38].

One possible candidate for a spatial soliton which may turn out to be an anyonic quasiparticle is the optical vortex soliton formed inside a cavity. It has been recently suggested that photons confined inside a Fabry–Perot resonator in the presence of a Kerr nonlinearity due to two-level atoms will form a photon superfluid, owing to their mutual interactions [36]. When the Landau critical velocity of the superfluid is exceeded, superfluid vortices are predicted to be generated [37]. The question arises: what kind of quantum statistics, if any, do these quasiparticles in the photon superfluid obey?

One approach to answering this question is to quantize the Kirchhoff equations of motion for point-like vortices [38]. Let us denote the positions of the vortices by (X_i, Y_i). Then the classical equations of motion governing these positions are

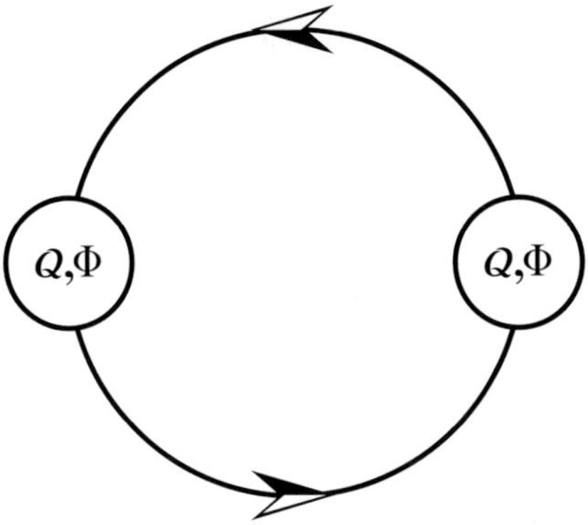

Fig. 5. Anyon interchange [34]. An anyon is a particle or a quasiparticle consisting of a bound state with charge Q and flux Φ. Two identical anyons pick up an Aharonov–Bohm phase upon interchange, indicating that there exists a novel kind of quantum statistics in two spatial dimensions which is neither Bose–Einstein nor Fermi–Dirac

$$\kappa_i \frac{\mathrm{d}X_i}{\mathrm{d}t} = +\frac{\partial H}{\partial Y_i} , \tag{32}$$

$$\kappa_i \frac{\mathrm{d}Y_i}{\mathrm{d}t} = -\frac{\partial H}{\partial X_i} , \tag{33}$$

$$H = -\frac{1}{4\pi} \sum_{i<j} \kappa_i \kappa_j \ln \left(\frac{R_{ij}}{r_0} \right)^2 , \text{ and} \tag{34}$$

$$R_{ij}^2 = (X_i - X_j)^2 + (Y_i - Y_j)^2 , \tag{35}$$

where $\kappa_i = \pm h/m_{\text{eff}}$ is the vorticity of the ith vortex, and r_0 is the vortex core radius. Note that the equations of motion are Hamiltonian in form, with X_i and Y_i being canonically conjugate variables. This immediately suggests a canonical quantization procedure, which starts by invoking the commutator

$$[X_i, Y_i] = \mathrm{i}C_i \delta_{ij} , \tag{36}$$

where $C_i = \hbar(\kappa_i N d)^{-1}$, with N being the number density of photons in a Fabry–Perot resonator of spacing d.

For the special case of two identical vortices with the same sign of topological charge (i.e. vorticity), the solution of these equations can be found exactly, and describes motion in a circular orbit of the two vortices around each other, just as in Fig. 5. This two-body problem reduces to a center-of-mass problem, in which the center of mass is a constant of motion, plus a one-body problem in the relative coordinates $X = X_1 - X_2$ and $Y = Y_1 - Y_2$.

Upon quantization, the relative coordinates for the two identical vortices obey the commutator

$$[X, Y] = 2\mathrm{i}C , \tag{37}$$

where $C = \hbar(\kappa N d)^{-1}$. This commutator leads to a quantization of the area of the circular orbit, which is proportional to X^2+Y^2, in units of $(n+1/2)\hbar$. The meaning of the fraction 1/2 is that it represents the zero-point motion of the two vortices, which is an irreducible consequence of the above commutator. Also, the meaning of the area of the classical orbit is that it represents the relative orbital angular momentum of the two-vortex system [38]. The quantization of this area implies that the relative angular momentum of the two vortices is quantized in *half-fractional* units of $\hbar$, with the fraction 1/2 originating from their relative zero-point motion. This implies that after one period of the classical orbit, or after two interchanges of their positions, the two-vortex system picks up a phase factor of $\mathrm{e}^{\mathrm{i}\pi} = -1$. Therefore, after only one interchange, the two-vortex system picks up a phase factor of

$$\mathrm{e}^{-\mathrm{i}\theta} = \pm\sqrt{\mathrm{e}^{\mathrm{i}\pi}} = \pm\mathrm{e}^{\mathrm{i}\pi/2} = \pm\mathrm{i}\,. \tag{38}$$

Thus we infer from (31) that the vortices obey "semionic", or "quarter-fractional", quantum statistics, which are halfway between Bose–Einstein and Fermi–Dirac statistics.

The physical origin of this anyonic effect for quantum vortex solitons is the zero-point motion of their vortex cores, which gives rise to a half-fractional unit of orbital angular momentum. In other words, this effect originates from the fact that the area, which is proportional to the value of X^2+Y^2 for the vortex core motion, must be a positive definite quantity, and from the fact that X and Y do not commute. This implies that the number of particles, i.e. photons, contained inside the vortex core is the fraction 1/2, while the phase picked up by a photon circulating around the vortex core is 2π. This is quite analogous to the case of the vortex-like quantum Hall, quasiparticle where the number of particles, i.e. electrons, contained inside the core is the fraction 1/3, while the phase picked up by an electron circulating around the integer flux quantum is also 2π (see Fig. 5). Furthermore, there may be yet another physical setting in which anyons can form: an atom trapped on an electromagnetic vortex might become an optical anyon [39].

In the case of vortices in superfluid helium thin films, the above conclusions were called into question on the grounds that the finite-sized core structure of the vortices might contribute ambiguities to the quantum statistical phase factor [40]. However, a specific microscopic model was used to calculate these ambiguous contributions, and it is unclear that this particular microscopic model actually applies to superfluid helium thin films. Since there is a much better-established microscopic theory for photons weakly interacting with each other via two-level-atom interactions than for helium atoms strongly interacting with each other in superfluid helium, there is hope that these ambiguities will eventually be resolved, but this theoretical work has not been done to date. Furthermore, it should be stressed that, ultimately, the most important direction for future research would be to perform an experiment to measure this quantum statistical phase factor. Optical experiments to measure this phase are especially called for, since there exist well-established techniques to measure this phase interferometrically. Analogous experiments to measure this phase in electron, helium, or Bose-condensed atomic systems are much more difficult.

Another situation in spatial-soliton physics where anyonic statistics might appear is for 1D spatial solitons. Both the sech and the kink or dark soliton are possible candidates, but for ease of visualization, let us think about the sech soliton only. It is well known that when two classical sech solitons collide, they pick up a phase factor. At the quantum level, when the number of photons inside each soliton is quantized, two quantum solitons with identical quantum numbers are again indistinguishable from each other, so that it makes sense again to ask the question of what kind of quantum statistics these solitons, viewed as identical "quasiparticles", obey. Invoking the correspondence principle, we infer that the classical phase factor picked up by two classical solitons, which is in general nonvanishing, will go continuously over to the quantum phase factor picked up by the two identical quantum solitons. Experiments have already indicated that the soliton–soliton interaction phase is nonvanishing [26]. By the correspondence principle, this phase should become the *anyonic* phase for two identical solitonic quasiparticles at the quantum level. One should be able to observe this phase by means of interference in a scattering experiment. For example, one could pattern the experiment after the one in nuclear physics used to measure the Bose–Einstein statistics of ^{12}C nuclei and the Fermi–Dirac statistics of ^{13}C nuclei, as discussed by Feynman [41]. A schematic of a possible experiment is shown in Fig. 6.

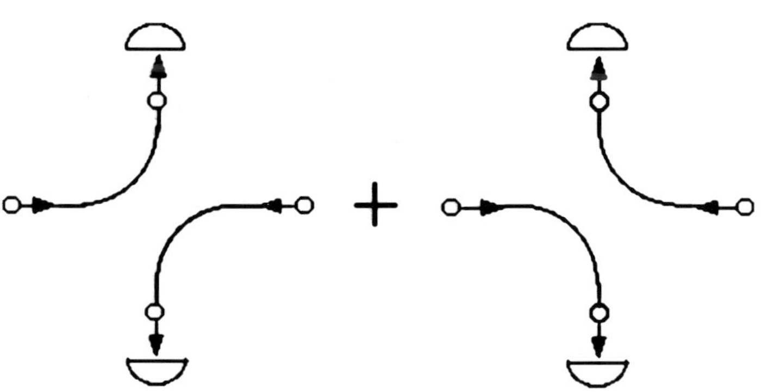

Fig. 6. Two contributing Feynman diagrams in an interference experiment which involves the superposition of two indistinguishable 90-degree scatterings of two indistinguishable quantum solitons. The *small open circles* on these paths denote the identical solitons, and the *D-shaped symbols* denote detectors, which are used in coincidence detection. Such an interference experiment would allow the measurement of the quantum statistical phase factor (possibly anyonic) for these identical solitons (adapted from [41])

In conclusion, the field of spatial solitons is a very rich one in terms of both applications and basic physics. Promising applications to optical communications and information processing are based on the idea of the control of light by light in solitons. At the quantum level, there are many avenues to explore, such as whether quantum spatial solitons can be viewed as "quasiparticles" in a "condensed-matter" system composed of interacting photons, and whether unusual quantum statistics exist for such optical solitons. Thus solitons may be fundamental for our understanding of the quantum interacting many-body problem, i.e. quantum field theories in lower dimensions.

Acknowledgments

This work, which was presented at the 85th birthday celebration in honor of Professor Charles Townes at Jackson Hole, Wyoming, was supported by the ONR and the NSF. I thank J. Boyce, I.H. Deutsch, J. Perez-Torres, A. Hansen, E. Garmire, J.C. Garrison, A.M. Moulthrop, C.H. Townes and E.M. Wright for their collaborative contributions to this work, and for many helpful discussions.

References

1. R.Y. Chiao, E. Garmire, C.H. Townes, Phys. Rev. Lett. **13**, 479 (1964).
1a. V.I. Talanov, Izvestia V.U.Z. Radiofyzika (U.S.S.R.) **7**, 564 (1964).
1b. G.A. Askar'yan, Zh. Éksp. Teor. Fiz. **42**, 1567 (1962) [Sov. Phys. JETP **15**, 1088].
2. R.Y. Chiao, PhD thesis, Massachusetts Institute of Technology, 1965.
3. P.L. Kelley, Phys. Rev. Lett. **15**, 1005 (1965).
4. V.E. Zakharov, A.B. Shabat, Sov. Phys. JETP **34**, 62 (1972).
5. O. Svelto, *Progress in Optics* Vol. 12, ed. by E. Wolf (Elsevier, Amsterdam, 1974), p. 3.
5a. M. Mlejnek, M. Kolesik, J.V. Moloney, E.M. Wright, Phys. Rev. Lett. **83**, 2938 (1999).
6. V.N. Lugovoi, A.M. Prokhorov, JETP Lett. **7**, 117 (1969); Y. R. Shen, Prog. Quantum Electron. **7**, 1 (1975); J.H. Marburger, Prog. Quantum Electron. **7**, 4 (1975).
7. J.E. Bjorkholm, A. Ashkin, Phys. Rev. Lett. **32**, 129 (1974).
8. Yu.N. Karamzin, A.P. Sukhorukov, JETP Lett. **20**, 339 (1974); Yu.N. Karamzin, A.P. Sukhorukov, Sov. Phys. JETP **41**, 414 (1975).
9. W.E. Torruellas, Z. Wang, D.J. Hagan, E.W. VanStryland, G.I. Stegeman, L. Torner, C.R. Menyuk, Phys. Rev. Lett. **74**, 5036 (1995).
10. S. Coleman, *Aspects of Symmetry* (Cambridge University Press, Cambridge, 1985), p. 194.
11. A. Barthélemy, S. Maneuf, C. Froehly, Opt. Commun. **55**, 201 (1985); S. Maneuf, R. Desailly, C. Froehly, Opt. Commun. **65**, 193 (1988); S. Maneuf, F. Reynaud, Opt. Commun. **66**, 325 (1988).

12. R. De La Fuente, A. Barthélemy, IEEE J. Quantum Electron. **QE-28**, 547 (1992).
13. A.W. Snyder, L. Poladian, D.J. Mitchell, Opt. Lett. **17**, 789 (1992).
14. J.S. Aitchison, A.M. Weiner, Y. Silberberg, M.K. Oliver, J.L. Jackel, D.E. Leaird, E.M. Vogel, P.W.E. Smith, Opt. Lett. **15**, 471 (1990).
15. J.S. Aitchison, A.M. Weiner, Y. Silberberg, D.E. Leaird, M.K. Oliver, J.L. Jackel, P.W.E. Smith, Opt. Lett. **16**, 15 (1990).
16. G. Khitrova, H.M. Gibbs, Y. Kawamura, H. Iwamura, T. Ikegami, J.E. Sipe, L. Ming, Phys. Rev. Lett. **70**, 920 (1993).
17. G.A. Swartzlander, Jr., C.T. Law, Phys. Rev. Lett. **69**, 2503 (1992); G.A. Swartzlander, Jr., in *Optical Vortices*, ed. by M. Vanetsov, K. Staliunas (Nova Science Publishers, New York, 1999), p. 107.
18. V. Tikhonenko, N.N. Akhmediev, Opt. Commun. **126**, 108 (1996).
19. V. Tikhonenko, J. Christou, B. Luther-Davies, Phys. Rev. Lett. **76**, 2698 (1996).
20. G. Molina-Terriza, L. Torner, D.V. Petrov, Opt. Lett. **24**, 899 (1999).
21. D.W. McLaughlin, J.V. Moloney, A.C. Newell, Phys. Rev. Lett. **51**, 75 (1983); A.C. Newell, J.V. Moloney, *Nonlinear Optics* (Addison-Wesley, Redwood City, 1992).
22. J. Boyce, J.P. Torres, R.Y. Chiao, Opt. Lett. **24**, 1850 (1999).
23. J.P. Torres, J. Boyce, R.Y. Chiao, Phys. Rev. Lett. **83**, 4293 (1999).
24. S. Trillo, S. Wabnitz, E.M. Wright, G.I. Stegeman, Opt. Lett. **13**, 672 (1988); W.J. Firth, A.J. Scroggie, Phys. Rev. Lett. **76**, 1623 (1996); R.A. Fuerst, D.-M. Baboiu, B. Lawrence, W.E. Torruellas, G.I. Stegeman, S. Trillo, S. Wabnitz, Phys. Rev. Lett. **78**, 2756 (1997).
25. M. Segev, G.C. Valley, B. Crosignani, P. Diporto, A. Yariv, Phys. Rev. Lett. **73**, 3211 (1994); D.N. Christodoulides, M.I. Carvalho, J. Opt. Soc. Am. B **12**, 1628 (1995); T.H. Coskun, A.G. Grandpierre, D.N. Christodoulides, M. Segev, Opt. Lett. **25**, 826 (2000).
26. S.R. Friberg, S. Machida, Y. Yamamoto, Phys. Rev. Lett. **69**, 3165 (1992).
27. Y. Lai, H.A. Haus, Phys. Rev. A **40**, 844 (1989); Phys. Rev. A **40**, 1138 (1989).
28. R.Y. Chiao, I.H. Deutsch, J.C. Garrison, Phys. Rev. Lett. **67**, 1399 (1991); I.H. Deutsch, R.Y. Chiao, and J.C. Garrison, Phys. Rev. Lett. **69**, 3627 (1992).
29. R.Y. Chiao, I.H. Deutsch, J.C. Garrison, E.W. Wright, in *Frontiers in Nonlinear Optics: the Serge Akhmanov Memorial Volume*, ed. by H. Walther, N. Koroteev, M.O. Scully (Institute of Physics, Bristol, 1993), p. 151.
30. M. Rosenbluh, R.M. Shelby, Phys. Rev. Lett. **66**,153 (1991).
31. H.B. Thacker, Rev. Mod. Phys. **53**, 253 (1981).
32. J.K. Boyce, PhD thesis, University of California Berkeley, 1998.
33. J.M. Leinaas, M. Myrheim, Nuovo Cimento B **37**, 1 (1977); G.A. Goldin, R. Menikoff, D.H. Sharp, J. Math. Phys. **22**, 1664 (1981).
34. F. Wilczek, Phys. Rev. Lett. **48**, 1144 (1982); Phys. Rev. Lett. **49**, 957 (1982).
35. D.C. Tsui, H.L. Störmer, A.C., Gossard, Phys. Rev. Lett. **48**, 1559 (1982); R.B. Laughlin, Phys. Rev. Lett. **50**, 1395 (1983).
36. R.Y. Chiao, J. Boyce, Phys. Rev. A **55**, 1431 (1999); R.Y. Chiao, Opt. Commun. **175**, 157 (2000).
37. E.L. Bolda, R.Y. Chiao, W. Zurek, Phys. Rev. Lett. **86**, 416 (2001).
38. R.Y. Chiao, A. Hansen, A.M. Moulthrop, Phys. Rev. Lett. **54**, 1339 (1984); A. Hansen, A.M. Moulthrop, and R.Y. Chiao, Phys. Rev. Lett. **55**, 1431 (1985).

39. E.M. Wright, R.Y. Chiao, J.C. Garrison, Chaos, Solitons and Fractals **4**, 1791 (1994).
40. F.D.M. Haldane, Y.-S. Wu, Phys. Rev. Lett. **55**, 2887 (1985).
41. R.P. Feynman, R.B. Leighton, M. Sands, *The Feynman Lectures on Physics*, Vol. 3 (Addison-Wesley, Reading, MA, 1965), p. 4-1.

Materials for Spatial Solitons

Barry Luther-Davies and George I. Stegeman

Summary. The basic properties of materials commonly employed in spatial soliton experiments are outlined and discussed. The requirements for different classes of materials used to observe Kerr-like, photorefractive, or parametric solitons are compared.

1 Introduction

Nonlinear optics has been a flourishing discipline since the 1960s, and a great deal of that time has been spent researching nonlinear materials. Literally thousands of materials have been reported, and yet when one scans the soliton literature it is clear that only a handful of these have been used for soliton experiments [1]. Clearly, the requirements for appropriate materials are demanding, and this aspect will be discussed in this chapter. Furthermore, soliton experimentalists are invariably practitioners of nonlinear optics, and usually not materials researchers. As a result, the focus in the spatial-soliton community has mostly been on standard, well-characterized, and readily available materials.

The material properties needed depend on the type of soliton being studied. In the temporal domain, the very low losses and large sample lengths available with glass fibers allow low-power temporal solitons to propagate many dispersion lengths [2]. For spatial solitons, the propagation distance is limited to only a few centimeters, either by fabrication technology or by material losses. This puts stringent conditions on the required nonlinear response, the damage threshold of the material, and the soliton width (since the diffraction length must be small compared with the propagation distance) for experimental work.

Fortunately, the variety of physical mechanisms which lead to self-trapping is much richer in the spatial than in the temporal domain (to date at least). Spatial solitons are currently classed into three distinct groups, based on the physics of the nonlinearity, namely Kerr and Kerr-like solitons, photorefractive solitons, and quadratic or parametric solitons [3]. For example, all Kerr and Kerr-like solitons rely primarily on any physical effect which produces an intensity-dependent change in refractive index. The origin can be electronic, or it can be due to carrier generation, thermal, etc. In general, Kerr solitons form in materials which have a local response, although in some cases, e.g. for a thermal nonlinearity, the response may be nonlocal. Photorefractive solitons utilize materials where a light-induced change of refractive index also occurs. However, in this case it is a DC electric-field distribution in a crystal that is affected by the optical field, and this in turn changes the refractive index via the electro-optic effect. Quadratic solitons also depend on second-order nonlinearities, however, in this case the response is at optical frequencies and the process generally involves energy exchange between different frequency components of the optical field.

The conditions for spatial solitons to form can be discussed in terms of a series of characteristic lengths. For example, for Kerr solitons, spatial diffraction is compensated by self-focusing (or self-defocusing in the case of dark solitons) when the diffraction length $L_{\mathrm{d}} \geq L_{\mathrm{nl}}$, where L_{nl} is the nonlinear length defined by the distance required for a nonlinear phase shift of $\pi/2$. For the soliton to maintain its form (which depends on its intensity), the loss must be small, i.e. $\alpha^{-1} > L_{\mathrm{d}}, L_{\mathrm{nl}}$, where α is the total loss. In fact, $\alpha L_{\mathrm{d}} \ll 1$ and $\alpha L_{\mathrm{nl}} \ll 1$ are the preferred conditions. These arguments can generally be extended to the other classes of solitons and directly define the desired material properties.

In the next sections we shall examine the material requirements, and the materials successfully used for spatial solitons to date.

2 Third-Order Kerr and Kerr-Like Saturating Nonlinearities

Ideal Kerr solitons, for which the change of refractive index is linearly related to the beam intensity ($\Delta n = n_2 I$, where I is the intensity) and the material response is instantaneous and local, are a very restricted class because of the requirements on geometry and nonlinearity. For example, it is well known that $(2+1)$D Kerr solitons exist only at one value of the beam power and are hence unstable. $(1+1)$D Kerr solitons are stable and correspond to creation of solitons which diffract in only one transverse dimension, for example in a planar or slab waveguide. In this case the balance of diffraction and self-focusing occurs only for the in-plane dimension. Since the index change for an ideal Kerr soliton should be proportional to the intensity at the same point in space, i.e. this relationship is strictly “local”, then the mathematical analysis of Kerr solitons is considerably simplified. However, “locality” in an experiment is defined in terms of the optical wavelength used, i.e. as long as the index change produced does not “diffuse” in space over a significant fraction of a wavelength, the interaction is effectively “local”. Furthermore, the theoretical description assumes that the temporal response of the medium is “instantaneous” (at least within the conventional slowly-varying-amplitude approximation). In practice, the required material response time depends on the time scale of the experiment. For example, for pulsed experiments, the response time required to obtain a good approximation to the ideal Kerr response should be much shorter than the pulse width. Even then, the experiment is complicated by the fact that the beam profile will vary during the pulse, and hence time resolution is required to observe Kerr solitons.

One should remember, however, that as the experimental conditions deviate increasingly from those assumed for the ideal Kerr model, the solitons will lose some of their model characteristics. For example, soliton collisions may become inelastic, resulting in energy loss to radiation fields [4]. It is perhaps fortunate that the limited physical scale of most experiments on spatial solitons results in observable behavior quite close to the Kerr case, even when the material response is far from ideal.

In general, the ideal Kerr response is best approached when the photon energy used for experiments is far from resonance with any single photon transition in the atoms and molecules making up the medium. Unfortunately, in such conditions the real and imaginary parts of the third-order nonlinearity are often rather small, making experimental investigations quite difficult! Complicating the situation further is the existence of multiphoton resonances that typically appear in spectral regions in which the linear absorption is negligible. There the absorption increases with intensity as $\Delta\alpha = \alpha_p I^{p-1}$, where $p = 2$ identifies two-photon absorption, usually the dominant nonlinear absorption mechanism. Since the index change $\Delta n = n_2 I$ required for soliton formation also depends linearly on the intensity, there is a direct trade-off which should be satisfied for useful Kerr soliton propagation, i.e. $T = \alpha_2 \lambda / n_2 < 1$ is required [5].

Much of the nonlinear-material characterization reported to date has been at least partially driven by a desire to understand the material properties in terms of the contributing electronic transitions and hence has been performed primarily in spectral regions near resonances. As a result, the nonresonant nonlinearities and two-photon coefficients are rarely known. It would be a boost to the spatial-soliton community if more materials characterization were carried out under nonresonant conditions. Since the ideal nonlinearity is nonresonant, the response is, to a first approximation, weakly dependent on the wavelength of the source.

An interesting question is whether there really exists a true "Kerr" material, that is, can the effect of resonances ever be neglected?

Resonances themselves modify the material response by breaking the linear relation between the change in refractive index and the intensity. This occurs because a resonance implies optical pumping between fixed energy levels in the material via the absorption of single or simultaneously of multiple photons from the incident field. The intensity dependence of the absorption depends on the number of absorbed photons. However, with increasing intensity, the transition will saturate (at an intensity I_s) as the population becomes excited from the lower to the upper energy level faster than the decay rate from the upper level. This results in bleaching and, via the Kramers–Kronig relation, to saturation of the light-induced change in refractive index. In general, in the presence of a resonance, the index change is no longer linearly related to intensity but takes on a more general form, proportional to I^p, where $p \neq 1$. As an additional point, it should be noted that the nonlinearity is generally complex, having both refractive and absorptive components. The presence of absorption due to a resonance can lead to a situation where the change in index is dominated by the imaginary (absorptive) component. In such conditions solitons or soliton-like structures may not exist at all.

Generally, I_s increases with increased detuning from the resonance and hence the higher-order ($p \geq 2$) contributions to Δn from that resonance decrease for fixed intensity. Therefore, whilst there will always be an intensity at which the material exhibits saturation, that intensity may well be above the material's damage threshold. As a result, the key to ensuring Kerr-law response is to operate in spectral regions optimally detuned from single- or multiple-photon resonances. In the final analysis, the best approach to finding Kerr-soliton-"friendly" materials and spectral regions is to experimentally characterize, over broad spectral ranges, the dependence of the index change on the local intensity for intensities up to a few times the intensity needed for soliton generation. The visible and infrared regions of the spectrum, including the communications bands, bound the usual wavelengths (500–1600 nm) in which solitons have been investigated. In materials with broad transparency (e.g. silica glass) this wavelength range is far from the primary UV and IR absorption bands, and hence one is perhaps safe to assume that resonances will have a negligible influence on the nonlinearity.

Propagation loss is an important practical issue. Establishing (and/or verifying) soliton characteristics requires propagation for at least a few diffraction lengths $L_d = \pi a^2 n_0/\lambda$, where a is the beam width, measured approximately from the peak intensity to where it decreases to 1/e of its peak value, and n_0 is the background (low-intensity) refractive index. As a "rule of thumb", the total loss α, including scattering losses and single- and multiple-photon absorption at the 1D soliton intensity (i.e. $I \approx \lambda^2/(2\pi^2 n_{\text{eff}} n_2 a^2)$), should satisfy $\alpha L_d \leq 0.3$. Table 1 lists a number of materials, for which the pertinent data are given in [6–13]. The first (1+1)D (i.e. in slab waveguides) solitons were observed in a multilayer structure consisting of liquid carbon disulphide confined between two glass plates. The electronic n_2 for CS_2 is quite small and the nonlinearity used here was the well-known re-orientational one, for which $n_2 \simeq 4 \times 10^{-14}$ cm^2/W [14]. The individual CS_2 molecules are linear and highly optically anisotropic, so that an incident optical field induces dipoles indexdipole primarily along the long molecular axis. Hence the incident field exerts a torque through these dipoles on these molecules, causing a partial alignment of the long molecular axes along the field direction. Thus the index increases parallel to the optical field and decreases in the plane normal to the field. The molecular orientational relaxation time τ is of the order of a few picoseconds, so the effective nonlinearity is constant

Table 1. Figures of merit of various materials. Intensity (assumed) = 1 GW/cm^2; $T = \lambda\alpha_2/n_2$ (goal: $T < 1$); "u" stands for "data unavailable"

Material	n_2 (cm^2/W)	α (cm^{-1})	T	λ (μm)
Semiconductors				
Kerr AlGaAs, $\lambda_{\text{gap}} = 0.79$ μm [6]	-4×10^{-12}	18	0.9	0.81
Kerr AlGaAs, $\lambda_{\text{gap}} = 0.75$ μm [7]	2×10^{-13}	0.1	< 0.1	1.55
AlGaAs amplifiers [8]	3×10^{-10}	u	u	0.90
Organics				
PTS (crystals) [9]	2.2×10^{-12}	0.8	< 0.1	1.6
DANS (side-chain polymer) [10]	8×10^{-14}	< 0.2	0.2	1.32
DEANST (20% solution) [11]	6×10^{-14}	$< 10^{-2}$	< 1	1.06
Glasses (fibers)				
SiO_2	2.4×10^{-16}	10^{-6}	$\ll 1$	> 1.06
$As_{0.38}S_{0.62}$ [13]	4.2×10^{-14}	0.02	< 2	> 1.32
RN [12]	1.3×10^{-14}	0.01	< 0.1	1.06

for pulse widths Δt longer than τ and decreases approximately as $\Delta t/\tau$ for shorter pulses. Note that one can define a saturation intensity in this case as approximately $n_2 I_s \simeq n_{\parallel} - n_{\perp}$, so that I_s is of the order of TW/cm^2. Larger nonlinearities can be obtained with liquid crystals. However, the reorientation time is also correspondingly longer and the "locality" of the response can extend to many wavelengths.

Kerr solitons have been observed in glasses at wavelengths in the red region of the spectrum [15]. These materials have transitions typically in the UV, so that even operating in the red is far away from one and two-photon resonances. Furthermore, in slab waveguides, the propagation losses are frequently limited by scattering due to limitations in the fabrication technology and not by absorption. Unfortunately, the nonlinearity $n_2 \simeq 3 \times 10^{-16}$ cm^2/W for fused silica is small, implying soliton intensities of a few TW/cm^2 for $\lambda = 1$ μm and $a \simeq 10$ μm. Other glasses with nonlinearities a few times larger than that of fused silica have been used [15]. Nevertheless, the soliton intensities required in these glasses are so high that femtosecond pulses needed to be used in order to prevent material damage. It is noteworthy, however, that there are other glasses with much higher nonlinearities and with small enough single- and multiple-photon absorption in the $\lambda > 1$ μm region of the spectrum to allow spatial-soliton generation at intensities of tens of GW/cm^2.

Other popular materials in which Kerr spatial solitons can be generated are semiconductors (e.g. AlGaAs), with photon energies below one-half of the semiconducting bandgap energy ($\hbar\omega < E_g/2$) [16]. For those photon energies, one-photon absorption is negligible, and two-photon absorption is in principle forbidden and in practice is too small to measure in high-quality samples just below $E_g/2$. Three-photon absorption is allowed but is very small for photon energies just less than one-half of the bandgap. Because the MBE fabrication techniques yield samples several centimeters long with scattering losses less than 0.5 dB/cm, attenuation is not an issue. Here the nonlinearity is local, has a femtosecond response, and $n_2 \simeq 2 \times 10^{-13}$ cm^2/W owing to virtual two-photon transitions. Since the typical soliton intensities are in the GW/cm^2 range, this system has proven to be a useful test bed for soliton physics.

What are the prospects of additional material systems for true Kerr nonlinearities suitable for spatial-soliton experiments? In general, "dielectric" materials have nonlinearities comparable to or smaller than that of liquid CS_2. Nonresonant nonlinearities in organic materials depend on the dominant physical mechanisms utilized. Two major classes of current interest, with potentially large nonlinearities, are charge transfer molecules and conjugated polymers. The first relies on specific chemical structures which exhibit large charge separation across the molecules. Although a great deal of characterization still needs to be done, it currently appears that nonresonant nonlinearities in the range 10^{-12} to 10^{-13} cm^2/W are possible, but with losses of a few dB/cm [10,11]. In the conjugated-polymer area, values in the range 10^{-11} to 10^{-12} cm^2/W have been reported in regions with acceptable absorption losses [9]. In fact, bright spatial solitons have been launched in a modified PPV. slab waveguide [17] Examples of each type of material are given in Table 1. The major problem is the fabrication of low-loss slab waveguides in such materials.

The preceding discussion of Kerr nonlinearities has concentrated on self-focusing nonlinearities, i.e. $n_2 > 0$. There have been, in fact, no demonstrations of dark or gray spatial solitons with nonresonant, self-defocusing nonlinearities. It is noteworthy that femtosecond temporal solitons have been demonstrated in waveguides of disperse "red one" (a charge transfer molecule) utilizing a value of $n_2 \simeq -10^{-20}$ m^2/V^2 (with positive group velocity dispersion), with acceptable losses, so this material could be a reasonable candidate [18]. In addition, some conjugated polymers, such as poly(p-phenylenevinylene), have been reported to have negative n_2 at 800 nm (between the single- and two-photon absorption edges), although to date the problems of fabricating low-loss waveguides in this material have prevented any observation of dark spatial solitons [19].

3 Higher-Order Kerr-Like Nonlinearities

The above discussion introduced the idea that resonances had a marked affect on the material's nonlinear response, modifying it from the ideal Kerr case. However, the influence of resonances should not be construed as necessarily negative from the experimental point of view, since the proximity of a resonance can cause very strong enhancement of the nonlinearity, lowering the threshold for soliton formation. Furthermore, the deviation from the Kerr response can result in saturation of the index change ($p < 1$), and this in fact is beneficial since it can stabilize $(2+1)$D solitons, allowing the creation of truly self-guided beams in isotropic media. However, as mentioned above, resonances inevitably introduce absorption (imaginary part of n_2), and hence propagation lengths can become limited and solitons may not form. Furthermore, since the resonance involves optical pumping between defined energy levels, the response time of the system is slow and determined by spontaneous decay from the upper energy level. In some instances this may result in the material response extending into the microsecond of millisecond range. In the case of optical pumping near a single-photon transition, the presence of resonance fluorescence may also introduce nonlocality into the nonlinearity.

A particularly interesting case of a strongly resonant material response being used for soliton studies is the case of metal vapors, such as rubidium. For this material, the laser is tuned close to the Rb D_2 line at 795 nm and hence the laser wavelength is critical on the 100 MHz scale. The transition can be considered as an approximation to a two-level atom. During optical pumping, atoms are excited from the $5S_{1/2}$ ground state to the $5P_{3/2}$ excited state. At sufficiently high pump levels the population of the lower level becomes bleached, reducing the material absorption. If the laser is detuned slightly from the line center, this absorption bleaching results in either a reduction (for low frequency detuning) or an increase (for high frequency detuning) of the refractive index of the material. Because absorption bleaching occurs, the material losses can drop to the point where a soliton can form and propagate through the medium, and if low vapor densities are used, the influence of resonance fluorescence can be made negligible.

It has been shown that the response of such a medium can be accurately described using a two-level-atom model with an inhomogeneously broadened transition in the presence of strong spectral hole-burning. [20] Saturation of both the absorption α and the refractive response $n_{\rm nl}$ under some conditions is anticipated as the populations in the upper and lower levels become equalized (taking the statistical weights into account), since in such conditions the absorption must drop to zero. Inasmuch as there is no simple analytic function describing that process, approximate functions of the following form have proved useful in beam propagation simulations:

$$
\begin{aligned}
n_{\mathrm{nl}} &= \frac{n_{\mathrm{s}}\, I}{1 + I/I_{s1}}\,, \\
\alpha &= \frac{\alpha_{\mathrm{s}}}{\sqrt{1 + I/I_{\mathrm{s}2}}}\,,
\end{aligned}
\tag{1}
$$

where $I_{\mathrm{s}1}$ ($\mathrm{W/cm^2} = 5$, $I_{\mathrm{s}2}$ $(\mathrm{W/cm^2}) = 0.07$, n_{s} $(\mathrm{W/cm^2}) = 0.9 \times 10^{-17} N_{\mathrm{Rb}}$ $(\mathrm{cm^{-3}})$, and α_{s} $(\mathrm{cm^{-1}}) = 0.2 \times 10^{-11} N_{\mathrm{Rb}}$ $(\mathrm{cm^{-3}})$, where N_{Rb} is the rubidium vapor density.

Notice that the different effective saturation intensities for the refractive and absorptive responses lead to strong absorption bleaching at intensities of only a few $\mathrm{W/cm^2}$, whilst preserving the effective nonlinearity. As a result, typical experimental conditions are $N_{\mathrm{Rb}} \simeq 5 \times 10^{11}$ $\mathrm{cm^{-3}}$, $I \simeq 5$ $\mathrm{W/cm^2}$, and propagation length $\sim$ 5–10 cm. Experiments can therefore be carried out using quite low-power CW lasers ($\sim$ 20 mW), which makes the experiments considerably less demanding than those on the Kerr materials. Rubidium is not the only material that has been used for soliton studies; in fact the first work on dark spatial solitons was performed in a Na vapor cell [21] although experimentally it is much less demanding to use Rb because of its low evaporation temperature. In addition, optical pumping in solid materials (such as doped laser crystals) can potentially be used to create solitons. In all cases the important influence of saturation allows (2 + 1)D structures to be studied, and in fact most recent experiments have been performed in saturating media (of which Kerr-like materials, photorefractive materials, and quadratic materials are all examples).

Self-focusing nonlinearities also occur in semiconductor optical amplifiers that are electrically or optically pumped [8]. For wavelengths below the transparency point, the index change produced by light injection is positive, and bright solitons exist. This concept has been used to guide CW bright solitons in AlGaAs with powers of 350 mW at 0.9 μm.

From the theoretical point of view, self-guided beams created in such saturating materials are soliton-like, rather than solitons, since their behavior can deviate substantially from the ideal Kerr case. As a result they cannot, in general, be described analytically (although there are some special cases of saturation where analytic solutions for single solitons do exist). Furthermore, collisions between these quasi-solitons are generally inelastic and result in radiative losses but also permit a range of rich phenomena such as soliton fusion, splitting, and reshaping. Since such phenomena may be useful in technologies such as optical switching, the additional behavior afforded by saturation generally outweighs those properties lost by comparison with the Kerr case.

4 Photorefractive Materials

Photorefractive crystals have been used widely in the past few years to create $(1+1)$D and $(2+1)$D spatial solitons. A local refractive-index change is produced in these media through the action of spatially dependent light-induced and externally applied electric fields together with the electro-optic response of the material. As a result, photorefractive materials must possess a second-order nonlinearity and are generally noncentrosymmetric inorganic crystals, such as strontium barium niobate, lithium niobate, or barium titanate.

It is well established that there are *three different types* of photorefractive response that can lead to self-trapping. In the first case, diffraction is suppressed through two-beam coupling involving refractive-index gratings created by on- and off-axis components of the beam in the presence of an externally applied electric field [22]. The phase of the grating is shifted by the external field to create *phase* coupling between the interacting waves, as is necessary to compensate for diffraction. The effect is transient, lasting for only a fraction of a second depending on the material properties, since the external field will eventually be screened by the photogenerated charges in the material. Note that in the absence of the external electric field, the photorefractive two-beam interaction results in *energy* coupling between the incident and scattered waves, and this cannot compensate diffraction. Furthermore, the response is nonlocal and can be changed from self-focusing to self-defocusing by changing the sign of the applied field. An important property of these solitons, which is shared by two-beam energy coupling in photorefractive materials, is their independence of the local light intensity (although the response time is intensity dependent).

The second and third types of response, which give rise to so-called *screening* [23] and *photovoltaic* [24] solitons, result from nonuniform screening of an external field (the former type) or from photovoltaic fields (the latter type) which appear in the crystal under steady-state conditions. Unlike the case for first kind of soliton, the effective nonlinearity in these cases is, to a good approximation, local, and hence the shape and width of the solitons depend on the light intensity. It is characteristic of such solitons that they can be treated using a conventional nonlinear-Schrödinger-equation model by including a saturating nonlinearity. As noted earlier, saturation is important in stabilizing $(2+1)$D spatial solitons against collapse, which occurs in media with a pure Kerr-like response. Hence $(2+1)$ solitons can be studied in photorefractive media.

Screening solitons have become the most commonly used in experiments, and hence it is worth dwelling on the physical processes involved in this nonlinearity (see Fig. 1). A strong external electric field is applied to the photorefractive crystal (usually strontium barium titanate) using electrodes attached to opposing crystal faces. In the absence of illumination, the field is distributed uniformly across the crystal, lowering (or raising) its average index of refraction via the Pockels effect. When a localized beam of light

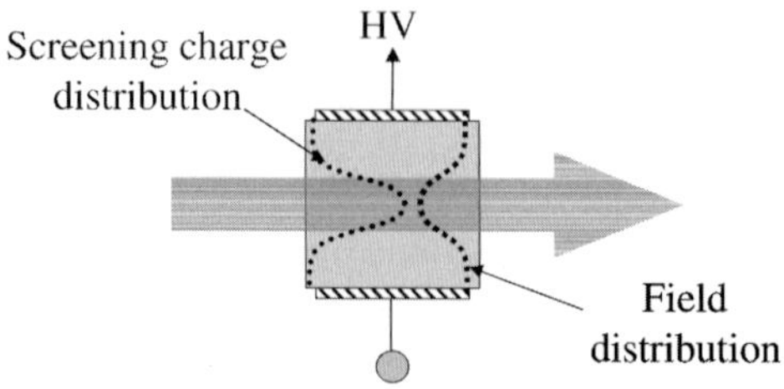

Fig. 1. Schematic of the creation of a screening spatial soliton in a photorefractive material. Optical excitation from the incident beam leads to the creation of a space charge near the beam. In the presence of an externally applied field, the charge screens the local field in the region of the beam and then, via the Pockels effects, this nonuniform field creates a local change in the refractive index

propagates through the crystal in a direction normal to the field, electric charges are excited from traps within the bandgap, increasing the charge density in the region of the beam. The presence of these charges modifies the conductivity of the crystal, and as a consequence the local field is screened. This modifies the local refractive index via the Pockels effect and can create the conditions for solitons to form. The largest changes in the field occur when the photoinduced charge in conductivity is of the order of the background value (which can be adjusted by uniformly illuminating the whole crystal with a background beam). The mechanism is clearly saturable, since a lower bound (zero) on the local field exists.

In the steady state, where a $(2+1)$D soliton has formed, the beam profile is invariant along the propagation direction (to first approximation), and the local field distribution (and hence index profile) is determined by the interaction of a two-dimensional charge distribution with a one-dimensional electric field. In general this leads to the creation of elliptical solitons, although in some cases the asymmetry can be small.

A general characteristic of photorefractive media is their slow response time but very high sensitivity. Laser powers in the mW range are adequate for most experiments, and photorefractive solitons have also been generated using low-power white-light sources, since the photorefractive response is generally broadband. Typically, solitons form on a timescale of several seconds and the charge distribution can remain "frozen" into the crystal for long periods, especially if it is stored in the dark. Whilst this can be useful in an experimental sense (especially in studies of the interaction of several solitons where it is convenient to be able to switch beams on and off without changing the index structure, to aid identification), for many practical applications a faster response is desired. To this end there has been increasing work on alternative materials such as Fe:InP.

Studies of the photorefractive response of semiconductors have appeared over the past two decades, with the high carrier mobility in these materials providing superior response speed compared with the oxide photorefractive crystals such as lithium niobate and strontium barium niobate. For example, a response time of 20 μs was reported for undoped GaAs [25] from two-beam coupling experiments, whilst in Fe:InP response times for photorefractive grating formation are in the microsecond range. A further advantage of the use of semiconductors is their long-wavelength response, and recently two-dimensional self-trapping of a 1.3 μm wavelength beam was reported in Fe:InP [26]. Thus semiconductors offer some possibility of response times in the microsecond or submicrosecond range, which would make devices using spatial solitons far more attractive from the practical point of view.

5 Second-Order Nonlinearities for Quadratic Solitons

The self-trapping mechanism which leads to spatial solitons in quadratic media is quite different from that used for the other solitons discovered to date [27]. As a result, with the exception of the diffraction and attenuation lengths, other characteristic lengths involved are different in this case. In the simplest physical embodiment, quadratic solitons may be created in an imperfectly phase-matched frequency-doubling crystal. The process, called "cascading", involves conversion first from the fundamental to a second-harmonic wave that travels at a different phase velocity from the fundamental. As a result of this imperfect phase matching, after some distance energy flows back from the second-harmonic wave to the fundamental (via parametric decay), but the wave so generated has its phase shifted relative to the unconverted fundamental. Through constructive interference, the phase of the combined fundamental wave becomes slightly altered. This phase shift is intensity dependent through the intensity sensitivity of the frequency-doubling process, and can result in a larger phase lag on the axis of a beam than in the off-axis regions, thereby counteracting diffraction. Because self-trapping here relies on rapid energy exchange with propagation distance, the characteristic nonlinear length here is the parametric gain length. For example, for a Type I Optical Parametric Generator process, $L_{\mathrm{nl}} = 2/[\omega\epsilon_0\sqrt{P(\omega_1)P(\omega_2)}d_{\mathrm{eff}}^{(2)}]$, so that $L_{\mathrm{nl}} < L_{\mathrm{d}}$ is required for soliton formation in the general sum frequency case. This example clearly works for positive phase matching, and the solitons degenerate into Kerr solitons for large phase mismatch.

A more general discussion of quadratic-soliton generation requires examination of the generating equations [27]. Again, for the simplest case of Type I second-harmonic generation, in a slab waveguide with diffraction along the x axis and waveguide confinement along the y-axis, we have, for propagation in the z direction,

$$\begin{aligned} -\mathrm{i}2k_1 \frac{\partial E(\omega_1;x,z)}{\partial z} + \frac{\partial^2 E(\omega_1;x,z)}{\partial x^2} &= -\Gamma E(\omega_2;x,z)E^*(\omega_1;x,z)\exp(\mathrm{i}\Delta kz), \\ -\mathrm{i}2k_2 \frac{\partial E(\omega_2;x,z)}{\partial z} + \frac{\partial^2 E(\omega_2;x,z)}{\partial x^2} &= -\Gamma E^2(\omega_1;x,z)\exp(-\mathrm{i}\Delta kz), \end{aligned} \tag{2}$$

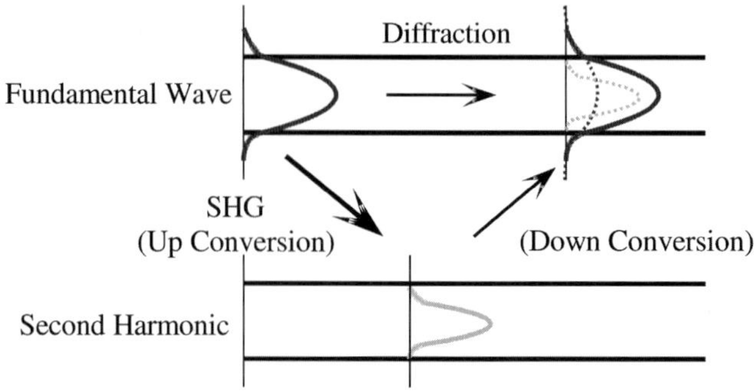

Fig. 2. Schematic representation of beam narrowing for both the fundamental and the harmonic components of a quadratic soliton. *Upper track*: fundamental. *Lower track*: harmonic. The *narrow dotted* line represents the down-converted fundamental, the *broad dotted* line represents the unconverted fundamental that has diffracted on propagation, and the *solid line* is the sum of the down-converted fundamental and the unconverted fundamental

where Δk is the wavevector mismatch and $\omega_2 = 2\omega_1$. Here the $E(\omega_{1,2}; x, z)$ are the complex, slowly varying amplitudes, the second term in each case describes diffraction in the x–z plane of the waveguide, and the nonlinear driving term Γ is proportional to the inverse of the parametric gain length L_{nl}. As shown schematically in Fig. 2, this term drives the growth of the harmonic $E(\omega_1)^2$ so that the generated harmonic is spatially narrower than the fundamental. Furthermore, the regenerated fundamental, $\propto E(\omega_1)^* E(2\omega_1)$, is also narrower than the original fundamental. Acting against this mutual beam narrowing is diffraction, and the robust balance between these mechanisms leads to a quadratic soliton. The key parameter is the parametric gain length, which must be shorter than the diffraction length.

It is clear that this process is not restricted to second-harmonic generation but can be equally applied to other wave-mixing processes that involve energy up-conversion and down-conversion in the presence of imperfect phase matching.

Because of the coherent interaction between the multifrequency waves involved in soliton propagation, the coherence length $L_{\mathrm{coh}} = \pi/(k_1 + k_2 - k_3)$ is an additional characteristic length. The known increase in the threshold intensity with detuning from the phase-matching condition implies that $L_{\mathrm{coh}} > L_{\mathrm{nl}}$ is probably required (the interplay between L_{coh} and L_{nl} has not been investigated). The existence of two or more waves at different frequencies in optically noncentrosymmetric media introduces the well-known problem of "walk-off" into many parametric mixing processes. Since all of the interacting waves must propagate together for a quadratic soliton to be formed, the coupling between the waves must be strong enough to overcome the walk-off. This means that $L_{\mathrm{wo}} > L_{\mathrm{nl}}$ is necessary.

A necessary criterion for quadratic-soliton formation is a material and wavelength combination which results in very efficient parametric mixing. Parametric processes involving second-order nonlinearities have been investigated and implemented in various frequency conversion devices since the early days of nonlinear optics. As a result there are many crystals and crystal geometries which have been developed, and these are all useful for quadratic solitons. In fact the first quadratic-soliton experiments were performed with a standard KTP doubling crystal at 1064 nm in a geometry in which the walk-off angles are a fraction of a degree [28]. Subsequent soliton experiments in bulk media all utilized standard materials, namely BBO, and LBO, $LiIO_3$, and QPM (quasi-phase-matched) $LiNbO_3$ [29–31].

The same comments are appropriate for quadratic solitons in waveguides, i.e. in the 1D case. The first (and only) experiments utilized noncritically, birefringently phase-matched $LiNbO_3$ waveguides first developed for optical parametric oscillators [32]. They required temperature tuning to achieve phase matching at the operating wavelength of 1.32 μm.

What can one expect with respect to different materials in the future? High-quality parametric processes have been reported in QPM KTP and $LiTaO_3$, both in waveguides and in bulk media, and these are prime candidates [33,34].

One of the barriers to concrete applications of quadratic solitons is that all of the materials discussed above have maximum phase-matchable nonlinearities of only 20 pm/V. This means that intensities of the order of hundreds of MW/cm^2 are needed for soliton formation. There are other materials which have larger nonlinearities, but at present they are not suitable for solitons. For example, single-crystal organics have values in the hundreds of pm/V.[1] However, there are two factors which are not advantageous in these materials. The transmission windows of these materials tend to be small because the larger the nonlinearity is, the longer the wavelength of the main absorption feature and the stronger this feature is. Furthermore, the large nonlinearities are accessed by electric fields along crystalline axes and are not phase-matchable, owing to both the large index dispersion and the birefringence. The phase-matchable geometries which have been found typically have very large walk-off angles [36]. Unless some processing mechanism can be found for implementing some form of quasi-phase matching along crystal axes, or some unique crystal with just the right optical properties is discovered, the prospects for these materials are not good.

Another material system with large-second order nonlinearities is the semiconductors. For example, GaAs has an off-diagonal nonlinearity of ~ 130 pm/V. However, phase matching is impossible in bulk materials because the material is optically anisotropic and the dispersion in the refractive index with wavelength is large. However, progress has been made in engineering phase matching in a waveguide geometry by two different techniques [37,38].

[1] See extensive material listing in [35].

Currently, the losses are too large for implementing any efficient parametric process, the first requisite for making quadratic solitons.

6 Conclusions

Self-trapped optical beams that display soliton-like behavior have been the subject of intense study in recent years, generally in materials whose nonlinear response deviates from the normal Kerr response. Whilst the range of materials that are available for experimental studies is much broader than in the temporal domain and has permitted a wide range of demonstration experiments, in all cases practical uses of these self-trapped beam are compromised by the material limitations. For Kerr solitons themselves, the third-order optical nonlinearity is generally too small, requiring impractically high beam intensities before solitons can be observed. Whilst the Kerr response of many materials has been measured, often the data fail to determine both the real and the imaginary part of the nonlinearity, or the reported values are such that the absorptive parts from linear or multiphoton absorption are too high for solitons to be created. At present one would have to be pessimistic about the chances of large Kerr nonlinearities being discovered in high-transparency optical materials suitable for the practical use of spatial solitons.

Saturating nonlinearities have been generally used in recent experiments, and here there are at least three classes of nonlinearity that have proved useful: Kerr-like behavior in materials pumped close to a single-photon resonance, photorefractive materials, and quadratic materials. In each case the required laser intensity is generally much smaller than that required for the Kerr materials: by around twelve orders of magnitude in the case of photorefractive materials! Nevertheless the materials remain a limitation. This may be because sensitivity has come at the price of poor temporal response, because of the limited crystal sizes that are available, because of inadequate second-order nonlinearity, because of unacceptable restrictions on operating wavelength, or for other reasons. As a result it is clear that if spatial solitons are to make the transition from a laboratory curiosity to an optical phenomenon that will be applied in an engineering sense, new materials will be needed. With this in mind, one can bracket the ideal response required for such a material. If an application in photonics is envisaged involving interconnections to conventional optical fibres, then the effective nonlinear index change must be ~ 0.005, a large value compared with that achieved in most experiments to date. This would need to be obtained at beam intensities less than $1\ \mathrm{W/cm^2}$, and at wavelengths useful to the telecommunications community, ~ 1.3 or $\sim 1.55\ \mu\mathrm{m}$. Whilst the response time could be in the millisecond or microsecond range to be useful for switching applications involving network reconfiguration, as time goes on, faster switching speeds will inevitably become necessary. The material would inevitably have to be a solid with good temperature stability, and be capable of being processed into a form convenient for interconnection to optical fibres. Let us hope that this does not define the fabled "unobtainium" of materials science!

References

1. *Nonlinear Optical Properties*, Handbook of Laser Science and Technology, Supplement 2: Optical Materials, ed. by M.J. Weber (CRC Press, Ann Arbor, 1995) pp. 269–333.
2. G.P. Agrawal, *Nonlinear Fiber Optics* (Academic Press, Boston, 1989).
3. M. Segev, Optical spatial solitons, J. Opt. Quantum Electron. **30**, 503–533 (1998).
4. G.I. Stegeman, M. Segev, Optical spatial soliton and their interactions: universality and diversity, Science **286**, 1518–1523 (1999).
5. S. Blair, K. Wagner, R. McLeod, Material figures of merit for spatial soliton interactions in the presence of absorption, J. Opt. Soc. Am. B **13**, 2141–2153 (1996).
6. M.J. LaGasse, K.K. Anderson, C.A. Wang, H.A. Haus, J.G. Fujimoto, Femtosecond measurements of the non-resonant nonlinear index in AlGaAs, Appl. Phys. Lett. **56**, 417–419 (1990).
7. G.I. Stegeman, A. Villeneuve, J. Kang, J.S. Aitchison, C.N. Ironside, K. Al-hemyari, C.C. Yang, C.-H. Lin, H.-H. Lin, G.T. Kennedy, R.S. Grant, W. Sibbett, AlGaAs below half bandgap: the silicon of nonlinear optical materials, Int. J. Nonlinear Optical Physics **3**, 347–371 (1994).
8. C. Kutsche, P. LiKamWa, J. Loehr, R. Kaspi, Quasi-CW self-guided optical beams in GaAs-AlGaAs double heterostructure slab waveguides, Electron. Lett. **34**, 906–907 (1998).
9. B. Lawrence, W. Torruellas, M. Cha, G.I. Stegeman, J. Meth, S. Etemad, G. Baker, Identification and role of two photon absorption in the p-conjugated polymer paratoluene-sulfonate, Phys. Rev. Lett **73**, 597–600 (1994).
10. D.Y. Kim, M. Sundheimer, A. Otomo, G.I. Stegeman, W.G.H. Horsthuis, G.R. Mohlmann, Third order nonlinearity of DANS waveguides at 1319 nm, Appl. Phys. Lett. **63**, 290–292 (1993).
11. H. Kanbara, H. Kobayashi, K. Kubodera, T. Kurihara, T. Kaino, Optical Kerr shutter using organic nonliner optical materials in capillary waveguides, IEEE Phot. Technol. Lett. **3**, 795–797 (1991).
12. J.U. Kang, T.D. Krauss, F.W. Wise, B.G. Aitken, N.F. Borerelli, Femtosecond measurement of enhanced optical nonlinearities of sulfide glasses and heavy-metal-doped oxide glasses, J. Opt. Soc. Am. B **12**, 2053–2059 (1995).
13. M. Asobe, K. Suzuki, T. Kanamori, K. Kubodera, Nonlinear refractive index measurement in chalcogenide-glass fibers by self-phase modulation, Appl. Phys. Lett. **60**, 1153–1154 (1992).
14. A. Barthelemy, S. Maneuf, C. Froehly, Propagation soliton et auto-confinement de faisceaux laser par non linearite optique de Kerr, Opt. Commun. **55**, 201–206 (1985).
15. J.S. Aitchison, A.M. Weiner, Y. Silberberg, M.K. Oliver, J.L. Jackel, D.E. Leaird, E.M. Vogel, P.W. Smith, Observation of spatial optical solitons in a nonlinear glass waveguide, Opt. Lett. **15**, 471–473 (1990).

16. J.U. Kang, G.I. Stegeman, A. Villeneuve, J.S. Aitchison, AlGaAs below half bandgap: a laboratory for spatial soliton physics, J. Eur. Opt. Soc., A **5**, 583–594 (1996).
17. U. Bartuch, U. Peschel, T. Gabler, R. Waldhaus, H.-H. Horhold, Experimental investigations and numerical simulations of spatial solitons in planar polymer waveguides, Opt. Commun. **134**, 49–54 (1997).
18. S. Yamakawa, K. Hamashima, T. Kinoshita, K. Sasake, Temporal solitary subpicosecond pulse propagation in a dye-doped polymer slab waveguide with a negative nonlinear refractive index, Appl. Phys. Lett. **72**, 1562–1564 (1998).
19. M. Samoc, A. Samoc, B. Luther-Davies, Z. Bao, L.Yu, B. Hsieh, U. Scherf, Femtosecond Z-scan and degenerate four wave mixing measurements of real and imaginary parts of the third-order nonlinearity of soluble conjugated polymers, J. Opt. Soc. Am. B **15**, 817–825 (1988).
20. V. Tikhonenko, J. Christou, B. Luther-Davies, Spiraling bright spatial solitons formed by the breakup of an optical vortex in a saturable self-focussing medium, J. Opt. Soc. Am. B **12**, 2046–2052 (1995).
21. G. Swartzlander, D. Andersen, J.J. Regan, H. Yin, A. Kaplan, Spatial dark-soliton stripes and grids in self-defocusing materials, Phys. Rev. Lett. **66**, 1581–1585 (1991).
22. M. Segev, B. Crosignani, A. Yariv, B. Fischer, Spatial solitons in photorefractive media, Phys. Rev. Lett. **68**, 923–926, (1992).
23. M. Segev, G.C. Valley, B. Crosignani, P. Di Porto, A. Yariv, Steady-state screening solitons in photorefractive materials with external applied field, Phys. Rev. Lett. **73**, 3211–3214 (1994); M.D. Iturbe-Castillo, P.A. Marquesz Aguilar, J.J. Sanchez-Mondragon, S. Stpeanov, V. Vysloukh, Spatial solitons in photorefractive $Bi_{12}TiO_{20}$ with drift mechanism of nonlinearity, Appl. Phys. Lett. **64**, 408–410 (1994).
24. M. Taya, M.C. Bashaw, M.M. Fejer, M. Segev, G.C. Valley, Observation of dark photovoltaic spatial solitons, Phys. Rev. A **52**, 3095–3100 (1995).
25. M.B. Klein, Beam coupling in undoped GaAs at 1.06 μm using the photorefractive effect, Opt. Lett **9**, 350–352 (1984).
26. M. Chauvet, S.A Hawkins, G.J. Salamo, M. Segev, D.F. Bliss, G. Bryant, Self-trapping of two-dimensional optical beams and light-induced waveguiding in photorefractive InP at telecommunications wavelengths, Appl. Phys. Lett. **70**, 2499–2501 (1997).
27. G.I. Stegeman, D.J. Hagan, L. Torner, $\chi^{(2)}$ cascading phenomena and their applications to all-optical signal processing, mode-locking, pulse compression and solitons, J. Opt. Quantum Electron. **28**, 1691–1740 (1996).
28. W.E. Torruellas, Z. Wang, D.J. Hagan, E.W. VanStryland, G.I. Stegeman, L. Torner, C.R. Menyuk, Observation of two-dimensional spatial solitary waves in a quadratic medium, Phys. Rev. Lett. **74**, 5036–5039 (1995).
29. P. Di Trapani, G. Valiulis, W. Chianglia, A. Adreoni, Two-dimensional spatial solitary waves from travelling wave parametric amplification of the quantum noise, Phys. Rev. Lett. **80**, 265–268 (1998).
30. X. Liu, L.J. Qian, F.W. Wise, Generation of optical spatiotemporal solitons, Phys. Rev. Lett. **82**, 4631–4634 (1999).
31. B. Bourliaguet, V. Couderc, A. Barthelemy, G.W. Ross, P.G.R. Smith, D.C. Hanna, C. De Angelis, Observation of quadratic spatial solitons in periodically poled lithium niobate, Opt. Lett. **24**, 1410–1412 (1999).

32. R. Schiek, Y. Baek, G.I. Stegeman, One-dimensional spatial solitons due to cascaded second-order nonlinearities in planar waveguides, Phys. Rev. E **53**, 1138–1141 (1996).
33. D. Eger, M. Oron, M. Katz, A. Zussman, Highly efficient blue light generation in $KTiOPO_4$ waveguides, Appl. Phys. Lett. **64**, 3208–3209 (1994).
34. S.-Y. Yi, S.-Y. Shin, Y.-S. Jin, Y.-S. Son, Second harmonic generation in a $LiTaO_3$ waveguide domain inverted by proton exchange and masked heat treatment, Appl. Phys. Lett. **68**, 2493–2495 (1996).
35. S. Singh, in *CRC Handbook of Laser Science and Technology, Supplement 2: Optical Materials*, ed. by M.J. Weber (CRC Press, Ann Arbor, 1995) pp. 147–266.
36. U. Meier, M. Bosch, C. Bosshard, F. Pan, P. Günter, Parametric interactions in the organic salt 4-N,N-dimethylamino-4'-N'-methyl-stilbazolium tosylate at telecommunication wavelengths, J. Appl. Phys. **83**, 3486–3489 (1998).
37. S.J.B. Yoo, Wavelength conversion technologies for WDM network applications, J. Lightwave Technol. **14**, 955–966 (1996)
38. P. Bravetti, A. Fiore, V. Berger, E. Rosencher, J. Nagle, O. Gauthier-Lafaye, "5.2–5.6 μm source tunable by frequency conversion in a GaAs-based waveguide, Opt. Lett. **23**, 331–333 (1998).

One-Dimensional Spatial Solitons in Kerr Media

Yaron Silberberg and George I. Stegeman

Summary. We give an overview of the experimental activity related to spatial solitons in Kerr or Kerr-like media. We review the results obtained in this area, starting from the early experiments up to the most recent achievements in semiconductors, discussing the formation, control, and application of solitons in one-dimensional structures.

1 Introduction

As outlined in the introductory chapter, optical solitons were discussed first for self-focusing Kerr media ($\Delta n = n_2 I$, where Δn is the index change induced optically by the intensity I) because they lead to the nonlinear Schrödinger equation (NLSE), which has analytical solutions in $(1+1)$D [1,2] (here the first "1" refers to the number of dimensions in which a beam spreads, and the second "1" to the propagation coordinate). As a result, the initial interest in solitons was more or less restricted to Kerr media in waveguide geometries. This included fibers, in which the temporal pulse spreading was associated with the spectral dispersion in the group velocity (group verlocity dispersion, GVD) [3]. In fact, much of what we know about Kerr solitons was first discussed and observed in fibers with temporal solitons [4]. When one is dealing with spatial solitons, the key characteristic is that a finite beam propagates in a slab waveguide without beam spreading in the plane of the waveguide. It is this latter case that is discussed in this chapter.

The initial problem was to find true Kerr materials, as discussed in the preceeding chapter, and to fabricate high-quality waveguides in them. Note that saturating nonlinearities will also give rise to self-trapped beams, i.e. spatial solitons, in slab waveguides, and having a Kerr material is not crucial to generating $(1+1)$D solitons. The focus on the Kerr case is a consequence of the desire to work with solitons whose properties and collisions can be described analytically [2]. In fact, most of the experiments on slab waveguides reported have utilized Kerr materials, or saturating materials in regions of operation in which they are effectively Kerr. In this chapter we shall discuss spatial solitons in $(1+1)$D waveguides, focusing on semiconductors. This will include materials with both passive and active nonlinearities. The bulk of the work reported to date has been on homogeneous waveguides of the type shown on the left-hand-side of Fig. 1. That is, all of the waveguiding

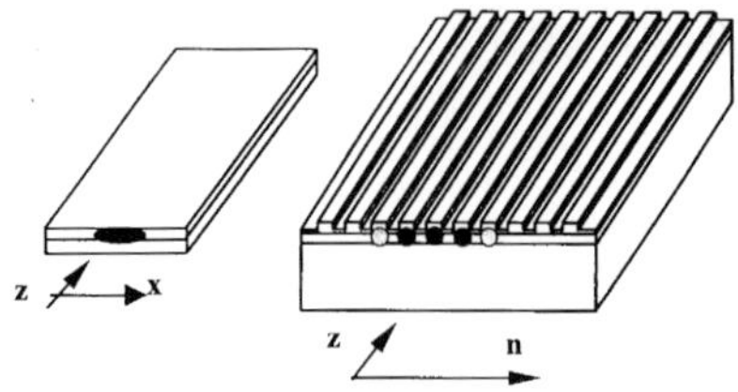

Fig. 1. Waveguide geometries for generating homogeneous spatial solitons (*left*) and discrete spatial solitons (*right*). Also shown are typical intensity distributions for the two cases

media have uniform optical properties. Very recently, spatial solitons have also been investigated in coupled nonlinear arrays. An example is shown in Fig. 1 (right). The added degree of freedom introduced by the discreteness leads to new soliton propagation properties. The corresponding theory is discussed in the chapter by Lederer, Darmanyan, and Kobyakov, and here we shall focus on the experiments only.

2 Early and Miscellaneous Experiments

The first experiments illustrating self-trapped beams were actually reported for bulk saturating media by Bjorkholm and Ashkin in 1974 [5]. These authors worked with sodium vapor near that atom's strong yellow absorption and emission line. Because there are no analytical solutions for this kind of self-trapped beam, these first experiments failed to generate a surge of interest in the soliton community. It was just over ten years later that the first experiments on beam trapping in one transverse dimension were reported by Froehly and coworkers in bulk liquid CS_2 [6]. They did not actually use a waveguide but were able to produce confinement in one dimension in a very

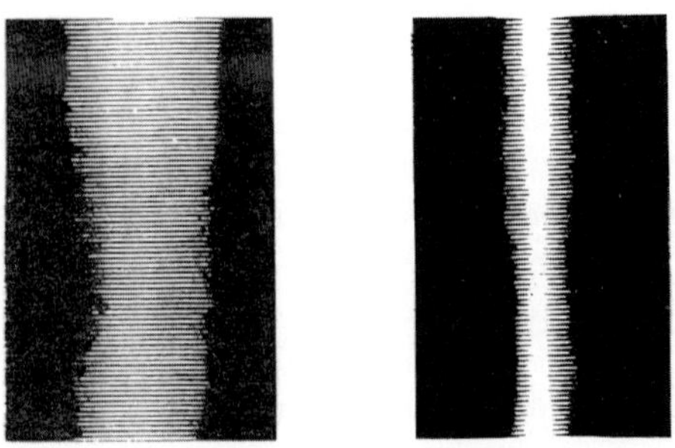

Fig. 2. Self-trapping of decoupled arrays of (1 + 1)D-like spatial solitons formed by interference patterns in liquid CS_2. (*Left*) diffracted interference pattern obtained at low input powers. (*Right*) self-trapped interference pattern at high input powers (reproduced from [6])

clever way. They formed an interference pattern in one dimension by interfering two high-power beams in CS_2 at a small crossing angle. The dark zones of the pattern do not allow light to diffract across the dimension orthogonal to the dark stripes, effectively creating a series of parallel waveguides. These workers observed self-trapping in the direction parallel to the bright stripes of the interference pattern. As shown in Fig. 2, the whole interference pattern narrows in the planes of the fringes and the full width returns to its input value. Thus each bright fringe behaves like a spatial soliton. In a subsequent experiment these authors reported a true waveguide demonstration of (1+1)D spatial solitons. The sample was liquid CS_2 sandwiched between two glass slides separated by a few microns and beam self-trapping in the plane of the slides was demonstrated [7]. These authors were able to show solitons for a number of different guided modes of the slab waveguide structure.

The next experiments dealt with glass waveguides [8]. The nonlinearity was over an order of magnitude less than for CS_2 and it was necessary to use 100 femtosecond pulses to avoid damage to the input facets. The first observation led soon to the demonstration of the collision of in-phase and out-of-phase solitons [9]. Two solitons were launched along parallel trajectories and the interaction between them was found to depend on their relative phase. It was demonstrated that in-phase solitons attract and out-of-phase solitons repel (see Fig. 3) [9]. Further collision experiments in CS_2 completed

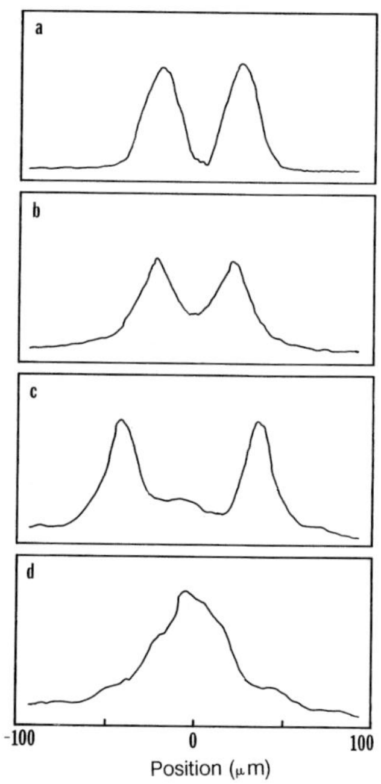

Fig. 3. Example of soliton collisions in a Kerr medium (glass waveguide). (**a**) The two input solitons. (**b**) The output intensity of the two solitons launched separately in time. (**c**), (**d**) The output fields obtained for two parallel input Kerr solitons in a glass waveguide: (**c**) in phase, (**d**) out of phase (reproduced from Ref. [9])

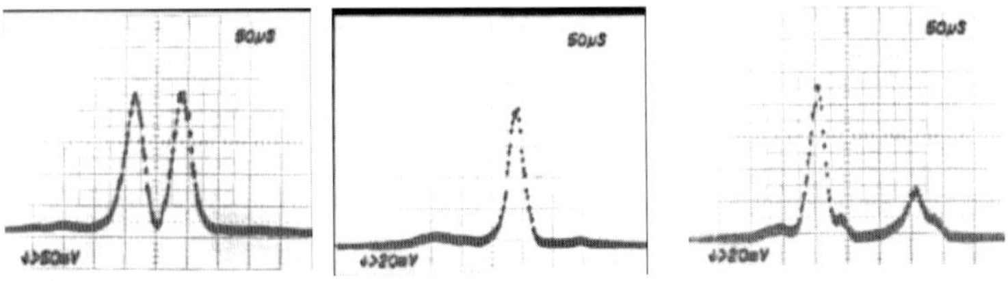

Fig. 4. Example of energy exchange in soliton collisions in a Kerr medium (CS_2). *Left*: the two input Kerr solitons in a CS_2 (1+1)D waveguide. *Center*: the output for a relative phase of $\pi/2$ between the solitons. *Right*: the output for a relative phase of $3\pi/2$ between the solitons (reproduced from [10]). In all cases, the horizontal scale is 50 µs/division, and the vertical scale is arbitrary

the picture of Kerr soliton interactions [10]. For relative phase angles of the two solitons of $\pm\pi/2$, energy flows from one soliton into the other (Fig. 4).

Chronologically, in the early 1990s and thereafter, there were a series of experiments on $(1 + 1)$D spatial solitons in AlGaAs waveguides. These will be discussed in a separate section. At this point, for completeness, we also mention that subsequently spatial solitons were excited in polymer and liquid-crystal waveguides. In the polymer case, the waveguide material was a film of DP-PPV/DP-PFV , a solution-processable derivative of poly-phenylene vinylene, which is a conjugated polymer [11]. The experiments were performed with a Ti:sapphire laser in a spectral region in which two-photon absorption was minimal. An advantage of using such polymeric materials is that they can be easily spin-coated onto glass substrates to produce high-quality waveguides. In the liquid-crystal case, there have been a number of reports over the years, but the principal impediment has been the large scattering losses usually obtained in liquid-crystal waveguides [12]. The most convincing work was recently published, in 2000 [13]. In that work solitons were propagated with minimum loss for distances of the order of a millimeter ($\alpha \simeq 5\ \text{cm}^{-1}$) [13]. Electric fields were used to align the nematic liquid crystal molecules at an angle to the propagation direction. The material used was the classical liquid crystal E7, and the power required to generate the solitons was only 4.2 mW.

3 Spatial Solitons in Homogeneous Semiconductor Waveguides

The largest body of work on spatial solitons in homogeneous waveguides has been done with semiconductors. This can be separated into two distinct areas. The first deals with AlGaAs waveguides operated at photon energies just below one-half of the semiconducting bandgap. Here spatial solitons have been studied in both homogeneous and discrete waveguide systems (see Fig. 1). The second area deals with semiconductor amplifiers.

3.1 Homogeneous AlGaAs Passive Waveguides

The $Al_xGa_{1-x}As/GaAs$ multilayer material system with an Al : Ga ratio of 0.18 : 0.82 for the waveguide core has turned out to be ideal for the investigation of spatial solitons in the 1.55 μm communications wavelength region. For photon energies less than one-half of the semiconductor bandgap and in the absence of impurity or excitonic states, two-photon absorption is forbidden and the nonlinear response of the waveguide is effectively Kerr [14]. The resulting nonlinearity, although much smaller than can be found near the bandgap, is self-focusing and still about 800 times that of fused silica [15]. Furthermore, the linear and nonlinear losses can be neglected over very useful ranges of input intensity for soliton generation [15]. High-quality waveguides have been made by MBE deposition of AlGaAs layers on a GaAs substrate. They typically consisted of a 1 μm thick guiding layer of $Al_{0.18}Ga_{0.82}$, a 4 μm thick lower cladding region of $Al_{0.4}Ga_{0.6}As$ (deposited on the GaAs substrate), and a 1.5 μm thick upper cladding region of $Al_{0.3}Ga_{0.7}As$. Etching of the upper cladding layer was typically used to define channel waveguides for the investigation of discrete solitons . The linear loss of these waveguides was measured to be less than 0.2 cm^{-1} at the operating wavelength. The measured waveguide nonlinearities are listed in Table 1. Note that the nonlinearities for the TE and TM polarizations are approximately equal, and that the ratio of the cross- to self-phase modulation coefficients is 0.94 ± 0.05, almost unity. This combination of values allows Manakov solitons to be excited.

The first experiments demonstrating that spatial solitons in AlGaAs could be successfully generated were reported by Aitchison et al. [18]. A cylindrical lens was used to shape the radiation, focused onto the end face of the slab waveguide, into a beam a few microns wide along the x axis and tens of millimeters wide along the y axis. Typically, the input beam at low powers diffracted at the output to 5–10 times the input size, i.e. the samples were 5–10 diffraction lengths long. Peak powers of $\sim$ 500 W were needed to excite spatial solitons whose beam width did not change with propagation distance. For higher powers, the output beam size was found to vary only weakly with

Table 1. Measured nonlinearities in cm^2/W for the AlGaAs waveguides discussed in the text; n_2(TE/TM) (= n_2(TM/TE)) is the cross-phase modulation coefficient in which the strong beam is TE or TM polarized and the index change is TM or TE polarized, respectively. Note that for the TE mode the electric-field vector is polarized parallel to [110], and for the TM mode it is parallel to [001].

n_2(TE)	n_2(TE)	n_2(TE/TM)	Reference
$1.5 \pm 0.2 \times 10^{-13}$	$1.4 \pm 0.2 \times 10^{-13}$	$1.4 \pm 0.2 \times 10^{-13}$	[16]
1.3×10^{-13}	1.0×10^{-13}	1.0×10^{-13}	[17]

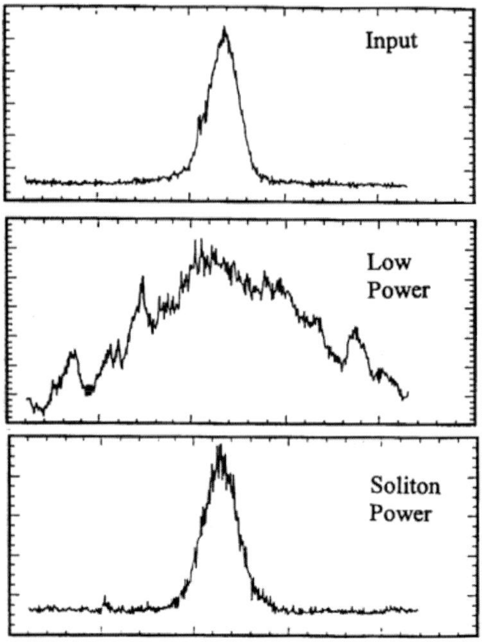

Fig. 5. The spatial variation in beam power along the x axis of the input beam (*top*), and at the output waveguide facet at a low input power (beam diffracts, *middle*) and at the soliton power (no diffraction, *bottom*)

input power. An example of the spatial profiles of the input beam and of the low- and high-power outputs is shown in Fig. 5.

The generation of a spatial soliton has been used to demonstrate a saturable absorber useful for mode locking of lasers [19]. The desired property is that the transmission of a device increases with increasing intensity, i.e. the throughput power increases faster than linearly with input power. A specially designed waveguide horn was fabricated so that at low powers the beam radiated by the waveguide horn diffracted, reducing the throughput through an output aperture. At higher powers, self-focusing took place and spatial solitons were generated. As a result, a larger fraction of the input power was transmitted. This was implemented successfully in an AlGaAs waveguide [19].

The near-ideal properties of these spatial solitons allowed the detailed investigation of soliton properties, interactions, and applications. Specifically, a variety of interactions were studied:

- the trapping of weak beams by spatial solitons;
- the dragging and trapping of spatial solitons;
- the generation of stable Manakov solitons, that is, solitons consisting of two coupled orthogonally polarized beams;
- the stability of the solitons for both polarizations;

- the electronic control of the propagation direction of spatial solitons leading to a demonstration of reconfigurable interconnects .

The propagation of two, orthogonally polarized, interacting beams can be written in approximation of the slowly varying amplitude and phase as

$$\begin{aligned}\frac{\partial v}{\partial z} &= \frac{\mathrm{i}}{2k_{\mathrm{TM}}}\frac{\partial^2 v}{\partial x^2} - \mathrm{i}k_0 n_2\left[(C|v|^2 + A|u|^2)v - Bu^2v^*\exp(-2\mathrm{i}\Delta kz)\right],\\ \frac{\partial u}{\partial z} &= \frac{\mathrm{i}}{2k_{\mathrm{TE}}}\frac{\partial^2 u}{\partial x^2} - \mathrm{i}k_0 n_2\left[(|u|^2 + A|v|^2)u - Bv^2u^*\exp(2\mathrm{i}\Delta kz)\right],\end{aligned} \tag{1}$$

for the TE (u) and TM (v) optical fields, polarized along the birefringence axes x and y, respectively. The first term on the right-hand side describes spatial diffraction in the plane, along the x axis. The second two terms introduce phase modulation, and the last term in each of the two equations introduces mixing between the two polarizations, a four-wave-mixing (FWM) term. Here k_0 is the propagation constant in vacuum, k_{TE} and k_{TM} are the waveguide propagation constants for polarizations along the two birefringent axes, $k_{\mathrm{TE}} \sim k_{\mathrm{TM}}$ so that $\Delta k = k_{\mathrm{TE}} - k_{\mathrm{TM}}$ is small, and n_2 is the nonlinear refractive index in the medium. C describes the anisotropy of self-phase modulation (SPM), and A is the ratio of cross-phase modulation (XPM) to SPM. B, the ratio of FWM to SPM, is responsible for coupling between the two polarizations. The values of the coefficients A, B, and C are determined by the characteristics of the medium. For the orientation of the AlGaAs waveguides used, $A \approx 1$, $B \approx 0.5$, and $C \approx 1$. Solving these equations, usually numerically, describes the coupled propagation of the two beams, including spatial-soliton beams.

One of the interesting features of this research was that it was possible to control the FWM term. This was achieved by beam-splitting the laser into two polarizations which were directed through various different optical components such as prisms, delay lines, etc. before being recombined at the sample input. The ultrashort, pulses of a few hundred femtoseconds were strongly affected by GVD in the optics, and hence the two polarizations were no longer coherent relative to each other at the input into the waveguides. As a result, the FWM term was effectively zero [20]. This clearly depends on the spectral width of the pulses, and, for longer pulses ($\sim$ 5 psec) with smaller spectral widths, the effect of the different GVD in the two optical paths was much less and the input beams were still coherent relative to each other.

3.1.1 Weak-Beam–Spatial-Soliton Interactions

A spatial soliton creates an index profile $\Delta n \propto n_2 I(x)$ in the medium for an orthogonally polarized beam owing to cross-phase modulation, i.e. the n_2(TE/TM) and the n_2(TM/TE) terms listed previously. Such a high-index region creates a channel waveguide for a weak orthogonally polarized beam. Therefore the trapping of a weak beam conceptually reduces to launching the weak probe beam into a channel waveguide with a solitonic-intensity index profile. In order to eliminate the FWM term, a time delay configuration was used with 600 fs pulses. Such soliton-induced guiding of signals can be used, for example, for reconfigurable interconnects [21]. The experiments were performed by directing the input pump beam (TE) normal to the end face of the sample and the weak beam at a small angle to the normal to the facet, the angle being chosen for the particular experiment being performed [22]. At low input powers, both the pump and the probe diffracted to 130 ± 10 µm at the output face, meaning that the sample was about eight diffraction lengths long. At a peak power of 550 W, the pump beam formed a spatial soliton and partially trapped the probe beam with an efficiency that depended on the relative incidence angle of the probe. The trapping efficiency versus angle is shown in Fig. 6. The width of the trapped component is exactly the same as that of the soliton. This is a direct consequence of the cross-phase modulation nonlinearity being equal to the self-phase nonlinearity, i.e. the equality of the self- and cross-phase modulations leads to the same induced index potential for both beams.

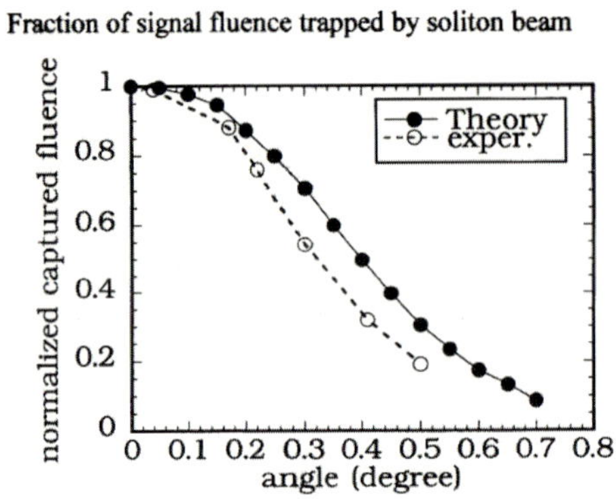

Fig. 6. Normalized signal beam fluence trapped by a spatial-soliton beam as a function of relative launching angle (reproduced from [22])

3.1.2 Dragging and Trapping of Orthogonally Polarized Spatial Solitons

Two orthogonally polarized spatial solitons, initially launched in slightly different directions, should lock together to copropagate in space by self-adjusting their "transverse velocities" (parallel to the input interface) to be equal, i.e. for equal input powers the "trapped" direction bisects the directions of the two input solitons. Such mutual soliton-trapping experiments were carried out in AlGaAs slab waveguides using a beam geometry for which

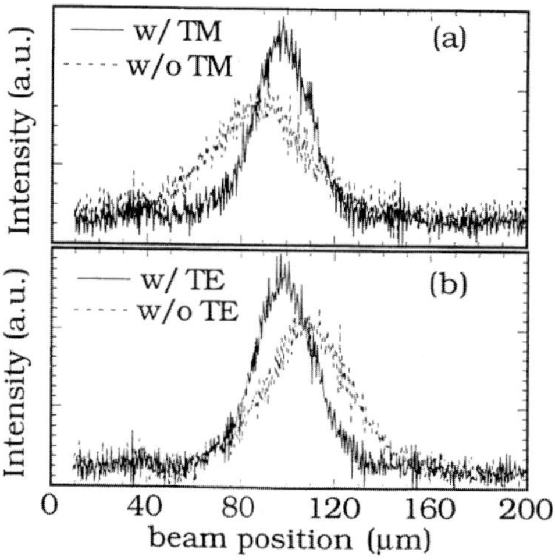

Fig. 7. (**a**) Output beam profile of TE solitons with and without the presence of a TM soliton. (**b**) Output beam profile of TM solitons with and without the presence of a TM soliton. The *solid* and *dotted lines* represent the output with and without the presence of the orthogonally polarized soliton, respectively. The solitons trap each other via cross-phase modulation and propagate together. The initial crossing angle was 0.10° (reproduced from [23])

$B = 0$ [23]. This means that the solitons interact only through the incoherent index change produced by the adjacent soliton. When only one soliton is present, it travels without change in shape from one end of the sample to the other, maintaining its original trajectory. For a relative launching angle of 0.1 degrees or less, the two beams self-trap, and overlap at the output. As shown in Fig. 7, the coupled solitons are spatially narrower than the individual solitons. In fact, they both narrow by approximately a factor of two, from 35 μm to 18 μm. The reason is clear from (1). Because $A = C = 1$, the index potential "seen" by each soliton is doubled for equal input powers. This leads to a reduction of the combined soliton width by a factor of two.

One of the properties of Kerr solitons is that the soliton number is conserved on collision, and this is apparently violated in this experiment. That is, two solitons enter the collision and one leaves. However, the soliton created is a Manakov soliton, which is an incoherent superposition of two single polarization Kerr solitons, so that in reality two solitons do leave the collision and the soliton number is conserved.

The ability of the solitons to trap each other depends on the relative launching angle. For larger relative angles, for example 0.2 degrees, as shown in Fig. 8, their output positions are shifted towards each other relative to the positions defined by the trajectories of the initial solitons when launched alone. Note that the trajectories at the output are parallel to those at the

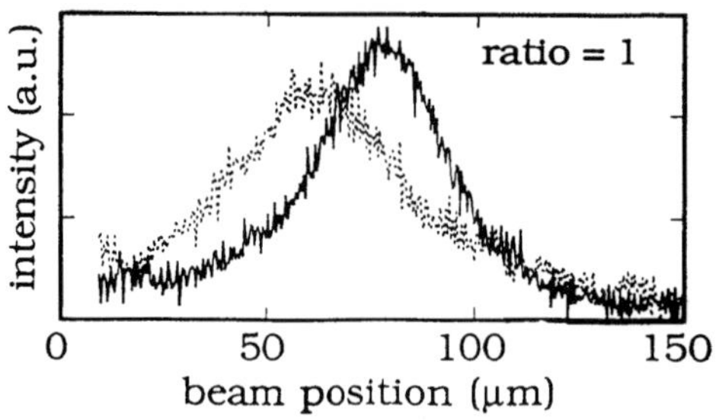

Fig. 8. Output beam profiles of TE solitons with and without the presence of TM soliton beams, for equal power inputs. The *solid* and *dotted lines* represent the TE soliton beam profile with and without the presence of the TM beam, respectively. The initial crossing angle was 0.2 degrees (reproduced from [23])

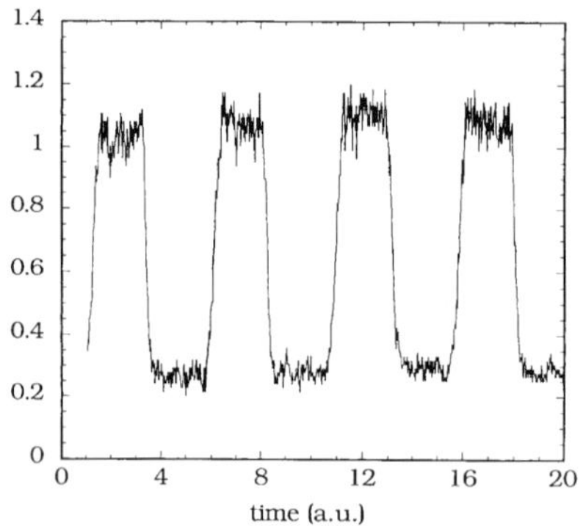

Fig. 9. An all-optically switched signal caused by soliton dragging (reproduced from [23])

input. The beams, however, are laterally shifted owing to the interaction. For equal power inputs, both spatial solitons are deflected (dragged) towards each other. Such soliton-induced lateral shifts have been used to implement all-optical switching based on appropriately placed detectors [23]. In this case a pinhole was used to optimize the switching function; the pinhole was placed at the peak of either the dragged or the undeflected soliton. To demonstrate switching, a slow mechanical chopper was inserted into the path of the pump beam, to simulate turning on and off of the control beam. The signal measured through the pinhole as a function of time is shown in Fig. 9. The maxima and minima represent the amount of signal when the pump beam is blocked and unblocked, respectively, by the chopper. This clearly shows the switching of a signal beam using an orthogonally polarized soliton beam.

3.1.3 Generation of Manakov Solitons

It was first noted by Manakov that two orthogonally polarized, mutually incoherent beams can combine to produce a stable soliton [24]. He found that it was possible to extend the classical inverse scattering technique to integrate (1) in the limit of no FWM term, $n_2(TE) = n_2(TM)$, and $C = 1$ (i.e. equality of the self- and cross-phase modulation coefficients). As discussed above, AlGaAs has the required special properties, and the FWM term can be effectively set to zero by using a specific dual-beam geometry. Therefore, planar AlGaAs slab waveguides provided a unique opportunity to excite Manakov solitons.

One of the properties of a Manakov soliton is that the soliton beam width is a constant, independent of the intensity ratio of the two orthogonal components as long as the total power is the same. Therefore, for a given power, the soliton beam width is the same for a Manakov spatial soliton as it is for a spatial soliton with only one polarization. Experiments were performed to verify this special property of Manakov spatial solitons [25]. First, a TE-polarized spatial soliton was excited with an input power of $\sim$ 550 W, and a beam width of 16 ± 2 μm. Then the same total input power was split in arbitrary intensity ratios between the TE- and TM-polarized beams. It was found that the two polarizations had identical shapes, and beam widths that were the same as those measured for a single-polarization soliton of the same power. This experiment was repeated for intensity ratios varying from 1 : 20 to 20 : 1 and identical results were obtained, as shown in Fig. 10. These results are all consistent with the generation and propagation of Manakov spatial soli-

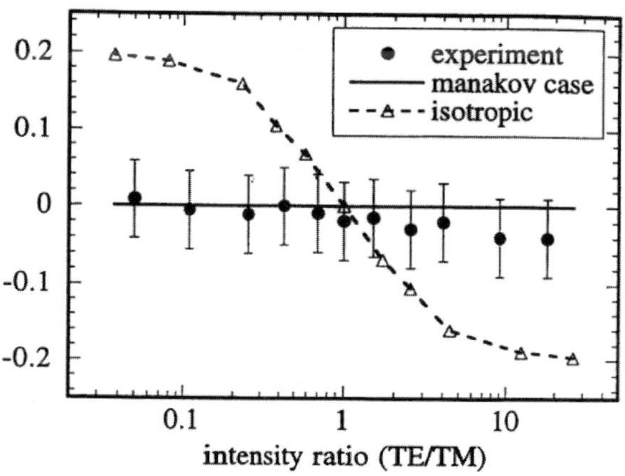

Fig. 10. The normalized difference (R $= (\omega_{\mathrm{TE}} - \omega_{\mathrm{TM}})/\omega_{\mathrm{TE}}$) between the output TE and TM beam widths (ω_{TE} and ω_{TM}, respectively) versus the intensity ratio $I_{\mathrm{TE}}/I_{\mathrm{TM}}$ of the two beams, colaunched at the input. The *solid line* is the Manakov prediction and the *dashed line with triangles* is for an isotropic medium, $I_{\mathrm{TE}} = I_{\mathrm{TM}}$ (reproduced from [25])

tons. In fact, the trapping result in Fig. 7 can be interpreted as the robust generation of a Manakov soliton.

3.1.4 Stability of Orthogonally Polarized Solitons

The existence of the FWM term (coherence between orthogonal polarizations) means that energy can be exchanged between the two orthogonal polarizations TE and TM. It also means that only one soliton is stable, the one with the lowest energy, and that the orthogonally polarized soliton will eventually decay into this stable soliton. This has been predicted theoretically by Akhmediev and collaborators, and verified experimentally [26,27]. In the experiment, either TE or TM spatial solitons were excited, along with a small fraction (2×10^{-3}) of the orthogonal (TM or TE) polarization owing to the (measured) interpolarization scattering in the waveguide [28]. This weak orthogonal polarization component acts as a "seed". If the "seed" polarization corresponds to the stable soliton, it will be amplified at the expense of its orthogonal polarization. The larger the input power, the faster the conversion into the stable mode. On the other hand, the "seed" will decay if the orthogonal polarization corresponds to the stable mode. In the sample investigated, the TE polarization corresponds to the "slow wave" and is the stable soliton. The results of the experiment are shown in Fig. 11. They show an increasing fractional conversion from the TM soliton into the TE polarization. They also show a power-independent fractional conversion of the TE into the TM polarization, which is interpreted as due to stray scattering. These results establish that the TE soliton (the slow soliton in this sample) is stable and the TM (the fast soliton) is unstable.

3.1.5 Soliton Reconfigurable Interconnect

The guiding of weak, orthogonally polarized signal beams by an index channel induced by a spatial soliton suggests that signal beams could be guided across a sample by an appropriately pointed soliton beam. Furthermore, the density of the output locations will not be limited by beam diffraction. The basic concept is summarized in Fig. 12, where the deflecting elements are prisms that can be turned on and off electronically. Thus the output destination is chosen electronically. The crossing angles are chosen to minimize the deflection of the solitons on crossing. The original plan was to use the electro-optic effect to turn the prisms on and off but it was found more effective to rely on carrier injection through triangular electrodes to change locally the refractive index and hence deflect the soliton (and signal).

The first experiment was to study the deflection properties of spatial solitons as they traversed a triangular region of modified refractive index. The geometry investigated is shown in Fig. 13a, and the results in Fig. 13b [29]. The key results are the following:

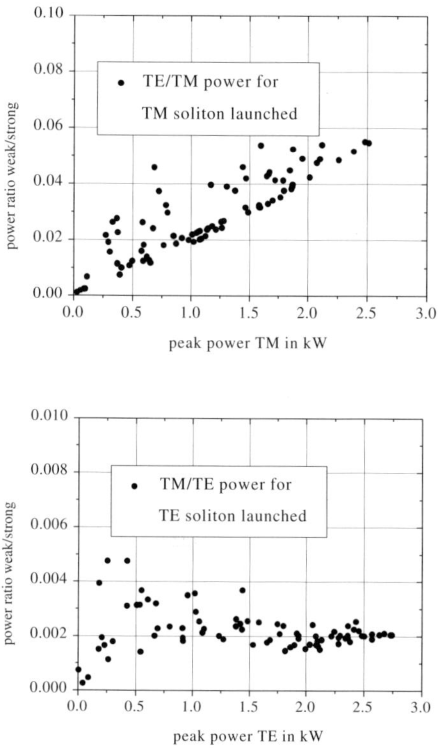

Fig. 11. Experimental data for the irradiance ratio of the weak to the strong polarization vs. total guided power for a TM (*top*) and TE (*bottom*) soliton with weak orthogonally polarized seeds launched into the waveguide (reproduced from [28])

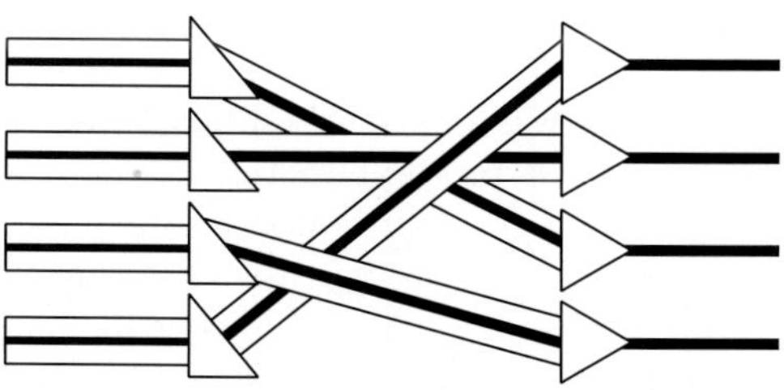

Fig. 12. Schematic of a proposed soliton-based reconfigurable interconnect. The *triangles* represent the deflection prisms, the *solid lines* represent the signal beams, and the *bands* around the solid lines represent the solitons

- The soliton survives transit through the triangular index pad without energy loss or breakup (I, II, and III in Fig. 13b).
- The net deflection varies smoothly with the fraction of the soliton that intercepts the triangular region [29].

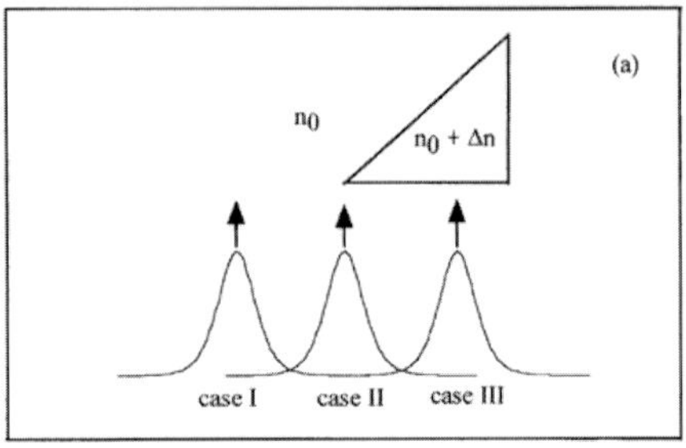

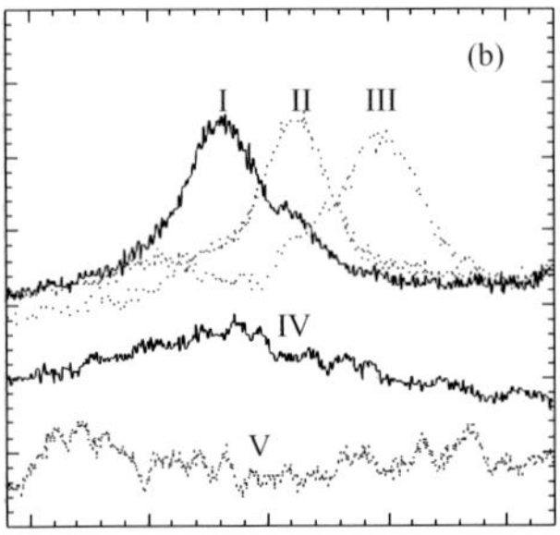

Fig. 13. (**a**) Schematic of the launching conditions relative to the wedge. (**b**) Experimental results showing the output beam profiles for the different launching cases shown in (**a**). Cases IV and V are for the same launching condition as cases I and II, respectively, but for a weak nonsoliton beam (reproduced from [29])

- Even the edges of the pad do not breakup the soliton, in comparison with a nonsoliton beam, which experiences classical diffraction by an edge (II and V).

These results showed that the spatial soliton was sufficiently "robust" to be used for soliton reconfigurable interconnects. A prototype interconnect was demonstrated, by the deflection of a signal-bearing soliton first into two and then into four well-defined output channels [30,31]. This increase was achieved by increasing the charge injection into the prism pad and optimizing the pad geometry. A major problem encountered was the absorption of the soliton and the signal by the injected electrons, which led to a decrease in the throughput with increasing deflection angle.

3.2 Spatial Solitons in Semiconductor Optical Amplifiers

Semiconductor optical amplifiers (SOAs) are waveguide devices in which inversion of the electron population is achieved between the bottom of the conduction band and the top of the valence band by either optical or electrical pumping [32]. Because of the inversion, an incident photon can be amplified, and the resulting induced refractive-index change it produces is positive. This is the opposite of what happens without inversion, in which case the photon is absorbed and the index change is negative. Therefore both gain and a self-focusing refractive-index change are available in such an inverted system. In fact, these self-trapping properties are well known in the semiconductor laser literature as "gain guiding". They make SOAs attractive for the generation and application of spatial solitons.

There have been two reports of spatial solitons in SOAs [33,34]. The first utilized an InGaAs multiple-quantum-well (MQW) laser amplifier operated at 1.32 μm. Khitrova et al. showed the formation of the two lowest order spatial solitons under gain conditions [33]. CW solitons were generated with input powers of 250 mW. In the second paper [34], the authors report that spatial solitons ~ 10 μm wide were generated in AlGaAs MQW amplifiers at 0.85 μm. The typical amplifier operating parameters and soliton parameters are shown in Fig. 14. Here about 350 mW was required for soliton generation.

One can anticipate increased activity in this area for two reasons. For one thing, these are inherently dissipative (and amplifying) systems, and new classes of solitons and collision phenomena have been predicted for this case [35]. Furthermore, the operating powers for soliton excitation are quite reasonable for realistic applications.

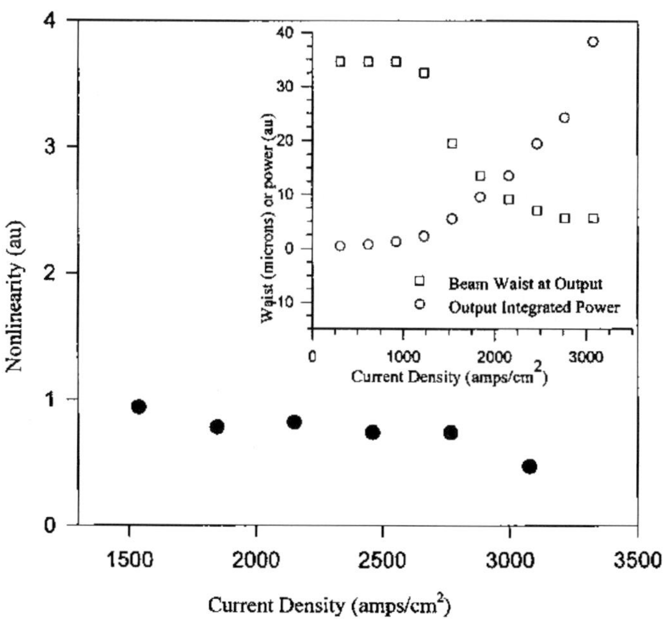

Fig. 14. Variation in the effective n_2 nonlinearity, soliton width, and soliton power with current density (reproduced from [34])

4 Discrete Solitons

There is one obvious practical difference between temporal solitons in fibers, where the nonlinearity counteracts dispersion, and spatial solitons in slab waveguides, where the nonlinearity counteracts diffraction. Dispersion is material dependent, it vanishes in vacuum, and its sign and strength can be chosen by proper choice of the waveguide material and the wavelength of the light. Diffraction, on the other hand, is a geometrical phenomenon: it is always present, and one has only limited control on its strength through the refractive index. One way to gain control over diffraction is to let the light propagate in a periodic medium, most simply in a linear array of coupled waveguides. Light expands then through "discrete diffraction" – by coupling outwards from waveguide to waveguide. It is obvious that the diffraction rate is controlled then by the strength of the coupling – light diffracts slowly in a weakly coupled array. It is less obvious that the strength and even the sign of diffraction can be chosen in a given array by controlling the input conditions [36]. When the light propagates in the direction of the array, or at a small angle to it, the diffraction is "normal", and a high-intensity beam will self-focus. However, if the light is coupled at an angle so that neighboring waveguides are excited out of phase, the diffraction is "anomalous", and a high intensity beam will self-defocus ! This behavior was observed experimentally just recently [37], as will be discussed below.

Discrete diffraction in a nonlinear Kerr medium leads to the formation of discrete solitons. These are solutions of the discrete NLSE (DNLSE) which keep a fixed spatial profile while propagating along the array. Discrete solitons have much in common with their continuous counterpart; however, they also differ in many details owing to the discrete nature of the governing equation. The theory of discrete systems is reviewed in the chapter by Lederer and coworkers, and here we shall concentrate on the experimental work.

All experiments to date have been performed in waveguide arrays made of AlGaAs, which has also been the main material system for spatial-soliton research in slab waveguides, as discussed above. Single-mode waveguides, typically 4 μm wide, have been defined by reactive-ion etching. The arrays typically contained 40–60 waveguides. The strength of coupling was controlled through the spacing between the waveguides, between 2 μm and 7 μm (see inset in Fig. 15).

The light source in these experiments was a synchronously pumped optical parametric oscillator (an OPAL system from Spectra-Physics) which emitted pulses in the 100–200 fs range, tunable near a wavelength of 1.5 μm, which is below half the bandgap of the semiconductor, hence reducing nonlinear absorption. The short pulses were needed in order to reach the required peak powers, and their maximum peak power was 1.5 kW.

The first experimental study reported the basic observation of the formation of discrete solitons in this system [38]. Light was launched into a single waveguide on the input side of a 6 mm long sample, and the light distribution

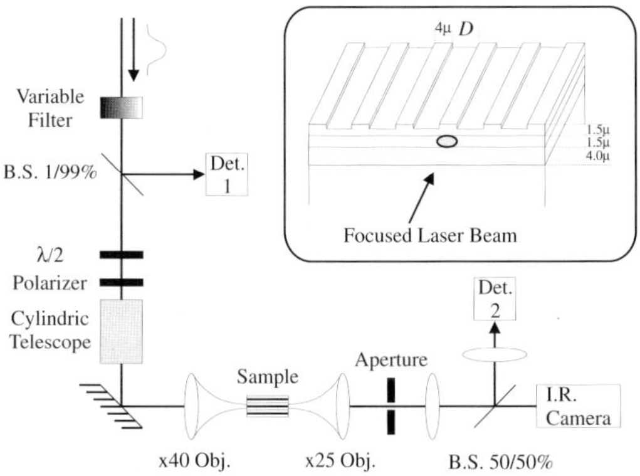

Fig. 15. The experimental setup for observation of discrete solitons. *Inset*: schematic drawing of the sample. The sample consists of an $Al_{0.18}Ga_{0.82}$ as core layer and $Al_{0.24}Ga_{0.76}As$ cladding layers grown on top of a GaAs substrate (reproduced from [38])

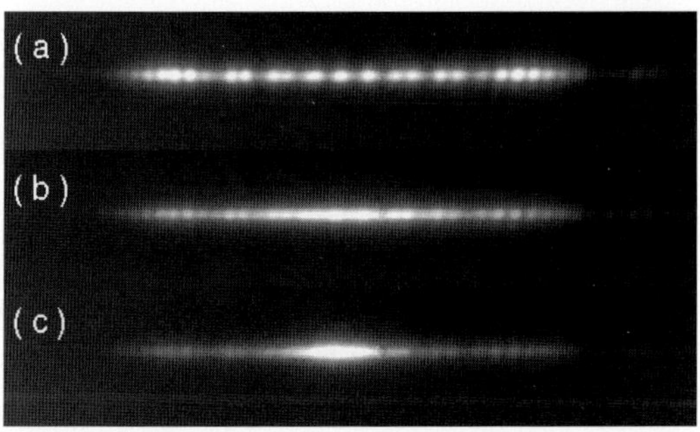

Fig. 16. Images of the output facet of a 6 mm long sample with 4 μm spacing between waveguides, for different powers. (**a**) Peak power 70 W. Linear propagation; light is diffracting in two main lobes and a few secondary peaks in between. (**b**) Peak power 320 W. Intermediate power, the distribution is narrowing. (**c**) Peak power 500 W. A discrete soliton is formed. (Reproduced from [38])

at the output end was recorded, as shown schematically in Fig. 15. At low light power, the propagation is linear, and the light expands over some 30 waveguides (top row in Fig. 16). From this distribution, it is possible to determine the coupling length, about 1.4 mm. As the input power is increased, the output distribution shrinks, as also shown in Fig. 16. At a power of 500 W, light is confined to about five waveguides around the input waveguide.

Simulations suggest that this distribution is achieved after propagating just 1 mm, and that from then on it propagates with little spatial change. This experiment was repeated with samples with different coupling strengths, yielding similar results. When light is launched into a single waveguide, the soliton that is eventually formed propagates along the array. However, it is important to realize that discrete solitons do not have to propagate in the direction of the waveguides. When the input is into several guides, if a linear phase slope is imposed at the input side, the transverse momentum leads to transverse motion of the soliton across the waveguides (see Fig. 17). However, while the slab soliton is rotationally and translationally invariant, the discrete soliton is not. It behaves differently when it is launched in different directions and from translated locations. A demonstration of these properties was reported in a subsequent study [39]. Here it was shown how the output location of the solitons can be shifted laterally quite drastically by imposing small changes on the input.

It is important to note that for each power level there are two types of soliton that propagate along the waveguides, one centered on a single waveguide and one centered in between two waveguides (see Fig. 17(C)).

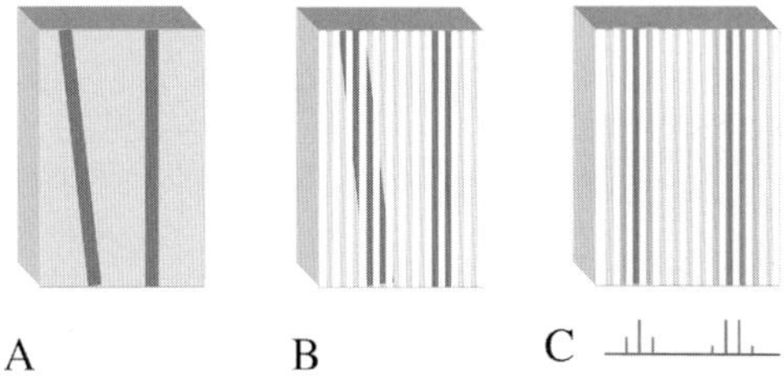

Fig. 17. Continuous slab solitons are invariant under translation and rotation (A), while discrete solitons are expected to exhibit different properties when launched with a transverse velocity component (B). In (C) the two families of discrete solitons, centered on and between waveguides, are shown schematically

Only the first type is stable. If a soliton is forced to move sideways, it has to jump from waveguide to waveguide, passing from a stable to an unstable configuration. The soliton has to overcome the so-called Peierls–Nabarro potential (PNP). For increasing power levels, the PNP increases, resulting in a strong localization of the soliton, mainly in a single waveguide which is effectively decoupled from the rest of the array. Note that in the corresponding continuous system of spatial solitons in a slab waveguide, any phase gradient imposed on the initial beam results in a corresponding tilt of the soliton motion.

In a study of soliton transverse motion [39], the output position of the soliton was monitored as the power or input location was varied. Figure 18 gives the output light distribution for an input distribution that was launched with a small tilt, as a function of input power. It shows how the output distribution, which initially was shifted considerably away from the region of the input waveguides, locks at higher powers to the input guides and propagates along the array.

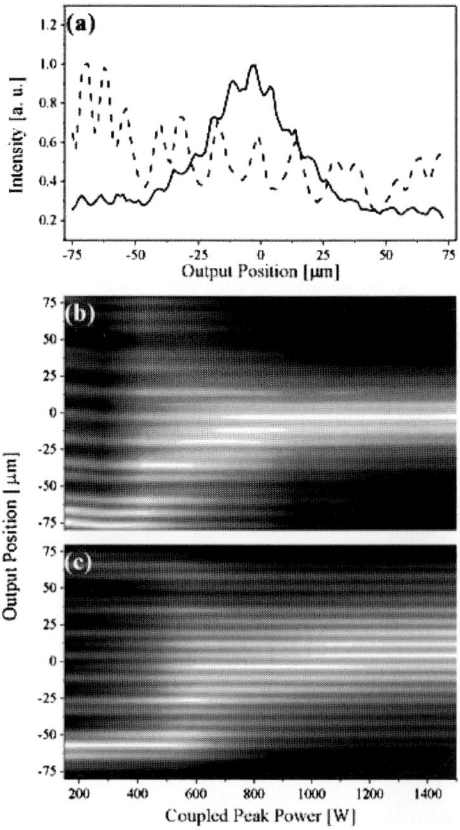

Fig. 18. Power-dependent soliton steering by an externally induced velocity. Output field distribution for an initial tilt of the beam of about 0.4 degrees. (**a**) Experimentally recorded power distributions for high (*solid line*) and low (*dashed line*) power. (**b**) Experimentally recorded field distribution at the output for varying power levels. (**c**) Simulation of (b) according to the DNLSE. (Reproduced from [39])

In a second experiment, solitons were steered sideways by means of asymmetry in the input coupling. The input distribution (this time without any phase tilt) was scanned across the input facet, and the output distribution was recorded. Whenever the input distribution was centered on a waveguide, or exactly between two waveguides, the soliton propagated in the waveguide direction without any transverse motion. However, when the input condition was not symmetric, a transverse motion was excited owing to a phase tilt that was induced through the Kerr effect. The periodic motion of the output location vs. the input location is clearly observable in Fig. 19. Also obvious is the higher sensitivity to the input location of the unstable soliton state (broken lines in Fig. 19).

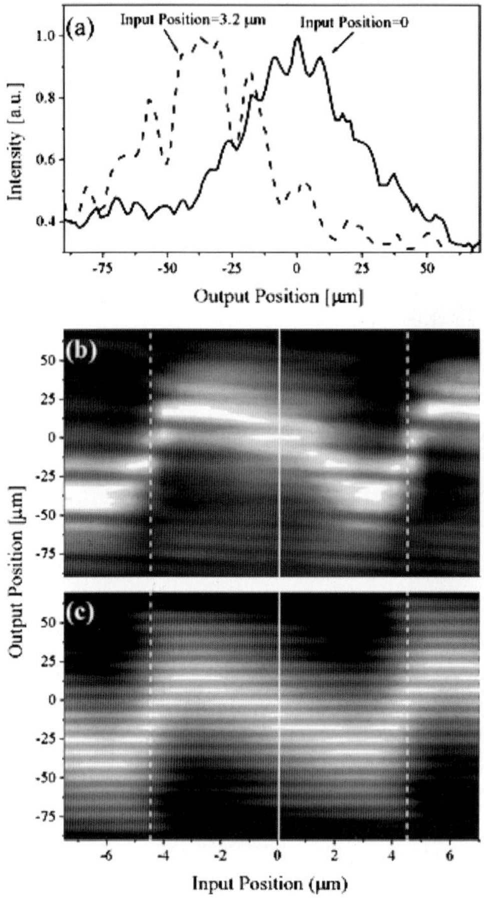

Fig. 19. Soliton steering by an internally induced velocity. Output field distribution for fixed input peak power $I_{\text{peak}} \sim 1500$ W as a function of the input beam position. (**a**) Experimentally recorded power distributions for two input positions relative to the central guide, (**b**) output field distributions for various positions of the initial excitation (*solid line*, beam centered on a waveguide; *dashed lines*, input beam centered in between two waveguides), (**c**) simulation of (**b**). (Reproduced from [39])

We discussed above how, through the choice of input condition, the diffraction can be controlled, and its sign can be reversed. Anomalous diffraction is a counterintuitive phenomenon: it means that components with a high (spatial) frequency have a slower transverse velocity than low-frequency components. In discrete diffraction, this happens when the input is excited with a tilt so that adjacent waveguides are excited out of phase. As far as linear optics is concerned, both "normal" and "anomalous" diffraction lead to broadening of a finite-width beam. However, while normal diffraction leads to self-focusing, anomalous diffraction leads to self-defocusing, that is, even faster broadening at high optical powers. This result was recently demonstrated [37] in a specially designed sample with 61 waveguides, each 3 μm wide, separated by 3 μm. The experimental results are given in Fig. 20.

As we have seen, waveguide arrays have proven to be a rich medium for nonlinear investigations, and discrete solitons are still being studied theoretically and experimentally. Current issues are interactions between discrete solitons, interactions between solitons and array defects, and various ways of achieving all-optical switching in waveguide arrays.

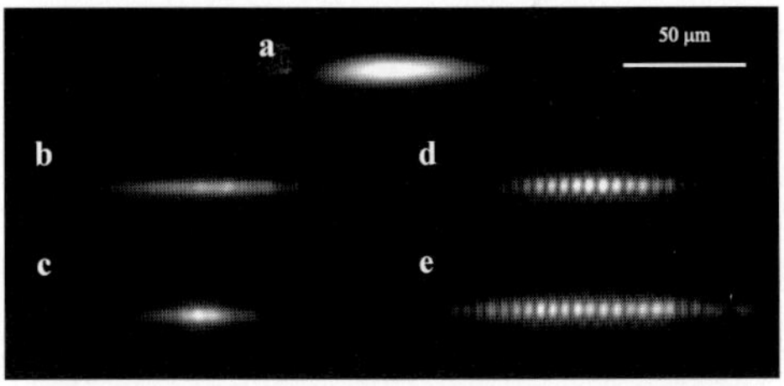

Fig. 20. Experimental results showing both nonlinear self-focusing and self-defocusing in the same array of waveguides, for slightly different initial conditions. (**a**) A micrograph of the input beam, ∼ 35 μm wide at FWHM. (**b**) Output light distribution for normal dispersion (i.e. light beam normal to the input facet) in the linear regime. The beam broadens slightly through discrete diffraction. (**c**) At high power ($I_{\text{peak}} \sim 150$ W), the field shrinks and evolves into a discrete bright soliton, as in Fig. 16. (**d**) For anomalous dispersion, i.e. a beam injected at an angle of 2.6 ± 0.3 degrees inside the array, the beam broadens slightly as in (b). Note the dark lines between the optical modes resulting from the p phase flips between adjacent waveguides. (**e**) When the power is increased ($I_{\text{peak}} \sim 100$ W), the distribution broadens significantly owing to self-defocusing

5 Summary

Slab waveguides have proven to be a versatile laboratory for studying the simplest kinds of spatial solitons in both homogeneous and discretized media. When Kerr media are involved, slab waveguides provide an opportunity to test many of the fundamental concepts of propagation and interaction in scenarios for which analytical theories are possible. This has been the case in both homogeneous and discrete array media.

Spatial solitons in (1 + 1)D have been investigated in a variety of media, including liquids, glasses, polymers and both passive and active semiconductors. The semiconductor medium Al–Ga–As has proven to be very valuable for these soliton studies because very high-quality, low-loss samples have been fabricated by MBE techniques, for both homogeneous and discrete waveguides. The nonlinearities are essentially Kerr over the intensity ranges needed, and multiphoton absorption is not a factor. Furthermore, some peculiarities in the nonlinearities have allowed special solitons such as Manakov solitons to be studied.

The work reported to date has by no means exhausted the number of interesting soliton phenomena to be investigated in slab waveguides. Many fascinating questions remain to be answered about discrete solitons, and about solitons in dissipative and amplifying media. Furthermore, the low excitation power levels for spatial solitons in semiconductor optical amplifiers offer hope for applications.

References

1. R.Y. Chiao, E. Garmire, C.H. Townes, Self-trapping of optical beams, Phys. Rev. Lett. **13**, 479 (1964); V.I. Talanov, Izv. Vysshikh Uchebn. Zavendii, Radiofizika, **42**, 1567 (1964) [Propagation of a short electromagnetic pulse in an active medium, Radiophysics **7**, 25 (1964)].
2. V.E. Zakharov, A.B. Shabat, Exact theory of two-dimensional self-focusing and one-dimensional self-modulation of waves in nonlinear media, Sov. Phys. JETP **34**, 62 (1972).
3. A. Hasegawa, F. Tappert, Transmission of stationary nonlinear optical pulses in dispersive dielectric fibers: I. Anomalous dispersion, Appl. Phys. Lett. **23**, 142 (1973).
4. G.P. Agrawal, *Nonlinear Fiber Optics* (Academic Press, New Yorkm, 1989); M.N. Islam, *Ultrafast Fiber Switching Devices and Systems* (Cambridge University Press, Cambridge 1992).
5. J.E. Bjorkholm, A. Ashkin, cw self-focusing and self-trapping of light in sodium vapor, Phys. Rev. Lett. **32**, 129 (1974).
6. A. Barthelemy, S. Maneuf, C. Froehly, Propagation soliton et auto-confinement de faisceaux laser par non linearite optique de Kerr, Opt. Commun. **55**, 201 (1985).
7. S. Maneuf, F. Reynaud, Quasi-steady state self-trapping of first, second and third order sub-nanosecond soliton beams", Opt. Commun. **66**, 325 (1988).
8. J.S. Aitchison, A.M. Weiner, Y. Silberberg, M.K. Oliver, J.L. Jackel, D.E. Leaird, E.M. Vogel, P.W. Smith, Observation of spatial optical solitons in a nonlinear glass waveguide, Opt. Lett. **15**, 471 (1990).
9. J.S. Aitchison, A.M. Weiner, Y. Silberberg, D.E. Leaird, M.K. Oliver, J.L. Jackel, P.W. Smith, Experimental observation of spatial soliton interactions, Opt. Lett. **16**, 15 (1991).
10. M. Shalaby, F. Reynaud, A. Barthelemy, Experimental observation of spatial soliton interactions with a $\pi/2$ relative phase difference, Opt. Lett. **17**, 778 (1992).
11. U. Bartuch, U. Peschel, T. Gabler, R. Waldhaus, H.-H. Horhold, Experimental investigations and numerical simulations of spatial solitons in planar polymer waveguides, Opt. Commun. **134**, 49 (1997).
12. M. Warengheim, J.F. Henninot, G. Abbate, Non-linearly induced self-waveguiding structure in dye doped nematic liquid crystals confined in capillaries, Opt. Express **2**, 483 (1998); M.A. Karpierz, M. Sierakowski, M. Swillo,

T. Wolinsky, Self-focusing in liquid crystalline waveguides, Mol. Cryst. Liq. Cryst. **320**, 157 (1998).
13. M. Peccianti, A. De Rossi, G. Assanto, A. De Luca, C. Umeton, I.C.Khoo, Electrically assisted self-confinement and waveguiding in planar nematic liquid crystal cells, Appl. Phys. Lett. **77**, 7 (2000).
14. K. DeLong, G.I. Stegeman, Dispersion of the two photon absorption parameter for all-optical switching, Appl. Phys. Lett. **57**, 2063 (1990).
15. G.I. Stegeman, A. Villeneuve, J.S Aitchison, C.N. Ironside, Nonlinear integrated optics and all-optical switching in semiconductors, in *Fabrication, Properties and Applications of Low-Dimensional Semiconductors*, ed. by M. Balkanski and I. Yancher, NATO ASI Series (Kluwer Academic, Dordrecht, pp. 415-449, 1995); G.I. Stegeman, A. Villeneuve, J. Kang, J.S. Aitchison, C.N. Ironside, K. Al-Hemyari, C.C. Yang, C.-H. Lin, H.-H. Lin, G.T. Kennedy, R.S. Grant, W. Sibbett, AlGaAs below half bandgap: the silicon of nonlinear optical materials, Int. J. Nonlinear Opt. Phys. **3**, 347 (1994)
16. A. Villeneuve, J.U. Kang, J.S. Aitchison, G.I. Stegeman, Cross-phase modulation in bulk AlGaAs and AlGaAs/GaAs multiple quantum wells, Appl. Phys. Lett. **67**, 760 (1995).
17. D.C. Hutchings, J.S. Aitchison, B.S. Wherrett, G.T. Kennedy, W. Sibbett, Polarization dependence of ultrafast nonlinear refraction in an AlGaAs waveguide at the half-band gap, Opt. Lett. **20**, 991 (1995).
18. J.S. Aitchison, K. Al-Hemyari, C.N. Ironside, R.S. Grant, W. Sibbett, Observation of spatial solitons in AlGaAs waveguides, Electron. Lett. **28**, 1879 (1992).
19. A. Villeneuve, J.S. Atichison, J.U. Kang, P.G. Wigley, G.I. Stegeman, Integrated ultrafast saturable absorber, Opt. Lett. **19**, 761 (1994).
20. J.S. Aitchison, J.U. Kang, G.I. Stegeman, Signal gain due to polarization coupling in an AlGaAs channel waveguide, Appl. Phys. Lett. **67**, 2456 (1995).
21. G.I. Stegeman, P. Mamyshev, W. Torruellas, A. Villeneuve, J.S. Aitchison, Photonic applications of spatial soliton switches, Proc. SPIE **2481**, 270 (1995).
22. J.U. Kang, G.I. Stegeman, J.S. Aitchison, Weak beam trapping by bright spatial solitons in an AlGaAs planar waveguide, Opt. Lett. **20**, 2069 (1995).
23. J.U. Kang, G.I. Stegeman, J.S. Aitchison, One-dimensional spatial soliton dragging, trapping and all-optical switching in AlGaAs waveguides, Opt. Lett. **21**, 189 (1996).
24. S.V. Manakov, On the theory of two-dimensional stationary self-focusing of electromagnetic waves, Sov. Phys. JETP **38**, 248 (1974).
25. J.U. Kang, G.I. Stegeman, J.S. Aitchison, N. Akhmediev, Observation of Manakov spatial solitons in AlGaAs planar waveguides Phys. Rev. Lett. **76**, 3699 (1996).
26. N.N. Akhmediev, J.M. Soto-Crespo, Dynamics of soliton-like pulse propagation in birefringent optical fibers, Phys. Rev. E **49**, 5742 (1994).
27. N.N. Akhmediev, A.V. Buryak, J.M. Soto-Crespo, D.R. Anderson, Phase-locked stationary soliton states in birefringent nonlinear optical fibers, J. Opt. Soc. Am. B **12**, 434 (1995).
28. L. Friedrich, G.I. Stegeman, J.M. Soto-Crespo, N.N. Akhmediev, J.S. Aitchison, Radiation related polarization instability of fast Kerr spatial solitons in slab waveguides, Opt. Commun. **186**, 1786 (2000). (2000).
29. J.U. Kang, G.I. Stegeman, G. Hamilton, J.S. Aitchison, Robustness of spatial solitons in AlGaAs waveguides. Appl. Phys. Lett. **70**, 1363 (1997).

30. L. Friedrich, J.S. Aitchison, P. Millar, G.I. Stegeman, Dynamic, electronically controlled, angle steering of spatial solitons in AlGaAs slab waveguides, Opt. Lett. **23**, 1438 (1998).
31. L. Friedrich, G.I. Stegeman, P. Millar, J.S. Aitchison, 1×4 optical interconnect using electronically controlled angle steering of spatial solitons, Phot. Technol. Lett. **11**, 988 (1999).
32. M.J. Adams, D.A.O. Davies, M.C. Tatham, M.A. Fisher, Nonlinearities in semiconductor laser amplifiers, Opt. Quantum Electron. **27**, 1 (1995); J. Mark, A. Mecozzi, Theory of the ultrafast optical response of active semiconductor waveguides, J. Opt. Soc. Am. B **13**, 1803 (1996).
33. G. Khitrova, H.M. Gibbs, Y. Kawamura, H. Iwamura, T. Ikegami, J.E. Sipe, L. Ming, Spatial solitons in self-focusing semiconductor gain medium, Phys. Rev. Lett. **70**, 920 (1993).
34. C. Kutsche, P. LiKamWa, J. Loehr, R. Kaspi, Quasi-CW self-guided optical beams in GaAs-AlGaAs double heterostructure slab waveguides, Electron. Lett. **34**, 906 (1998).
35. J.M. Soto-Crespo, N. Akhmediev, A. Ankiewicz, Pulsating, creeping and erupting solitons in dissipative systems, Phys. Rev. Lett. **85**, 2937 (2000).
36. H.S. Eisenberg, Y. Silberberg, R. Morandotti, J.S. Aitchison, Diffraction management, Phys. Rev. Lett. **85**, 1863 (2000).
37. R. Morrandotti, H.S. Eisenberg, Y. Silberberg, M. Sorel, J.S. Aitchison, Self-focusing and defocusing waveguide arrays, submitted Phys Rev. Lett (2000).
38. H.S. Eisenberg, Y. Silberberg, R. Morandotti, A.R. Boyd, J.S. Aitchison, Discrete spatial optical solitons in waveguide arrays, Phys. Rev. Lett. **81**, 3383 (1998).
39. R. Morandotti, U. Peschel, J.S. Aitchison, H.S. Eisenberg, Y. Silberberg, Dynamics of Discrete Solitons in Optical Waveguide Arrays, Phys. Rev. Lett. **83**, 2726 (1999)

Photorefractive Spatial Solitons

Eugenio DelRe, Bruno Crosignani, and Paolo Di Porto

Summary. In this chapter we discuss the basic phenomenology and theoretical framework associated with photorefractive spatial solitons. Our attention is mainly centered on explaining the fundamental processes that allow self-trapping of visible beams. In doing this, we trace the principal steps of progress in the field, from the first prediction of a photorefractive soliton at the beginning of the '90s, to the more recent experiments that promise useful applications.

1 Introduction

Photorefractive crystals are electro-optic dielectrics that host a small amount of photosensitive impurities. Light propagation leads to the generation of out-of-equilibrium mobile charge, which, in order to reach a stable electrostatic configuration, redistributes throughout the crystal. The ensuing space-charge field, modifying electro-optically the index of refraction of the crystal, changes the trajectory of the ionizing light distribution, altering, in turn, the original charge equilibrium conditions [1]. This feedback mechanism gives rise to a variety of nonlinear effects that go under the generic term of "photorefractive nonlinear optics" [2–5]. For confined optical beams, the nonlinear beam dynamics lead to two basic, qualitatively different phenomena: beam fanning and self-lensing. These are connected, to the two basic charge transport mechanisms, diffusion and drift, respectively. Photorefractive spatial solitons, whose investigation began with the seminal work of Segev et al. in 1992 [6], emerge when beam self-focusing exactly balances diffraction, and are thus generally connected to regimes in which charge drift plays a fundamental role [6–9]. In the picture of a nonlinear beam, a spatial beam self-traps when the light–space-charge feedback mechanism finds its dynamic equilibrium point in a nondiffracting slab or of light, corresponding to an appropriate waveguide-like refractive-index distribution. In what follows, we discuss such nonlinear phenomena, concentrating in particular on the basic theory and phenomenology.

2 Photorefractive Beam Nonlinearity

Photorefractive beam dynamics are the result of the combined effect of optical wave propagation and the electro-optic response to the electric field generated by the displacement of photoexcited charge. The description starts from two basic issues. (1) Given an optical distribution, what is the induced equilibrium charge displacement and hence the photoinduced electric field? (2) How does this electric field modify, through the electro-optic response, optical propagation?

2.1 Space-Charge Field

The basic model that describes charge separation in a photorefractive material was progressively formulated in the second half of the 1970s [10]. Although it makes use of a number of approximations, it incorporates all the basic ingredients that allow for the prediction and description of photorefractive self-trapping, at least in the case of one-dimensional beams, that is, beams that are confined in only one transverse dimension. The light–matter interaction reflects the typical band structure of a lightly doped dielectric. In particular, the structure can normally be approximated by considering two intraband levels: one donor and one acceptor. For n-type photorefractives, donor sites can be optically ionized by light of an appropriate wavelength, generally in the visible, depending on the given impurity. Furthermore, the concentration of donor sites N_d ($N_\mathrm{d} \approx 10^{18}$–$10^{19}\,\mathrm{cm}^{-3}$) is much greater than that of acceptor impurities N_a (i.e. $\alpha \equiv N_\mathrm{d}/N_\mathrm{a} \gg 1$). In the absence of ionizing illumination and thermal excitation, equilibrium is reached when all the acceptor sites are ionized by donors, that is, the concentration of ionized donors $N_\mathrm{d}^+ = N_\mathrm{a}$. For a given optical intensity distribution $I(x, y, z)$, charge equilibrium is reached when mobile-charge generation and recombination exactly balance, that is, $(\beta + sI)(N_\mathrm{d} - N_\mathrm{d}^+) = \gamma N N_\mathrm{d}^+$, where β is the thermal excitation rate, s is the photoexcitation cross section, γ is the electron recombination rate, and N is the local concentration of conduction electrons. In general, the nonequilibrium condition is described by the rate equation

$$\frac{\partial}{\partial t} N_\mathrm{d}^+ = (\beta + sI)(N_\mathrm{d} - N_\mathrm{d}^+) - \gamma N N_\mathrm{d}^+ \,. \tag{1}$$

Electrons move in the crystal under the influence of drift, diffusion, and, in some noncentrosymmetric samples, the photovoltaic effect [11], giving rise to a current density $\boldsymbol{J}$ described by

$$\boldsymbol{J} = q\mu N \boldsymbol{E} + k_\mathrm{b} T \mu \boldsymbol{\nabla} N + \beta_\mathrm{ph}(N_d - N_d^+) I \boldsymbol{c} \,, \tag{2}$$

where $-q$ is the electron charge, μ is the charge mobility, k_b is the Boltzmann constant, T is the crystal temperature, and β_ph is the component of the photovoltaic tensor along the optical axis $\boldsymbol{c}$. The model is completed by imposing charge continuity and Poisson's equation, along with appropriate boundary conditions:

$$\frac{\partial}{\partial t}\rho + \nabla \cdot \boldsymbol{J} = 0 \,, \tag{3}$$

$$\nabla \cdot \boldsymbol{E} = \frac{\rho}{\epsilon} , \tag{4}$$

where $\rho = q(N_\mathrm{d}^+ - N_\mathrm{a} - N)$, and ϵ is the crystal dielectric constant. For conditions in which the electro-optic sample is biased by a constant applied voltage V, the condition

$$V = -\int_a^b \boldsymbol{E} \cdot \mathrm{d}\ell \tag{5}$$

holds, where the line integral goes from one electrode (a) to the other (b). Finally,

$$\nabla \times \boldsymbol{E} = 0 . \tag{6}$$

This system of equations can be cast into a single nonlinear differential equation relating $\boldsymbol{E}$ to I [12].

Two fundamental photorefractive time scales emerge: the charge recombination time (or charge lifetime) $\tau_\mathrm{r} = 1/(N\gamma)$ (see (1)), and the dielectric relaxation time $\tau_\mathrm{d} = \epsilon/(\mu N q)$ (see (3)). For most configurations of interest, $\tau_\mathrm{r} \ll \tau_\mathrm{d}$, and the study of space-charge evolution starts by considering adiabatically $\partial N_\mathrm{d}^+/\partial t = 0$ in (1), along with the generally valid assumptions that $N \ll N_\mathrm{a}$ and that $\alpha \gg 1$ as mentioned. The resulting nonlinear equation relating $\boldsymbol{E}$ to I is

$$\nabla \cdot \left[\frac{\gamma\epsilon}{q\mu s\alpha}\frac{\partial \boldsymbol{E}}{\partial t} + \boldsymbol{E}(\beta/s + I)\frac{1-\epsilon\nabla\cdot\boldsymbol{E}/\alpha N_\mathrm{a} q}{1+\epsilon\nabla\cdot\boldsymbol{E}/N_\mathrm{a} q} \right. \\ \left. + \frac{k_\mathrm{b} T}{q}\nabla\cdot\left((\beta/s + I)\frac{1-\epsilon\nabla\cdot\boldsymbol{E}/\alpha N_\mathrm{a} q}{1+\epsilon\nabla\cdot\boldsymbol{E}/N_\mathrm{a} q}\right)\right] = 0 . \tag{7}$$

With appropriate approximations, this equation allows one to calculate $\boldsymbol{E} = \boldsymbol{E}(I)$.

2.2 Beam Propagation

The space-charge field $\boldsymbol{E}$ influences beam propagation through the electro-optic modulation of the index of refraction. As in standard electro-optic samples, the index modulation can be phenomenologically described by the relation

$$\Delta n_{ij} = -\frac{1}{2}n^3 r_{ijk} E_k - \frac{1}{2}n^3 \epsilon_0^2(\epsilon_r - 1)^2 g_{ijkl} E_k E_l , \tag{8}$$

where n is the unperturbed crystal index of refraction, r_{ijk} and g_{ijkl} are the linear and quadratic electro-optic tensors, respectively, and $\boldsymbol{E} = (E_x, E_y, E_z)$. For a noncentrosymmetric sample, the quadratic term is generally irrelevant, whereas for centrosymmetric samples the linear response is absent.

For a monochromatic paraxial beam, propagation is described by the parabolic equation

$$\left[\frac{\partial}{\partial z} - \frac{\mathrm{i}}{2k}\nabla_\perp^2\right] A_i(x,y,z) = -\frac{\mathrm{i}k}{n}\Delta n_{ij} A_j(x,y,z) , \tag{9}$$

where $k = 2\pi n/\lambda$ is the wavevector, A_x and A_y are the transverse components of the slowly varying optical field, i.e. $\boldsymbol{E}_{\mathrm{op}}(\boldsymbol{r},t) = \boldsymbol{A}(\boldsymbol{r})\exp(\mathrm{i}kz - \mathrm{i}\omega t)$, $\omega = 2\pi c/n\lambda$, and $I = |\boldsymbol{A}(\boldsymbol{r})|^2$. Through (8), and (7), (9) becomes the general nonlinear equation that describes almost all photorefractive nonlinear beam dynamics, and, in particular, solitons.

Photorefraction is thus a consequence of an interplay between a small absorbed (i.e. nonpropagating) part of an optical beam and the remnant propagating light, which, in fact, propagates *linearly* in the medium (as in thermo-optic effects). This fact, which distinguishes photorefraction from the fundamental nonlinear optical phenomena connected with direct optical self-action, is useful in understanding both the strengths and the limits of photorefraction. The strengths lie in the fact that the response is due to a temporally extended buildup process, allowing the observation of intense beam self-action even with low optical intensities. This temporally "nonlocal" response, however, limits the attainable dynamics. Thus, photorefractive phenomena are sometimes referred to as *beam* nonlinearities instead of *optical* nonlinearities, this characteristic being also the basis of phenomena connected with the self-trapping of time-varying light beams.

3 Self-Trapping Mechanisms

The highly nonlinear system described above gives rise to a host of different phenomena associated with beam self-trapping , which have been the object of intense investigation during the past decade. The starting point of this scientific effort was the idea, formulated in 1992 [6,13] and later confirmed in pioneering experiments [14], that, in biased drift-dominated photorefractives, self-lensing could lead to trapping of confined optical beams. In what follows we shall describe the various different types of photorefractive self-trapping mechanisms, i.e. quasi-steady-state solitons, steady-state screening solitons, photovoltaic solitons, semiconductor screening solitons, and centrosymmetric self-trapping. Attention is concentrated on scalar configurations, although effects connected to the tensorial nature of (8) have also been investigated [15,16].

3.1 Quasi-Steady-State Solitons

Quasi-steady-state solitons are beams that self-trap in a biased photorefractive crystal during a finite time window, and subsequently undergo beam diffraction, breakup and fanning [14]. They contain, in an embryonic form,

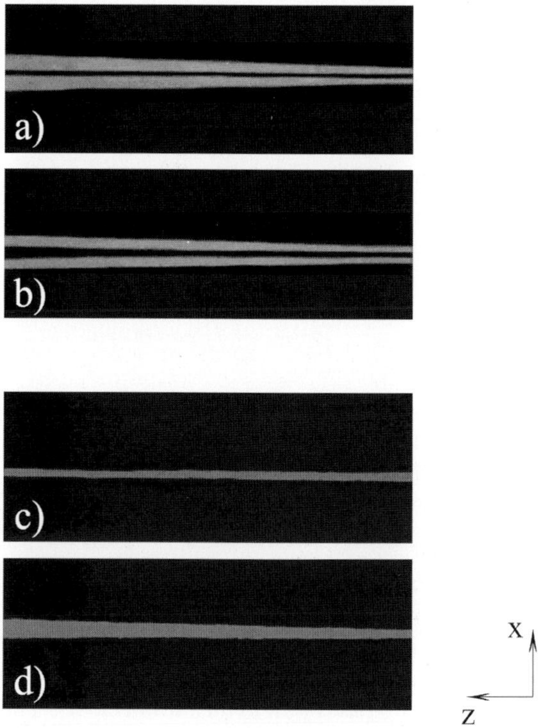

Fig. 1. Top-view photographs of (**a**) a dark $(1+1)$D quasi-steady-state 12 μm soliton at $\lambda = 457$ nm, (**b**) a linearly diffracting dark notch with drift nonlinearity deactivated, (**c**) a guided $\lambda = 632$ nm He–Ne beam, and (**d**) the same beam undergoing diffraction when no soliton is present (from [24])

most of the ingredients that lead to the formation of stable self-trapping. They were the first type of photorefractive spatial soliton to be predicted [6,13] and observed, both as bright and dark self-trapped beams in one and two transverse dimensions [14,17,18], and are still today a subject of investigation [19,20]. Given their transient nature, they have been mostly associated with studies on the time evolution of photorefractive beam dynamics [19,21–23], taking into account the full time-dependent model of (1)–(6). Their potential for passive optical guiding in bulk media was recognized early on [24], in an experiment that represents the first reported application of photorefractive solitons. Here, a dark photorefractive quasi-steady-state soliton was made to guide a nonphotoactive read beam, as shown in Fig. 1.

Although the theoretical description of these self-trapped beams is rather intricate, the physical mechanism in the slab soliton case is intuitive. As schematically illustrated in Fig. 2, bright slab quasi-steady-state solitons are generated when a continuous-wave visible photoactive laser beam (i.e. a beam

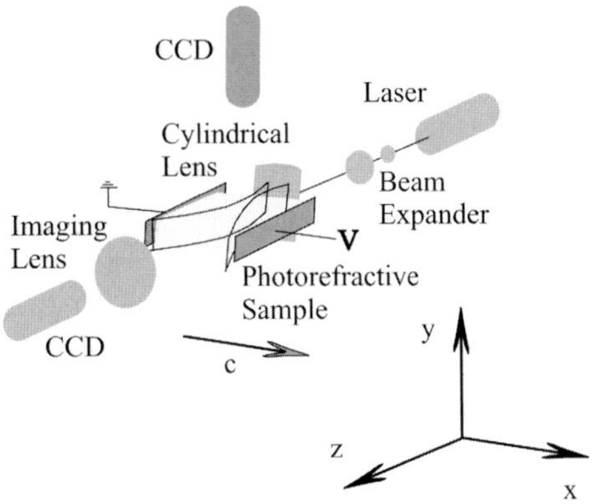

Fig. 2. Schematic setup allowing the generation and observation of quasi-steady-state photorefractive bright slab solitons. Dark solitons are generated with an appropriate phase mask at the input, and (2 + 1)D needle solitons are generated when the cylindrical lens is replaced by a spherical one

capable of ionizing photorefractive impurities) is focused onto the input facet of a zero-cut photorefractive sample (for example, a doped SBN crystal).

The beam is made to propagate along an ordinary principal axis (the z axis) and is itself extraordinarily polarized (along the x axis). Electrodes on the x facets of the crystal allow the application of an external bias field. The beam is confined in the x dimension, whereas it is extended (nondiffracting) in the second transverse direction, y. During the initial stages of beam evolution, for times shorter than τ_{d}, mobile charges are photoexcited and drift in the external field. For n-type samples, such as SBN, mobile electrons statistically drift towards the positive electrode, and the ensuing charge separation from the fixed ionized impurities gives rise to a double layer that screens the external field. The electro-optic response of (8) is reduced to the scalar equivalent

$$\Delta n = -\frac{1}{2} n^3 r_{33} E \, , \tag{10}$$

where $i = j = k = 3$ (x axis), and r_{33} is the contracted form of r_{333}. For an appropriate arrangement, $r_{33}E > 0$, and the unscreened region of the crystal suffers a global decrease of index of refraction. In the screened region, this effect is weaker, and the net result is a higher index of refraction in the illuminated region, giving rise to self-lensing. During the screening stages, if the applied field is sufficiently high, the charge pattern passes through a self-lensing regime such as to trap the diffracting beam into a slab soliton. However, the propagating light continues to generate mobile charges, and these keep drifting towards the positive pole until the actual charge separation reaches saturation, when screening is total. This leads to a highly insensitive saturated response in *all* illuminated regions, giving rise to a gen-

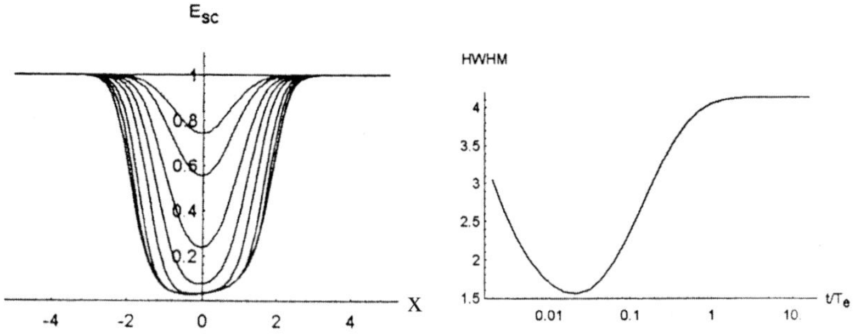

Fig. 3. Results of numerical simulation for the evolution of normalized space-charge field E_{sc}, for $t/T_e = 0, 0.1, 0.2, 0.5, 1, 2, 5, 10, \infty$, where $T_e = \gamma\epsilon/q\mu s\alpha$ (*left*) and a typical evolution of the half-width-half-maximum (HWHM, in normalized units) for a (1 + 1)D soliton configuration (from [21])

erally widened index modulation (a wide step-index "waveguide") that does not correspond to a self-trapping index structure. This simple picture holds for beam intensities much higher than typical dark-equivalent illuminations, i.e. for $I \gg I_\mathrm{d} \equiv \beta/s$ (of the order of 10^{-3}–10^{-6} W/cm^2), and for applied external fields much higher than typical diffusion fields (~10^2 V/cm for a 10 μm beam). The quasi-steady-state self-trapping regime has been analytically described in [6,13,25] by means of a phenomenological model, and numerical investigations of the entire transient regime in both (1 + 1)D and (2 + 1)D can be found in [19,21–23]. An analytical expression for the time-dependent space-charge field can be found in [12,26]. Both theory and experiment agree on two main features: (1) formation of quasi-steady-state solitons is not dependent on the intensity of the propagating beam (as long as $I \gg I_\mathrm{d}$), and (2) quasi-steady-state solitons are characterized by a whole *region* of existence, forming, for a given input beam size, over a range of external bias fields (always such that $V/L \gg E_\mathrm{d}$) [27,28].

As can be seen in Fig. 3 (right), the quasi-steady-state regime occurs over a relatively extended plateau (note the logarithmic time scale in Fig. 3). The plateau arises when an intermediate light-trapping space-charge field E forms, in a process that can be qualitatively explained as follows. Given that the time dynamics are locally proportional to the light intensity, once trapping occurs, space-charge broadening is temporarily slowed down by the absence of light in the immediate vicinity of the screening region. As trapping is weakened, the time dynamics speed up and reach a time-behaviour similar to the buildup process.

3.2 Screening Solitons

The most widely studied self-trapping mechanism in photorefractives is without doubt the so-called screening nonlinearity. The reason for this lies in the fact that the resulting self-trapped beams are steady-state, that is, they do not form and breakup as a transient but, rather, persist in time as long as the laboratory parameters maintain their values. Furthermore, in their one-dimensional manifestation as (1+1)D slab solitons, the model described below leads, with appropriate approximations, to an explicit, saturable, Kerr-like nonlinear equation (in a merely formal analogy) that allows direct qualitative and quantitative prediction and interpretation, as opposed to either merely phenomenological models or purely numerical strategies. Screening solitons are supported by a slightly modified configuration of the kind that supports quasi-steady-state solitons and, like these, exist also in the higher-dimensional needle case. The higher dimensionality, of central importance for beam-steering applications, fundamentally complicates the theoretical description and raises issues connected to the anisotropy of the underlying physical mechanism.

The discovery of steady-state photorefractive screening solitons began with the preliminary observation that response saturation, which leads to soliton annihilation in quasi-steady-state configurations, can be inhibited by making use of an artificial background illumination [29]. It was soon shown that this could indeed lead to stable self-trapped photorefractive beams [30–32]. The basic idea is the following. Since the runaway charge separation leading to decay of quasi-steady-state solitons is due to the inevitable charge accumulation at the edge of the propagating beam, this being a direct consequence of charge recombination and low dark conductivity, then if the global crystal conductivity is artificially increased with a constant optical illumination of the whole sample, charge can move through the equivalent circuit formed by the crystal and the voltage supply, thereby avoiding unrestrained buildup. The final charge separation depends on the ratio of the intensity of the propagating beam to that of the background illumination, and the resulting electric field in the crystal follows qualitatively that of a series of resistors to which a constant voltage V is applied, resulting in a lower field in the beam region, and thus, as in the previous quasi-steady-state case, giving rise to self-lensing [33]. Since the mechanism is based on a dynamic equilibrium, stable steady-state spatial self-trapping occurs when this mechanism generates a self-consistent charge separation that exactly balances optical diffraction. This occurs for a precise set of physical parameters and gives rise to what is generally referred to as the "soliton" existence curve [34], making screening steady-state solitons very different from their transient counterparts.

Screening solitons have been observed in a number of different configurations, attesting to their relatively general nature. They have been observed as bright $(1+1)$D slab solitons [29], dark $(1+1)$D slabs [30,35], as bright $(2+1)$D needle solitons [36,37], and even as dark needles [39]. They have been detected in several different crystals, such as SBN [33,36,37], BSO and BGO [29,38], the semiconductor InP [40,41], and $BaTiO_3$ [42], and have been pre-

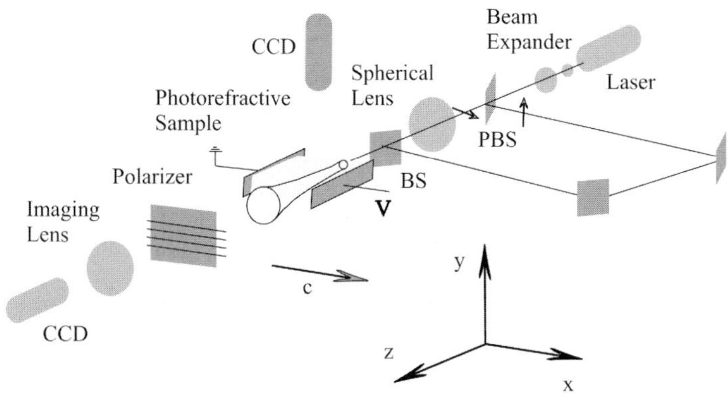

Fig. 4. Schematic setup allowing the generation and observation of steady-state photorefractive bright, screening, needle solitons. Note the introduction of a background illumination and a selector before the CCD camera on the z axis

dicted and observed in ferroelectrics in the high-symmetry paraelectric phase [43–45].

The experimental apparatus that allows the formation of screening solitons is similar to that needed to observe quasi-steady-state solitons illustrated in Fig. 2, with the addition of a background illumination, as indicated in Fig. 4. In the most appropriate configuration, the background illumination is obtained by illuminating the entire crystal with a plane wave of the same wavelength λ as that of the soliton beam; this plane wave is polarized orthogonal to the soliton beam polarization, i.e. orthogonal to the crystal c axis (ordinary beam), and is made to copropagate along the z axis with the soliton beam itself. Use of the same wavelength allows the use of a single CW laser for the entire setup, and makes the theoretical interpretation independent of the sample-specific optical photorefractive cross section $s = s(\lambda)$. The ordinary polarization avoids the possible coupling of background light into the soliton-supporting index pattern, given that the off-axis tensorial terms of the electro-optic response described in (8) are generally weaker than the diagonal ones (i.e. $r_{13} < r_{33}$). Finally, copropagation makes the ratio of the peak soliton beam intensity I_0 to the background illumination intensity, I_b constant for each value of z along the propagation direction, provided that the absorption is the same for the two beams. This last condition is particularly important because soliton formation strongly depends on the value of this ratio $u_0^2 = I_0/I_\mathrm{b}$, known as the intensity ratio, which must be independent of z. Screening solitons have been used to observe a number of startling nonlinear effects, such as soliton spiraling [46], fusion [47], and soliton annihilation, [48] to realize new optical devices, such as a reconfigurable direction coupler [49] and an enhanced second-harmonic generation system [50,51], and to investigate hitherto unexplored phenomena, such as self-trapping of incoherent light beams, discussed in another chapter in this book.

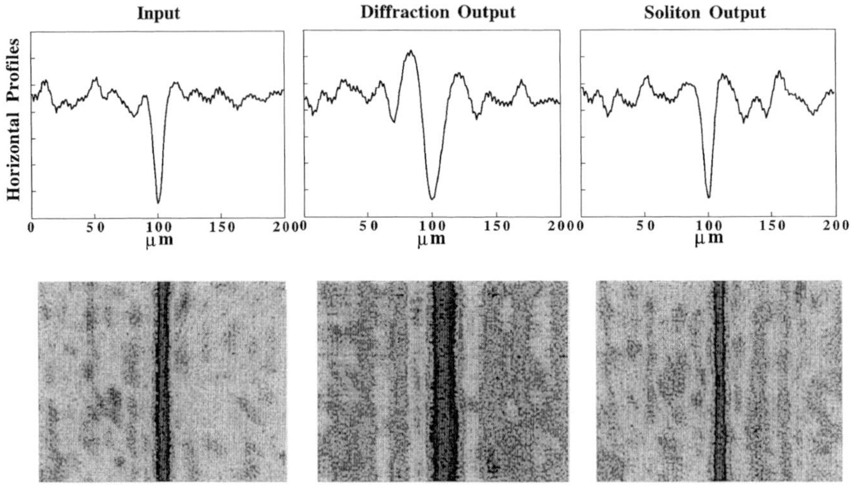

Fig. 5. Dark slab screening solitons. Beam profiles and photographs of the input, normally diffracting output, and soliton output beams after propagation in a 5 mm SBN crystal. Taken from [35]

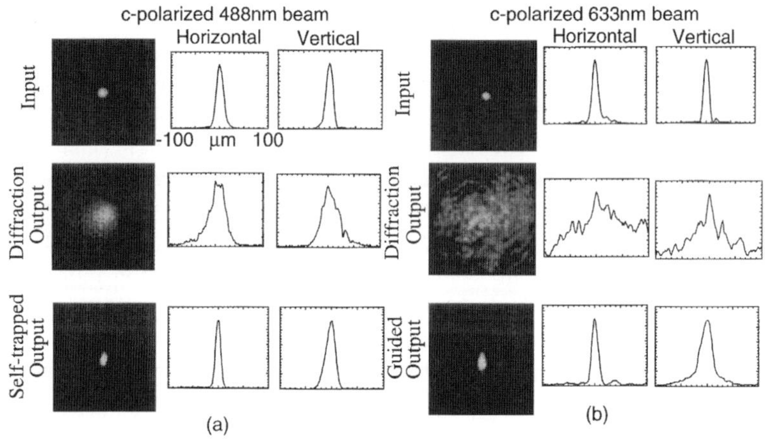

Fig. 6. Needle soliton: photographs and beam profiles (cross sections) of (**a**) a c-polarized 488 nm needle soliton beam and (**b**) a c-polarized 632.8 nm 2D non-photoactive probe beam. The photographs and sections at the *top* show the input beams, those in the *middle* show the diffracted beams at zero voltage, and those at the *bottom* show the slightly distorted needle soliton output and the corresponding passively guided probe beam. Taken from [51]

The theoretical description of slab screening solitons [30] as shown in Fig. 5 is to be considered one of the most important successes of the Kukhtarev model described above, whereas the description of needle solitons as shown in Fig. 6 is still subject of investigation, and the strikingly symmetric self-trapping that emerges from a highly asymmetric physical process baffles more than one researcher [52–55].

3.2.1 Screening Slab-Solitons

A description of screening slab solitons has been formulated on the basis of an approximate reduction of the full Kukhtarev model [30–34]. Consider the reduced (1 + 1)D system associated with steady-state slab solitons. This reduces (7) to a partial differential equation in $\boldsymbol{A}(x,z)$ and $I(x,z) = |\boldsymbol{A}(x,z)|^2$. Furthermore, the typical values of the physical quantities involved are such that $\varepsilon\nabla \cdot E/\alpha N_\mathrm{a} q \ll 1$, and in most cases $\gamma \equiv \varepsilon\nabla \cdot E/N_\mathrm{a} q \ll 1$ is a much stronger, but still relatively plausible, assumption. For example, in SBN, a 10 μm screening soliton is formed with an electric-field scale $|\boldsymbol{E}| \sim 10^2$ kV/m, $N_\mathrm{a} \approx 10^{22}$ m^{-3}, and $\gamma \sim 0.1$. Neglecting γ with respect to one (and even more so γ/α), the nonlinear differential equation relating the optical field to the space-charge field reads

$$\frac{\partial}{\partial x}\left(E\left(I_\mathrm{b}+I\right)+\frac{k_\mathrm{b}T}{q}\frac{\partial}{\partial x}\left(I_\mathrm{b}+I\right)\right)=0\,, \tag{11}$$

where the z derivatives have been neglected with respect to the x derivatives, a valid assumption for paraxial cases and, in particular, if we are looking for soliton solutions. The second term on the left-hand side comes directly from the diffusion term in (2), assuming $\beta_\mathrm{ph} \equiv 0$. We can neglect it in a drift-dominated regime, since in general, $k_\mathrm{b}T/(q\ell) \ll |E|$, where ℓ is the transverse spatial soliton scale (10 μm in the above-mentioned case). The space-charge field thus approximately obeys the relationship

$$E=\frac{\delta}{I_\mathrm{b}+I}\simeq-\frac{V}{L}\frac{1}{1+I/I_\mathrm{b}}\,, \tag{12}$$

where δ is a constant that is fixed by the boundary condition of (5). Given the fact that $\ell \ll L$, where L is the distance between the crystal electrodes (the sample size in the transverse x direction), $\delta \simeq -I_\mathrm{b}V/L$. Given the scalar electro-optic response of (10) in a ferroelectric (neglecting the quadratic term), the nonlinearity is of the type $\Delta n \sim 1/\left(1+I/I_\mathrm{b}\right)$, i.e. a saturated Kerr-like nonlinearity. The nonlinear propagation equation is obtained by substituting (12) into (10), and then substituting this expression for Δn directly into the parabolic equation (9). The final nonlinear scalar propagation equation reads

$$\frac{\mathrm{d}^2u(\xi)}{\mathrm{d}\xi^2}=\pm\left(\frac{\Gamma}{b}-\frac{1}{1+u(\xi)^2}\right)u(\xi)\,, \tag{13}$$

where we have imposed self-consistently the scalar solitary-wave solution form $A(x,z) = u(x)\mathrm{e}^{\mathrm{i}\Gamma z}\sqrt{I_\mathrm{b}}$, and have normalized the transverse spatial scale to the so-called nonlinear length scale $d = (\pm 2kb)^{-1/2}$, i.e. $\xi = x/d$, where $b = (1/2)kn^2 r_{33}(V/L)$. The plus sign corresponds to $b > 0$ and leads to self-focusing and bright spatial solitons, whereas the minus sign corresponds to $b < 0$ and describes self-defocusing and dark spatial solitons. Equation (13) lacks a first derivative and can be integrated once, giving the relationship $\Gamma/b = \log\left(1 + u_0^2\right)/u_0^2$ for bright beams, and $\Gamma/b = 1/\left(1 + u_\infty^2\right)$ for dark beams, where $u_\infty = u(\infty) = -u(-\infty)$. It is not solvable analytically, unless $u_0 \ll 1$, where it reduces to a Kerr nonlinearity but refers to a situation in which solitons are not observable. In general, the equation can be solved numerically, and examples of profiles of self-trapped beams are reported, for example, in [12]. The profiles are more similar to a hyperbolic secant function than to a Gaussian, meaning that in experiments, in an initial evolution, the Gaussian laser beam is adiabatically transformed into a stable soliton profile. One important issue is the so-called soliton existence curve [34]. For a given value of u_0, the self-trapped profile has a given width, in terms of the normalized spatial coordinate ξ (see Fig. 7). This means, experimentally, that for a given input Gaussian FWHM and a given value of u_0, there is a value of V that allows the observation of a stable screening soliton [33].

Whereas the qualitative success of this treatment is evident, obtaining quantitative agreement is far from trivial. Experiments aimed at drawing a quantitative comparison between theory and experiment indicate that the agreement is not full, and this is generally attributed to a series of factors [34]. First of all, the theory is an analytical approximation to the photorefractive process. Secondly, there is evidence that the small tensorial electro-optic coupling of nondiagonal terms (see (8)) cannot be wholly neglected, and some of the background illumination interacts with the soliton-supporting pattern,

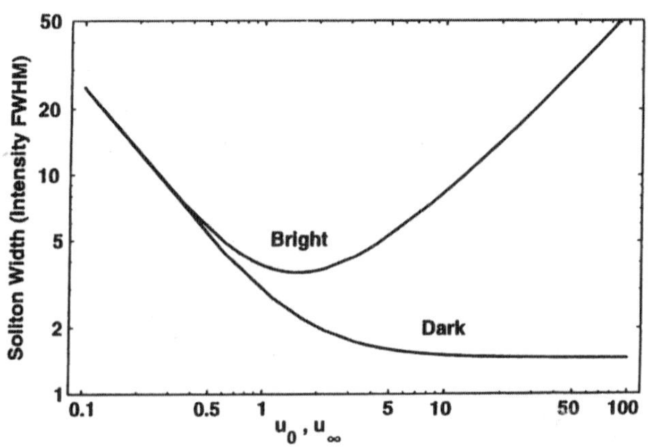

Fig. 7. Calculated existence curve for slab solitons. Width (FWHM of the profile of the intensity u^2) of a bright soliton (*upper curve*) as a function of u_0, and of a dark soliton (*lower curve*) as a function of u_∞. Taken from [33]

changing the beam-trapping conditions. Lastly, as occurs for most experiments involving photorefractive ferroelectrics , the electro-optic parameters vary from sample to sample, depending even on the type of crystal configuration, making comparison arduous [2,3]. Yet another complication, or rather an area of research still to be explored, is represented by transverse instability [56].

3.2.2 Screening Needle Solitons

One of the major breakthroughs in nonlinear photorefractive beam dynamics is the steady-state self-trapping of two-dimensional beams ($(2+1)$D), leading to diffractionless propagation of micron-size light needles [36,37] or, more generically, two-dimensional waves, such as vortices [39]. These have been widely documented in various crystals [43,51] and configurations, in strict analogy to their one-dimensional counterparts. Most results indicate that self-trapped beams that originate from the focused fundamental, circularly symmetric TEM_{00} mode of a laser are to a good approximation themselves circularly symmetric, and that a similar symmetry conservation seems to hold also for dark solitons. This experimental fact does, however, represent a theoretical riddle [26]. Even from only an intuitive point of view, there is no apparent reason why a circularly symmetric beam should self-focus while maintaining this symmetry, when the very nonlinearity that seems to allow self-trapping is altogether not symmetric. There are two main asymmetries embedded in the physical process. The first is the electro-optic response, which is highly anisotropic. The second is the direction of the bias electric field, which is directed initially along the x direction, but undergoes inevitable distortion during charge separation as the full 2D transverse photorefractive mechanism operator. At present, our understanding can be summarized as follows. Whereas the optical beam manifests, under some conditions, a circularly symmetric profile, the underlying space-charge field is highly anisotropic [57,58]. When single, isolated solitons are considered, this puzzling situation is of no direct import. Its main influence is on soliton–soliton interaction, as discussed below [59–62]. The difficulty in understanding the intuitive basis of needle self-trapping is not fully mitigated by any sort of quasi-analytical approach, as in the reduced slab soliton case. In fact, in the $(1+1)$D case, the propagation equation is *local* (see (13)), the nonlocal nature being contained only in the imposition of the global voltage drop across the crystal (see (5)). The full $(2+1)$D propagation equation is highly nonlocal, in the sense that the nonlinear local (symmetric) and nonlocal (asymmetric) interaction terms are comparable. In the Kerr limit of $u^2 \ll 1$ already mentioned, the local term dominates over the nonlocal one, and it is possible to prove the existence of circularly symmetric bright and dark solitons and to derive the corresponding existence curve [26]. This regime, however, cannot be experimentally explored, the accessible regime being that for which $u^2 > 1$. In this important, saturated regime, the analytical approach becomes extremely complicated and the nonlinear wave equation does not lend itself to any analytical solution or approach. All that can really be said is that, from a purely mathematical point of view, exact circularly symmetric solutions do not exist [55], in direct contrast to some experimental evidence. One viable alternative to address the issue of the interpretation of needle solitons is to test experimental findings using purely numerical simulations. Even this approach, however, is extremely difficult because, apart from the complexity of the higher-dimensional system, in analogy to slab solitons and in agreement with experiments, needle solitons also exist, when the beam parameters, which provide the boundary conditions for the propagation problem, are chosen so as to lie as close as possible to the existence curve (which is, of course, a priori unknown) [54]. Recent numerical and experimental results seem to indicate that it is the very nonlocal terms that make the problem intractable, that allow needle formation even in the presence of a highly anisotropic nonlinearity [63]. More fundamentally, one cannot completely rule out the possibility that some of the approximations introduced in the quasi-analytical slab soliton model may not be valid in the $(2+1)$D case, or that even the simple Kukhtarev model may not be able to describe the physical situation in a satisfactory way.

Formally, the theoretical "problem" arises when an external bias field is introduced, and is strictly connected with the requirement that the space-charge field is to be conservative. In the $(2+1)$D case, the approximated equation relating the space-charge field to the optical field (analogous to (11)) reads

$$\nabla \cdot \left(\boldsymbol{E}\,(I_\mathrm{b}+I) + \frac{k_\mathrm{b}T}{q}\nabla \cdot (I_\mathrm{b}+I) \right) = 0\,, \tag{14}$$

and its most general solution can be written in the form

$$\begin{aligned}\boldsymbol{E}(x,y,z) = {} & \frac{\delta\,[1+I(x=\pm\infty)/I_\mathrm{b}])}{1+I/I_b}\boldsymbol{e}_x \\ & - \frac{k_\mathrm{b}T}{q}\nabla\cdot\ln(1+I/I_\mathrm{b}) + \frac{\delta\,[1+I(x=\pm\infty)/I_\mathrm{b}]}{1+I/I_\mathrm{b}}\nabla\times\boldsymbol{v}\,,\end{aligned} \tag{15}$$

where $\boldsymbol{e}_x$ is a unit vector in the x direction, δ is associated with the boundary condition of (5), and $\boldsymbol{v}(x,y,z)$ is an arbitrary vector field determined by imposing the condition of (6). Looking for z-independent intensity profiles associated with soliton-like propagation, we can take $\boldsymbol{v}=(0,0,f(x,y))$, where f obeys

$$\nabla_\perp^2 f - \nabla_\perp f \cdot \nabla_\perp \ln(1+I/I_\mathrm{b}) = \frac{k_\mathrm{b}T}{q}\frac{\partial}{\partial y}(1+I/I_\mathrm{b})\,, \tag{16}$$

where $\nabla_\perp = (\partial/\partial x, \partial/\partial y)$ [26].

3.2.3 Self-bending

Self-bending and self-deflection are a generic signature of system asymmetry in photorefractive self-trapping [64–68], yet, given the significant advances in the understanding of screening solitons, the study of self-bending and self-deflection has been mainly associated with steady-state solitons. As mentioned above, in conventional optical propagation in a photorefractive material, in the absence of an external applied field and photovoltaic effects, spontaneous beam instability leads to beam fanning. This fanning is a distinct signature of space-charge fields induced by charge diffusion, the only charge-separating mechanism in the absence of drift, and originates from spurious inhomogeneities in the intensity distribution, due to scattering from crystal imperfections. In the presence of a self-trapped micron-sized beam, however, things are quite different. The marked intensity inhomogeneity of the nondiffracting beam itself engenders a small, but not negligible, symmetric charge distribution that gives rise to an asymmetric field component, and an associated small asymmetric component of the pattern in the self-lensing structure, along the *entire* propagation trajectory (as opposed to what occurs for a diffracting beam, where self-bending occurs only in the initial stages of propagation). Although self-bending can be neglected in most configurations of interest, it does play a fundamental role in limiting the maximum attainable solitary-wave propagation in a photorefractive material. This is because self-bending increases nonlinearly along the propagation axis and, for a long enough propagation distance inevitably leads to soliton annihilation. More importantly, soliton annihilation is not a merely geometrical limitation, since it depends, like diffraction, on the transverse spatial scale, being stronger for smaller beams.

3.3 Photovoltaic Solitons

As mentioned earlier, some photorefractive noncentrosymmetric crystals, such as $LiNbO_3$, $BaTiO_3$, and $LiTaO_3$, manifest the so-called photovoltaic, effect [69] which can be generally described by introducing a photoinduced current component (the last term of (2)) in the Kukhtarev model [11]. In most situations, the net effect can be reduced to that of a conventional biased nonphotovoltaic material by introducing an effective bias field, the so-called photovoltaic field. This allows, in analogy to screening solitons, photovoltaic solitons [68,70–72]. Thus, although the underlying driving mechanism is physically different (giving rise, in some cases, to peculiar phenomenology, as for example in [73]), their description follows the steps indicated above for screening solitons. Experimental proof of bright, dark, and vortex solitons, in both the lower-dimensional slab and higher-dimensional needle cases, has been reported [74–76].

3.4 Self-Trapping in Semiconductors

An interesting and potentially useful extension of the photorefractive self-trapping phenomenology is represented by the observation of slab and needle photorefractive screening solitons in semiconducting iron-doped indium phosphide (InP) [40,41]. These solitons can be interpreted in much the same manner as their ferroelectric counterparts described above, although they do show some major differences. First of all, they can be formed with infrared beams, at typical telecommunications wavelengths. Secondly, they are characterized by shorter response times than for their ferroelectric counterparts.

3.5 Self-Trapping in Paraelectrics

Photorefractives, like a large proportion of electro-optic crystals, are ferroelectrics. As such, they manifest spontaneous polarization below the critical Curie temperature T_c, passing from their high-temperature centrosymmetric phase to a noncentrosymmetric phase. Although most electro-optic, and consequently photorefractive, experiments are carried out with ferroelectrics in the lower-symmetry phase, since the strong spontaneous polarization of poled samples allows considerable responsivity, some research has been carried out with samples in the higher-symmetry phase, also referred to as paraelectrics. To enhance the electro-optic response, the crystal is brought close to the phase-transition temperature T_c, where the dielectric response is very strong.

Photorefractive screening solitons have been predicted and observed in paraelectric KLTN near the transition [43–45], in analogy to screening solitons in noncentrosymmetric photorefractives. In this case, the main difference from the description given above is that the electro-optic response is purely quadratic, and is generally described by the scalar relationship

$$\Delta n = -\frac{1}{2} n^3 \epsilon_0^2 (\epsilon_r - 1)^2 g_{11} E \ , \tag{17}$$

where, in analogy to the noncentrosymmetric case, the electric field E is applied along a given principal axis (the x direction), the beam is polarized along this axis, and $g_{11} = g_{xxxx}$. KLTN undergoes a structural phase transition at room temperature, but for a range of crystal temperatures T above the transition, ϵ_r takes values of the order of 10^3–10^4. This makes index modulations sufficient to significantly modify beam propagation attainable with reasonably low applied electric fields. The resulting approximate slab soliton theory is similar to the one described above for screening slab solitons, and the final propagation equation reads

$$\frac{d^2 u(\xi)}{d\xi^2} = \pm \left[\frac{1}{1+u_0^2} - \left(\frac{1+u_\infty^2}{1+u(\xi)^2} \right)^2 \right] u(\xi) \ , \tag{18}$$

leading to spatial, bright slab self-trapping when $u_\infty^2 = 0$ and $u(0) = u_0$, and the minus sign is used, and to dark self-trapping when $u(0) = u_0 = 0$, and the plus sign is used. In analogy to the noncentrosymmetric slab soliton description, we have imposed self-consistently the scalar solitary-wave solution form $A(x,z) = u(x)\mathrm{e}^{\mathrm{i}\Gamma z}\sqrt{I_\mathrm{b}}$ and have normalized the transverse spatial scale to the so-called nonlinear length scale $d = (\pm 2k\mathrm{b})^{-1/2}$, i.e. $\xi = x/d$, where $b = (1/2)kn^2 g_{11}\epsilon_0^2(\epsilon_\mathrm{r} - 1)^2(V/L)^2$. The minus sign corresponds to bright solitons, and the plus sign corresponds to dark self-trapping.

As in the noncentrosymmetric case, needle solitons have been documented as being strikingly symmetric, and present the same theoretical riddles as their lower-symmetry counterparts.

Major differences between photorefractive beam dynamics in ferroelectrics and paraelectrics arise in the mechanisms associated with charge diffusion, a relatively marginal process in trapping in noncentrosymmetric materials [77]. In noncentrosymmetric materials, diffusion basically plays the role of an asymmetric seed that eventually leads to appreciable beam bending, and finally to beam annihilation. In a paraelectric, on the other hand, this is not so. The quadratic response makes the index modulation associated with the intrinsically asymmetric diffusion fields symmetric again, thus leading to self-lensing in the absence of external bias [78].

Starting from (7), making the approximations indicated in Sect. 3.2, and imposing the condition that the applied voltage be null (in (5)), i.e. $\boldsymbol{J} = 0$, gives an approximate expression for the internal field $\boldsymbol{E} = -(k_\mathrm{b}T/q)[\nabla \cdot I/(I_\mathrm{b}+I)]$. When this is inserted into (8) and in (9), it gives the final nonlinear propagation equation

$$\left(\mathrm{i}\frac{\partial}{\partial Z} + \nabla_\perp^2\right)u + \left[\gamma_1\left(\frac{\partial\left|u\right|^2/\partial X}{\left|u\right|^2+1}\right)^2 + \gamma_2\left(\frac{\partial\left|u\right|^2/\partial Y}{\left|u\right|^2+1}\right)^2\right]u = 0\ , \quad (19)$$

where $u = A_x I_\mathrm{b}^{-1/2}$, $(X,Y) = 2^{1/2}(kx, ky)$, $Z = kz$, $\nabla_\perp^2 = \partial^2/\partial X^2 + \partial^2/\partial Y^2$, $\gamma_1 = -k^2n^2\epsilon_0^2(\epsilon_\mathrm{r} - 1)^2 g_{11}(K_\mathrm{b}T/q)^2$, and $\gamma_2 = -k^2n^2\epsilon_0^2(\epsilon_\mathrm{r} - 1)^2 g_{12}(K_\mathrm{b}T/q)^2$. If I_b can be neglected in comparison with the beam intensity $I(X,Y)$, this equation admits exact analytical solutions, a remarkable fact in itself, given the almost absence of exact solutions in nonlinear propagation problems [79]. These solutions describe a number of interesting optical phenomena, such as self-modified optical diffraction and beam aspect ratio locking, of which partial experimental observation has been reported [79]. Furthermore, (19) supports a class of Gaussian and non-Gaussian self-trapped solutions in the form of noncircular spatial solitons, which, for the currently available values of γ, are outside the reach of observation.

4 Material Nonlinearities and Solitons

Photorefraction is generally associated with a light–crystal interaction that does not lead to actual material distortion and can be described by a linear relationship between the photoinduced electric field $\boldsymbol{E}$ and the resultant static polarization $\boldsymbol{P}$ of the crystal. Ferroelectrics, in a more general context, respond to local electric fields in a much more complicated fashion, allowing local reorientation of the spontaneous polarization and hence leading to a series of complex domain phenomena. This highly nonlinear crystal susceptibility has been used in standard holographic configurations for the permanent fixing of index patterns. In relation to photorefractive spatial solitons, two different phenomena have been investigated. The first is the fixing of spatial solitons, leading to the permanent imprinting of guiding structures in bulk samples of SBN. [80] A more complicated mechanism can be observed in a paraelectric undergoing a structural phase-transition. In this case, light induced diffusion charge fields pin down and seed a self-trapping guiding structure, giving rise to spontaneous self-trapping [81].

5 Nonlinear Beam Interactions

As is true for a large variety of nonlinear waves in physics, some of the most interesting, counterintuitive, and useful phenomenology is encountered when two or more nonlinear beams are made to interact or collide, and photorefractive self-trapping is no exception. In fact, in the last decade, photorefractive solitons, in particular screening solitons, have played a leading role in nonlinear collisional studies. This is mainly connected with the fact that photorefractive solitons are supported by a saturated nonlinearity that presents a more varied phenomenology than do more traditional Kerr-like waves. Thus, for example, photorefractive self-trapping occurs for both slab and needle beams in bulk environments, this greatly increases the degrees of freedom at work. Soliton phenomenology has been investigated in SBN for incoherent slab and needle screening solitons, with the important documentation of beam attraction and ultimately fusion [67,82–84]. Phase-dependent attraction and repulsion of coherent, parallel, nonlinear screening needle beams has been observed in both BTO [85] and SBN, along with phase-dependent interaction, fusion, and birth [86,87]. Interaction has been used to investigate symmetry and asymmetry in the formation of needle solitons as mentioned above [59,62]. The most important associated result is that the strong anisotropy in the space-charge field supporting a needle soliton can give rise to soliton–soliton interactions that bear little resemblance to conventional Kerr-like behavior. For example, incoherent solitons can repel each other [61]. Exotic phenomenology directly associated with the higher dimensionality of the system has allowed the documentation of soliton spiraling [46], in which two needle solitons spiral so as to conserve angular momentum, and of a collision

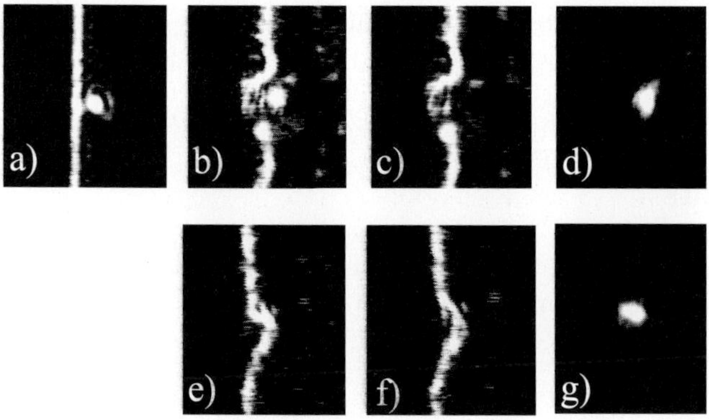

Fig. 8. Coherent interaction of a slab soliton and a needle screening soliton in a photorefractive sample of KLTN: (**a**) input; (**b**) output in the repulsive case $\Delta\phi_0 = \pi$; (**c**), (**d**) as in (b), with the needle and the stripe, repectively, blocked; (**e**) output in the attractive case $\Delta\phi_0 = 0$; (**f**), (**g**) as in (e) with the needle and the stripe, respectively, blocked. Taken from [88]

and interaction of a needle soliton with a slab soliton, in which the different dimensionalities of the waves induce intrinsically inhomogeneous interaction forces [88] (see Fig. 8).

6 Applications

The intense scientific effort associated with photorefractive spatial solitons has, to date, been mainly directed towards the observation and documentation of the diverse effects connected with the general characteristics of soliton physics. Regarding applications, pioneering experiments have demonstrated the basic but important beam-steering capabilities of single photorefractive solitons [24]. The self-induced waveguide formed at a photoactive visible wavelength can be used to guide a longer-wavelength, nonphotorefractively active beam, such as an infrared signal, through an otherwise bulk environment [89]. Actual beam rerouting, however, is not feasible, given the temporal restraints of the photorefractive response; however, these can be loosened using intense laser pulses [90]. Notwithstanding this limitation, two conceptual applications that are not directly hampered by slow response have been experimentally demonstrated. The first is a reconfigurable directional coupler, obtained simply by launching two independent parallel solitons and using the double waveguide structure to observe a directional coupler at a longer, infrared wavelength [49]. The second is connected to enhanced second-harmonic generation through self-induced phase matching of a micron-sized beam throughout a sample of $KNbO_3$, this being a direct consequence of beam self-trapping (similar to second-harmonic generation in waveguides) [50]. More recently, direct electro-optic beam handling in paraelectrics through soliton beams has been demonstrated, and promises a less ambitious electro-optic, as opposed to all-optical, beam-handling functionality [91].

Acknowledgments

Eugenio DelRe is partly supported by an agreement between the Fondazione Ugo Bordoni and the Italian Communications Administration.

References

1. A. Ashkin, G. Boyd, J. Dziedzic, R. Smith, A. Ballman, J. Levinstein, K. Nassau, Optically-induced refractive index inhomogeneities in $LiNbO_3$ and $LiTaO_3$, Appl. Phys. Lett. **9**, 72–74 (1966).
2. P. Yeh, *Introduction to Photorefractive Nonlinear Optics* (Wiley, New York, 1993).
3. L. Solymar, D.J. Webb, A. Grunnet-Jepsen, *The Physics and Applications of Photorefractive Materials* (Clarendon, Oxford, 1996).
4. P. Gunter, J.-P. Huignard (eds.), *Photorefractive Materials and Their Applications*, Vol. 1 (Springer, Berlin, Heidelberg, 1988).
5. P. Gunter, J.-P. Huignard (eds.), *Photorefractive Materials and Their Applications*, Vol. 2 (Springer, Berlin, Heidelberg, 1989).
6. M. Segev, B. Crosignani, A. Yariv, B. Fischer, Spatial solitons in photorefractive media, Phys. Rev. Lett. **68**, 923–926 (1992).
7. G.I. Stegeman, M. Segev, Optical spatial solitons and their interactions: universality and diversity, Science **286**, 1518–1523 (1999).
8. M. Segev, Optical spatial solitons. Opt. Quantum Electron. **230**, 503–533 (1998).
9. M. Segev, G. Stegeman, Self-trapping of optical beams: spatial solitons, Phys. Today **51**, 42–48 (1998).
10. V.L. Vinetskii, N.V. Kukhtarev, Phys. Solid State **16**, 2414 (1975).; N. Kukhtarev V., S.G. Odoulov, Wave front convolution in 4-wave interaction in media with nonlocal nonlinearity, Sov. Phys. JETP Lett. **30**, 6–11 (1979).
11. B.I. Sturman, V.M. Fridkin, *The Photovoltaic and Photorefractive Effects in Noncentrosymmetric Materials* (Gordon and Breach, Philadelphia, 1992).
12. B. Crosignani, P. Di Porto, M. Segev, G. Salamo, A. Yariv, Nonlinear optical beam propagation and solitons in photorefractive media, Riv. Nuovo Cimento **21**, (6), 1–37 (1998).
13. B. Crosignani, M. Segev, D. Engin, P. Di Porto, A. Yariv, G. Salamo, Self-trapping of optical beams in photorefractive media, J. Opt. Soc. Am. B **10**, 446–453 (1993).
14. G.C. Duree, J.L. Shultz, G.J. Salamo, M. Segev, A. Yariv, B. Crosignani, P. Di Porto, E.J. Sharp, R.R. Neurgaonkar, Observation of self-trapping of an optical beam due to the photorefractive effect, Phys. Rev. Lett. **71**, (1993). 533–536
15. M. Segev, G.C. Valley, S.R. Singh, M.I Carvalho, D.N. Christodoulides, Vector photorefractive spatial solitons, Opt. Lett. **20**, 1764–1766 (1995).

16. S.R. Singh, M.I. Carvalho, D.N. Christodoulides, Vector interactions of steady-state planar solitons in biased photorefractive media, Opt. Lett. **20**, 2177–2179 (1995).
17. G. Duree, M. Morin, G. Salamo, M. Segev, B. Crosignani, P. Di Porto, E. Sharp, A. Yariv, Dark photorefractive spatial solitons and photorefractive vortex solitons, Phys. Rev. Lett. **74**, 1978–1981 (1995).
18. M.D. Castillo, J.J. Sanchez-Mondragon, S.I. Stepanov, M.B. Klein, B.A. Wechsler, Probe beam wave-guiding induced by spatial dark solitons in photorefractive BTO crystal, Rev. Mex. Fis. **41**, 1–10 (1995).
19. N. Fressengeas, D. Wolfersberger, J. Maufoy, G. Kugel, Build up mechanisms of (1+1)-dimensional photorefractive bright spatial quasi-steady-state and screening solitons, Opt. Comm. **145**, 393–400 (1998).
20. D. Kip, M. Wesner, V. Shandarov, P. Moretti, Observation of bright spatial photorefractive solitons in a planar strontium barium niobate waveguide, Opt. Lett. **23**, 921–923 (1998).
21. N. Fressengeas, J. Maufoy, G. Kugel, Temporal behavior of bidimensional photorefractive bright spatial solitons, Phys. Rev. E **54**, 6866–6875 (1996).
22. N. Fressengeas, D. Wolfersberger, J. Maufoy, G. Kugel, Experimental study of the self-focusing process temporal behavior in photorefractive $Bi_{12}TiO_{20}$, J. Appl. Phys. **85**, 2577–2586 (1999).
23. J. Maufoy, N. Fressengeas, D. Wolfersberger, G. Kugel, Simulation of the temporal behavior of soliton propagation in photorefractive media, Phys. Rev. E **59**, 6116–6121 (1999).
24. M. Morin, G. Duree, G. Salamo, M. Segev, Wave-guides formed by quasi-steady-state photorefractive spatial solitons, Opt. Lett. **20**, 2066–2068 (1995).
25. M. Segev, B. Crosignani, P. Di Porto, A. Yariv, G. Duree, G. Salamo, E. Sharp, Stability of photorefractive spatial solitons, Opt. Lett. **19**, 1296–1298 (1994).
26. B. Crosignani, P. Di Porto, A. Degasperis, M. Segev, S. Trillo, (1997). Three-dimensional optical beam propagation and solitons in photorefractive crystals, J. Opt. Soc. Am. B **14**, 3078–3090
27. G.Q. Zhang, S.M. Liu, J.J. Xu, Q. Sun, G.Y. Zhang, Optical dark and bright spatial solitons in photorefractive media, Phys. Lett. A **204**, 146–150 (1995).
28. G. Duree, G. Salamo, M. Segev, A. Yariv, B. Crosignani, P. Di Porto, E. Sharp, Dimensionality and size of photorefractive spatial solitons, Opt. Lett. **19**, 1195–1197 (1994).
29. M. Castillo, P. Aguilar, J. Sanchez-Mondragon, S. Stepanov, V. Vysloukh, Spatial solitons in photorefractive $Bi_{12}TiO_{20}$, Appl. Phys. Lett. **64**, 408–410 (1994).
30. M. Segev, G.C. Valley, B. Crosignani, P. Di Porto, A. Yariv, Steady-state spatial screening solitons in photorefractive materials with external applied-field, Phys. Rev. Lett. **73**, 3211–3214 (1994).
31. D.N. Christodoulides, M.I Carvalho, Bright, dark, and gray spatial soliton states in photorefractive media, J. Opt. Soc. Am. B **12**, 1628–1633 (1995).
32. S.R. Singh, D.N. Christodoulides, Evolution of spatial optical solitons in biased photorefractive media under steady-state conditions, Opt. Commun. **118**, 569–576 (1995).
33. M. Segev, M. Shih, G.C. Valley, Photorefractive screening solitons of high and low intensity, J. Opt. Soc. Am. B **13**, 706–718 (1996).
34. K. Kos, H.X. Meng, G. Salamo, M. Shih, M. Segev, G.C. Valley, One-dimensional steady-state photorefractive screening solitons, Phys. Rev. E **53**, R4330–R4333 (1996).

35. Z.G. Chen, M. Mitchell, M. Shih, M. Segev, M.H. Garrett, G.C. Valley, Steady-state dark photorefractive screening solitons, Opt. Lett. **21**, 629–631 (1996).
36. M. Shih, M. Segev, G.C. Valley, G. Salamo, B. Crosignani, P. Di Porto, Observation of 2-dimensional steady-state photorefractive screening solitons, Electron. Lett. **31**, 826–827 (1995).
37. M. Shih, P. Leach, M. Segev, M. Garrett, G. Salamo, G.C. Valley, Two-dimensional steady-state photorefractive screening solitons, Opt. Lett. **21**, 324–326 (1996).
38. M. Castillo, J. Sanchez-Mondragon, S. Stepanov, M. Klein, B. Wechsler, (1+1)-dimensional dark spatial solitons in photorefractive $Bi_{12}TiO_{20}$ crystal, Opt. Commun. **118**, 515–519 (1995).
39. Z.G. Chen, M.F. Shih, M. Segev, D.W. Wilson, R.E. Muller, P.D. Maker, Steady-state vortex-screening solitons formed in biased photorefractive media, Opt. Lett. **22**, 1751–1753 (1997).
40. M. Chauvet, S. Hawkins, G. Salamo, M. Segev, D. Bliss, G. Bryant, Self-trapping of planar optical beams by use of the photorefractive effect in InP:Fe, Opt. Lett. **21**, 1333–1335 (1996).
41. M. Chauvet, S. Hawkins, G. Salamo, M. Segev, D. Bliss, G. Bryant, Self-trapping of two-dimensional optical beams and light-induced waveguiding in photorefractive InP at telecommunication wavelengths, Appl. Phys. Lett. **70**, 2499–2501 (1997).
42. J.A. Andrade-Lucio, M.D. Iturbe-Castillo, P.A. Marquez-Aguilar, R. Ramos-Garcia, Self-focusing in photorefractive $BaTiO_3$ crystal under external DC electric field, Opt. Quantum Electron. **30**, 829–834 (1998).
43. E. DelRe, M. Tamburrini, M. Segev, E. Refaeli, A.J. Agranat, Two-dimensional photorefractive spatial solitons in centrosymmetric paraelectric potassium-lithium-tantalate-niobate, Appl. Phys. Lett. **73**, (1998). 16–18
44. E. DelRe, B. Crosignani, M. Tamburrini, M. Segev, M. Mitchell, E. Refaeli, A.J. Agranat, One-dimensional steady-state photorefractive spatial solitons in centrosymmetric paraelectric potassium lithium tantalate niobate, Opt. Lett. **23**, 421–423 (1998).
45. M. Segev, A.J. Agranat, Spatial solitons in centrosymmetric photorefractive media, Opt. Lett. **22**, 1299–1301 (1997).
46. M. Shih, M. Segev, G. Salamo, Three-dimensional spiraling of interacting spatial solitons, Phys. Rev. Lett. **78**, 2551–2554 (1997).
47. W. Krolikowski, S.A. Holmstrom, Fusion and birth of spatial solitons upon collision, Opt. Lett. **22**, 369–371 (1997).
48. W. Krolikowski, B. Luther-Davies, C. Denz, T. Tschudi, Annihilation of photorefractive solitons, Opt. Lett. **23**, 97–99 (1998).
49. S. Lan, E. DelRe, Z.G. Chen, M.F. Shih, M. Segev, Directional coupler with soliton-induced waveguides, Opt. Lett. **24**, 475–477 (1999).
50. S. Lan, M.F. Shih, G. Mizell, J.A. Giordmaine, Z.G. Chen, C. Anastassiou, J. Martin, M. Segev, Second-harmonic generation in waveguides induced by photorefractive spatial solitons. Opt. Lett. **24**, 1145–1147 (1999).
51. S. Lan, M.F. Shih, M. Segev, Self-trapping of one-dimensional and two-dimensional optical beams and induced waveguides in photorefractive $KnbO_3$, Opt. Lett. **22**, 1467–1469 (1997).
52. A.V. Mamaev, M. Saffman, A.A. Zozulya, Propagation of dark stripe beams in nonlinear media: snake instability and creation of optical vortices, Phys. Rev. Lett. **76**, 2262–2265 (1996).

53. A.V. Mamaev, M. Saffman, D.Z. Anderson, A.A. Zozulya, Propagation of light beams in anisotropic nonlinear media: from symmetry breaking to spatial turbulence, Phys. Rev. A **54**, 870–879 (1996).
54. S. Gatz, J. Herrmann Anisotropy, nonlocality, and space-charge field displacement in (2 + 1)-dimensional self-trapping in biased photorefractive crystals, Opt. Lett. **23**, 1176–1178 (1998).
55. M. Saffman, A.A. Zozulya Circular solitons do not exist in photorefractive media, Opt. Lett. **23**, 1579–1581 (1998).
56. E. Infeld, T. Lenkowska-Czerwinska, Analysis of stability of light beams in nonlinear photorefractive media, Phys. Rev. E **55**, 6101–6106 (1997).
57. A.A. Zozulya, D.Z. Anderson, Propagation of an optical beam in a photorefractive medium in the presence of a photogalvanic nonlinearity or an externally applied electric field, Phys. Rev. A **51**, 1520–1531 (1995).
58. C.M. Gomez-Sarabia, P.A. Marquez Aguilar, J.J. Sanchez Mondragon, S. Stepanov, V. Vysloukh, Dynamics of photoinduced lens formation in a photorefractive $Bi_{12}TiO_{20}$ crystal under an external dc electric field, J. Opt. Soc. Am. B **13**, 2767–2774 (1996).
59. H.X. Meng, G. Salamo, M. Segev, Primarily isotropic nature of photorefractive screening solitons and the interactions between them, Opt. Lett. **23**, 897–899 (1998).
60. S. Gatz, J. Herrmann, Propagation of optical beams and the properties of two-dimensional spatial solitons in media with a local saturable nonlinear refractive index, J. Opt. Soc. Am. B **14**, 1795–1806 (1997).
61. A. Stepken, F. Kaiser, M.R. Belic, Anisotropic interaction of three-dimensional spatial screening solitons, J. Opt. Soc. Am. B **17**, 68–77 (2000).
62. W. Krolikowski, M. Saffman, B. Luther-Davies, C. Denz, Anomalous interaction of spatial solitons in photorefractive media, Phys. Rev. Lett. **80**, 3240–3243 (1998).
63. E. DelRe, A. Ciattoni, A.J. Agranat, Anisotropic charge displacement supporting isolated photorefractive needles, Opt. Lett. **26**, 908–910 (2001).
64. D.N. Christodoulides, M.I. Carvalho, Compression, self-bending, and collapse of gaussian beams in photorefractive crystals, Opt. Lett. **19**, 1714–1716 (1994).
65. M.I. Carvalho, S.R. Singh, D.N. Christodoulides, Self-deflection of steady-state bright spatial solitons in biased photorefractive crystals, Opt. Commun. **120**, 311–315 (1995).
66. J. Petter, C. Weilnau, C. Denz, A. Stepken, F. Kaiser, Self-bending of photorefractive solitons, Opt. Commun. **170**, 291–297 (1999).
67. V.A. Aleshkevich, V.A. Vysloukh, Y.V. Kartashov, Formation and interaction of spatial solitons in a photorefractive medium with drift and diffusion components of the nonlinear response, Quantum Electron. **29**, 621–625 (1999).
68. J.S. Liu, K.Q. Lu, Screening-photovoltaic spatial solitons in biased photovoltaic–photorefractive crystals and their self-deflection, J. Opt. Soc. Am. B **16**, 550–555 (1999).
69. A.M. Glass, D. von der Linde, T.J. Negran, High-voltage bulk photovoltaic effect and the photorefractive process in $LiNbO_3$, Appl. Phys. Lett. **25**, 233–235 (1974).
70. G.C. Valley, M. Segev, B. Crosignani, A. Yariv, M.M. Fejer, M. Bashaw, Dark and bright photovoltaic spatial solitons, Phys. Rev. A **50**, R4457–R4460 (1994).
71. M. Taya, M.C. Bashaw, M.M. Fejer, M. Segev, G.C. Valley, Observation of dark photovoltaic spatial solitons, Phys. Rev. A **52**, 3095–3100 (1995).

72. M. Segev, G.C. Valley, M.C. Bashaw, M. Taya, M.M. Fejer, Photovoltaic spatial solitons, J. Opt. Soc. Am. B **14**, 1772–1781 (1997).
73. C. Anastassiou, M.F. Shih, M. Mitchell, Z.G. Chen, M. Segev, Optically induced photovoltaic self-defocusing-to-self-focusing transition, Opt. Lett. **23**, 924–926 (1998).
74. M. Taya, M.C. Bashaw, M.M. Fejer, M. Segev, G.C. Valley, Y junctions arising from dark-soliton propagation in photovolatic media, Opt. Lett. **21**, 943–945 (1996).
75. Z. Chen, M. Segev, D.W. Wilson, R.E. Muller, P.D. Maker, Self-trapping of an optical vortex by use of the bulk photovoltaic effect, Phys. Rev. Lett. **78**, 2948–2951 (1997).
76. W.L. She, K.K. Lee, W.K. Lee, Observation of two-dimensional bright photovoltaic spatial solitons, Phys. Rev. Lett. **83**, 3182–3185 (1999).
77. E. DelRe, A. Ciattoni, B. Crosignani, P. Di Porto, Nonlinear optical propagation phenomena in near-transiton centrosymmetric photorefractive crystals, J. Nonlinear Opt. Phys. **8**, 1–20 (1999).
78. B. Crosignani, E. DelRe, P. Di Porto, A. Degasperis, Self-focusing and self-trapping in unbiased centrosymmetric photorefractive media, Opt. Lett. **23**, 912–914 (1998).
79. B. Crosignani, A. Degasperis, E. DelRe, P. Di Porto, A.J. Agranat, Nonlinear optical diffraction effects and solitons due to anisotropic charge-diffusion-based self-interaction, Phys. Rev. Lett. **82**, 1664–1667 (1999).
80. M. Klotz, H.X. Meng, G.J. Salamo, M. Segev, S.R. Montgomery, Fixing the photorefractive soliton, Opt. Lett. **24**, 77–79 (1999).
81. E. DelRe, M. Tamburrini, M. Segev, R. Della Pergola, A.J. Agranat, Spontaneous self-trapping of optical beams in metastable paraelectric crystals, Phys. Rev. Lett. **83**, 1954–1957 (1999).
82. Z.G. Chen, M. Segev, T.H. Coskun, D.N. Christodoulides, Y.S. Kivshar, Coupled photorefractive spatial-soliton pairs, J. Opt. Soc. Am. B **14**, 3066–3077 (1997).
83. M.F. Shih, M. Segev, Incoherent collisions between two-dimensional bright steady-state photorefractive spatial screening solitons, Opt. Lett. **21**, 1538–1540 (1996).
84. M.F. Shih, Z.G. Chen, M. Segev, T.H. Coskun, D.N. Christodoulides, Incoherent collisions between one-dimensional steady-state photorefractive screening solitons, Appl. Phys. Lett. **69**, 4151–4153 (1996).
85. G.S. GarciaQuirino, M.D. IturbeCastillo, V.A. Vysloukh, J.J. SanchezMondragon, S.I. Stepanov, G. LugoMartinez, G.E. TorresCisneros, Observation of interaction forces between one-dimensional spatial solitons in photorefractive crystals, Opt. Lett. **22**, 154–156 (1997).
86. W. Krolikowski, S.A. Holmstrom, Fusion and birth of spatial solitons upon collision, Opt. Lett. **22**, 369–371 (1997).
87. H.X. Meng, G. Salamo, M.F. Shih, M. Segev, Coherent collisions of photorefractive solitons, Opt. Lett. **22**, 448–450 (1997).
88. E. DelRe, S. Trillo, A.J. Agranat, Collisions and inhomogeneous forces between solitons of different dimensionality, Opt. Lett. **25**, 560–562 (2000).
89. M.F. Shih, Z.G. Chen, M. Mitchell, M. Segev, H. Lee, R.S. Feigelson, J.P. Wilde, Waveguides induced by photorefractive screening solitons, J. Opt. Soc. Am. B **14**, 3091–3101 (1997).

90. K. Kos, G. Salamo, M. Segev, High-intensity nanosecond photorefractive spatial solitons, Opt. Lett. **23**, 1001–1003 (1998).
91. E. DelRe, M. Tamburrini, A.J. Agranat, Soliton electro-optic effects in paraelectrics, Opt. Lett. **25**, 963–965 (2000).

Incoherent Solitons: Self-Trapping of Weakly Correlated Wavepackets

Mordechai Segev and Demetrios N. Christodoulides

Summary. Incoherent Solitons are self-trapped wave-packets upon which the phase distribution is random. In this chapter we describe this new area of science, from the first observation of solitons made of light from an incandescent light bulb, to the recent discoveries of pattern formation and modulation instabilities in weakly-correlated nonlinear wave systems.

1 Introduction, First Experiments, and Intuition

Solitons, or localized wavepackets that do not broaden in a dispersive environment, have been demonstrated to arise as a result of a variety of nonlinear self-trapping mechanisms and in many branches of physics. They exhibit particle-like behavior that is manifested in a richness of interaction effects [1]. Yet solitons are a universal phenomenon and, despite their diversity, share many common features [2]. In their best-known realization, these particle-like wavepackets are fully coherent entities. In other words, given the phase at a given location on the soliton and the frequency of the carrier wave, one can predict the phase anywhere (in space and time) on that self-trapped pulse. Wavepackets, however, do not necessarily need to be coherent. For example, one can focus into a narrow spot a light beam from a natural source such as the sun or from an incandescent light bulb. Can such a partially incoherent beam self-trap in a nonlinear medium? In a more general context, one can construct an ensemble of weakly correlated particles or a partially coherent wavepacket. Can this incoherent ensemble of particles form a self-trapped entity? And if it does self-trap, how does the correlation information (coherence function) influence the trapping? How do these complex entities interact with one another? Are they stable? In this chapter, we provide an updated overview of this new and exciting branch of solitons that just five years ago seemed to be no more than an oxymoron: incoherent solitons. As we show in this chapter, these partially coherent self-trapped entities, by their nature, exhibit a host of unique properties that have no counterpart whatsoever in the coherent regime.

Until 1995, all soliton experiments, in all branches of science, employed a coherent "pulse". But in 1996, Mitchell et al. at Princeton demonstrated self-trapping of beams (spatial "pulses") in which the phase varied randomly in space/time across any plane [3]. In that experiment, a quasi-monochromatic,

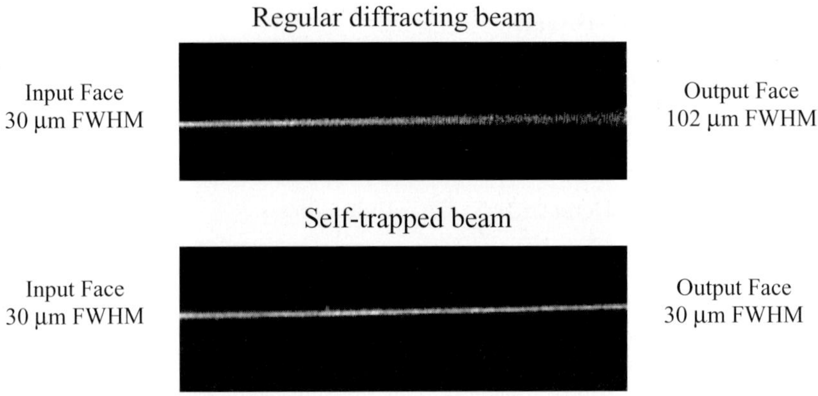

Fig. 1. Top-view photographs of normally diffracting (*top*) and self-trapped (*bottom*) beams

partially spatially incoherent light beam was employed. The beam originated from a laser and passed through a rotating diffuser that introduced a new random phase pattern every microsecond. The beam was launched into a slowly responding photorefractive crystal and, under appropriate conditions, the envelope of this beam self-trapped into a single nondiffracting narrow filament. The fact that the self-trapped beam was indeed a partially incoherent entity was manifested in its diffraction properties. The input beam was 30 µm wide FWHM (full-width-half-maximum) and, in the absence of self-trapping, it diffracted to 102 µm after 6 mm of propagation in the medium. Had this beam been fully coherent, it would had diffracted to only 35.75 µm in the same material and after the same propagation distance. To a large extent, this partially incoherent beam was a quasi-homogeneous spatially incoherent beam: the ratio between the beam diameter and the correlation distance across the beam was roughly a factor of eight. Yet this partially incoherent wavepacket, which exemplifies an ensemble of weakly correlated particles, self-trapped to form a localized nondiffracting beam, an incoherent soliton, when an appropriate nonlinearity was employed. In the particular experiment of [3], the nonlinearity used was of the photorefractive screening type [4,5]. The application of 550 volts between the electrodes, separated by 6 mm, resulted in self-trapping of this partially coherent beam, which maintained a constant width of 30 µm throughout its propagation (Fig. 1). This is how the first incoherent soliton was born. In a subsequent experiment, Mitchell and Segev have demonstrated that an incoherent white-light beam, i.e. a "pulse" that is both temporally and spatially incoherent, can also self-trap [6]. In that experiment the self-trapped beam originated from a simple incandescent light bulb which emitted light between wavelengths of 380 and 720 nm (Fig. 2). Soon afterwards, two different theories were suggested. First, the coherent-density theory was developed [7]. This theory

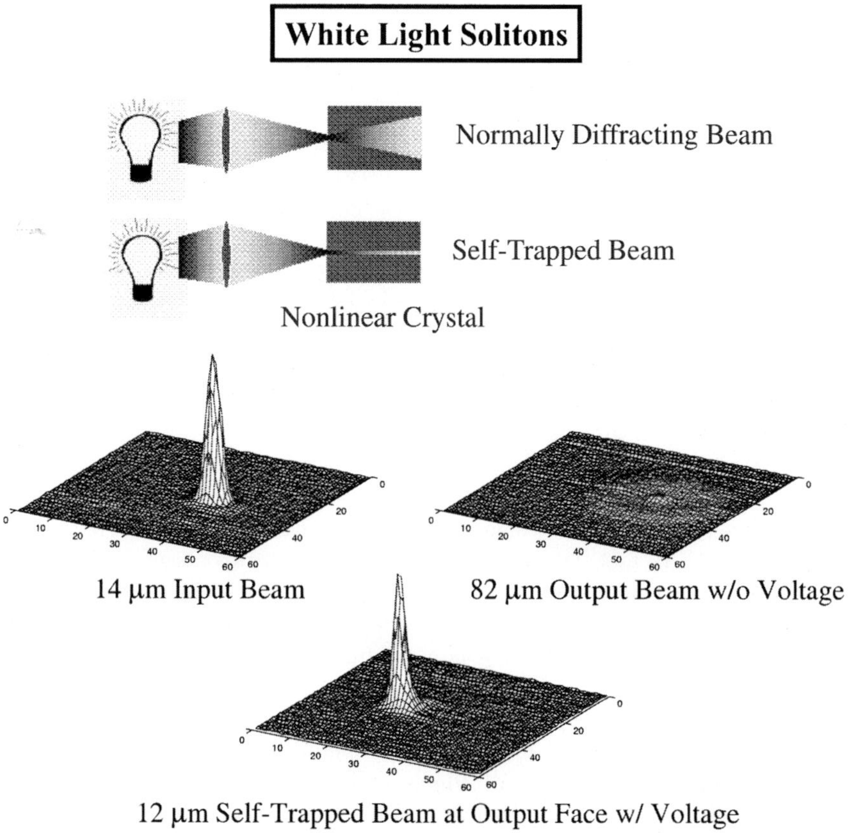

Fig. 2. Self-trapping of an incoherent "white" light beam

is, by its nature, a dynamic approach. Using this method, the first analytical solution describing an incoherent soliton (a Gaussian–Schell soliton) was obtained (in saturable nonlinear systems of the logarithmic type) [8]. Several months later, the modal theory [9] was developed, and, by virtue of its inherent simplicity, it became the method of choice for identifying incoherent solitons, their range of existence, and their correlation properties. One year later, yet another theory was proposed: the theory describing the propagation of mutual coherence [10].[1] Interestingly enough, even though at first sight these three theoretical approaches seem to be dissimilar, they are in fact formally equivalent to one another [12]. They are simply different representations of the same propagation process, each having its strengths and weaknesses. In addition to these exact theories, another, more simplified, ray optics approach was suggested [13]. The ray optics formulation of incoherent

[1] The propagation equation describing the evolution of the mutual coherence function in inertial nonlinear media was first derived by Pasmanik [11].

solitons almost fully coincides with earlier studies on random-phase solitons in plasmas [14]. By its nature, however, ray optics can provide only simple and intuitive information about incoherent solitons, because all phase information is absent. For example, it fails to describe the coherence properties of a partially coherent soliton, since it views such an entity as a bundle of completely uncorrelated rays. The real success of the three rigorous theories explaining incoherent solitons, namely the coherent density [7], the modal theory [9], and the mutual coherence theory [10], was in their ability to come up with exciting new predictions, some of which suggested truly fascinating phenomena. It became very clear right from the start that incoherent solitons are not some esoteric creatures specifically related to photorefractives, but rather form a general and rich new class of solitons, whose existence is relevant to diverse fields, even beyond nonlinear optics. For example, and as will be discussed later on, incoherent modulational instability and incoherent pattern formation actually relate to many systems in nature: from clustering in a cooled atomic gas to self-supported "stripes" of electrons in semiconductors, from high-T_c semiconductors to structures in fluids and gravitational effects. In fact, the underlying physics relates to almost any weakly correlated wave system which has a noninstantaneous nonlinearity. In spite of this generality, however, all the experiments have thus far been carried out in photorefractives, primarily as a matter of convenience. Only recently have such experiments started to be done in other optically nonlinear media, such as nematic liquid crystals. It is correct to say, however, that no optical nonlinearity is truly instantaneous. This includes materials with ultrafast Kerr nonlinearities, the origin of which is electronic.[2] For these reasons, a meaningful criterion as to whether or not an incoherent soliton can exist in a particular nonlinear medium is provided by comparing the response time of the nonlinearity with the characteristic time for random phase fluctuations across an incoherent multimode beam.

To understand the new ideas involved, some aspects of incoherent light have to be explained first. A spatially incoherent beam is nothing but a multimode (a so-called "speckled") beam whose structure varies randomly with time. The beam consists of many tiny bright and dark "patches" (thus the term "multimode") that are caused by a random phase distribution, which varies rapidly with time. The envelope of this beam is defined by the time-averaged intensity. To illustrate how a spatially incoherent beam is understood, consider a detector array (e.g. a human eye) which monitors the beam. When this detector responds much more slowly than the characteristic phase fluctuation time, all it will "see" is the time-averaged envelope. Such

[2] Even in the fastest nonlinear medium, the shortest possible response time is the dephasing time or the lifetime (whichever is the shorter), when at least one of the optical transitions is into a resonant (real) energy level, or $\hbar/\Delta E$ when all the optical transitions are into virtual levels (e.g. for the optical Kerr effect), ΔE being the energy difference from the nearest real energy level.

an incoherent beam diffracts much more than a coherent beam of the same beam width, since each tiny "patch" (speckle) contributes to the diffraction of the (time-averaged) envelope of the beam. In the limiting case in which the speckles are much smaller than the beam size, diffraction is dominated by the degree of coherence, i.e. the size of the speckle, rather than the diameter of the beam's envelope. In this case, the beam is said to be a quasi-homogeneous spatially incoherent beam.

Instantaneous nonlinearities *cannot* self-trap such a beam. If an incoherent beam is launched into a self-focusing nonlinear medium that responds "instantaneously" (e.g. an optical Kerr medium), each small speckle forms a small "positive lens" and captures a small fraction of the beam. These bright–dark features in the beam change very fast throughout propagation, and the tiny induced waveguides intersect and cross each other in a random manner. The net effect is beam breakup into small fragments, and self-trapping of the beam envelope does not occur. It is therefore obvious that only noninstantaneous nonlinear media can support incoherent solitons. In fact, this is exactly how the (noncoincidental) idea of incoherent solitons arose [1]: the slow response of the photorefractive nonlinearity at low light intensity hinted that the rapidly varying *spatial* information (phase and amplitude) carried in an optical beam was averaged out, and that the nonlinear medium responded only to the time-averaged intensity of the beam. This said, the noninstantaneous nature of the underlying nonlinearity is only a prerequisite for incoherent solitons. It is a necessary but not a sufficient condition.

Altogether, for self-trapping of an incoherent beam to occur, several conditions must be satisfied. First, the response time of the nonlinear medium must be much longer than the random-fluctuation time across the incoherent beam. Such a nonlinearity responds to the time-averaged envelope and not to the instantaneous "speckles" that constitute the incoherent beam [1]. Second, the multimode (speckled) beam should be able to induce a multimode waveguide via the nonlinearity. Otherwise, if the induced waveguide is able to support only a single guided mode, the incoherent beam will simply undergo spatial filtering, radiating all of its power except for the small fraction that coincides with that guided mode. Third, as with all solitons, self-trapping requires self-consistency: the multimode beam must be able to guide itself in its own induced waveguide [9]. This means that the time-averaged intensity of the beam must correspond to a superposition of time-averaged populations of the guided modes in the (multimode) waveguide induced by the total time-averaged intensity of the beam itself. Furthermore, because the waveguide is self-induced (via the nonlinearity), the time-averaged modal populations must be commensurate with the relative proportions of the modes that make up the time-averaged intensity of the beam and induce the waveguide. In other words, the time-averaged modal populations must have the right proportion to close a self-consistency loop [9].

In the next sections of this chapter, we describe this rapidly advancing new branch of soliton science: *incoherent solitons*, or, in a more general sense, the *self-trapping of incoherent wavepackets*.

2 The Modal Theory of Incoherent Solitons

The first theory of incoherent solitons was actually not the modal theory, but the coherent-density, theory [7] which relies on a nonlinear-optics extension to the Van Cittert–Zernike, theorem. However, here we shall begin with the modal theory [9], which is, by its nature, easier to understand intuitively. This theory is well suited for identifying incoherent solitons, their structure, their coherence properties, and the range in parameter space that supports them. To explain the modal theory in a simple way, consider first a one-dimensional ((1 + 1)D) partially spatially incoherent optical beam which is quasi-monochromatic. The notation (1 + 1)D stands for one direction of propagation, z, and one transverse direction, x, in which the beam diffracts or self-traps. Let the electric-field amplitude of such a beam be $E(x,z,t) = U(x,z,t)\exp[\mathrm{i}(kz-\omega t)]$, where $k = 2\pi n_0/\lambda$ is the wavenumber in the medium, λ being the vacuum wavelength and n_0 the background refractive index. Since the beam is incoherent, it is a multimode (speckled) beam and it can thus be represented as a superposition of m modes, $U = \sum_m c_m(t)U_m(x)\exp(\mathrm{i}\tilde{\beta}_m z)$, where the $c_m(t)$ are the relative weights (populations) of the modes. In general, the relative weights $c_m(t)$ are complex random variables whose random phase/amplitude fluctuations occur within a characteristic fluctuation time t_c. The multimode structure and the random phase terms are what make this beam spatially incoherent. (It is still quasi-monochromatic as long as $t_\mathrm{c} \gg \omega^{-1}$.) We are seeking stationary self-trapped beams, that is, beams for which the intensity structure does not vary with the propagation direction z. Thus, these modes correspond to the (properly weighted) guided modes of the waveguide induced by the total time-averaged intensity of the beam $I = \langle |U|^2\rangle$. Here the angle brackets $\langle\rangle$ denote a time average taken over the response time of the medium, τ, which has to be much longer than the characteristic fluctuation time t_c. The optical field (containing all these modes) obeys the usual paraxial nonlinear wave equation, which can be separated (by matching the random phase factors) into the modal equations

$$-\tilde{\beta}_m U_m + \frac{1}{2k}\frac{\partial^2 U_m}{\partial x^2} + \frac{k}{n_0}\Delta n(I)\,U_m = 0\,, \tag{1}$$

where $\Delta n(I)$ is the nonlinear change in the refractive index, as a function of the (time-averaged) beam intensity I. In the most general case, $\Delta n(I)$ can be written as $\Delta n(I) = \Delta n_0\,F(I)$, where Δn_0 scales the magnitude of the nonlinearity and $F(I)$ is a normalized function. For example, $F(I) = I$ for Kerr media, $F(I) = -1/(1+I)$ for the photorefractive screening nonlinearity (where I is in units of the dark irradiance or the background illumination) [4], $F(I) = I/(1+I)$ for a homogeneously broadened two-level atomic system (where I is in units of the saturation intensity), etc. In dimensionless units (1) leads to the following normalized equation:

$$-\beta_m U_m + \frac{\partial^2 U_m}{\partial \xi^2} + F(I)\,U_m = 0\,, \tag{2}$$

where $\xi = x/x_0 = 2\pi(x/\lambda)\sqrt{2n_0\,\Delta n_0}$, $\zeta = z/z_0 = 2\pi\,\Delta n_0(z/\lambda)$, and β_m is the propagation constant of mode U_m in dimensionless units. The intensity I can be calculated from the optical field:

$$I(\xi) = \langle \sum_m c_m U_m(\xi)\exp(\mathrm{i}\beta_m\zeta) \sum_{m'} c^*_{m'} U^*_{m'}(\xi)\exp(-\mathrm{i}\beta_{m'}\zeta)\rangle\,. \tag{3}$$

The time averaging is performed over $\tau \gg t_\mathrm{c}$, and therefore

$$\begin{aligned} I(\xi) &= \langle \sum_m \sum_{m'} c_m U_m(\xi) c^*_{m'} U^*_{m'}(\xi)\exp[\mathrm{i}(\beta_m - \beta_{m'})\zeta]\rangle \\ &= \sum_m \langle |c_m(t)|^2\rangle |U_m(\xi)|^2 \equiv \sum_m d_m^2 |U_m(\xi)|^2\,, \end{aligned} \tag{4}$$

where d_m^2 is the time-averaged population of mode m. In deriving (4) we have made use of the fact that under incoherent excitation, $\langle c_m c^*_{m'}\rangle \propto \delta_{mm'}$, i.e. the statistical correlation between different modes is zero. (To understand this, let us assume that $c_m(t)$ contains a phase factor $\theta_m(t)$, which represents random phase fluctuations with a characteristic fluctuation time t_c. The fluctuations in one mode are completely uncorrelated with those in another mode, and thus the time average of the cross-interference terms between the modes is zero.) The modal wavefunctions $U_m(\xi)$ and the propagation constants β_m can be found by solving the set of (2) in a self-consistent fashion. The self-consistency loop is illustrated in Fig. 3. The intensity I gives rise to $\Delta n(I)$,

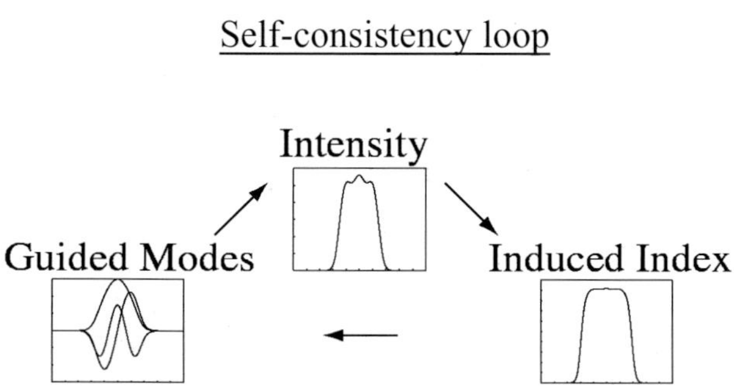

Fig. 3. Illustration of the "self-consistency loop" between the intensity, induced index change, and modal constituents of the induced waveguide

which can guide a discrete set of N guided modes. If a subset M of these guided modes coincide with the set of modes $\{U_m\}$ that make up the incoherent beam, then the beam can be guided by its own self-induced waveguide. For this to happen, however, the modal (time-averaged) populations d_m^2 must be chosen properly, because the populations affect the total intensity and thus the structure of the induced waveguide (for example, an incoherent beam that is made of antisymmetric modes only is not likely to exist in any isotropic nonlinear medium, because the singularity and the phase jump of π at the center of the beam tear the beam in two). The problem of solving for $\{U_m(\xi)\}$, d_m^2, and β_m becomes an eigenvalue problem, and in the most general case it is solved numerically through relaxation methods [9]. In some cases, however, one can find closed-form solutions for particular families of solitons in specific nonlinear media. The first closed-form modal solution for incoherent solitons was found [15] in nonlinear media where $F(I) = \log(1+I) \sim \log(I)$ for $I \gg 1$, where the intensity I is in units of the saturation intensity. In this case, if the total intensity of the beam has a Gaussian shape, then the modes are Gauss–Hermite functions and the modal population is described by a Poisson distribution function [15]. Another elegant example occurs in Kerr media, for which $F(I) = I$, where closed-form solutions can be found for the family of solitons whose time-averaged intensity is $I = I_0 \ \mathrm{sech}^2(\xi)$ [16,17]. In this case, (2) yields

$$\frac{\mathrm{d}^2 U_m}{\mathrm{d}\xi^2} + \left[\alpha^2 \mathrm{sech}^2(\xi) - 2\beta_m\right] U_m = 0 \,. \tag{5}$$

The total electric field of the incoherent soliton can be obtained through a superposition of all the modes involved, that is, $U \propto \sum_m c_m U_m(\xi) \exp(\mathrm{i}\beta_m \zeta)$. The transformation $\mu = \tanh \xi$ leads to an equation that yields solutions of the form $U_m(\xi) = P_N^m(\tanh \xi)$, $P_N^m(y)$ being the associated Legendre functions of the first kind, and $\beta_m = m^2/2$. The value of the integer N (associated with α^2, or the magnitude of nonlinearity) represents the number of allowed modes in the self-induced waveguide, and m is the mode index number. The lowest-order mode occurs when $m = N$, and the highest occurs at $m = 1$. For a given value of N, the statistically varying optical field of the incoherent spatial soliton is given by $U(\xi, \zeta) \propto \sum_{m=1}^{N} c_m P_N^m(\tanh \xi) \exp(\mathrm{i} m^2 \zeta/2)$. Returning to the self-consistency condition, reveals that the mode occupancy coefficients also depend on the number of modes N. The time-averaged intensity of this beam is $I = \langle |U|^2 \rangle = \sum_{m=1}^{N} \langle |c_m|^2 \rangle [P_N^m(\tanh \xi)]^2$ since, under incoherent excitation, the time average of the cross-interference terms among different modes is zero ($\langle c_m c_{m'}^* \rangle \propto \delta_{mm'}$). The mode occupancy coefficients c_m are found by employing self-consistency: the total intensity must be $I = I_0 \ \mathrm{sech}^2(\xi)$. The functional form of the normalized eigenfunctions is given in Appendix A of [16]. With this method, one can have three-component ($N = 3$) incoherent solitons, four-component solitons ($N = 4$), etc.

The Kerr nonlinearity actually allows closed-form solutions for solitons that have an *asymmetric* intensity structure [18]: an unusual feature that

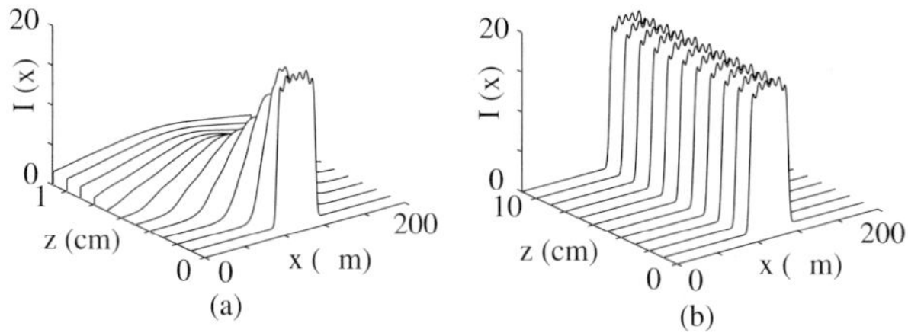

Fig. 4. The propagation dynamics of a six-mode incoherent beam: (**a**) during linear diffraction; (**b**) during self-trapping

typically cannot exist for coherent solitons in isotropic nonlinear media. In the general case, however, the modal constituents of an incoherent soliton must be found numerically through relaxation codes [9]. Using the modal theory, one can identify the incoherent solitons and use the solutions to simulate beam propagation and interactions with other (coherent or incoherent) beams, using (numerical) beam propagation methods. An example showing the propagation dynamics of a six-mode partially incoherent beam for a saturable nonlinearity $F(I) = -1/(1+I)$, which is the (1 + 1)D photorefractive screening nonlinearity [4,5,9], is shown in Fig. 4, where Figs. 4a and 4b show the linear diffraction of the beam (when the nonlinearity is set to zero) and its self-trapped propagation, respectively.

With the tools provided by the modal theory, one can identify the range of parameters in which incoherent solitons exist. Once incoherent solitons have been identified in terms of their modal constituents and their corresponding distribution function (modal population), it is useful to cast the solutions into some parameter space and identify the range in which soliton solutions exist: the so-called *existence range*. Before discussing incoherent solitons in parameter space, we would like to recall some of the properties of coherent solitons. Single-mode coherent solitons are conveniently described by an "existence curve" that relates the normalized width of a soliton, $\Delta\xi$, to the peak amplitude of the soliton, U_0. In normalized units, a scalar soliton is described by a curve that links U_0 to the normalized soliton width $\Delta\xi$.[3] Such solitons form a one-parameter family: for every U_0 there is a single $\Delta\xi$ that supports a (fundamental or lowest-order) soliton. Solitons that are made of more than one mode have many more degrees of freedom. For example, bimodal solitons exist in a two-dimensional region of parameters called the "existence range" [9], because, in addition to U_0 and $\Delta\xi$, they are defined by the relative weight of modes one and two, i.e. they posses two degrees of freedom. Along the same lines, trimodal solitons have five degrees of freedom [9]. In fact, the

[3] The normalized soliton width is related to the V-number of the waveguide induced by the soliton. The concept of "soliton existence curve" was first introduced in [4], in the context of the photorefractive screening nonlinearity, but it is a general concept and can be applied to all solitons. For example, it applies also the thresholding nonlinearity [19].

general trend, at least for the Kerr nonlinearity, is known [18]: N-mode solitons possess $2N-1$ degrees of freedom, and thus give rise to an existence range of $2N-1$ dimensions that defines the range in which such soliton solutions exist. This generalization and the concept of "soliton existence range" is valid for both incoherent solitons and multimode solitons, which are both described by the same range.

Here is the place to emphasize the fundamental difference between incoherent solitons and multimode solitons. Multi-mode (or composite) solitons are solitons in which two (or more) optical fields jointly self-trap in a nonlinear medium (through the waveguide induced by the total intensity of the beam), and in which the fields belong to different modes of their jointly induced waveguide (see [2] and references therein). The field components of multimode solitons must be such that the interference cross terms between fields do not contribute to Δn. There are several methods to achieve this: one is based on having the components in different polarizations, a second method is based on having the fields at different (widely spaced) frequencies, and a third method relies on mutual incoherence of the modes. In this latter method, each mode is populated by a single, fully coherent optical field, but the fields that populate two different modes are incoherent with respect to one another [20]. In other words, each optical field has a *deterministic* population that *does not vary with time*: $c_m^2 = d_m^2$, but the phase differences between fields are random or stochastic so that the interference cross terms average out to zero. Thus, when the response time of the nonlinear medium is slower than the characteristic fluctuation time, the interference contribution to Δn averages out. Because the theory of multimode solitons resembles the modal theory of incoherent solitons, and because the use of the mutual-incoherence method became so fashionable in launching multimode solitons experimentally, there is some confusion in the literature between these two fundamentally different entities. Theory-wise, if the nonlinear material is chosen such that it "has no memory", i.e. $\tau \gg t_c$, the underlying light–matter interaction system is fully ergodic and the theories of incoherent solitons and multimode solitons coincide with one another. However, *for incoherent solitons the actual populations vary with time* (c_m is a stochastic function of time), whereas *for a multimode soliton the modal populations remain constant in time.* (One may expect a substantial difference between the two cases when τ is on the order of t_c.) Experimentally, however, the difference between incoherent and multimode solitons is fundamental. Multi-mode solitons have to be engineered by splitting a coherent laser beam and delaying the components with respect to one another by an optical path that exceeds the coherence length of the laser [20]. However, each component remains fully coherent at all times and its intensity is constant with time. Incoherent solitons, on the other hand, are "naturally" generated either by using quasi-thermal spatially incoherent light (by passing a laser beam through a fast-rotating diffuser [3]) or by using an incoherent source, such as an incandescent light bulb [6] or

LEDs. The implication is that for incoherent solitons the modal populations vary with time, and only the time-averaged population is fixed, whereas for multimode solitons the populations themselves are fixed. What this means is that multimode solitons are actually "engineered": both the wavefunctions and the modal populations are prepared and launched together to jointly form a multimode soliton. Incoherent solitons, on the other hand, cannot be engineered, because the coherence function of incoherent solitons, as shown in Sect. 5 below, depends not only on the coordinate difference (as is the case for all naturally incoherent sources), but also on the absolute coordinates within the beam. One can control the correlation distance of the input beam, but it is extremely difficult to realize a beam whose coherence function varies locally from one point to another. In other words, the only coherence quantity that can be controlled rather well is the degree of coherence (correlation distance) of the entire beam; it is very hard (except for some very special examples) to actually realize experimentally a coherence function that varies in space. However, perhaps the most elegant proof that incoherent solitons and multimode solitons are not the same entity is provided by the work on antidark solitons. As explained later on in Sect. 6, antidark solitons that reside in an incoherent background are stable, whereas antidark solitons that are merely a special case of multimode solitons are fundamentally unstable.

To conclude this issue, it is essential to realize that *incoherent solitons are fundamentally different from multimode solitons.*

In summary of this section, we have reviewed the modal theory of incoherent solitons, providing a general formalism for identifying soliton solutions as an ensemble of modes of their jointly induced waveguide, for which the time-averaged modal population is commensurate with their time-averaged intensity distribution and the underlying nonlinearity. The advantages of the modal theory are in its ability to actually identify the soliton solutions, either in closed form or as a solution obtained by a numerical relaxation code. Moreover, through the modal theory one can calculate the parameter range in which such solitons exist and use this knowledge for launching incoherent solitons experimentally.

3 The Coherent-Density Theory and the Propagation of Mutual Coherence

The coherent-density theory represents an alternative avenue via which one can explore the dynamics of partially incoherent optical fields in noninstantaneous nonlinear media [7]. With this method, the intensity and the correlation statistics of a partially incoherent beam can be monitored during propagation using a modified version of the Van Cittert–Zernike theorem. In this approach, the incoherent input beam is decomposed into infinitely many "coherent fragments", all of which happen to be mutually incoherent with respect to each other. The initial relative weight of each such component is determined from the angular power spectrum of the source. Each fragment propagates in a coherent fashion in the nonlinear optical material and the total intensity of the partially coherent field is found by superimposing the intensities of these coherent components. Before we describe this method, perhaps it is instructive to consider the diffraction (linear) dynamics of a partially incoherent beam. After all, the main difference between a coherent and a partially coherent field is in the way it diffracts. Let a quasi-monochromatic optical wave be expressed in terms of a slowly varying envelope $\phi = \phi(x, y, z)$, i.e., $E = \phi(x, y, z)\exp(ikz)$. In the linear regime (under the influence of diffraction), the envelope ϕ evolves according to

$$\phi = \frac{1}{(2\pi)^2}\int_{-\infty}^{+\infty}\int_{-\infty}^{+\infty} \mathrm{d}k_x\,\mathrm{d}k_y\,\Phi(k_x, k_y)\mathrm{e}^{\mathrm{i}(k_x x + k_y y) - \mathrm{i}(z/2k)(k_x^2 + k_y^2)}\,, \qquad (6)$$

where $\Phi(k_x, k_y)$ is the Fourier transform of the optical field right at the input $z = 0$. Now let the optical field at the origin be written as $\phi(x, y, z = 0) = m(x, y)\phi_0(x, y)$, where $m(x, y)$ is a complex modulation function and $\phi_0(x, y)$ is the field before modulation, which implicitly contains all the spatial statistical properties of the source. If we assume that the source fluctuations constitute a stationary random process, the statistical autocorrelation function of $\phi_0(x, y)$ is given by $\langle\phi_0(x, y)\phi_0^*(x', y')\rangle = R(x - x', y - y')$. From here, the autocorrelation of the source spectrum can be determined, that is $\langle\Phi_0(k_x, k_y)\Phi_0^*(k_x', k_y')\rangle = (2\pi)^2\delta(k_x - k_x')\delta(k_y - k_y')G(k_x', k_y')$, where $\Phi_0(k_x, k_y)$ and $G(k_x, k_y)$ are the Fourier transforms of $\phi_0(x, y)$ and $R(x, y)$, respectively. Physically, the function $G(k_x, k_y)$ represents the angular power spectrum of the partially incoherent source and is always real, since $R(\boldsymbol{r}_1 - \boldsymbol{r}_2) = R^*(\boldsymbol{r}_2 - \boldsymbol{r}_1)$. From the frequency convolution theorem, one can then easily find that

$$\begin{aligned}&\langle\Phi(k_x, k_y)\Phi^*(k_x', k_y')\rangle\\ &\quad = \frac{1}{(2\pi)^2}\int\!\!\int \mathrm{d}\eta\,\mathrm{d}\xi\,G(\eta, \xi)M(k_x - \eta, k_y - \xi)M^*(k_x' - \eta, k_y' - \xi)\,, \qquad (7)\end{aligned}$$

where $M(k_x, k_y)$ is the Fourier transform of the spatial modulation function $m(x, y)$. Given that the intensity of this incoherent wavefront is given by $I = \langle|\phi|^2\rangle$ we finally obtain the following result [7]:

$$\begin{aligned}I(x, y, z) &= \int\!\!\int \mathrm{d}\theta_x\,\mathrm{d}\theta_y\,G_N(\theta_x, \theta_y)\\ &\times\left|\frac{1}{(2\pi)^2}\int\!\!\int \mathrm{d}k_x\,\mathrm{d}k_y\,M(k_x, k_y)\mathrm{e}^{\mathrm{i}[k_x(x - \theta_x z) + k_y(y - \theta_y z) - (z/2k)(k_x^2 + k_y^2)]}\right|^2, \qquad (8)\end{aligned}$$

where $G_N(\theta_x, \theta_y)$ has been normalized, and $\theta_x = k_x'/k$, $\theta_y = k_y'/k$ represent angles in radians with respect to the propagation (z) axis. It is interesting to note that the intensity term $|*|^2$ that appears in (8) represents in fact a solution of the coherent paraxial equation of diffraction

$\mathrm{i}(f_z + \theta_x f_x + \theta_y f_y) + (1/2k)\nabla^2 f = 0$. In other words, the diffraction behavior of a partially incoherent beam can be effectively described by adding the intensity contributions of all the components involved, each propagating at angles θ_x, θ_y. Normalization requires that at the origin, $f(x, y, z = 0) = \sqrt{G_N(\theta_x, \theta_y)}m(x, y)$. The auxiliary function f is what we refer to here as the *coherent density*. This result can now be generalized in the presence of nonlinearity. If we now assume that the refractive index of the nonlinear material varies according to $n^2 = n_0^2 + 2n_0\,\Delta n(I)$, then in this case the coherent density evolves according to [7]

$$\mathrm{i}\left(\frac{\partial f}{\partial z} + \boldsymbol{\theta}\cdot\boldsymbol{\nabla}_T f\right) + \frac{1}{2k}\nabla_T^2 f + k_0\,\Delta n(I)\,f = 0\,. \tag{9}$$

Here $\boldsymbol{\theta} = \theta_x \boldsymbol{x}_0 + \theta_y \boldsymbol{y}_0$, $\boldsymbol{\nabla}_T$ is the $\boldsymbol{\nabla}$ operator in directions normal to z, and $\Delta n(I)$ represents the nonlinear functional dependence of the refractive index on intensity. The overall intensity I is obtained via $I = \int |f(x, y, z, \boldsymbol{\theta})|^2 \mathrm{d}^2\boldsymbol{\theta}$. From the coherent density f, the mutual coherence function J_{12} can then be obtained. Using similar arguments as before, one can show that

$$J_{12}(\boldsymbol{r}_1, \boldsymbol{r}_2, z) = \int \mathrm{d}^2\boldsymbol{\theta}\; f_1 f_2^* \exp[\mathrm{i}k\boldsymbol{\theta}\cdot(\boldsymbol{r}_1 - \boldsymbol{r}_2)]\,, \tag{10}$$

where $f_j = f(\boldsymbol{r}_j, z, \boldsymbol{\theta})$, and $\boldsymbol{r} = x\boldsymbol{x}_0 + y\boldsymbol{y}_0$. Note that the intensity $I_j = J_{jj} = \int \mathrm{d}^2\boldsymbol{\theta}|f_j|^2$, i.e. is in agreement with previous statements. Equation (10) represents a modified version of the Van Cittert–Zernike theorem . [21] Together with (9), it describes the dynamical evolution of the mutual coherence function J_{12} in inertial nonlinear media. Consequently, the complex coherence factor can be obtained from J_{12}, i.e. $\mu_{12} = J_{12}/\sqrt{J_{11}J_{22}}$, from which the correlation distances can be determined [21].

In the special case of biased photorefractives, the nonlinearity used in (9) is of the type $\Delta n = -\Delta n_0/(1 + I_N)$, where I_N is a normalized intensity [4,5]. Figure 5 depicts the evolution of a one-dimensional partially coherent Gaussian beam when it propagates in a biased SBN:75 crystal. In these examples the normalized angular power spectrum associated with this Gaussian beam was also taken to be Gaussian, that is, $G_N = (\sqrt{\pi}\theta_0)^{-1}\exp(-\theta^2/\theta_0^2)$, where θ_0 represents the width of the angular spectrum. Figure 5 shows what happens to this beam when the bias voltage is below that required for self-trapping. As expected, in this case the beam tends to expand during propagation. When the bias field is 917 V/cm, stable self-trapping occurs and the beam behaves as an incoherent quasi-soliton [7]. Above this field (at 1667 V/cm) the beam starts to exhibit considerable spatial compression. In all cases, $\theta_0 = 9.56$ mrad and the normalized intensity ratio is $r = 3$. These results are in excellent agreement with the experimental observations reported in [3].

The equivalence of the coherent density method to the other two theoretical approaches can be established in a straightforward fashion. By differen-

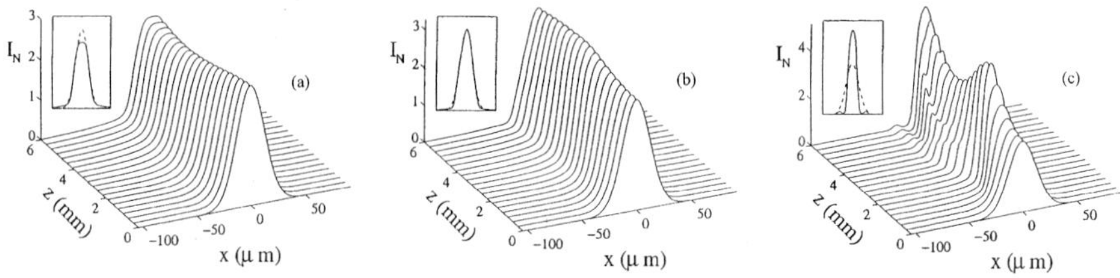

Fig. 5. Evolution of the normalized intensity profile of a partially coherent beam propagating in a photorefractive SBN:75 crystal. The initial (Gaussian) beam has a width of 30 µm FWHM, the width of the angular spectrum is 9.56 mrad, and the intensity ratio is 3. The three figures display the propagation dynamics for three different applied fields: (**a**) 667 V/cm, (**b**) 917 V/cm, and (**c**) 1667 V/cm. The *insets* show the intensities at the input (*dashed curves*) and at the output at $z = 6$ mm (*solid curves*)

tiating (10) we find that

$$\frac{\partial J_{12}}{\partial z} = \int \mathrm{d}^2\boldsymbol{\theta}\ \mathrm{e}^{\mathrm{i}k\boldsymbol{\theta}\cdot(\boldsymbol{r}_1-\boldsymbol{r}_2)} \left(f_2^* \frac{\partial f_1}{\partial z} + f_1 \frac{\partial f_2^*}{\partial z} \right) . \tag{11}$$

From this, one can easily deduce that

$$\mathrm{i}\frac{\partial J_{12}}{\partial z} = k_0 \left[\Delta n(I_2) - \Delta n(I_1)\right] J_{12} + \int \mathrm{d}^2\boldsymbol{\theta} \mathrm{e}^{\mathrm{i}k\boldsymbol{\theta}\cdot(\boldsymbol{r}_1-\boldsymbol{r}_2)} \times \left[\left(-\mathrm{i} f_2^* \boldsymbol{\theta}\cdot\nabla_{T1} f_1 - \frac{1}{2k} f_2^* \nabla_{T1}^2 f_1 \right) + \left(-\mathrm{i} f_1 \boldsymbol{\theta}\cdot\nabla_{T2} f_2^* + \frac{1}{2k} f_1 \nabla_{T2}^2 f_2^* \right) \right] . \tag{12}$$

Since, on the other hand,

$$\frac{1}{2k}\nabla_{T1}^2 J_{12} = \int \mathrm{d}^2\boldsymbol{\theta} \mathrm{e}^{\mathrm{i}k\boldsymbol{\theta}\cdot(\boldsymbol{r}_1-\boldsymbol{r}_2)} f_2^* \left(\frac{1}{2k}\nabla_{T1}^2 f_1 + \mathrm{i}\boldsymbol{\theta}\cdot\nabla_{T1} f_1 - \frac{k}{2}\theta^2 f_1 \right) , \tag{13}$$

$$\frac{1}{2k}\nabla_{T2}^2 J_{12} = \int \mathrm{d}^2\boldsymbol{\theta} \mathrm{e}^{\mathrm{i}k\boldsymbol{\theta}\cdot(\boldsymbol{r}_1-\boldsymbol{r}_2)} f_1 \left(\frac{1}{2k}\nabla_{T2}^2 f_2^* - \mathrm{i}\boldsymbol{\theta}\cdot\nabla_{T2} f_2^* - \frac{k}{2}\theta^2 f_2^* \right) , \tag{14}$$

the following result holds true:

$$\mathrm{i}\frac{\partial J_{12}}{\partial z} + \frac{1}{2k}\left(\nabla_{T1}^2 - \nabla_{T2}^2\right) J_{12} + k_0 \left[\Delta n(I_1) - \Delta n(I_2)\right] J_{12} = 0 . \tag{15}$$

Equation (15) represents the evolution equation for the mutual coherence and is exactly that derived earlier by Pasmanik [11]. This latter result clearly demonstrates that the coherent density formalism (described by (9) and (10)) is fully equivalent to (15). The equivalence of the modal theory to the other two can be established via Karhunen–Loeve expansions [12]. In reality these three approaches constitute different representations of the same propagation process. Their difference is manifested in the way in which they treat initial conditions. More details regarding this issue can be found in [12].

4 Dark Incoherent Solitons

Following the observations of bright incoherent solitons, a natural question came up: can such incoherent beams also support dark solitons ? This seems indeed a natural question to ask, yet the answer to this question is far from trivial. On the basis of the knowledge of coherent dark solitons (see [22] for a review), fundamental 1D coherent dark solitons require a transverse phase shift of π at the center of the dark stripe, whereas an initially uniform-transverse phase leads to a Y-junction soliton. Furthermore, 2D coherent dark solitons (vortex solitons) require a helical $2m\pi$ transverse-phase structure (where m is an integer). Extending the idea of dark coherent solitons to dark incoherent solitons raises several difficult questions. If dark incoherent solitons are to exist, is their phase structure important (as for coherent dark solitons) or is it irrelevant (as for bright incoherent solitons, in which the phase is fully random)? And, if the phase does play a role, how can it be "remembered" by these incoherent entities throughout propagation? Although the existence of bright incoherent solitons was well established (both experimentally [3,6] and theoretically [7–10,13]), the possibility of dark incoherent solitons was not at all clear at the time. The existence of dark incoherent solitons was predicted theoretically in [23]. In that paper, the coherent-density approach revealed that when an arbitrary dark-stripe-bearing incoherent beam is launched into a noninstantaneous self-defocusing medium, the beam undergoes considerable evolution but eventually stabilizes, with some small oscillatory "breathing", around a self-trapped solution. Surprisingly, the simulation showed that a single dark incoherent soliton requires an initial transverse phase jump of π and that the dark incoherent soliton is always gray.[4] These theoretical results [23], although not providing detailed answers to the questions raised above, did suggest that dark incoherent solitons should exist, and indeed their existence was demonstrated experimentally [24–26].

The experiments [24] employed the rotating-diffuser method to generate partially spatially incoherent light. After the rotating diffuser, the soliton-forming beam was reflected from a phase mask (a $\lambda/4$ step mirror), which generated a dark notch on a broad partially spatially incoherent background. The notch-bearing beam was launched into a biased photorefractive crystal. Self-trapping of the dark notch was observed at a suitable bias field that introduced the necessary screening nonlinearity so that diffraction was bal-

[4] It is fair to say that Hasegawa, in his paper [14] on random phase solitons in plasmas, suggested that dark incoherent solitons do exist, in the limit where the transverse correlation distance is assumed to be zero (or at least much smaller than the width of the beam). However, as we show later on, this pioneering suggestion is immaterial because, for dark incoherent solitons, the transverse-phase shift, right at the very center of the singularity, plays a crucial role. In addition, in this case the soliton-induced waveguide is typically single-moded. Thus, unlike bright incoherent solitons, dark incoherent solitons cannot be described by transport (ray optics) theories, in any limit or approximation.

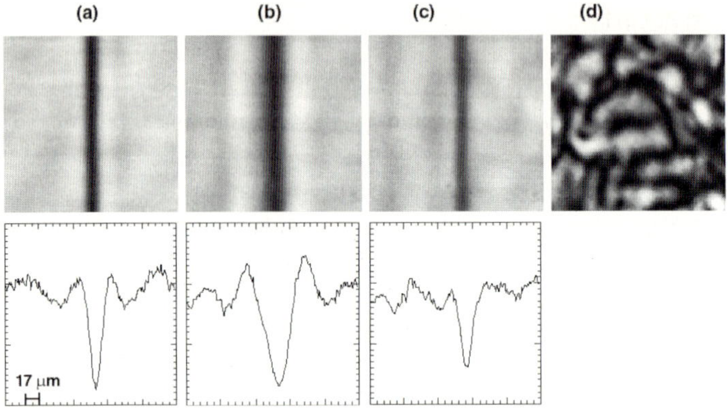

Fig. 6. Self-trapping of a dark stripe carried by a partially-incoherent beam. Shown are photographs and beam profiles of (**a**) the input beam, (**b**) the diffracted output beam, and (**c**) the self-trapped output beam. The last photograph (**d**) shows the output beam with the nonlinearity "on" when the diffuser was stationary, illustrating the fragmentation of an incoherent dark stripe in a fully instantaneous self-defocusing medium

anced by self-defocusing. Typical experimental results on self-trapping of a 1D dark incoherent beam are shown in Fig. 6. In agreement with the prediction [23], the observed fundamental incoherent dark soliton was always gray. Thus, unlike coherent dark solitons, which can be either black or gray, dark incoherent solitons are always gray ($\sim 40\%$ grayness in Fig. 6c). Another striking difference between incoherent and coherent dark (or bright) solitons is the nature of the temporal response of the nonlinearity. Coherent spatial solitons can occur in either instantaneous or noninstantaneous nonlinear media, but incoherent spatial solitons require a noninstantaneous response. For example, Fig. 6d shows what happens when the rotation of the diffuser is stopped and the material nonlinearity is allowed to reach steady state. The self-defocusing medium responds to the "instantaneous" (now stationary) speckles, fragmenting the beam and prohibiting self-trapping of the dark notch. Self-trapping of an incoherent dark notch depends critically on the degree of coherence. When the beam is more spatially incoherent, the self-trapped notch becomes grayer and a higher nonlinearity is needed for trapping [24]. The effect of the initial phase distribution at the center of a dark incoherent soliton was investigated in subsequent experimental and theoretical studies [23,25]. This research revealed that the initial phase at the center is crucial to the evolution of a dark incoherent soliton. If the phase at the center of the input beam goes through a jump of π, then a single dark (gray) incoherent soliton emerges [23,24]. But if the phase is continuous across the input dark notch, then *two* dark incoherent solitons emerge in the form of a Y [25]. The emergence of this Y-splitting is very similar to the occurrence of Y-splitting for coherent dark solitons [22], which also requires a uniform input phase across the input notch. This behavior of incoherent Y-splitting persists for a wide range of parameters: for example, the splitting angle remains almost the same even though the transverse correlation distance may vary by orders of magnitude.

To highlight the mystery involved in dark incoherent solitons, let us first recall the *linear* propagation of a *coherent* notch-bearing beam. When the beam is fully coherent, there is a marked difference between the diffraction of a notch with a π phase jump and a notch with a uniform phase: the former has zero amplitude at the center, at all propagation distances, whereas for the latter, diffraction gives rise to a nonzero amplitude at the center of the notch. Therefore, one might certainly expect that the *nonlinear* propagation of a *coherent* notch-bearing beam would also show a marked difference between uniform-phase and π-phase beams: the former gives rise to a single dark soliton (or, if the nonlinearity is high enough, to a sequence of an odd number of dark solitons), whereas the latter gives rise to Y-splitting (or to an even number of dark solitons) [27]. All of this is true for coherent beams. When the beam is partially spatially incoherent, the *linear* propagation dynamics of a uniform-phase and a π-phase-jump notch-bearing beam become almost indistinguishable from one another after a very short distance of propagation, and in both the notch diminishes rapidly, irrespective of the phase at its center [25]. This happens because the independent speckles that make up the beam radiate in a noncorrelated manner and diffuse all the phase information content. It is therefore quite surprising that in the midst of statistical phase fluctuations, the incoherent notch-bearing beam "remembers" its initial phase imprint, and evolves according to it into a single fundamental soliton or into a Y-type soliton doublet [23,25]. It is apparent that some unexpected "phase memory" effect is taking place during the nonlinear propagation dynamics of such incoherent beams, which is absent during linear propagation [25]. A complete (analytic) explanation of these phase-memory effects is still not fully available, but the importance of the phase at the center of the notch is now well understood from the modal theory of dark incoherent solitons [28].

The principles of the modal theory of dark incoherent solitons [28] are identical to those for bright incoherent solitons, with one major difference. For bright solitons, only guided modes (bound states) are self-trapped, whereas dark incoherent solitons involve a belt of radiation modes (unbound states). Moreover, in certain cases, dark incoherent solitons can be composed of radiation modes alone. In general, solving for dark incoherent solitons requires a numerical relaxation code. However, in some particular nonlinear media, a closed-form solution can be found for specific families of solitons. One such example arises in self-defocusing Kerr media when the time-averaged intensity profile of the fundamental dark incoherent soliton (actually gray soliton) is of the form $I(\xi) = I_0\,[1 - \epsilon^2\,\mathrm{sech}^2(\xi)]$. The parameter $\epsilon^2 \leq 1$ is associated with the grayness of the soliton.

Proceeding along the lines of the modal theory, the modes (bounded and unbounded) must satisfy the self-consistency condition. Thus, given the intensity profile and the resulting induced waveguide (induced potential), one seeks modal compositions where the time-averaged intensity due to radiation and bound modes is equal to the original assumed intensity profile (the profile that characterizes this family of solutions). We use again (2), with $F(I) = -I$. The modes (eigenfunctions) of this equation can be written as $U_m(\xi)\exp(\mathrm{i}\beta_m\zeta)$. Guided modes (bound states) are possible whenever $\beta_m + I_0 \equiv q^2 > 0$, and radiation modes (unbound states) when $\beta_m + I_0 \equiv -Q^2 < 0$. As stated earlier, incoherent dark solitons typically involve both unbound and bound states.[5] This is because for dark solitons (coherent or incoherent), the intensity at "infinity" (far away from the center of the dark notch) is nonzero. This in turn implies that guided modes alone cannot make up an incoherent dark soliton, because the amplitudes of all such bound states decay (typically exponentially) at large distances from the center. Radiation modes, on the other hand, always possess nonzero amplitude at infinity, and thus every dark incoherent soliton must contain a belt of radiation modes that participate in the self-trapping process.[6] Proceeding along the lines of the self-consistency principle, one can solve for the modes: find the values of the propagation constants and the modal wavefunctions. The simplest example is when the soliton involves of a single guided mode along with a belt of radiation modes. In this case, $I_0\epsilon^2 = 2$, and the single bound mode at $\beta_\mathrm{b} + I_0 = 1$ is given by $U_\mathrm{bound} = \mathrm{sech}\,\xi$. There are two types of radiation modes: even, $U_\mathrm{r,e} = Q\cos(Q\xi) - \sin(Q\xi)\tanh\xi$, and odd, $U_\mathrm{r,o} = Q\sin(Q\xi) + \cos(Q\xi)\tanh\xi$. The radiation modes form a continuous set, and are thus weighted by a (continuous) weight function $c_\mathrm{e,o}(Q)$. The total electric field is therefore

$$U = c_\mathrm{b}U_\mathrm{bound}(\xi)\exp(\mathrm{i}\beta_\mathrm{b}\zeta) + \int_0^\infty \mathrm{d}Q\,[c_\mathrm{e}(Q)U_\mathrm{r,e}(\xi,Q) + c_\mathrm{o}(Q)U_\mathrm{r,o}(\xi,Q)]\exp[\mathrm{i}\beta_\mathrm{r}(Q)\zeta]\,. \tag{16}$$

The upper limit of the integral over the radiation modes is taken to infinity, but in practice the modal weight (population) function decays rapidly, and thus the paraxial approximation is still valid. Under incoherent excitation, the following relationships hold: $\langle c_{\mathrm{b},m}c^*_{\mathrm{b},n}\rangle \propto \delta_{mn}$, $\langle c_{\mathrm{b},m}c^*_{\mathrm{e,o}}\rangle = 0$,

[5] In contradistinction, bright solitons in their "usual" sense, whether coherent or incoherent, are localized wavepackets for which the total intensity at "infinity" (far away from their center) is zero. Therefore, coherent bright solitons are made of a single bound state that is guided in its own self-induced waveguide, whereas incoherent and multimode bright solitons involve multiple bound states (guided modes) , but in both cases no radiation modes (unbound states) participate in the self-trapping. The exception is bright solitons that reside in a uniform background – the so-called antidark solitons [37]. Thus far, the only such antidark solitons that have been observed experimentally are those of [36].

[6] Unlike those coherent dark solitons that are actually the second guided mode at the cutoff frequency [22].

$\langle c_o c_e^* \rangle = 0$, and $\langle c_e(Q) c_e^*(Q') \rangle = \langle c_o(Q) c_o^*(Q') \rangle \propto \delta(Q - Q')$, i.e. the statistical time expectation value of the field coefficients is zero between different bound and radiation modes. In other words, the time-averaged value of the interference between different modes is zero. The intensity profile of the dark incoherent soliton goes to a constant value I_0 far away from the notch. Since the bound mode is of finite extent, the intensity far from the notch is made up from radiation modes alone. A close inspection of the profiles of the even and odd radiation modes reveals that the only way these modes can add up to give a constant intensity profile (at infinity) is if they have the same weight function. When even and odd modes of the same "order" Q share the same weight function (the incoherent source shows no preference for either odd or even radiation modes), then they add up to $|U_{r,e}|^2 + |U_{r,o}|^2 = Q^2 + \tanh^2 \xi$ which approaches a constant value $Q^2 + 1$ as $\xi \to \infty$. We can therefore write the total intensity of the dark incoherent soliton as

$$I(\xi) = A^2 \operatorname{sech}^2 \xi + \int_0^\infty \mathrm{d}Q\, D(Q) \left(Q^2 + \tanh^2(\xi)\right) , \tag{17}$$

where $\langle |c_b|^2 \rangle \propto A^2$, and $D(Q)$ is the distribution function of the radiation modes: $\langle c_e(Q) c_e^*(Q') \rangle = \langle c_o(Q) c_o^*(Q') \rangle \propto D(Q) \delta(Q - Q')$. From the self-consistency principle, we require that the intensity given by (17) is identical to the assumed time-averaged intensity $I(\xi) = I_0(1 - \epsilon^2 \operatorname{sech}^2 \xi)$. This happens when $I_0 = \int_0^\infty \mathrm{d}Q\, D(Q)(Q^2 + 1)$ and $A^2 = \int_0^\infty \mathrm{d}Q\, D(Q)[1 - \epsilon^2(1 + Q^2)]$. These two conditions, combined with the modal wavefunctions given above, are the analytic solution for the simplest dark incoherent soliton. The distribution function $D(Q)$ is by no means unique, but it must be chosen so that it decays fast enough with increasing Q. A natural choice would be $D(Q) = D_0 \exp(-Q/Q_0)$, i.e. a Boltzmann-like distribution function (where Q_0 represents the width of the distribution). The exponentially decreasing nature of $D(Q)$ reflects the fact that the angular power spectrum of the incoherent beam must decrease with the launch angle [7,23], i.e. more power is expected to couple into spatial (plane wave) components that are propagating at shallow angles (small Q) with respect to the propagation axis. For this particular choice of $D(Q)$, the above conditions yield $D_0 = I_0/[Q_0(2Q_0^2 + 1)]$ and $A^2 = I_0[1 - \epsilon^2(2Q_0^2 + 1)]/[2Q_0^2 + 1]$. A numerical test of the solution (with a beam propagation code employing the coherent-density approach) confirms that this is indeed a valid and stable solution. This example of a first-order incoherent dark soliton is illustrated in the eigenvalue diagram of Fig. 7.

The exponential choice for the radiation mode distribution function actually allows a variety of solutions, each characterized by a choice of variables, I_0, Q_0, and ϵ. It can be readily shown that solutions exist whenever $\epsilon^2(2Q_0^2 + 1) \le 1$. Similarly, one can obtain dark incoherent solitons that consist of two or more bound states (guided modes), in addition to the continuous belt of radiation modes [28]. Another interesting possibility is a dark incoherent soliton that consists of radiation modes only and in which the bound state is empty: $I_b = 0$; that is, although a bound state inevitably exists, it

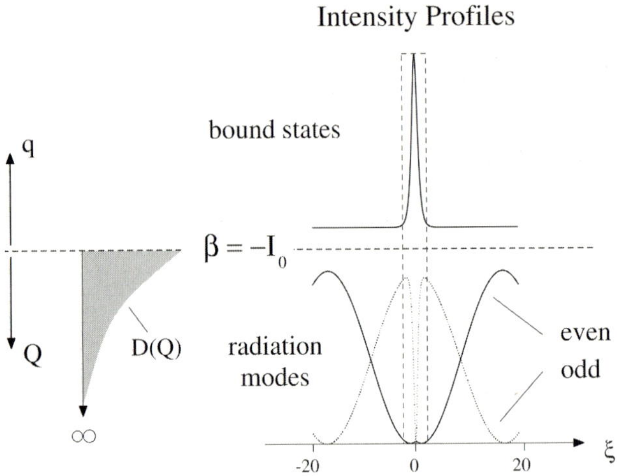

Fig. 7. Eigenvalue diagram associated with a first-order incoherent dark soliton. The bound-state intensity and the intensities of the even and odd radiation modes are also depicted. The stripe delineated by *dashed lines* in the *right-hand part* shows the spatial extent of the soliton-induced waveguide

is not populated. This "radiation-only" dark incoherent soliton occurs when $\epsilon^2(2Q_0^2+1)=1$. In fact, from the correlation length traces, one can speculate that the gray fundamental solitons encountered in the theoretical study of [23] were of this type (radiation-only strucures) or almost of this type, and possibly this was the case in the experimental observations of [24].

This particular family of closed-form solutions for dark incoherent solitons provides considerable insight and answers some of the questions raised above related to the essence of such solitons. First, one can readily observe that dark incoherent solitons must be gray, owing to the presence of both even and odd radiation modes at the center of the notch. Second, and much more importantly, this family of solutions explains the necessity to superimpose a phase jump of π at the dark notch of the incoherent beam, as found numerically and experimentally [23–26]. From Fig. 7, it is clear that the odd radiation modes dominate (possess a higher amplitude) within the soliton-induced waveguide, whereas the even radiation modes have diminishingly small amplitudes inside the waveguide. This is apparent from consideration of $U_{\mathrm{r,e}}$ and $U_{\mathrm{r,o}}$ near $\xi=0$ when the radiation modes are confined to a narrow belt around $Q\sim 0$ (which is always the physical case for such solitons, which obey the paraxial approximation). Thus, in order to launch such a soliton effectively, the phase must be properly manipulated to favor the odd radiation modes, which is exactly what the phase jump of π at the center of the notch does. This is why a phase jump of π at the center facilitates the observation of such incoherent dark solitons.

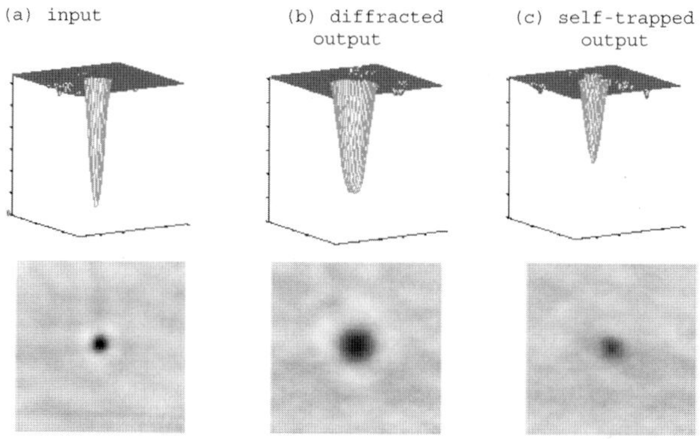

Fig. 8. Self-trapping of an optical "vortex" carried by a partially incoherent beam. Shown are 3D intensity plots and photographs of (**a**) the input beam, (**b**) the diffracted output beam, and (**c**) the self-trapped output beam. For better visualization, the 3D intensity plots have been truncated to remove the background noise away from the "hole"

Along with the first experimental observation of $(1+1)$D incoherent dark solitons, self-trapping of a $(2+1)$D incoherent dark "beam" (a two-dimensional void in an incoherent beam) has also been demonstrated [24]. By analogy with $(1+1)$D dark incoherent solitons, which required the launch of a notch with a phase jump of π at its center, a $(2+1)$D incoherent beam was superimposed on a vortex-type beam. In other words, a vortex (from a helicoidal mask) was nested in an incoherent beam launched into the nonlinear crystal. At a sufficiently high self-defocusing nonlinearity, the two-dimensional dark "beam" self-trapped, as shown in Fig. 8. As for the 1D dark incoherent soliton, the more incoherent the beam is, the larger the nonlinearity required for trapping is and the grayer the dark "hole" is.

5 Coherence Properties of Incoherent Solitons

Perhaps the most fascinating feature of incoherent solitons is their coherence structure. The very fact that the phase at any given point in such a soliton is not deterministic but random, and the way in which it correlates with the phase at any other point in this beam are described by a distribution function: the spatial coherence function, or the correlation function. This stands in sharp contradistinction to coherent solitons, in which all points are fully correlated at all times and the phases at all points vary in unison. The coherence properties of incoherent solitons can be calculated theoretically in any of the three approaches described above.[7] Interestingly, when two (or more)

[7] Note that the ray optics theory for describing incoherent solitons cannot, fundamentally, describe the coherence properties because it is based on a transport approach in which all phase information is washed out.

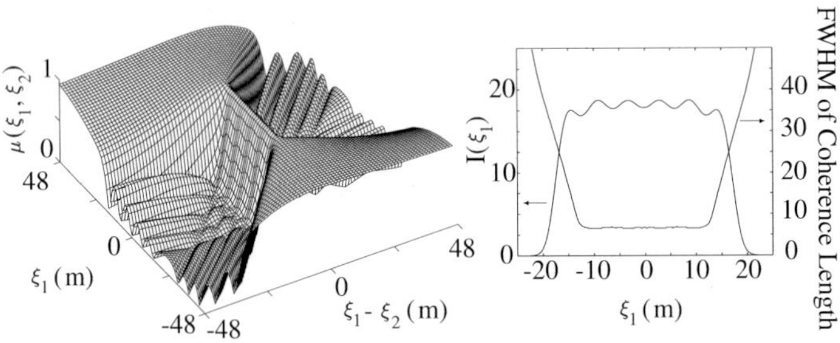

Fig. 9. The spatial coherence function of the six-mode beam of Fig. 4. *Left*: 3D plot of the coherence function as a function of coordinate and coordinate difference. *Right*: intensity profile and correlation distance as a function of position

incoherent solitons interact with one another, their coherence properties are affected. It is therefore possible to utilize interactions between (and among) incoherent solitons, and interactions between incoherent and coherent solitons to manipulate (in space) the coherence properties of the optical field. The coherence properties of incoherent solitons can be readily found through the modal theory [9,16]. The one-dimensional spatial coherence (correlation) function is defined as [21]

$$\mu_{1,2}(\xi_1,\xi_2) = \frac{\langle U(\xi_1)U^*(\xi_2)\rangle}{\sqrt{|U(\xi_1)|^2|U(\xi_2)|^2}} , \tag{18}$$

This definition can be readily extended to two transverse dimensions and to include the evolution with propagation of the correlation function. In the context of the modal theory, in the (1 + 1)D case, the correlation function becomes

$$\mu_{1,2}(\xi,\xi+\delta) = \sum_{m=1}^{N} \frac{U_m(\xi)U_m^*(\xi+\delta)}{\sqrt{|U(\xi)|^2|U(\xi+\delta)|^2}} , \tag{19}$$

where $|U(\xi)|^2 = \sum_{m=1}^{N} |U_m|^2 = I(\xi)$. The correlation function reveals several important features that are unique to incoherent solitons. For example, Fig. 9 (left) depicts the correlation function of the six-mode incoherent soliton shown earlier in Fig. 4. It is obvious that the correlation function is not only a function of the coordinate difference (as it is for many natural light sources, e.g. the sun), but actually varies with the absolute coordinate. That is, μ_{12} is not a function of δ alone, but a function of both ξ and δ. To illustrate this fact, it is convenient to examine the correlation distance, which is the characteristic maximum distance between two points that are still correlated, given by $l_c(\xi) = \int_{-\infty}^{+\infty} |\mu_{12}(\xi,\xi+\delta)|^2 \, d\delta$. Figure 9 (right) shows the correlation distance and the total intensity, both as a function of the coordinate ξ. It is apparent that in the wide central region of the bright incoherent soliton (where the intensity is fairly large), the correlation distance is rather low. But

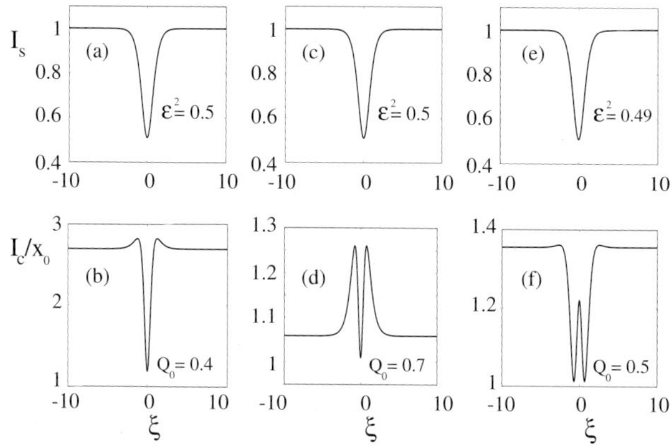

Fig. 10. Intensity profiles and the corresponding correlation distances of a first-order dark incoherent soliton: (**a**), (**b**), when the bound mode is populated, (**c**), (**d**) when the bound mode is almost empty; (**e**), (**f**) show the same information for a second-order dark incoherent soliton, when both bound modes are populated

at the margins of the soliton, where the local intensity decays exponentially, the correlation distance increases monotonically. This is because a bright incoherent soliton is made up from guided modes (bound states) only. In the center, many guided modes contribute to the local field, and thus, because the modes are uncorrelated with one another, the local correlation distance is short. At the margins of the soliton, however, the guided modes decay exponentially. The farthest-reaching mode is the highest guided mode. Thus, far away from the center of the soliton, only the highest mode survives and the correlation distance is infinite, that is, the beam becomes fully spatially coherent: $|\mu_{12}(\xi, \xi + \delta)| \to 1$ as $\xi \to \infty$ (while keeping δ finite). Altogether, for all bright incoherent solitons, the tendency is for a short l_c in the central region of the beam and an increasingly longer l_c at the beam margins, where l_c eventually goes to infinity.[8]

It is also possible to extract the coherence properties of dark incoherent solitons from the modal theory [28]. For Kerr incoherent dark solitons where closed-form solutions exist, the coherence function $\mu_{12}(\xi_1, \xi_2) = \langle U(\xi_1)U^*(\xi_2)\rangle / \sqrt{|U(\xi_1)|^2 |U(\xi_2)|^2}$ for the lowest-order dark incoherent soli-

[8] In retrospect, the fact that l_c is roughly constant and is low at the center of the soliton is what allows the approximation that treats an incoherent soliton as a bundle of completely uncorrelated rays [13,14] (in the limit of incoherent solitons whose width is much greater than l_c). This approximation does not work for few-mode solitons, nor does it work for any incoherent solitons at distances large enough from the center.

ton, can be obtained from

$$\langle U(\xi_1)U^*(\xi_2)\rangle \propto A^2 \operatorname{sech}\xi_1 \operatorname{sech}\xi_2 + \int_0^\infty \mathrm{d}Q D(Q)\left[U_{\mathrm{r,e}}(\xi_1)U_{\mathrm{r,e}}(\xi_2) + U_{\mathrm{r,o}}(\xi_1)U_{\mathrm{r,o}}(\xi_2)\right] . \tag{20}$$

From this, the correlation distance can be evaluated by calculating the corresponding integral. The results are shown in Fig. 10, which displays intensity profiles and the corresponding correlation distances as a function of the coordinate. From this figure, it is obvious that the correlation distance varies dramatically inside and near the dark notch, whereas it attains a constant value far from the notch. In other words, the coherence properties of dark incoherent solitons are very much "opposite" to those of their bright counterparts. The reason for that is due to the way radiation and bound states are spatially distributed within the dark notch. In the vicinity of the notch, the correlation distance depends on the coordinate. On the other hand, far away from the notch, only radiation modes contribute. The even and odd radiation modes of order Q sum to a constant value far away from the notch. Thus, far away from the notch, the superposition of all guided modes attains a constant composition that does not depend on the coordinate. Finally, we would like to point out that correlation traces of both bright and dark incoherent solitons can also be dynamically monitored (via (10)) using the coherent-density method. In [23], the evolution of the correlation statistics of bright and dark soliton was investigated when these entities were launched from stationary (Schell) sources. The solitons were found to adjust not only their intensities but also their coherence properties, in agreement with what one would have expected from the modal theory.

From the analyses of the coherence properties of both bright and dark incoherent solitons, it is apparent that the self-trapping process reshapes the statistics of the incoherent beam. The most dramatic coherence property of such solitons is the coordinate-dependent correlation distance. Unlike most incoherent light sources (Schell-like sources, e.g. the sun), which have non-localized statistics (their correlation distance does not depend on the absolute location), for incoherent solitons the correlation distance depends on the actual coordinate in the beam. With this understanding, it is natural to ask whether it is possible to "engineer" the coherence properties of an incoherent beam through the self-trapping process or via the interaction between solitons. The answer is a definite yes. In a recent paper [29], it has been demonstrated that the spatial coherence of a partially incoherent light beam can be greatly enhanced through an energy-conserving interaction (with no absorption/gain taking place) with an incoherent or coherent dark spatial soliton. The interaction itself is rather simple: a broad, partially incoherent, bright "signal" beam copropagates with a dark (coherent or incoherent) soliton and interacts with it during propagation. Computer simulations have shown that during the interaction process, a portion of the incoherent beam

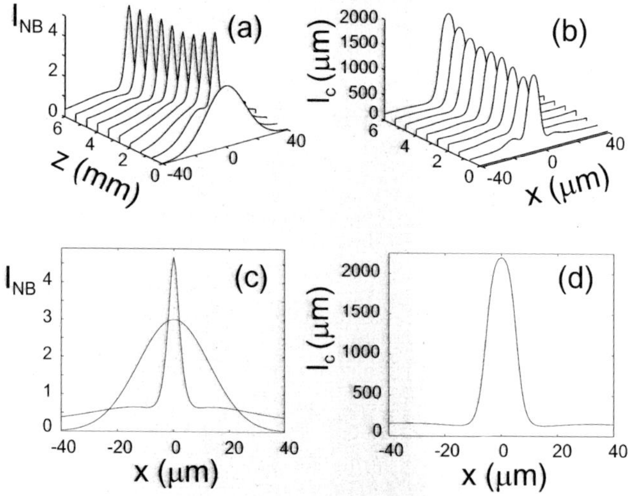

Fig. 11. (**a**) Evolution of the intensity profile of a bright signal beam and (**b**) the corresponding correlation length when the beam copropagates with a dark coherent spatial soliton in a biased photorefractive crystal. (**c**) The intensities at the input (*wide curve*) and at the output at $z = 6$ mm (*narrow curve*) of an incoherent bright beam, and (**d**) its correlation length after 6 mm of propagation

is trapped within the dark notch of the dark soliton, thus forming a sharp intensity spike. In this region, the correlation distance is dramatically increased by at least two orders of magnitude. Thus, incoherent light can be effectively "cooled" (its entropy is being reduced) at any arbitrarily chosen point on a partially incoherent wavefront by using either a coherent or an incoherent dark spatial soliton. This is the only passive system known to exhibit an increase in both the local intensity and the local correlation distance, simultaneously. The outcome of such an interaction process is depicted in Fig. 11, where the intensity spike is apparent, and, even more interestingly, where the coherence of this sharp spike has also increased dramatically (by a factor of 240) after some propagation distance. The simulations suggest that the coherence of this intensity spike can be further increased at longer propagation distances.

The explanation of the intensity spike is straightforward: the dark soliton induces a narrow waveguide and traps part of the incoherent light in it. The coherence enhancement is more complicated, but it can be intuitively understood by considering the modal properties of the waveguide induced by a dark coherent soliton. As previously shown, such a waveguide is single-mode. As a result, in the absence of any modal competition, the spatial coherence of the trapped light in this waveguiding structure is expected to increase. In essence, the lowest mode alone (almost) is trapped in the waveguide induced by the dark soliton, and this mode is fully coherent with itself. The reason for the finite value of the coherence length at the output is the small contribution from radiation modes that have not yet escaped. In fact, it is expected that as $z \to \infty$, the coherence length will increase to infinity and the signal beam will become fully coherent at the center. Note that this coherence enhancement or "cooling" has been carried out entirely passively, that is, without any power transfer between the two beams. Moreover, this kind of local increase in coherence (local cooling) can be performed at any arbitrarily chosen point on a partially incoherent wavefront, depending on the position of the control dark-soliton beam. Nevertheless, one cannot collectively enhance the coherence of the entire "signal beam" and at the same time increase the intensity, since the brightness of a beam is preserved in a medium with no gain or loss. This is in accordance with the law of brightness, which can be alternatively deduced from thermodynamic arguments. "Cooling" the entire signal beam will therefore require power exchange between the bright and dark beams (e.g. two-wave mixing effects, etc.), which does not occur in a "conventional" self-focusing or defocusing system.

At this point, it is already clear that incoherent solitons and incoherent self-focusing processes provide a means of control of the correlation (coherence) function of optical beams. It is therefore very interesting to note the strong analogy between this pronounced trapping/correlation enhancement and the process of Bose–Einstein condensation as demonstrated through evaporative cooling [30].

6 Modulational Instability and Pattern Formation with Incoherent Wavepackets

Modulational instability (MI) is a universal process that appears in most nonlinear wave systems in nature. Because of MI, small amplitude and phase perturbations (from noise) grow rapidly under the combined effects of nonlinearity and diffraction (or dispersion, in the temporal domain). As a result, a broad optical beam (or a quasi-CW pulse) tends to disintegrate during propagation [31,33], leading to filamentation [33] or to breakup into pulse trains [31].[9] In general, MI appears in the same parameter region as where bright solitons occur. In fact, the relation between MI and solitons is rather intuitive. A soliton forms when the localized wavepacket induces (via the nonlinearity) a potential and "captures" itself in it, thus becoming a bound state in its own induced potential. In the spatial domain of optics, a spatial soliton forms when a very narrow optical beam induces (through self-focusing) a waveguide structure and guides itself in its own induced waveguide. The relation between MI and solitons is best manifested in the fact that the filaments (or the pulse trains) that emerge from the MI process are actually trains of almost ideal solitons. Therefore, MI is largely considered to be a precursor

[9] For a review of MI in the temporal domain, see [32].

to soliton formation. Over the years, MI has been systematically investigated in connection with numerous nonlinear processes. However, it was always believed that MI was inherently a coherent process and thus it could only appear in nonlinear systems with a perfect degree of spatial/temporal coherence. Following the discovery of incoherent solitons, it was natural to wonder whether MI also existed for incoherent light beams. Or, in a broader physical sense, will patterns form in a nonlinear, weakly correlated, multiparticle system? Early in 2000, this question was answered in a positive way [34]: MI does occur in such systems, but only if the "strength" of the nonlinearity exceeds a specific threshold that is set by the spatial correlation (coherence) function. But the development of the ideas of MI and of pattern formation in such systems took several intriguing, and at the same time unexpected, turns, especially with the first experimental observation of this phenomenon later on that year [35]. It appears that the occurrence of incoherent MI actually reflects on many other nonlinear systems well beyond optics: it implies that patterns can form spontaneously (from noise) in nonlinear many-body systems involving weakly correlated particles, such as, electrons in semiconductors in the vicinity of the quantum Hall regime, high-T_c superconductors, and atomic gases at temperatures in the vicinity of the Bose–Einstein condensation temperature. The ideas of incoherent MI and of pattern formation in nonlinear weakly correlated wave systems are very new, and much progress is being made as this chapter is being written. In this section, we describe the essence of the physical processes involved, along with the intuitive insight about what is yet to come.

The key to the existence of incoherent solitons is the noninstantaneous nature of the nonlinearity, which responds only to the beam's time-averaged intensity structure and not to the instantaneous highly speckled and fragmented wavefront. In other words, the response time of the nonlinear medium must be much larger than the average time of phase fluctuations across the beam. Thus, the time-averaged intensity induces, through the nonlinearity, a multimode waveguide structure (a potential well that can bind many states), whose guided modes are populated by the optical field with its instantaneous speckled structure. With this noninstantaneous nature of the nonlinearity in mind, the motivation was to find out if patterns can form spontaneously in a partially coherent uniform beam, through the interplay between nonlinearity and diffraction, in a such a random-phase wavefront of uniform intensity. The first step was theoretical, and it has been shown [34] that a uniform but partially incoherent wavefront is unstable in such a nonlinear medium, provided that the nonlinearity exceeds a well-defined threshold that is set by the coherence properties. Above that threshold, MI should occur, and patterns should form.

The theory of incoherent MI is conveniently derived from the equation describing the propagation of mutual coherence [10], although the coherent-density theory [7] yields the very same results (which is by no means surprising, because the theories are equivalent [12]). Given the complex amplitude representing the electric field of the optical wave, $E(x,z,t) = U(x,z,t) \times \exp[\mathrm{i}(kz-\omega t)]$, $U(x,z,t)$ being the slowly varying amplitude, we recall the quantity $J_{12}(x_1,x_2,z) = \langle U(x_1,z,t)U^*(x_2,z,t)\rangle$ or, in convenient dimensionless units, $J_{12}(\xi_1,\xi_2,\zeta) = \langle U(\xi_1,\zeta,t)U^*(\xi_2,\zeta,t)\rangle$ (as shown above, J_{12} is actually associated with the correlation function $\mu_{12}(\xi_1,\xi_2,\zeta)$). At this point it is convenient to use $J_{12}(r,\delta,\zeta)$, where $r = (\xi_1+\xi_2)/2$ and $\delta = \xi_1 - \xi_2$ are the midpoint and the difference (normalized) coordinates, respectively. To shorten the notation, from this point and on we refer to J_{12} as J.

In the paraxial approximation, J satisfies the equation [10]

$$\frac{\partial J}{\partial \zeta} - \mathrm{i}\frac{\partial^2 J}{\partial r \partial \delta} = \mathrm{i}(F[\xi_1,\zeta] - F[\xi_2,\zeta])J\,, \tag{21}$$

where $F[r,\zeta] = F[I(r,\zeta)]$ is the normalized nonlinear index change, which depends only on the time-averaged *local* intensity $I(r,\zeta)$. In this notation, the time-averaged intensity is $I(r,\zeta) = J(r,\delta=0,\zeta)$. The definition of J yields $J(r,\delta,\zeta) = J^*(r,-\delta,\zeta)$.

To study MI, one assumes that the incident light intensity is uniform in space, except for small intensity perturbations that depend on r and ζ. Thus, J can be written as $J(r,\delta,\zeta) = J^{(0)}(\delta) + J^{(1)}(r,\delta,\zeta)$, where $J^{(1)} \ll J^{(0)}$, $J^{(0)}$ representing the background of uniform intensity $I_0 = J^{(0)}(\delta=0)$. The dependence of the normalized index change $F[I(\xi,\zeta)]$ on r comes from $J^{(1)}$, and thus to lowest order, $F[\xi_1,\zeta] - F[\xi_2,\zeta] = \kappa[J^{(1)}(r_1,\delta=0,\zeta) - J^{(1)}(r_2,\delta=0,\zeta)]$, where $\kappa \equiv \mathrm{d}F[I]/\mathrm{d}I$ evaluated at I_0 is the marginal nonlinear index change. For example, in Kerr media $F[I] = \gamma I$ (where γ is some constant), so that $\kappa = \gamma$. This procedure results in a linearized equation for the perturbation

$$\frac{\partial J^{(1)}}{\partial \zeta} - \mathrm{i}\frac{\partial^2 J^{(1)}}{\partial r \partial \delta} = \mathrm{i}\kappa\left[J^{(1)}\left(r+\frac{\delta}{2},0,\zeta\right) - J^{(1)}\left(r-\frac{\delta}{2},0,\zeta\right)\right]J^{(0)}\,. \tag{22}$$

The eigenmode solutions of this equation are of the form $J^{(1)}(r,\delta,\zeta) = \exp(g\zeta + \mathrm{i}\alpha r)L(\delta) + \exp(g^*\zeta - \mathrm{i}\alpha r)L^*(-\delta)$, where g is associated with the MI gain, α (always real) is the transverse wavenumber, and L is some complex function of δ. Because (22) depends on $J^{(0)}$, the evolution of $J^{(1)}$ depends on the coherence function of the input (uniform) intensity. In the general case, this equation requires numerical solution. However, in the specific case when $J^{(0)}(\delta)$ has a Lorentzian-shaped transverse k spectrum, it is possible to obtain closed-form results [34]. It is natural to go to k space and define $\tilde{J}^{(0)}(q) = (2\pi)^{-1}\int_{-\infty}^{+\infty}\mathrm{d}\delta\, J^{(0)}(\delta)\exp(\mathrm{i}q\delta)$. After some algebra [34], it is possible to show that in the Lorentzian case of $\tilde{J}^{(0)} \propto (q^2+q_0^2)^{-1}$, the ζ dependence of $J^{(1)}$ is characterized by a growth coefficient g (as a function of α) that can be calculated directly, given the parameters q_0 and κI_0:

$$g(\alpha) = -q_0|\alpha| + |\alpha|\sqrt{\kappa I_0 - (\alpha/2)^2}\,. \tag{23}$$

Having found $g(\alpha)$, one can easily determine the intensity of the perturbation $I_1(r,\zeta) = J^{(1)}(r,\delta = 0,\zeta)$. The variable q represents the transverse k distribution of the light beam, and thus q_0 is proportional to the characteristic angular width, θ_0, of the angular power spectrum of the partially coherent beam [34]. When $g > 0$, the perturbation $J^{(1)}$ grows exponentially, that is, MI occurs. On the other hand, when $g < 0$, the perturbation decays and no patterns grow on top on the uniform input intensity. In the coherent limit of $q_0 \to 0$, the growth coefficient g is *always* positive (MI occurs) and correctly coincides with that of coherent MI [36,37]. For an incoherent beam, however, MI occurs *only if* $\kappa I_0 > q_0^2$, whereas when $\kappa I_0 < q_0^2$, MI is *entirely eliminated.* Thus, the more incoherent the beam is (q_0 large), the larger the marginal index change required to induce MI is. If we recall that for fully coherent beams, MI always occurs, irrespective of the magnitude of the nonlinearity, the implication of the new result is far-reaching: *MI in incoherent nonlinear wave systems has a specific threshold that depends solely on the correlation function.* Computer simulations and further analytical work have shown [34] that this is a universal trend, and a threshold for MI exists for any input correlation function.

At this point it is instructive to seek an intuitive explanation for the presence of a threshold for incoherent MI. The explanation, after the fact, is rather simple. Consider first a fully coherent and linear system, and imagine a periodic perturbation superimposed on a uniform-intensity beam. Because the system is linear and coherent, the modulation depth (visibility or contrast) of this perturbation neither grows nor decays during propagation, because all points in the beam are fully correlated with one another (i.e. vary in unison with time). Consequently, if some, no matter how weak, self-focusing nonlinearity is added to the system, then every maximum in the perturbation induces a positive index change in its vicinity. Thus, the intensity maxima narrow (self-focus), the modulation depth of the perturbation grows, and MI occurs. Coherent MI occurs even if the magnitude of the nonlinearity is vanishingly small, because there is no mechanism that can suppress the MI growth. This is why coherent MI has no threshold for its existence. Now, consider a partially incoherent, linear wave system, and again imagine a perturbation superimposed on a uniform-intensity beam. In such a linear but partially incoherent system, the perturbation naturally decays, because different points in the beam behave like independent sources and the modulation depth of the perturbation diminishes with propagation. Adding some self-focusing nonlinearity to this system provides a means of increasing the modulation depth of the perturbation. Whether or not the modulation depth of the perturbation will increase or decrease during propagation depends upon the relative rates of the two conflicting effects: the "washout" effect that results from incoherence, and the self-focusing effects that tend to amplify the perturbation. When the two effects exactly balance, the perturbation maintains its initial modulation depth. This is the threshold point. If the self-focusing tendency is stronger than the effect of incoherence, then the perturbation grows and MI occurs. This is the intuitive explanation for the presence of a threshold for incoherent MI and for its dependence on the correlation function. Interestingly, this seems to be the only known MI process that has any threshold in a one-way propagation scheme (as distinct from similar threshold effects that occur in cavities or for cavityless counterpropagating waves).

To summarize the theoretical findings related to incoherent MI obtained thus far [34], the main predictions are as follows:

- There exists a sharp threshold for the nonlinear index change, below which perturbations (noise) on top of a uniform input beam decay and above which a quasi-periodic pattern forms.
- The threshold depends upon the coherence properties of the input beam: the threshold increases with decreasing correlation distance (decreasing spatial coherence).
- Saturation alone, while keeping the maximum index change and correlation distance fixed, arrests the growth of the MI and can decrease the rate to below the MI threshold.

The first theory of incoherent MI [34] was soon followed by experimental work on this subject. The first experimental observation of effects related to incoherent MI [35] indeed confirmed many of the predictions, including the very pronounced threshold effects. But at the same time the experiments revealed much intriguing, and completely unpredicted, new physics. It was confirmed that incoherent MI occurs above a specific threshold that depends on the beam's coherence properties (correlation distance), and leads to a periodic train of one-dimensional filaments. At a higher value of nonlinearity, and very much unexpectedly, incoherent MI displays a two-dimensional instability and leads to the appearance of self-ordered 2D lattices of light spots. A typical experimental result depicting incoherent MI is shown in Fig. 12. The experiment was carried out in a photorefractive crystal, employing the screening nonlinearity [4,5]. When an external voltage with a magnitude large enough to allow for MI is applied to the nonlinear crystal, the previously homogeneous light distribution at the output face of the sample becomes periodically modulated and starts to form 1D filaments of incoherent light. Figure 12 shows the beam intensity at the output plane of the nonlinear crystal. The coherence length of the incoherent light was $l_c = 17.5\,\mu\text{m}$ and the intensity ratio $I_o/I_{sat} = 1$. Figure 12a depicts the output intensity without nonlinearity. Figures 12b, c, d correspond to a value of the nonlinearity just below the threshold for 1D incoherent MI, at threshold, and just above the threshold. This shows beyond any doubt (i) the existence of incoherent MI, and (ii) that incoherent MI occurs only when the nonlinear index change exceeds a well-defined threshold. In particular, Fig. 12c shows a mixed state exactly at threshold, in which order and disorder coexist. This is a clear indication that the nonlinear interaction undergoes an order–disorder phase

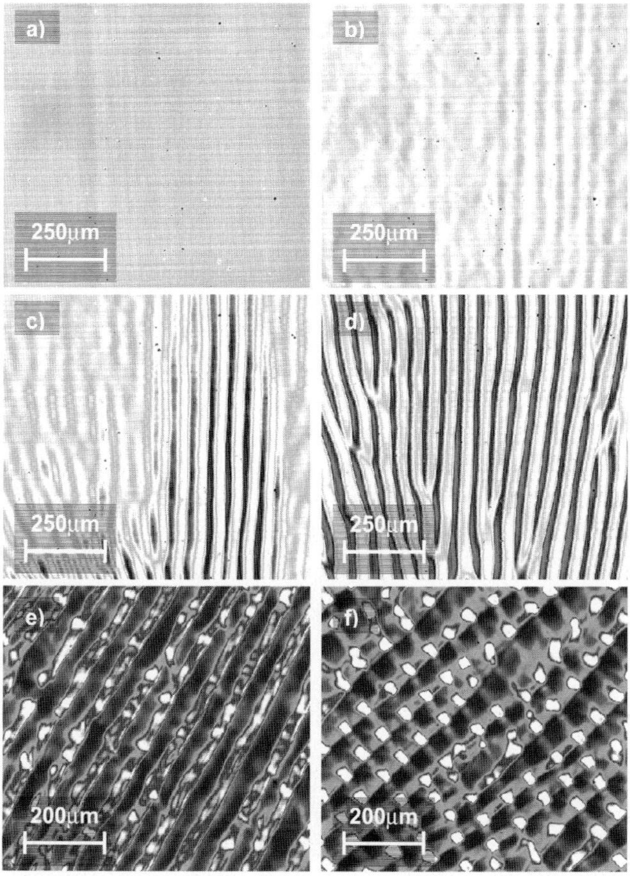

Fig. 12. The intensity structure of a partially spatially incoherent beam at the output plane of a nonlinear crystal. The sample was illuminated homogeneously with partially spatially incoherent light with a coherence length of 17.5 μm. The displayed area is $1.0 \times 1.0\ \mathrm{mm}^2$ in (**a**)–(**d**) and $0.8 \times 0.8\ \mathrm{mm}^2$ in (**e**) and (**f**). The size of the nonlinear refractive-index change of the crystal was successively increased from (**a**) $\Delta n_0 = 0$ (the linear case), to (**b**) 3.5×10^{-4}, (**c**) 4.0×10^{-4}, (**d**) 4.5×10^{-4}, (**e**) 9×10^{-4}, and (**f**) 1×10^{-3}. The plots (**b**)–(**d**) show the cases just below threshold (no features), at threshold (partial features), and just above threshold (features everywhere) for 1D incoherent MI, which leads to 1D filaments. Far above this threshold, at a much higher value of the nonlinearity, the 1D filaments become unstable (**e**), and finally become ordered in a regular two-dimensional pattern (**f**)

transition. These phenomena were predicted by the incoherent-MI theory [34]. But the experiments, as often happens, revealed new surprises. When the nonlinearity is further increased, a second threshold is reached: the filaments become unstable (Fig. 12e) and start to break into *an ordered array of spots (2D filaments)* as shown in Fig. 12f. We emphasize that, in all the pictures displayed in this figure, the correlation distance is much shorter than the distance between two adjacent stripes or filaments. Thus, this work is a clear demonstration that patterns can form in weakly correlated, nonlinear multiparticle systems. The dependence of the MI threshold on the coherence properties of the beam was also studied experimentally [35]. The experiments showed that, for the case of a fully coherent input beam, MI occurs even at a vanishingly small nonlinearity. This is because coherent MI has no threshold. When the correlation distance is reduced, however, a well-defined threshold is observed: the jump from very low visibility to a large visibility is always abrupt, because for every beam with a finite correlation length there is always a threshold for MI. The experiments have clearly demonstrated that the MI threshold shifts towards a higher value of nonlinearity with decreasing correlation distance l_c. A different set of experiments [35], has confirmed another prediction, about the dependence of incoherent MI on the degree of saturation in saturable nonlinear media. On the basis of the theory of 1D incoherent MI [34], one expects that saturation of the optical nonlinearity should arrest the growth of MI. Experimentally, this was tested by launching a "flat top" beam that served as a "quasi-uniform beam" in its flat top. When this beam is launched into a nonlinear crystal with a high enough nonlinearity, patterns form in several regions in the beam, as shown in Fig. 13. In the flat top of the beam, only low-visibility stripes appear. In this region, the nonlinearity is above threshold but is in rather deep saturation, so the growth of MI is suppressed. At the margins of the beam, on the other hand, where the nonlinearity is far from saturation, high-visibility stripes appear. In this region the nonlinearity is above threshold and at the same time it is not saturated, so the MI growth rate is large. Finally, in the far margins of the beam, the local nonlinearity is below threshold, because $I(r) \ll I_{\text{sat}}$. A nice by-product of this particular experiment is the clear evidence (in Fig. 13) that the 1D stripes emerge at different orientations, and are not affected much by the local noise in the crystal (striations). It is expected that similar experiments with incoherent MI in the saturated regime with high nonlinearities that lead to 2D "lattices" of filaments will reveal a wealth of new features, because the "lattice" will form features of varying order and of varying scale in different regions of the beam. Certainly, this fascinating topic merits further research.

The possibility of suppressing MI via coherence control has opened up new avenues for investigating soliton structures that would have been otherwise impossible to observe. Such entities include, for example, the so-called antidark solitons, which have been experimentally observed for the first time only recently [36]. Antidark solitons are self-trapped "bright" (localized)

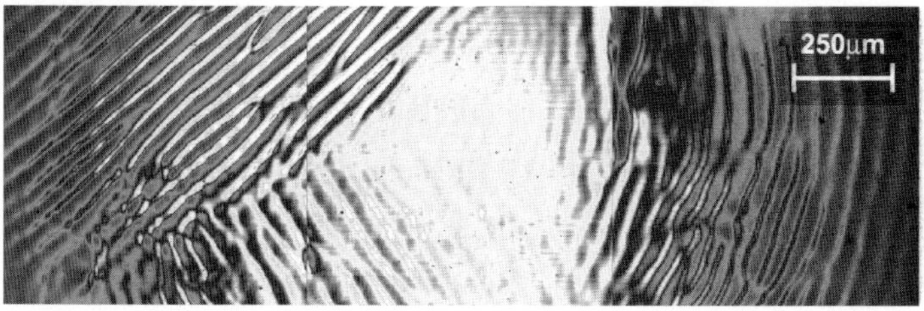

Fig. 13. Suppression of incoherent MI due to saturation of the nonlinearity. The intensity structure of a finite signal beam (Gaussian beam with a width (FWHM) of 1 mm) at the output plane of the crystal. The intensity ratio (peak intensity of beam to background/saturation intensity) is $I_0/I_{\mathrm{sat}} = 3$. Without nonlinearity ($\Delta n_0 = 0$), the output beam shows no features. The photograph was taken at $\Delta n_0 = 6 \times 10^{-4}$. The saturable nature of the nonlinearity clearly suppresses MI in the center of the beam, whereas strong modulation and filaments of random orientation occur in the margins of the beam

wavepackets that reside in a nonzero background [37]. Because these solitons are bright-like, they occur only in self-focusing media. In the coherent limit, antidark solitons tend to disintegrate rapidly during propagation since the nonzero background happens to be modulationally unstable.[10] On the other hand, if the background is made sufficiently incoherent, such that the nonlinearity is below the MI threshold, the antidark soliton as a whole is stable [36]. As a result such a soliton propagates indefinitely without suffering from MI. Thus far, these are the only antidark solitons ever observed experimentally.

Before closing this section, we would like to relate nonlinear optical systems to other nonlinear systems of weakly correlated particles. The theoretical predictions and experimental observations imply that in all such systems patterns will form spontaneously, provided that the nonlinearity is larger than a threshold value, which in turn is set by the correlation distance. For example, it is expected that 1D and 2D patterns will form in an atomic gas at, or slightly above, the Bose–Einstein condensation temperature, when the atoms possess independent degrees of freedom (and cannot be described by a single wavefunction, as in the Bose–Einstein-condensate state) but they are still weakly correlated. At least for atoms with attractive collision forces (a negative scattering length) [40], such patterns should form, depending on the dimensionality of the problem describing the atoms in the trap. The equation governing the evolution of the "mean field" of an atomic gas is the so-called Gross–Pitaevskii equation (see, e.g. [41]), which is identical to the nonlinear

[10] A bimodal antidark soliton with a coherent background was investigated by Chen et al. [38]. As shown there, theoretically and experimentally, such an antidark soliton is unstable. Later on, this family of solutions was generalized to a multicomponent antidark soliton [39]. Both the bimodal and the multicomponent antidark soliton are unstable.

wave equation in nonlinear optics that gives rise to (1 + 1)D Kerr solitons. In other areas of physics, there are already at least some hints that such patterns do exist in disordered many-body nonlinear systems. To be specific, several experimental papers have reported a large anisotropy in the resistivity of a two-dimensional electron system with weak disorder [42]. The observed anisotropy is now attributed to the combination of nonlinear transport and weak disorder [43], which is exactly the transport equivalent of the combination of an optical nonlinearity and incoherence in optical systems such as ours. Theoretical work predicts the existence of 1D stripes (electron stripes) of charge density waves [43]. Spontaneous formation of stripes has also been predicted and observed in high-T_c superconductors (see [44] and references therein), which are again nonlinear weakly correlated many-body systems. Finally, as discussed in [34], spontaneously forming patterns are also known in at least one system of classical particles: in a gravitational system. The spontaneous emergence of patterns in all of these diverse fields of science indicates that pattern formation in nonlinear weakly correlated systems is a universal property. It is a gift of nature that in optics we can study this effect directly, visualizing every little detail of the physics involved, and at the same time are able to isolate the underlying effects and to develop a theory that captures the core effects that are indeed universal. The hope is to be able to understand such patterns in a higher dimensionality, now that they have been observed in the laboratory [35], and to draw new exciting conclusions that have implications for other fields of nature.

7 Effects of Anisotropic Coherence and Propagation Dynamics

Thus far in all the sections above, it has been implicitly assumed that all the nonlinear beam interactions take place in a fully isotropic partially coherent environment. But, it is quite possible that the statistical properties of the partially incoherent optical field may be *anisotropic*. In other words, the correlation distance in the x direction may be different from that along y. This in turn implies that at the input, the angular power spectrum is also anisotropic and therefore exhibits two different angular widths θ_{0x} and θ_{0y}. The possible effects of anisotropic incoherence on incoherent spatial solitons were first investigated in nonlinear systems of the logarithmic type [15]. In this study it was found that *elliptical incoherent solitons* are possible, provided that their coherence properties are anisotropic. These elliptical soliton states are possible even though the material nonlinearity itself is fully isotropic. Depending on the degree of coherence anisotropy, the shape of these entities can be circular or, at the other extreme, stripe-like. It is worth emphasizing that this property is entirely unique to incoherent solitons. As shown earlier [45], a coherent elliptical self-trapped beam never displays stationary propagation in a fully isotropic nonlinear medium. Instead, such a coherent beam undergoes oscillations, with two periods of oscillation, that depend on the two different beam diameters [45].[11] This is because the nonlinearity, being isotropic, cannot support self-trapping at two different transverse scales simultaneously. The intuitive explanation is simple: the magnitude of the nonlinearity needed to compensate for diffraction is set by the diffraction properties, which, for a fully coherent beam, are determined by the ratio between the beam width and the optical wavelength. Therefore, a beam that has different diffraction properties in its two transverse directions can never reach stationary propagation if the underlying nonlinearity is the same for both directions. As a result the beam oscillates periodically [45]. However, if the beam is partially spatially incoherent, its diffraction properties are determined by *both* its diameter *and* its correlation distance [7,9]. In particular, an elliptical beam can be made to possess isotropic diffraction properties by appropriately introducing anisotropic coherence. In other words, one can manipulate the coherence properties of such an elliptical beam so that it has a shorter l_c in the direction where its diameter is larger and, at the same time, a longer l_c in the direction where its diameter is shorter. Such a beam, having isotropic diffraction properties in spite of its ellipticity, can be self-trapped by a fully isotropic nonlinearity, and displays stationary propagation [15]. In a subsequent study [46], elliptical incoherent solitons were identified numerically in saturable nonlinear systems. The nonlinearity of these systems was taken to be of the form $\Delta n = -\Delta n_0/(1 + I)$, which closely resembles that in photorefractives. Again, these solitons were found to be possible as long as their diffraction characteristics (because of incoherence) were isotropic [46]. The dynamics and collisions of these elliptical solitons were also investigated in these systems. The interaction properties of elliptical incoherent solitons led to another surprise. As shown theoretically, when two elliptic incoherent solitons with initial trajectories that lie in the same plane collide, they tend to spiral about one another [46]. This is very much unexpected because, the input beam configuration does not seem to carry angular momentum, and, in a fully isotropic system that lacks angular momentum , spiraling "cannot" occur, not even when two coherent solitons interact with one another in an incoherent fashion [47]. Evidently, this statement is true and accurate for coherent solitons, but not for incoherent solitons. In the case of a collision between two elliptical incoherent solitons, the solitons rotate about one another in opposite directions, so as to conserve the zero total angular momentum. However, unlike the almost indefinite spiraling of incoherently interacting coherent solitons [47], the spiraling rotation of elliptical incoherent solitons ends after several rotations, and the solitons coalesce [46].

Another example where anisotropic incoherence has been used is in the recent observation of stable $(1 + 1)$D solitons in bulk Kerr media [48]. These solitons are perfectly coherent in one transverse dimension (say, x), which is the direction of self-trapping, and at the same time they are made sufficiently

[11] Such self-trapped beams are sometimes called "mighty morphic solitons".

incoherent in the other direction (y) – so much so that the transverse MI (in y) is below the MI threshold and all perturbations in y are suppressed [34]. These are the only known solitons in a self-focusing medium that are stable (in the *absolute* sense) in a higher dimension.

Another manifestation of the intriguing instability properties of incoherent solitons has to do with the (2+1)D configuration. As shown recently [49], (2+1)D partially incoherent beams can undergo catastrophic collapse in (noninstantaneous) Kerr media, much like their coherent counterparts. However, the critical power for this instability increases with increasing incoherence, that is, the shorter the correlation distance, the higher the critical power. The self-focusing collapse dynamics of such incoherent entities is fascinating, and will undoubtedly lead to new features.

Last and certainly not least is the interaction dynamics between incoherent solitons [18]. This is another feature that exemplifies the strong relation between the coherence properties of such solitons and their dynamics. The multimode nature of incoherent solitons suggests that their collision properties should be very different from those of coherent solitons. Thus far, this interesting direction has already led to the discovery of three new features of the interaction. The first one has to do with coherence enhancement through soliton interactions [29], and was described in a previous section. Another new feature is shape transformation upon collision between incoherent (or between multimode) solitons. In this process, two solitons (typically each is asymmetric) collide, and two new solitons of new shapes emerge from the collision process. Such interesting propagation dynamics were predicted in [18], but thus far have been demonstrated only with multimode solitons [50], not with incoherent solitons. The collision properties of two-dimensional elliptic incoherent solitons have recently been investigated in [46].

8 Prospects

The rapid progress in this new area of incoherent solitons has led to many interesting fundamental ideas and possible applications. In this chapter, we have concentrated on the fundamentals, but one can envision numerous applications related to incoherent solitons. The exciting ability to obtain self-trapped light beams from incoherent sources, such as light emitting diodes, for reconfigurable optical interconnects and beam steering is just one possibility. Self-trapping of incoherent wavepackets is a research area that has been around for barely four years. It is a very young area, nonetheless, it is already evident that the phenomena in this area are related to many other areas of physics, in which nonlinearities, stochastic behavior, and statistical (ensemble) averaging are involved. The closest immediate example is cooled atomic gases, where the nonlinearity results from head-on collisions (and thus the ensemble average arises) of atoms [41]. In a nutshell, the self-trapping effects described in this chapter relate to any noninstantaneous nonlinear system of weakly correlated waves or particles. We believe that, as has happened so many times in scientific history, the best is yet to come.

Dedication and Acknowledgments

We would like to dedicate this chapter to Shlomo Segev, the father of Moti Segev. Shlomo Segev was a gentle and noble man, who passed away on January 6, 2000. Shlomo Segev was born in Dorohoi, Romania, in 1932 and went through concentration camps, torture, and starvation during the Holocoast. After the Second World War, he was drafted into the Romanian army and was not allowed to emigrate to Israel for eleven years. In 1961 he eventually fled to Israel, with a wife and a three-year-old Moti. All his life, Shlomo was a Zionist, and has educated his children for honesty and integrity. We will always remember him for what he was: a gentle and noble man with zero respect for money, which he never had anyway. In Hebrew, *YEHI ZIKHRO BARUCH.*

The work on incoherent solitons was carried out, to a large extent, by our former and current talented graduate students and postdoctoral fellows. It is their work that made the area of incoherent solitons possible. In particular, we wish to emphasize the milestones set by Dr. Matt Mitchell, in doing the very first experiments demonstrating incoherent solitons and in making many other discoveries from that point on. We gratefully acknowledge the contributions of all our former and current graduate students and postdoctoral fellows involved in this research: Prof. Zhigang Chen, Prof. Tamer Coskun, Dr. Marin Soljacic, Prof. Detlef Kip, Prof. Ming-feng Shih, Dr. Eugenia Eugenieva, Charalambos Anastassiou, Alexandra Grandpierre, and Dr. Ziad Musslimani. We are privileged to have worked with such a group of talented people. Finally, we gratefully acknowledge the continuing support of several funding agencies. We wish to express our gratitude, more than to anybody else, to Dr. Mikael Ciftan of the US Army Research Office, who believed in our work and supported us when nobody else did. We value his honesty and friendship.

The research at the Technion was supported by the Israel Science Foundation and by the Israeli Ministry of Science. The research at Princeton and at Lehigh is part of the MURI program on optical spatial solitons and is supported by the US Army Research Office, the National Science Foundation, and the Air Force Office of Scientific Research.

References

1. M. Segev, G.I. Stegeman, Phys. Today **51**, 42 (1998).
2. G.I. Stegeman, M. Segev, *Optical Spatial Solitons and Their Interactions: Universality and Diversity*, Special Issue on Frontiers in Optics, Science **286**, 1518 (1999).

3. M. Mitchell, Z. Chen, M. Shih, M. Segev, Phys. Rev. Lett. **77**, 490 (1996).
4. M. Segev, G.C. Valley, B. Crosignani, P. Di Porto, A. Yariv, Phys. Rev. Lett. **73**, 3211 (1994); D.N. Christodoulides, M.I. Carvalho, J. Opt. Soc. Am. B **12**, 1628 (1995); M. Segev, M. Shih, G.C. Valley, J. Opt. Soc. Am. B **13**, 706 (1996).
5. M. Shih, M. Segev, G.C. Valley, G. Salamo, B. Crosignani, P. Di Porto, Electron. Lett. **31**, 826 (1995); Opt. Lett. **21**, 324 (1996).
6. M. Mitchell, M. Segev, Nature **387**, 880 (1997).
7. D.N. Christodoulides, T. Coskun, M. Mitchell, M. Segev, Phys. Rev. Lett. **78**, 646 (1997).
8. D.N. Christodoulides, T.H. Coskun, R.I. Joseph, Opt. Lett. **22**, 1080 (1997).
9. M. Mitchell, M. Segev, T. Coskun, D.N. Christodoulides, Phys. Rev. Lett. **79**, 4990 (1997).
10. V.V. Shkunov, D.Z. Anderson, Phys. Rev. Lett. **81**, 2683 (1998).
11. G. A. Pasmanik, Sov. Phys. JETP **39**, 234 (1974).
12. D.N. Christodoulides, E. Eugenieva, T. Coskun, M. Segev, M. Mitchell, Phys. Rev. E **63**, R35601 (2001).
13. A.W. Snyder, D.J. Mitchell, Phys. Rev. Lett. **80**, 1422 (1998).
14. A. Hasegawa, Phys. Fluids **18**, 77 (1975); Phys. Fluids **20**, 2155 (1977).
15. D.N. Christodoulides, T.H. Coskun, M. Mitchell, M. Segev, Phys. Rev. Lett. **80**, 2310 (1998).
16. M.I. Carvalho, T.H. Coskun, D.N. Christodoulides, M. Mitchell, M. Segev, Phys. Rev. E **59**, 1193 (1999).
17. V.A. Vysloukh, V. Kutuzov, V.M. Petnikova, V.V. Shuvalov, Quantum Electron. **27**, 843 (1997).
18. N.N. Akhmediev, W. Krolikowski, A.W. Snyder, Phys. Rev. Lett. **81**, 4632 (1998); A. Ankiewicz, W. Krolikowski, N.N. Akhmediev, Phys. Rev. E **59**, 6079 (1999).
19. Z.H. Musslimani, M. Segev, D.N. Christodoulides, M. Soljacic, Phys. Rev. Lett. **84**, 1164 (2000).
20. D.N. Christodoulides, S.R. Singh, M.I. Carvalho, M. Segev, Appl. Phys. Lett. **68**, 1763 (1996); Z. Chen, M. Segev, T.H. Coskun, D.N. Christodoulides, Opt. Lett. **21**, 1436 (1996); M. Mitchell, M. Segev, D.N. Christodoulides, Phys. Rev. Lett. **80**, 4657 (1998).
21. L. Mandel, E. Wolf, *Optical Coherence and Quantum Optics* (Cambridge University Press, New York, 1995).
22. Y.S. Kivshar, B. Luther-Davies, Phys. Rep. **298**, 81 (1998).
23. T.H. Coskun, D.N. Christodoulides, M. Mitchell, Z. Chen, M. Segev, Opt. Lett. **23**, 418 (1998).
24. Z. Chen, M. Mitchell, M. Segev, T. Coskun, D.N. Christodoulides, Science **280**, 889 (1998).
25. T. Coskun, D.N. Christodoulides, Z. Chen, M. Segev, Phys. Rev. E **59**, R4777 (1999).
26. Z. Chen, M. Segev, D.N. Christodoulides, R.S. Feigelson, Opt. Lett. **24**, 1160 (1999).
27. Z. Chen, M. Segev, S.R. Singh, T. Coskun, D.N. Christodoulides, J. Opt. Soc. Am. B **14**, 1407 (1997).
28. D.N. Christodoulides, T. Coskun, M. Mitchell, Z. Chen, M. Segev, Phys. Rev. Lett. **80**, 5113 (1998).
29. T. Coskun, A.G. Grandpierre, D.N. Christodoulides, M. Segev, Opt. Lett. **25**, 826 (2000).

30. M.H. Anderson, J.R. Ensher, M.R. Matthews, C.E. Wieman, E.A. Cornell, Science **269**, 198 (1995)
31. V.I. Bespalov, V.I. Talanov, JETP Lett. **3**, 307 (1966); V.I. Karpman, JETP Lett. **6**, 277 (1967); G.P. Agrawal, Phys. Rev. Lett. **59**, 880 (1987); S. Wabnitz, Phys. Rev. A **38**, 2018 (1988). A. Hasegawa, W.F. Brinkman, J. Quantum Electron. **16**, 694 (1980); K. Tai, A. Hasegawa, A. Tomita, Phys. Rev. Lett. **56**, 135 (1981).
32. G.P. Agrawal, *Nonlinear Fiber Optics*, 2nd ed. (Academic Press, San Diego, 1995), Chap. 5.
33. E.M. Dianov, P.V. Mamyshev, A.M. Prokhorov, S.V. Chernikov, Opt. Lett. **14**, 1008 (1989); P.V. Mamyshev, C. Bosshard, G.I. Stegeman, J. Opt. Soc. Am. B **11**, 1254 (1994); M.D. Iturbe-Castillo, M. Torres-Cisneros, J.J. Sanchez-Mondragon, S. Chavez-Cerda, S.I. Stepanov, V.A. Vysloukh, G.E. Torres-Cisneros, Opt. Lett. **20**, 1853 (1995); M.I. Carvalho, S.R. Singh, D.N. Christodoulides, Opt. Commun. **126**, 167 (1996).
34. M. Soljacic, M. Segev, T. Coskun, D.N. Christodoulides, A. Vishwanath, Phys. Rev. Lett. **84**, 467 (2000).
35. D. Kip, M. Soljacic, M. Segev, E. Eugenieva, D.N. Christodoulides, Science **290** 495 (2000).
36. T. Coskun, D.N. Christodoulides, Y. Kim, Z. Chen, M. Soljacic, M. Segev, Phys. Rev. Lett. **84**, 2374 (2000).
37. Y.S. Kivshar, Phys. Rev. A **43**, 1677 (1991).
38. Z. Chen, M. Segev, T.H. Coskun, D.N. Christodoulides, Y.S. Kivshar, V.V. Afanasjev, Opt. Lett. **21**, 1821 (1996).
39. N. Akhmediev, A. Ankiewicz, Phys. Rev. Lett. **82**, 2661 (1999).
40. C.A. Sackett, J.M. Gerton, M. Welling, R.G. Hulet, Phys. Rev. Lett. **82**, 876 (1999); S.L. Cornish, N.R. Claussen, J.L. Roberts, E.A. Cornell, C.E. Wieman, Phys. Rev. Lett. **85**, 1795 (2000).
41. T. Busch, J.R. Anglin, Phys. Rev. Lett. **84**, 2298 (2000).
42. M.P. Lilly, K.B. Cooper, J.P. Eisenstein, L.N. Pfeiffer, K.W. West, Phys. Rev. Lett. **82**, 394 (1999); Phys. Rev. Lett. **83**, 824 (1999); W. Pan, R.R. Du, H.L. Stormer, D.C. Tsui, L.N. Pfeiffer, K.W. Baldwin, K.W. West, Phys. Rev. Lett. **83**, 820 (1999); M. Shayegan, H.C. Manoharan, preprint [cond-mat/9903405] (1999).
43. A.A. Koulakov, M.M. Fogler, B.I. Shklovskii, Phys. Rev. Lett. **76**, 499 (1996); A.H. MacDonald, M.P.A. Fisher, preprint [cond-mat/9907278], Phys. Rev. B, in press; H. MacDonald, M.P.A. Fisher, preprint [cond-mat/0001021] (2000).
44. V.J. Emery et al., Proc. Natl. Acad. Sci. USA **96**, 8814 (1999), and references therein.
45. B. Crosignani, P. Di Porto, Opt. Lett. **18**, 1394 (1993); A.W. Snyder, D.J. Mitchell, Opt. Lett. **22**, 16 (1997); V. Tikhonenko, Opt. Lett. **23**, 594 (1998)..
46. E.D. Eugenieva, D.N. Christodoulides, M. Segev, Opt. Lett. **25**, 972 (2000).
47. M. Shih, M. Segev, G. Salamo, Phys. Rev. Lett. **78**, 2551 (1997); A. Buryak, Y.S. Kivshar, M. Shih, M. Segev, Phys. Rev. Lett. **82**, 81 (1999).
48. C. Anastassiou, M. Soljacic, M. Segev, D. Kip, E. Eugenieva, D.N. Christodoulides, Z.H. Musslimani, Phys. Rev. Lett. **85**, 4888 (2000).
49. O. Bang, D. Edmundson, W. Krolikowski, Phys. Rev. Lett. **83**, 5479 (1999).
50. W. Krolikowski, N. Akhmediev, B. Luther-Davies, Phys. Rev. E **59**, 4654 (1999).

Quadratic Solitons

William Torruellas, Yuri S. Kivshar, and George I. Stegeman

Summary. In this chapter we review the state of knowledge from both the theoretical and experimental perspectives of quadratic spatial solitons. This class of spatial solitons offers a wide parameter space to be investigated in addition to effects not seen in Kerr or Photorefractive spatial solitons. In many ways they offer an ideal testbed to investigate coherent interactions and mutual trapping of mulit-wavelength optical fields. We introduce the chapter with a historical perspective followed by an overview of the current state of understanding of the theory of quadratic solitons. Experimental results demonstrating the existence of this soliton family are presented for the (1+1)D and (2+1)D cases. Finally we present results of spatial modulation instability of these indeed intriguing and fascinating solitons.

1 Introduction and Historical Remarks

One of the classical concepts that have pervaded optical-spatial-soliton physics since its earliest days has been that optically self-induced changes in the refractive index modify the propagation properties of the beam that has caused those changes [1]. This approach has led to many insights and simplifications in our understanding of spatial solitons in photorefractive, Kerr, and Kerr-like nonlinear media (see, e.g., [2,3]). Parametric solitons in a quadratic medium, or *quadratic solitons*, are a genre that does not fall neatly into this simplification, but does follow similar historical lines. In 1967, Lev Ostrovskii [4] showed that self-induced changes in the phase front of an optical beam propagating in a crystal, seemingly unrelated to refractive-index changes, might occur in a noncentrosymmetric nonlinear medium when multifrequency components were present and interacted parametrically. In the mid 1970s, Karamzin and Sukhorukov [5,6] predicted the existence of *two-wave parametric quadratic solitons* in both the waveguide geometry and the bulk, and even more general three-wave parametric solitons were investigated theoretically in the literature [7,8].

After those early days there were sporadic further theoretical developments, mainly in the Russian literature, well documented in the book by Sukhorukov [9], published in Russian. This field was revitalized theoretically as a part of the general interest in spatial solitons which followed their experimental discovery in Kerr and Kerr-like media in the late 1980s and early 1990s. Unfortunately, it took until the mid 1990s for the first experiments on quadratic solitons to be reported. The main reason for this delay was that high-quality materials did not become available until this time; in addition, the advantages afforded by quadratic parametric interactions for all-optical applications, such as light guiding light, were not realized until then. Experimentally, the ideas evolved naturally from the rediscovery of self-action effects in quadratic media and the availability of high-damage-threshold optical materials with long enough propagation lengths. This has allowed, in just a few years, the demonstration of (1+1)-dimensional spatial solitons in a mode-matched planar $LiNbO_3$ waveguide, spatial solitons with two transverse dimensions in bulk KTP crystals, solitons in traveling-wave optical parametric amplifiers; the demonstration of the temporal compression of pulses, implying the existence of temporal solitons; (1+1)-dimensional temporal solitons in a KTP optical parametric oscillator (OPO) with additional intracavity dispersion, and quasi-"optical bullets", in which light is self-trapped in both time and one spatial dimension in a bulk medium. (As in a perturbed Kerr-type medium, the evolution of solitary waves in (3+1) dimensions, in which light is confined in both time and two spatial dimensions, needs to be experimentally demonstrated and seems elusive.) The intimate connection between transverse (or modulational) instabilities and solitons has also been demonstrated experimentally. With the present and future advances in the growth and engineering of quadratic materials, rapid progress will undoubtedly occur, leading to further demonstrations and implementation of concepts that take full advantage of the solitonic nature of the optical parametric interactions discussed here (see also the recent review papers [10–12]).

This chapter presents a brief overview of the recent results on parametric spatial solitons in optical materials with a quadratic (or $\chi^{(2)}$) nonlinear response. As is normally the case, theoretical developments have outstripped experiments in this field, and in Sect. 2 we deal with the fundamental models for quadratic solitons and also present an overview of some basic results of the underlying theory, for different dimensions and geometries. Section 3 gives a brief summary of experimental results reported to date, and the chapter is concluded by Sect. 4, where we summarize our presentation.

2 Summary of Theoretical Results

2.1 Basic Models for Quadratic Solitons

The conventional theory of spatial optical solitons is based on the concept of the nonlinearity-induced change of the refractive index of an optical material that may occur in Kerr (or Kerr-like) materials with a cubic (or the so-called $\chi^{(3)}$) nonlinear response. In the case of the quadratic (or so-called $\chi^{(2)}$) nonlinearity of noncentrosymmetric nonlinear media, the light does not change the refractive index, and nonlinear effects can be observed when the fundamental frequency becomes phase-matched with one of its harmonics. The simplest effect of such a resonant interaction is the generation of a second-harmonic wave with a frequency 2ω by a fundamental wave with a frequency ω, a particular case of a more general process of *three-wave mixing*. Such an interaction is efficient when the matching condition between the wave propagation constants is satisfied. This is well known in the theory of type I and type II second-harmonic generation (SHG) (see, e.g., [13,14]). For example, in an anisotropic dielectric medium, for any wavevector direction $\boldsymbol{k}/k$, two *different* corresponding values of $k(\omega)$ can be found. In other words, for any direction of propagation there are two normal waves, called the *ordinary* ('o') and *extraordinary* ('e') waves, which have different polarizations and different phase velocities. For the ordinary wave, the direction of the wavevector $\boldsymbol{k}$ coincides with the direction of the Poynting vector $\boldsymbol{S}$ (i.e. with the direction of energy flow), whereas for the extraordinary wave the directions of $\boldsymbol{k}$ and $\boldsymbol{S}$ do not coincide.

To derive the most general model of three-wave mixing in a diffractive quadratic medium, we consider parametric interaction between three stationary quasi-plane monochromatic waves with envelopes E_j (where $j = 1, 2, 3$) and assume the resonant condition, $\omega_1 + \omega_2 = \omega_3$, where the corresponding wavevectors satisfy $k_1(\omega_1) + k_2(\omega_2) - k_3(\omega_3) = \Delta k$, where $\Delta k \ll k_j$. If all three vectors $\mathbf{k}_j$ have the same direction, the phase velocity walk-off is absent. However, if any of the three waves are extraordinary, then their energy flows diverge. Choosing the z axis as the direction of $\boldsymbol{k}_j$, and the x axis to be in the plane defined by $\boldsymbol{k}_j$ and the direction of the energy walk-off, we consider the electric field as a sum of three fields at the resonantly interacting frequencies. Then, in the approximation of slowly varying envelopes, we can derive a system of three coupled equations that describes *spatial solitons due to type II SHG* in a bulk medium,

$$2\mathrm{i}k_1 \frac{\partial E_1}{\partial z} + \nabla_\perp^2 E_1 + \frac{8\pi\omega_1^2}{c^2}\chi_1^{(2)} E_3 E_2^* \mathrm{e}^{-\mathrm{i}\Delta k z} = 0\,,$$

$$2\mathrm{i}k_2 \left(\frac{\partial E_2}{\partial z} - \rho_\omega \frac{\partial E_2}{\partial x}\right) + \nabla_\perp^2 E_2 + \frac{8\pi\omega_2^2}{c^2}\chi_2^{(2)} E_3 E_1^* \mathrm{e}^{-\mathrm{i}\Delta k z} = 0\,,$$

$$2\mathrm{i}k_3 \left(\frac{\partial E_3}{\partial z} - \rho_{2\omega} \frac{\partial E_3}{\partial x}\right) + \nabla_\perp^2 E_3 + \frac{8\pi\omega_3^2}{c^2}\chi_3^{(2)} E_1 E_2 \mathrm{e}^{\mathrm{i}\Delta k z} = 0\,, \qquad (1)$$

where $\nabla_\perp = (\partial/\partial x, \partial/\partial y)$, c is the speed of light, ρ_ω and $\rho_{2\omega}$ are the walk-off parameters, and $\chi_j^{(2)}$ are the elements of the second-order susceptibility tensor. Equations (1) describe the case when the spatial walk-off of all waves occurs in the same plane. Formally, this is true only for single-axis crystals, but the generalization is simple. In a slab waveguide, the structure of the linear guided modes in the direction of the modal trapping provided by the waveguide is known. Then, using a separation of variables, $E_j(x, y, z) = F_j(y)\mathcal{E}_j(x, z)$ and integrating out the dependences on y, we can obtain a similar system of nonlinear coupled equations but with normalized (scaled) coefficients.

In one special case, when $\omega_1 = \omega_2 = \omega_3/2$, only one characteristic frequency $\omega_1 \equiv \omega_0$ is involved. This requires only one source of coherent radiation at the fundamental frequency ω_0, and then a wave at the double frequency $2\omega_0$ is generated owing to type I SHG. In this case, we put $E_1 = E_2$ and $\rho_\omega = 0$, so that *spatial solitons due to type I SHG* in a $\chi^{(2)}$ slab waveguide are described by the system of two coupled equations

$$2\mathrm{i}k_0 \frac{\partial E_1}{\partial z} + \frac{\partial^2 E_1}{\partial x^2} + \frac{8\pi\omega_0^2}{c^2}\chi_1^{(2)} E_3 E_1^* \mathrm{e}^{-\mathrm{i}\Delta k z} = 0 \,,$$

$$4\mathrm{i}k_0 \left(\frac{\partial E_3}{\partial z} - \rho_{2\omega}\frac{\partial E_3}{\partial x}\right) + \frac{\partial^2 E_3}{\partial x^2} + \frac{32\pi\omega_0^2}{c^2}\chi_3^{(2)} E_1^2 \mathrm{e}^{\mathrm{i}\Delta k z} = 0 \,, \tag{2}$$

where ω_0, k_0, and E_1 are the frequency, wavenumber and electric-field intensity, respectively, of the fundamental harmonic; E_3 and $\rho_{2\omega}$ are the electric-field intensity and walk-off angle for the second-harmonic wave; and $\Delta k = 2k_0 - k_3$ is the wavevector mismatch.

Similar equations, but more general than (2), describe *two distinct physical situations* (see, e.g., [15–19]): (i) *spatial beams* in a slab waveguide exhibiting diffraction in the transverse direction x, in which case the modal diffraction ratio $\sigma = k_3/k_1 \approx 2$, and (ii) *temporal pulses*, where x stands for the retarded time, in which case the group velocity dispersion ratio is arbitrary, $\sigma > 0$. In the latter case, the second-order derivative in time (instead of x in (2)) accounts for the modal dispersion, which can be positive or negative.

We can normalize these equations by measuring the transverse coordinate (or time) x in units of the input beam size (or pulse duration) r_0, and the propagation coordinate z in units of the diffraction (or dispersion) length $z_\mathrm{d} \equiv r_0^2 k_2$. Looking for stationary solutions of those equations in the form $E_1 = \tilde{E}_1 \mathrm{e}^{\mathrm{i}\beta z}$ and $E_3 = \tilde{E}_3 \mathrm{e}^{\mathrm{i}(2\beta+\Delta)z}$, we can finally derive a dimensionless system of two coupled equations for the normalized field envelopes v and w (see [15] for details);

$$\mathrm{i}\frac{\partial v}{\partial z} + r\frac{\partial^2 v}{\partial x^2} - v + wv^* = 0 \,,$$

$$\mathrm{i}\sigma\frac{\partial w}{\partial z} - \mathrm{i}\delta\frac{\partial w}{\partial x} + s\frac{\partial^2 w}{\partial x^2} - \alpha w + \frac{1}{2}v^2 = 0 \,. \tag{3}$$

Here $\Delta \equiv z_\mathrm{d}\,\Delta k$, $\alpha = 2\sigma + \sigma\Delta/\beta$, δ accounts for walk-off and is proportional to $\rho_{2\omega}$, and $r = \pm 1$ and $s = \pm 1$ take into account different types of the modal dispersion and nonlinear mismatch. In this notation, the dimensionless parameter β is proportional to the nonlinearity-induced phase velocity shift, and the sign functions r and s are both positive or both negative (two cases) for the spatial case, or arbitrary (four cases) for the temporal case.

2.2 (1 + 1)-Dimensional Quadratic Solitons

2.2.1 Solitons in the Cascading Limit

In the simplest case of type I SHG without walk-off, the two-wave solitons are described by a stationary form of the normalized equations (3) at $\delta = 0$, reduced to a system of real equations

$$rv'' - v + wv = 0 \,,$$

$$sw'' - \alpha w + \frac{1}{2}v^2 = 0 \,. \tag{4}$$

Equations (3), and their stationary form (4), represent a generic model of two-wave $\chi^{(2)}$ solitons in the absence of walk-off. Their solutions have been analyzed by many authors, in the (1 + 1)-dimensional case [5,15–31] and in the more general (2 + 1)-dimensional case [5,6,8,27,32,33]. In the case of more general nondegenerated three-wave mixing, solitary waves have been also investigated (see, e.g., [34,35]).

It is straightforward to show why one should expect to find spatially localized solutions of (3). Indeed, for $\alpha \gg 1$, i.e. for large positive Δk, the second equation of (3) reduces to a simple algebraic relation, $w \approx v^2/2\alpha$, and the first equation of (3) becomes the standard nonlinear Schrödinger (NLS) equation ,

$$\mathrm{i}\frac{\partial v}{\partial z} + r\frac{\partial^2 v}{\partial x^2} - v + \frac{1}{2\alpha}|v|^2 v = 0 \,, \tag{5}$$

which possesses stable *bright* (for $r = +1$) or *dark* (for $r = -1$) soliton solutions. We call the limit of large α *the cascading limit* . In this limit, the effective Kerr-like behavior due to cascaded $\chi^{(2)}$ effects is recovered.

Using this simple reduction to the NLS equation, Buryak and Kivshar [16] and Steblina et al. [36] looked for localized solutions of the stationary equations (4) in the form of an asymptotic series in α. They obtained, for *bright solitons* ($r = +1$),

$$v(x) = 2\alpha^{1/2}\text{sech}\, x + 4s\alpha^{-1/2}\tanh^2 x\,\text{sech}\, x + \dots \,,$$

$$w(x) = 2\,\text{sech}^2 x + s\alpha^{-1}(16\,\text{sech}^2 x - 20\,\text{sech}^4 x) + \dots \,, \tag{6}$$

and, for *dark solitons* ($r = -1$) ,

$$v(\zeta) = \sqrt{2}\alpha^{1/2}\tanh\zeta + \sqrt{2}s\alpha^{-1/2}(\zeta\,\text{sech}^2\zeta - \tanh\zeta\,\text{sech}^2\zeta) + \dots \,,$$

$$w(\zeta) = \tanh^2\zeta + s\alpha^{-1}(2\zeta\tanh\zeta\,\text{sech}^2\zeta - 4\,\text{sech}^2\zeta + 5\,\text{sech}^4\zeta) + \dots \,, \tag{7}$$

where $\zeta \equiv x/\sqrt{2}$.

Since the properties of the Kerr solitons of (5) are well known, the existence of the asymptotic solutions (6) and (7) suggests that for $\alpha \gg 1$ the system (3) should have bright solitons for $r = +1$, $s = \pm 1$, and dark solitons for $r = -1$, $s = \pm 1$, similarly to (5). However, this conclusion is not satisfactory, because (i) the formal localized solutions (6) and (7) can be nonstationary (radiative) for the system (3) owing to their resonance with linear waves; and (ii) in the case of dark solitons, the solutions (7) can also be unstable owing to *parametric modulational instability* , as was first demonstrated by Ferro and Trillo [26,37], Buryak and Kivshar [16], and He et al. [38].

2.2.2 Two-Wave Bright Solitons

As follows from the NLS limit and the analysis of modulational instability, bright solitons of (3) exist for $r = s = +1$ and $\alpha > 0$ in the form of one-hump localized profiles for the real functions $v(x)$ and $w(x)$. Families of such two-wave parametric solitons have been found numerically by Buryak and Kivshar [24,15] and Torner [27].

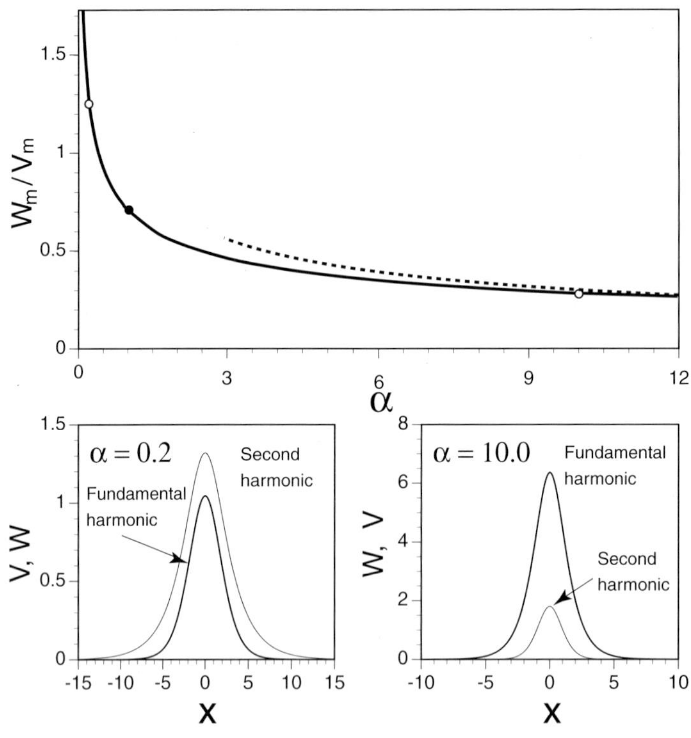

Fig. 1. Parameters of the family and of two characteristic profiles (*open circles*) of two-wave bright solitons in the model (3) for $r = s = +1$. The *filled circle* corresponds to an exact analytical solution at $\alpha = 1$, and the *dashed curve*, showing $w_{\mathrm{m}}/v_{\mathrm{m}} = 1/\sqrt{\alpha}$, to the asymptotic NLS-like solution (from [15])

Examples of two-wave localized solutions of (4) are presented in Fig. 1 for $\alpha = 0.2$ and $\alpha = 10.0$. For $\alpha \gg 1$ the maximum amplitude of the fundamental component $v_{\rm m}$ is much larger than the corresponding value $w_{\rm m}$ for the second-harmonic component, and this case corresponds to the asymptotic solution (6), $v \approx \pm 2\sqrt{\alpha}\,\mathrm{sech}\, x$, $w \approx 2\,\mathrm{sech}^2 x$. The ratio $w_{\rm m}/v_{\rm m}$ characterizing the whole family is plotted in Fig. 1, where the filled circle corresponds to the analytical solution at $\alpha = 1$ [5],

$$v(x) = \sqrt{2}\, w(x) = \frac{3\sqrt{2}}{2}\mathrm{sech}^2(x/2)\,, \tag{8}$$

and the dashed curve corresponds to the NLS limit given by the asymptotic dependence $w_{\rm m}/v_{\rm m} \approx 1/\sqrt{\alpha}$.

Approximate analytical forms of two-wave parametric quadratic solitons can be found by means of the variational approach, as first suggested by Steblina et al. ([36]; see also [28,39]). Such an approach provides a good approximation for all integral characteristics of solitary waves; however, it fails to take into account either the variation of the harmonic profiles or the soliton tails. Very recently, Sukhorukov [40] suggested an elegant method, based on the scaling properties of quadratic solitons, that allows to find an approximate solution of (4) in the form

$$v(x) = v_{\rm m}\mathrm{sech}^p(x/p)\,, \quad w(x) = w_{\rm m}\mathrm{sech}^2(x/p)\,, \tag{9}$$

where the parameters are defined by the following expressions:

$$v_{\rm m}^2 = \frac{\alpha w_{\rm m}^2}{w_{\rm m}-1}\,, \quad p = \frac{1}{w_{\rm m}-1}\,, \quad \alpha = \frac{4(w_{\rm m}-1)^3}{2-w_{\rm m}}\,. \tag{10}$$

These determine $w_{\rm m}$ for arbitrary α as a solution of a cubic equation, thus defining all other parameters as functions of α. For $0 < \alpha < \infty$, the parameters change monotonically in the range $0 < v_{\rm m} < \infty$, $1 < w_{\rm m} < 2$, and $\infty > p > 1$. At $\alpha = 1$, the values are $v_{\rm m} = 3/\sqrt{2}$, $w_{\rm m} = 3/2$, $p = 2$, and the solution (9), (10) reduces to the exact analytical solution (8). For $\alpha \gg 1$, the solution (9), (10) describes correctly the asymptotic NLS-type approximation as well.

Analysis of the stability of the bright soliton family based on the direct integration of (3) and an eigenvalue analysis of the corresponding linearized problem has shown that both stable and unstable solitons exist, depending on the values of the system parameters α and σ [41,42]. In the cascading limit ($\alpha \gg 1$), the solitons shown in Fig. 1 are stable, whereas in the other limit ($\alpha \to 0$) the solitons become unstable (e.g. the soliton at the bottom left of Fig. 1 is unstable for $\sigma = 2$). In spite of the instability for $\alpha \to 0$, for $\sigma < \sigma_{\rm cr}$ (including $\alpha \sim 1$) the parametric solitons are stable even under the action of very strong perturbations. These two-wave solitons can be generated from a rather broad class of initial conditions. Figure 2 shows the generation of

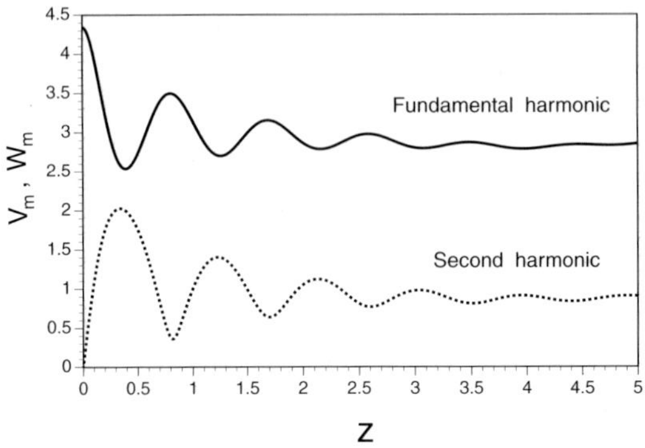

Fig. 2. Self-trapping of a sech-profile input pulse into a two-wave parametric bright soliton. Shown are the peak intensities for both harmonics (from [15])

a soliton from a first-harmonic sech-form input beam. Owing to diffraction, the input beam becomes broader but it also generates a second harmonic, and, after a rather short transition period, a two-component bright soliton forms. This kind of behavior is possible only because of the existence of the continuous family of stable bright solitons.

In addition to the fundamental (one-hump) localized solutions described above, the numerical analysis revealed the existence of continuous (in α) families of two-hump (and multihump) bright solitons, which can be treated as bound states of one-hump solitons [28–31,43]. Numerical stability analysis indicates that all these multihump bright solitons are unstable [44,45] and either split into partial, stable solitons or disintegrate completely; the latter scenario occurs for sufficiently small values of α that *stable* single solitons do not exist.

In spite of the fact that in the cascading limit the effective NLS equation (5) does not depend on the sign s, the localized solutions of (4) are very different for $s = +1$ and $s = -1$. Although stationary one-hump solitons do not exist, owing to resonance with linear waves, numerical analysis of (4) still allows one to find *discrete sets of bound states* of single solitons. For such bound states, linear radiation is suppressed outside, but it exists between the solitons in the form of a trapped standing wave.

In the case $r = +1$ and $s = -1$, one such solution is known in an explicit analytical form [23]. It exists at $\alpha = 2$ and has the form

$$v(x) = 6\sqrt{2}\ \tanh x\ \mathrm{sech}\, x\ , \quad w(x) = 6\ \mathrm{sech}^2 x\ . \tag{11}$$

The solution (11) is the simplest of the discrete set of two-soliton bound states due to trapped radiation, but they are all found to be unstable.

2.2.3 Two-Wave Dark Solitons

Following the preliminary results in the cascading limit, one can expect to find dark solitons in the case $r = -1$, $s = +1$, which corresponds to a defocusing

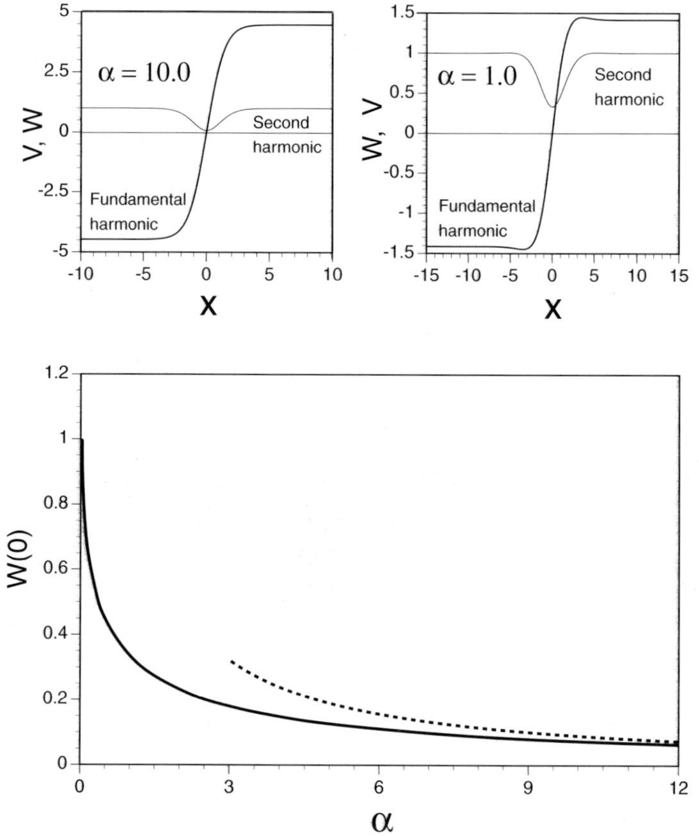

Fig. 3. *Top*: characteristic profiles of two-wave dark solitons in the model (3) at $r = -1$ and $s = +1$. *Bottom*: minimum intensity of the second-harmonic component vs. α. The *dashed curve* corresponds to the NLS solution (from [15])

effective cubic nonlinearity. Indeed, the numerical results obtained by Buryak and Kivshar [15,17] indicate that single, dark, radiationless solitons exist for $r = -s = -1$ as localized solutions of (4). In the case $r = -s = -1$, a continuous family of parametric dark solitons exists for $0 < \alpha < \infty$, and in the interval $0 < \alpha < 8$ these solitons have nonmonotonic radiationless oscillatory tails. Examples of these two-wave dark solitons are presented in Fig. 3 for $\alpha = 1.0$ (nonmonotonic tails) and $\alpha = 10.0$ (monotonic tails).

In the cascading limit ($\alpha \gg 1$) the solution can be presented in the asymptotic form (7). When α is not large, the asymptotic solution does not describe oscillatory tails for $\alpha < 8$, but the dark soliton family still exists for $\alpha > 0$, and it can be characterized by the minimum amplitude of the second harmonic $w(0)$, for example. The dependence of $w(0)$ on α is shown in Fig. 3. For large α it approaches the asymptotic dashed curve $w(0) \approx 1/\alpha$, which corresponds to the NLS soliton in the cascading limit. The dark solitons of Fig. 3 have modulationally stable backgrounds in a certain domain of the parameter plane (α, σ). For $\sigma \sim 1$, the soliton backgrounds are modulationally stable for $2 < \alpha < \infty$; this range of α does not depend on σ.

Owing to the existence of decaying radiationless oscillating tails, a dark soliton can trap another soliton to form a bound state, *a twin-hole dark soliton* [16]. This mechanism is known for other types of solitons, but the $\chi^{(2)}$ twin-hole and multihole dark solitons are the first examples of continuous families of *stable bound states of dark solitons*. As $\alpha \to 8$, the distance between the neighboring dark solitons in a bound state increases to infinity.

The dark solitons discussed above exist in (3) for $r = -s = -1$. Recently He et al. [30] found other continuous families of dark solitons for $r = -s = +1$ in the interval $0 < \alpha < 2$, i.e. in the second window of the parameter space where the resonance with linear waves is absent. These dark solitons possess radiationless oscillating tails and thus can also form bound states.

Similarly to the case of bright solitons, single dark radiationless solitons do not exist for $r = s = -1$ in (3), and only discrete sets of two- (and multi-) soliton radiationless bound states can be found [16]. Moreover, at $\alpha = 1$ there exists an exact analytic solution [46]

$$v(x) = \sqrt{2}\, w(x) = \sqrt{2} \left[1 - \frac{3}{2} \mathrm{sech}^2(x/2) \right], \tag{12}$$

which is a two-soliton bound state of the first order. Analytical results [16,37] indicate that all such bound states are unstable owing to the development of parametric modulational instability of the background beam.

2.3 (2 + 1)-Dimensional Quadratic Solitons

2.3.1 General Remarks

Similarly to other types of solitons, parametric solitary waves can become unstable in higher dimensions (see, e.g., [47] and references therein). In both the spatial and the temporal domains, the symmetry-breaking instability can generate solitary waves *self-trapped in higher dimensions*. Development of a soliton theory for higher dimensions (e.g. for the spatial (2 + 1)-dimensional geometry, which means two transverse and one longitudinal coordinates) is a conceptually a nontrivial task, since in a Kerr medium a self-trapped beam (a spatial soliton of circular symmetry) described by the (2+1)-dimensional NLS equation (see, e.g., [48]) is *unstable* and displays a catastrophic collapse, where the beam width tends to infinity at a certain finite propagation distance (see, e.g., [49] and also the review paper [50]). Several physical mechanisms which can suppress or even eliminate such collapse-type instabilities are known (see, for example, the recent review paper by Kivshar and Pelinovsky [47]).

In a medium with a quadratic nonlinear response, parametric interaction between the harmonics has been shown to support stable solitary waves even in higher dimensions [8,51], and the transverse instability does not lead to wave collapse dynamics.

Multidimensional spatial solitons have been discussed, first for parametric two-wave or type I SHG processes [5,6] and then for three-wave parametric interaction [8] in the special case of zero phase matching. Kanashov and Rubenchik [8] demonstrated the possibility of noncollapsing stable solitary waves in higher dimensions (e.g. light bullets) in the case of quadratic media.

The next effort to analyze multidimensional solitary waves was attempted by Hayata and Koshiba [20], who rediscovered the exact analytical solution of Karamzin and Sukhorukov [5] and also applied a special kind of Hartree-like approach to construct an approximate multidimensional solitary wave of radial symmetry for a special value of the phase-matching parameter. Numerical simulations demonstrating the generation of two-dimensional beams due to parametric interaction at a nonzero phase mismatch and in the presence of walk-off have been reported by Torner et al. ([32]; see also [52,53]), whereas beam collisions and steering were analyzed much later [54–56].

Buryak et al. [33] and Torner et al. [57] demonstrated the existence of a two-parameter class of $(2+1)$-dimensional stable solitary waves in a bulk quadratic medium; the parameters are the phase mismatch and the amplitude of one of the fields. Importantly, the approximate solution found by Hayata and Koshiba [20] was shown to be just a single point of this family of localized waves found numerically. Two-dimensional solitary waves have also been found by means of the variational approach [36], and their existence and properties have been independently confirmed by other researchers.

Recently, the theory of multidimensional solitary waves has been advanced by the study of beam self-focusing in the presence of both cubic and quadratic nonlinearities [58], of parametric solitons in a defocusing Kerr medium [59,60], and also of walking solitons and light bullets in a bulk medium [61–63].

2.3.2 Fundamental Cylindrical Solitons

To discuss $(2+1)$-dimensional solitary waves, we use the following form of the normalized equations:

$$\mathrm{i}\frac{\partial V}{\partial z} + \nabla_{\perp}^2 V - V + WV^* = 0\,,$$

$$\mathrm{i}\sigma\frac{\partial W}{\partial z} + \nabla_{\perp}^2 W - \mathrm{i}\delta\frac{\partial W}{\partial x} - \alpha W + \frac{V^2}{2} = 0\,, \tag{13}$$

where all parameters have the same definitions as in Sect. 2, and the variable y is normalized in the same way as x. The circularly symmetric stationary solitary waves of (13) without walk-off are described by the localized solutions of the following system of ordinary differential equations;

$$\frac{\mathrm{d}^2V}{\mathrm{d}r^2} + \frac{1}{r}\frac{\mathrm{d}V}{\mathrm{d}r} - V + WV = 0\,,$$

$$\frac{\mathrm{d}^2W}{\mathrm{d}r^2} + \frac{1}{r}\frac{\mathrm{d}W}{\mathrm{d}r} - \alpha W + \frac{V^2}{2} = 0\,, \tag{14}$$

where $r \equiv \sqrt{x^2+y^2}$ is a radial coordinate.

In the system (14), the functions $V(r)$ and $W(r)$ can be treated as two Cartesian coordinates in a corresponding mechanical problem of the motion of a particle in the effective potential $U(W,V) = (1/2)(V^2W - \alpha W^2 - V^2)$, where the first-order derivative plays the role of an effective "anisotropic" dissipation. Then, similarly to the case of the $(2+1)$-dimensional NLS equation, spatially localized solutions of (14) correspond to special (separatrix) trajectories in the phase space $(V, W, \mathrm{d}V/\mathrm{d}r, \mathrm{d}W/\mathrm{d}r)$, which start at the point $(V_{\max}, W_{\max}, 0, 0)$ at $r = 0$ and approach asymptotically the point $(0,0,0,0)$ as $r \to \infty$. Buryak et al. [33] (see also [30,42]) have found these separatrix trajectories and the corresponding soliton profiles for (13) numerically by employing the shooting technique. These results are summarized in Fig. 4, which presents a family of $(2+1)$-dimensional solitary waves for which the functions $V(r)$ and $W(r)$ decrease monotonically for $0 < r < \infty$.

Analytical expressions for the beam profiles for any α can be found approximately by means of the variational approach. We note that the stationary solutions of (14) can be treated as extrema of a Lagrangian with the density $\mathcal{L} = (1/2)[(\nabla_\perp W)^2 + (\nabla_\perp V)^2 + W^2 + \alpha V^2 - W^2 v]$, i.e. as solutions of the variational problem $\delta L = 0$, where $L = \int\int \mathcal{L}\, \mathrm{d}x'\mathrm{d}y'$. To apply the variational approach, Steblina et al. [36] approximated the soliton family by a simple ansatz in the form of two Gaussian functions,

$$W(r) = A\mathrm{e}^{-\rho r^2}\,, \quad V(r) = B\mathrm{e}^{-\gamma r^2}\,, \tag{15}$$

where A, B, ρ, and γ are unknown functions of α. Substituting the ansatz (15) into the Lagrangian L, we obtain a system of four equations for A, B, ρ, and γ from the corresponding variational problem $\partial L/\partial a_j = 0$, where $a_j \equiv \{A, B, \rho, \gamma\}$. The result for ρ is given by the equation

$$32\rho^3 + 2\alpha\rho - \alpha = 0\,, \tag{16}$$

and γ and the peak amplitudes $A \equiv w_0$ and $B \equiv v_0$ are defined as $\gamma = 4\rho^2$, $B = (1+2\rho)^2$, and $A^2 = \alpha(1+2\rho)^4/8\rho^2$. Therefore, (16) determines the function $\rho = \rho(\alpha)$, which, in turn, defines the dependences $\gamma(\alpha)$, $A(\alpha)$, and $B(\alpha)$ characterizing the soliton family. Steblina et al. [36] demonstrated good agreement of the integral characteristics calculated from numerical and variational solutions. A further extension of the variational procedure was developed by Malomed et al. [64] and Skryabin and Firth [65,66] for the analysis of optical light bullets, described by a $(3+1)$-dimensional system of

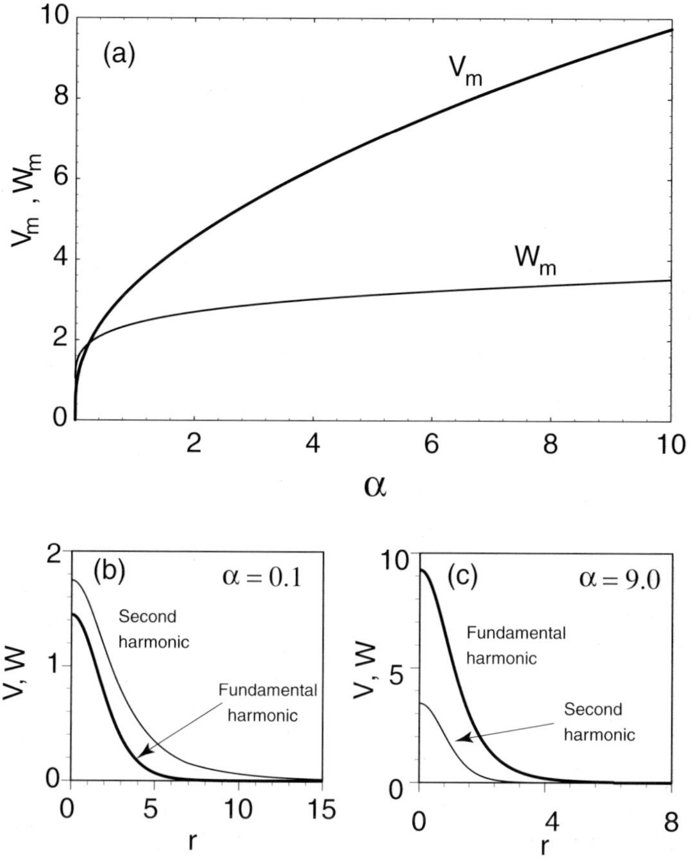

Fig. 4. (**a**) Family of circularly symmetric two-wave solitons, characterized by the maximum amplitudes of the harmonic beams. (**b**), (**c**) Characteristic profiles of two-wave solitary waves at $\alpha = 0.1$ and $\alpha = 9.0$, respectively (from [33])

coupled equations of the type (14). Finally, we wish to note that Sukhorukov [40] developed another analytical approach based on the scaling properties of quadratic solitons, which provided even better approximations for the profiles of the $(2+1)$-dimensional quadratic solitons of radial symmetry.

2.3.3 Soliton Scattering and Fusion

In general, the system (13) does not possess translational (Galilean) invariance. However, for $\sigma = 2$ (i.e. for spatial solitons) (13) has the property of Galilean invariance, and moving solitons (in fact, walking solitons) can be obtained by means of the vector gauge transformation

$$V(\boldsymbol{r}, z) = V_{\mathrm{s}}(\boldsymbol{r} - \boldsymbol{C}z)\ \exp\left[\mathrm{i}(\boldsymbol{C} \cdot \boldsymbol{r})/2 - \mathrm{i}\boldsymbol{C}^2 z/4\right] ,$$

$$W(\boldsymbol{r}, z) = W_{\mathrm{s}}(\boldsymbol{r} - \boldsymbol{C}z)\ \exp\left[\mathrm{i}(\boldsymbol{C} \cdot \boldsymbol{r}) + \mathrm{i}\delta C_x z - \mathrm{i}\boldsymbol{C}^2 z/2\right] , \tag{17}$$

where $\boldsymbol{r} = (x, y)$ is the transverse coordinate vector, and the "velocity" parameter $\boldsymbol{C} = (C_x, C_y)$ characterizes the beam displacement in the transverse plane.

Using the transformation (17), it is easy to investigate numerically different types of collisions between (2+1)-dimensional solitons in the framework of (13). In the (1 + 1) geometry, soliton collisions were investigated by Werner and Drummond [22], Baboiu et al. [67], Etrich et al. [68], Haelterman et al. [31], Clausen et al. [69], and Baboiu and Stegeman [70], who revealed that these collisions depend strongly on the initial relative phase between the solitons. A very similar interaction was observed for head-on collisions of (2 + 1)-dimensional solitons, as reported by Buryak et al. [33] (see also [54,55]). When the relative phase of two colliding solitons is zero, the solitons attract each other and finally fuse into a single two-dimensional soliton of a larger amplitude. The amplitude of this "fused" soliton oscillates, indicating the excitation of a *soliton internal mode*, as was pointed out by Etrich et al. [71]. However, when the two interacting solitons are significantly out of phase ($\pi/4 < \Delta\psi < 7\pi/4$), the interaction between them can be repulsive, and both solitons, after exchanging some energy, still survive the collision.

Recent experimental discoveries of stable (2 + 1)-dimensional solitons in various nonlinear media initiated the experimental study of fully three-dimensional interactions between solitary beams. Steblina et al. [72] developed the theory of nonplanar (i.e. fully three-dimensional) collisions of (2+1)-dimensional $\chi^{(2)}$ solitons. These authors presented an analytical model describing the scattering, spiraling and fusion of solitons. This mechanical model corresponds to a conservative system with an effective Lagrangian

$$E = \frac{1}{2} M_R \dot{R}^2 + \frac{1}{2} M_\psi \dot{\psi}^2 - U(R, \psi) , \tag{18}$$

where $R \equiv \sqrt{X^2 + Y^2}$ is the relative distance between the interacting beams, X and Y being the separations between the solitons in the x and y directions, and ψ is the relative phase between the solitons. The elements of the effective-mass matrix can be calculated explicitly: $M_R \equiv \pi \int\int_{-\infty}^{\infty} [|v(r)|^2 + 2\sigma|w(r)|^2] r\, \mathrm{d}r$, and $M_\psi \equiv -2\partial M_R/\partial\beta$, where $r \equiv \sqrt{x^2 + y^2}$ and the integrand is calculated on the family of radially symmetric one-soliton solutions known numerically [33]. The effective potential energy is defined as

$$U(R, \psi) = \frac{M_R s^2 C^2}{4R^2} + U_1(R) \cos\psi + U_2(R) \cos(2\psi) ,$$

where the functions U_1 and U_2 are expressed in terms of the soliton overlap integrals. The impact parameter s defines the distance between the trajectories of noninteracting solitons, and $C \equiv \dot{R}_0$ is the relative velocity between the solitons prior to their interaction.

When the distance between interacting solitons is large, the soliton interaction is determined by the tail asymptotics, given by $v(r) \sim \exp(-\sqrt{\beta}r)/\sqrt{r}$ and $w(r) \sim \exp[-\sqrt{\sigma(2\beta + \Delta)}r]/\sqrt{r}$. Therefore, asymptotic expressions for the energies U_1 and U_2 can also be estimated: $U_1(R) = -A\exp(-\sqrt{\beta}R)/\sqrt{R}$ and $U_2(R) = -B\exp[-\sqrt{\sigma(\Delta + 2\beta)}R]/\sqrt{R}$, where A and B are positive constants which are determined numerically.

As was shown by Steblina et al. [72], the effective mechanical model defined by the Lagrangian (18) can be used for predicting the outcome of soliton collisions. Importantly, the soliton interaction forces depend strongly on the relative phase ψ. In the case of out-of-phase collisions ($\psi = \pi$), the "centrifugal force", defined by the first term of the effective potential U, and the direct interaction force, defined by the second term $U_1(R)\cos\psi$, are both *repulsive*. Therefore, the effective particle cannot fall onto the force centre, i.e. the solitons cannot fuse. The interaction scenario is very different for in-phase soliton collisions ($\psi = 0$). For low relative velocities (or sufficiently large s) the solitons cannot overcome the centrifugal potential barrier, and they spiral about each other. At higher velocities (or smaller s), the solitons fuse.

2.4 Important Generalizations and Related Problems

2.4.1 Walking Solitons

Both (1 + 1)- and (2 + 1)-dimensional spatial solitons, since they are excited in anisotropic media, should experience *spatial walk-off* due to different propagation directions of the energy and phase fronts, the effect being described by the first-order derivative terms in (1). In particular, beam walk-off is always present in experiments in which the birefringence-tuning phase-matching technique is used.

The first study of the effect of spatial walk-off on quadratic solitons was presented by Karamzin et al. [7] for the case of three-wave interaction in a waveguide geometry. Later, Torner et al. [73] provided a detailed analysis of (1 + 1)-dimensional two-wave spatial solitons in the presence of Poynting vector beam walk-off (or different group velocities for the two waves), and demonstrated that these solitons may exist in the form of so-called *walking solitons* , which "walk" at a certain angle to the propagation direction. Such solitary waves have a nontrivial phase-front tilt and a more complicated structure, and they are stable below a certain critical value of the walk-off parameter [74]. Some generalizations to the case of type II SHG and three-wave parametric solitons in the presence of walk-off were later reported by Mihalache et al. [35].

The fundamental (2 + 1)-dimensional solitary waves and light bullets can also be generalized by taking into account the spatial walk-off effect between the harmonics. Such generalizations were considered by Mihalache et al. [61,63], who studied mutual trapping between the fundamental and second-harmonic beams in a bulk $\chi^{(2)}$ medium in the presence of walk-off, and found two-parameter families of solitary waves with nontrivial phase dependence, the multidimensional walking solitons. Such solitary waves are not circular but *elliptical*, with an axis of the ellipse along the propagation direction [63]; the asymmetry appears because of the walk-off. Similarly to the $(1+1)$-dimensional walking solitons [73], these localized waves propagate at a certain angle to the direction selected by the walk-off effect, developing a stationary phase chirp (see [61,63]). In contrast to the $(1+1)$-dimensional walking solitons, the family of stationary fundamental (one-hump) $(2+1)$-dimensional walking solitons was found by solving a system of partial differential equations, which is a more complicated task. Some numerical results obtained by Mihalache et al. [61], e.g. a nonmonotonic dependence of the power of the solitary waves on the propagation constant, indicate that the walking solitons may become unstable, and that the instability domain depends on two parameters, the phase mismatch and the walk-off. However, a rigorous stability analysis of this class of solitary waves still remains to be done.

An interesting concept of a so-called *guided-center walking soliton* was introduced by Torner [75], who considered self-trapping of light in periodic quadratic nonlinear media with a large walk-off. Torner [75] analyzed soliton formation in structures with quasi-phase-matching (QPM) where compensation for the large material walk-off is achieved by periodically reversing the sign of the walk-off. In each domain of such a structure, the walk-off can be far too large to permit soliton formation but, as was shown by Torner [75], an average guiding-center soliton might exist in the form of a spatially localized two-frequency beam with periodically varying parameters and a slow drift.

2.4.2 Three-Wave Quadratic Solitons

As was explained in Sect. 2.1, the most general situation of three-wave mixing in an optical medium with a quadratic nonlinear response should be described by the three coupled equations (1), and it includes also the case of type II SHG. Correspondingly, the three-wave solitary waves of the model (1) provide an important generalization of the two-wave parametric solitons analyzed above. Starting from the classic work by Karamzin and Sukhorukov [6], the model of three-wave solitary waves was studied in detail, in the case of zero mismatch [8], to find some particular solutions for three-color solitons [76] and the families of three-component stationary localized modes [34,77,78], to study their generation and switching [55,56,79,80], to study their stability [34,78,81], and to investigate other aspects.

In many cases, the three-wave solitons due to type II SHG behave similarly to their two-wave relatives. However, some of the soliton properties become more complicated, owing to the existence of one additional parameter – the power imbalance between two polarization components of the fundamental beam. The power imbalance allows many new effects, such as polarization switching control [56]. Owing to the existence of two independent parameters, the study of the stability of three-wave solitary waves goes beyond the standard Vakhitov–Kolokolov stability theory and, for example, the marginal stability line is defined by the condition

$$\frac{\partial(P_v, P_u)}{\partial(\beta_v, \beta_u)} = \frac{\partial P_v}{\partial \beta_v}\frac{\partial P_u}{\partial \beta_u} - \frac{\partial P_v}{\partial \beta_u}\frac{\partial P_u}{\partial \beta_v} = 0\,,$$

where β_v and β_u are two independent propegation constants, and P_v and P_u are two energies (Manley–Rowe invariants), which define the total power, $P_{\mathrm{tot}} = P_u + P_v$, and the power imbalance, $P_{\mathrm{imb}} = P_v - P_u$.

2.4.3 Competing Nonlinearities

The nonlinear response of any $\chi^{(2)}$ material always includes a contribution from the next-order *cubic* (or $\chi^{(3)}$) nonlinearity, which, under certain conditions, might become important and strongly compete with the $\chi^{(2)}$ nonlinearity. The influence of cubic nonlinearity on SHG and nonlinear switching has been studied for more than 15 years (see, e.g., [82–84]) and is known to lead to such effects as distortion of the SH spectrum and effective saturation of the SHG conversion efficiency.

There exist several physical mechanisms for such competing nonlinearities to occur. First of all, any $\chi^{(2)}$ material has an *inherent cubic nonlinearity* that may become important at high powers or when the fundamental wave and its SH are not closely phase-matched. Another mechanism is related to the average beam propagation in a $\chi^{(2)}$ crystal with a grating, i.e. in QPM media. Indeed, as has been demonstrated by Clausen et al. [85], if the $\chi^{(2)}$ nonlinearity varies periodically along the direction of propagation, the *effective averaged dynamical equations* describing the beam propagation in such a QPM medium also include an effect of *induced cubic nonlinearity*, as a result of an incoherent coupling between the wave at the main spatial (QPM) frequency with higher-order modes.

In general, effective cubic nonlinear terms may appear in the conventional model of $\chi^{(2)}$ two-wave mixing, owing to an incoherent coupling of the two main interacting modes with other modes or higher-order cascading effects. This may happen, for example, in the simple case of SHG in a waveguide which is single-moded at the fundamental frequency ω but supports two modes at 2ω. A similar situation may occur in the problem of *multistep cascading*, when the influence of third and fourth second-order processes, involving both sum and difference frequency mixing, are taken into account [86,87]. If only one of the processes is nearly phase-matched, the other one can be treated in the cascading-limit approximation, leading again to an effective cubic nonlinearity. Thus, a mismatched parametric coupling between modes is a common physical mechanism that induces an effective cubic nonlinearity in the equations for parametrically coupled harmonics. As a result, competition between quadratic and cubic nonlinearities is a very general physical phenomenon, and it occurs in many types of SHG processes.

The topic of solitary-wave dynamics in media with competing quadratic and cubic nonlinearities is fairly new, although about a dozen papers related to this subject have already been published [19,58–60,88–95].

2.4.4 Doughnut Modes and Vortex Solitons

One more concept that has recently attracted the attention of several research groups is the existence and stability of higher-order solitary waves of circular symmetry in a medium with a quadratic nonlinearity. In a cubic medium, this kind of higher-order self-trapping has been discussed for modes without angular dependence [96], and also for the spiral propagation of optical beams (see, e.g., [97]). Such kind of beams are naturally expected to exist in quadratic media, as *doughnut modes* or *vortex-like structures* consisting of two mutually coupled harmonic waves.

To describe modes with a nontrivial angular dependence, one should look for solutions of radial symmetry in the $(2+1)$-dimensional case in the form

$$v(x,y;z) = V(r)\exp[\mathrm{i}(\kappa z + l\phi)]\ , \quad w(x,y;z) = W(r)\exp[2\mathrm{i}(\kappa z + l\phi)]\ ,$$

where $r = \sqrt{x^2+y^2}$ is the radial coordinate, ϕ is the polar angle, and the integer number l characterizes the order of the solution. The majority of the results presented below are devoted to the case of the fundamental mode, $l = 0$. However, localized solutions with $l \neq 0$ may exist as well, provided both V and W vanish as $r \to 0$. This kind of radially symmetric ring-like solitary wave was investigated by Firth and Skryabin [98], who called them "*doughnut solitons*". Such higher-order solitons have finite orbital angular momentum, but they have been found to be unstable and to fragment into simple solitons flying out tangentially from the initial ring, behaving just like free Newtonian particles. Importantly, a similar scenario of instability and filamentation of ring-like solitary waves was reported for a non-Kerr medium with nonlinearity saturation [98], the problem is very closely related to the recent experiments done in Rb vapors by Tikhonenko et al. [99,100] who created a ring-like intensity distribution by applying a special kind of phase mask to produce a helical phase-ramp dislocation.

Similar results have been reported by Torner and Petrov [101,102] (see also [103–105]), who called these higher-order ring-type solitary waves *vortex solitons* owing to the presence of a phase dislocation carried by this localized mode, in an analogy with the terminology adopted earlier by Tikhonenko et al. [99,100]. In numerical simulations, Torner and Petrov [101] used just a single input beam with a phase dislocation, $v(z=0) = Ar^{|l|}\mathrm{e}^{\mathrm{i}l\phi}\mathrm{e}^{-r^2/w^2}$, where r is the cylindrical coordinate of the beam, and they observed an azimuthal beam breaking into spatial solitons. Decay of nonlinear higher-order modes, similar to that reported by Firth and Skryabin [98], has been observed in numerical simulations performed by Torres et al. [103], who also revealed that all these localized states are *azimuthally unstable*, decaying into a set of stable $(2+1)$-dimensional fundamental solitons. Torres et al. [103] reported that, depending on the type of the initial perturbation, the decay of ring-type solitons with charges $l = 2$ or 3 may occur into $N = 4$ or 5 fundamental solitons.

Generalization of the concept of vortex-like modes and an analysis of their azimuthal instability in the case of type II interaction (i.e. nondegenerate three-wave mixing) was carried out by Torres et al. [104] and Torner et al. [105], who found higher-order stationary solutions for the three-component model with an orbital angular momentum. Similarly to the case of two-wave mixing, all such solitons are unstable, and they decay into a set of three-wave fundamental solitons of circular symmetry. This research led to some suggestions for optical devices operating with the orbital angular momentum of the light beams that would process information by mixing topological wavefront charges and producing a certain number of spatial solitons (see [105] for details).

Vortex solitons on a nonvanishing background, a $(2+1)$-dimensional analogue of dark solitons, were first analyzed by Alexander et al. [94], who predicted the existence of two-wave parametrically coupled vortex modes where the fundamental-harmonic field has a phase twist of 2π corresponding to a phase twist of 4π in the second-harmonic field. In the other work, Alexander et al. [95] predicted different types of such parametric vortices, e.g. a "halo-vortex" consisting of a two-component vortex core surrounded by a bright ring made of its harmonic field, and a "ring-vortex soliton", which is a harmonic field that guides a ring-like localized mode of the fundamental-frequency field. The important feature of all such parametric vortices is a requirement for a nonvanishing contribution from a higher-order, *defocusing cubic* nonlinearity that stabilizes the otherwise unstable two-mode background beam (see [94,95] for details).

2.4.5 Quadratic Light Bullets

In spite of the fact that this book is mainly devoted to spatial solitons, it is worth mentioning some very recent developments in the theory of the so-called *light bullets*, self-trapped states localized in both the temporal and the spatial domains. The light bullets were introduced in the context of nonlinear pulse propagation in a diffractive and dispersive medium, in the framework of the generalized NLS equation [106], where such states can be simply described by radially symmetric solutions of a higher-dimensional NLS equation. Later on, light bullets were extensively analyzed by Edmundson and Enns, in the framework of the $(3+1)$-dimensional generalized NLS equation with transiting [107] and saturable [108] nonlinearities.

Generalization of the concept of light bullets to the case of spatiotemporal two-frequency parametric solitons in $\chi^{(2)}$ media is far from being straightforward. The main reason for this is the dependence of the group velocity dispersion (GVD) on the carrier-wave frequency. As a result, similarly to (2+1)-dimensional walking solitons, even the fundamental stationary $\chi^{(2)}$ light bullets without walk-off are described by a system of partial, rather than ordinary, differential equations. Such multidimensional solitons are, in general, *radially asymmetric*, and the asymmetry depends on the ratio $\delta_\omega \equiv d_\omega/d_{2\omega}$ of the second-order GVD coefficients.

The first indication of the possibility of stable spatiotemporal $\chi^{(2)}$ solitons was obtained by Kanashov and Rubenchik [8] for the very special case of $\delta_\omega = 1$ and the exact phase matching, $\Delta k = 0$. Later, further particular results were presented by Hayata and Koshiba [20] and Bergé et. al. [51]. Only recently have Malomed et al. [64] (see also [65,66]) tackled a more general formulation of this problem, by applying a version of the variational technique, developed earlier by Steblina et al. [36], in order to find approximate profiles of radially symmetric (2+1)-dimensional $\chi^{(2)}$ optical bullets. For the special case of $\delta_\omega = 1$, these two groups have also found radially symmetric numerical solutions, which have then been compared with profiles obtained by variational analysis, obtaining a reasonably good agreement. Malomed et al. [64] and Skryabin and Firth [65] also presented some estimations of the stability of the solutions on the basis of the modified Vakhitov–Kolokolov criterion and direct numerical simulations. It was shown that stable $\chi^{(2)}$ light bullets are only possible if both the fundamental and the second-harmonic waves are in the anomalous-dispersion regime. However, when only the fundamental wave has anomalous dispersion, but the absolute value of the parameter δ is small enough, then the existence of quasi-stable soliton-like propagation has also been demonstrated [64]. More detailed theoretical analysis of two-frequency $\chi^{(2)}$ light bullets, both in $(2+1)$- and $(3+1)$-dimensional models, including a more realistic asymmetric case, was reported by Mihalache et al. [62,109].

2.4.6 Discrete $\chi^{(2)}$ Systems and Nonlinear Photonic Modes

For many years, there have been attempts to prepare molecular-crystalline multilayer structures analogous to inorganic superlattices and quantum well structures. Recently, such structures have been prepared by molecular-beam-deposition methods and are undergoing intense experimental study by several leading experimental groups around the world (see, e.g., [110–113]). Recent progress in molecular-beam-deposition techniques has led to the preparation of organic multilayered structures of high optical quality [112,113]. These novel engineered materials have opened up a new field of research, which is very promising from the technological as well as the fundamental point of view.

In particular, recent theoretical analysis has shown that the various interactions taking place at the interfaces are of great importance and can be responsible for the appearance of new linear and nonlinear optical effects. More importantly for the subject of the present review, new localized states, *Fermi resonance interface modes* [114–117] and Fermi resonance interface solitons [39,118,119], have been predicted. The existence of these states may give rise to a resonant enhancement of the quadratic and cubic nonlinear optical susceptibilities [115,116]. As a matter of fact, the phenomenon of the Fermi resonance in molecular systems is very similar to SHG in optics, and the Fermi resonance solitary waves are basically the same as the two-wave quadratic solitons described above.

It particular, this physical problem can be reduced to the same equations as those which occur in the theory of two-wave quadratic optical solitons. As a result, many of the results obtained for these molecular multilayer systems are essentially the same as those described above. In particular, Agranovich and Kamchatnov [118] rediscovered the exact two-wave solution known in nonlinear optics since 1974 [5,6], and Agranovich et al. [39,118] developed a variational formulation for the localized modes, an approach similar to that developed in nonlinear optics [36].

We would like to emphasize, however, that one of the specific and most interesting features of these kinds of localized interface Fermi resonance modes is the discreteness of the original nonlinear lattice model, an issue that has been only very recently addressed [120–125].

A novel type of such a problem, but similar to the basic concept of Fermi resonance solitons in organic superlattices discussed above, is that of a nonlinear lattice with quadratic nonlinearities, each element of which consists of the fundamental mode resonantly interacting with its second harmonic (see, e.g., [122–126]). This kind of *nonlinear* $\chi^{(2)}$ *lattice* has a physical motivation different from the theory of Fermi resonance modes, and describes an array of weakly interacting identical optical waveguides where nonlinearity is produced by a phase-matched SHG effect. The basic model of such a $\chi^{(2)}$ lattice is described by the simplest discretization of the fundamental equations for quadratic solitons,

$$\mathrm{i}\frac{\mathrm{d}v_n}{\mathrm{d}z} + \eta_v(v_{n+1} + v_{n-1}) + v_n^* w_n = 0 \, ,$$

$$\mathrm{i}\frac{\mathrm{d}w_n}{\mathrm{d}z} + \eta_w(w_{n+1} + w_{n-1}) + \frac{1}{2}v_n^2 = 0 \, , \qquad (19)$$

where $n = [1, N]$, N being the total number of sites, and η_w and η_v determine the strength of the coupling between the fields at neighboring sites. The model (19) can be considered as a generalization of the discrete NLS equation. In application to optics, the discrete NLS equation has been employed in the theory of arrays of nonlinear waveguides (see, e.g., [127] and references therein), and its simple version of two coupled waveguides is well known as a model of a *nonlinear directional coupler* [128].

Unlike the theory of nonlinear couplers with cubic nonlinearity [128], (19) allows exact analytical solutions only for $N = 1$ (this is just the standard SHG model), and becomes nonintegrable for $N = 2$, the so-called $\chi^{(2)}$ dimer, when the model may display chaotic dynamics [122]. However, an analysis of stationary modes of the $\chi^{(2)}$ dimer model can be easily performed, including a stability analysis. Because for large values of the effective mismatch parameter the model (19) for $N = 2$ reduces to an integrable model of a nonlinear coupler [128], Bang et al. [122] used this nearly integrable limit to analyze how the stationary modes of the discrete NLS model, when used as initial conditions in the $\chi^{(2)}$ system, develop a gradual transition to chaotic dynamics.

The case of larger N is more complicated and less studied. Dubovsky and Orlov [120,121], motivated by the problem of Fermi resonance modes, found analytically some stationary solutions of the model (19) for $N = 2$, 4, and 6. The set of real solutions they found is definitely incomplete; for example, for $N = 2$ they find only two of the seven different classes analyzed by Bang et al. [122].

However, in the case of many sites ($N \gg 1$), the model (19) has been shown to support strongly localized self-trapped states of different (odd and even) configurations, *discrete quadratic solitons*, which involve only a few neighboring sites [123,125]. In the small-amplitude limit, such modes are very close to the two-wave quadratic solitons of the continuum model, and they can move through the lattice, fusing, annihilating, or passing through each other when colliding.

Recently, Sukhorukov et al. [129] analyzed nonlinear localized modes in a photonic structure with quadratic nonlinearities where the linear refractive index varies periodically, and quadratic nonlinearity is introduced through thin layers embedded periodically in the linear structure. By considering the standing waves in such a structure, Sukhorukov et al. [129] derived a system of effective equations for the wave amplitudes in the nonlinear layers that reduce to the stationary form of (19). Different kinds of nonlinear modes in such structures were analyzed, and the case of a single nonlinear layer in a linear periodic structure was studied in detail. Some of the properties of the two-frequency nonlinear localized modes were found to resemble closely the properties of quadratic solitons excited in a homogeneous medium (see also [130] for the case of a homogeneous medium with a single $\chi^{(2)}$ layer).

3 Summary of Experimental Results

The experimental work on quadratic spatial solitons to date can be separated thematically into *four areas*. The first area corresponds to the observation of freely propagating (1 + 1)- and (2 + 1)-dimensional spatial quadratic solitons and the exploration of their properties such as suppression of walk-off, dragging, etc. The second area is that of self-trapping in time, including temporal compression that leads to temporal and spatiotemporal solitons. The third distinct area corresponds to the recent development of quadratic solitons in cavities. Finally, universal features of solitons, such as collisions and instabilities, have been investigated, as the fourth area.

3.1 Freely Propagating Spatial Solitons

In contrast to the case of optical fibers, where propagation distances can be tens or hundreds of dispersion lengths, spatial-soliton experiments are typically limited to propagation over distances of less than five diffraction lengths. This is sufficient to establish the formation of solitons, but is too short to follow collisions, etc. to their completion. As discussed in the chapter by Luther-Davies and Stegeman, it is necessary to be near phase-matching to generate spatial solitons at reasonable powers. As a result, the materials used for quadratic-soliton studies had for the been previously used as frequency doublers.

3.1.1 (1 + 1)-Dimensional Spatial Solitons

In a classic experimental demonstration, Schiek and coworkers [131] have been able to achieve, for the first and at this point only time, $(1+1)$-dimensional spatial-soliton formation in a 5 cm long $LiNbO_3$ waveguide (see also [132]). In their case, phase-matching was achieved by temperature control of the mode matching in appropriately engineered waveguides with in-diffused titanium. A flattened temperature profile over most of the crystal length was achieved, allowing second-harmonic conversion efficiencies in excess of 50%. At input powers in excess of 1 kW with a 1320 nm Nd:YAG mode-locked, Q-switched laser, soliton self-trapping of the fundamental beam, induced by the presence of a generated *weak second-harmonic beam* in the positive-phase-mismatch region was demonstrated. Under this condition, the quadratic interaction acts as an effective *third-order* nonlinear process, commonly referred to as a *cascading process*. The induced cascaded self-focusing nonlinearity counteracts diffraction in the plane where the waveguiding confinement is not present. Schiek et al. [131] were also able to demonstrate trapping, even close to the phase-matching temperature with large pump depletion. In Fig. 5, we present the results of the self-trapping achieved by increasing the pump power at a fixed temperature of the oven.

In the SHG process, one (type I) or two (type II) input fundamental fields mix via a parametric $\chi^{(2)}$ process to generate a second-harmonic wave. Because efficient conversion requires wavevector conservation between the interacting beams, dispersion in the refractive index frequently results in different group velocity directions for the interacting beams, leading to beam walk-off and reduced SHG efficiency. The sample geometry for the experiments described above was type I, with propagation along the x axis, the y axis normal to the waveguide surfaces, and the z axis in the plane. For the

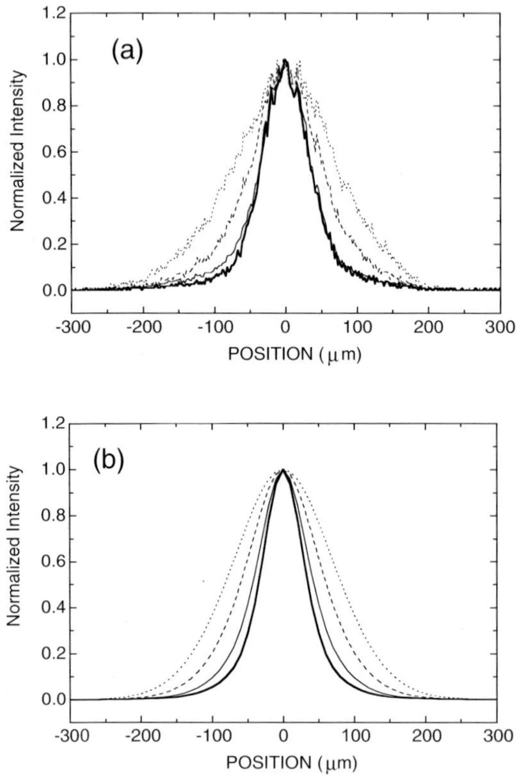

Fig. 5. Normalized output beam profiles at a temperature of 335.05 °C for beams propagating in a $LiNbO_3$ waveguide with total peak powers of 320 W, 760 W, 1.5 kW, and 1.9 kW: (**a**) experiment, (**b**) numerical simulations (from [132])

TM (fundamental) polarization, the group and phase velocities are always *collinear*, but this is not true for the TE (second-harmonic) wave for propagation at some angle to the x axis. Off the x axis, *walking solitons* are formed in which the fundamental and second-harmonic beams are locked together, having the same spatial group velocity direction, so in fact the walk-off is suppressed by nonlinearity. This unique feature of quadratic solitons has also been demonstrated experimentally [133].

3.1.2 (2 + 1)-Dimensional Spatial Solitons

The formation of stable spatial solitons in two transverse dimensions, i.e. in a bulk medium, requires some form of saturating nonlinearity. In 1995, it was shown experimentally that strong nonlinear coupling can lead to the formation of quadratic solitons in bulk media, balancing simultaneously the effects of diffraction and walk-off and hence allowing the production of clean, diffraction-free beams with enhanced peak output intensities [134–136]. In those experiments, Gaussian-shaped pulses at 1064 nm with an FWHM duration of approximately 35 ps and a repetition rate of 10 Hz were focused onto a KTP crystal with a diameter of 20 (or 40) μm, corresponding to a Rayleigh range (diffraction length) of approximately 0.5 (or 2) mm. That is, for the two beam diameters used, the KTP crystal was 20 or 5 diffraction lengths long. Figure 6 (left) clearly shows that a 20 μm input beam diffracts within the 1 cm length of the KTP crystal when below a certain threshold power. However, above a threshold power (Fig. 6 (right)), a clean, cylindrically symmetric beam is generated for both the fundamental and the second-harmonic beams. These experiments were performed with a KTP crystal cut for type II phase-matching along the XY plane. In such a geometry, one of the input fundamental fields is an ordinary field polarized along the Z axis, while the second input fundamental field is an extraordinary field with a polarization in the XY plane at the theoretically predicted phase-matching angle (26°). Energy depletions of the fundamental field exceeding 50% at phase-matching were observed.

The experimental situation just described is the most complex possible for SHG, and the governing equations for this type of three-wave interaction are given by (1), where $E_{1,2}$ are the envelopes of the two orthogonally polarized fundamental fields and E_3 is the envelope of the second-harmonic field; ρ_ω and $\rho_{2\omega}$ are, respectively, the walk-off angles of the extraordinary fundamental and the second harmonic with an extraordinary polarization, equal to 0.19° and 0.28°; the nonlinear coupling coefficient was calculated to be 6 cm^{-1} for an input intensity of 1 GW/cm^2; and $\Delta k\, L$ is the phase mismatch parameter. In the strong-coupling regime, the three-wave mixing process is therefore different from the well-known cubic nonlinearity driving the cubic NLS equation. As shown in Fig. 6, the numerical predictions are in excellent agreement with the experimental measurements of the beam waist at the output of the 1 cm long KTP crystal, both qualitatively and quantitatively. Linear and nonlinear absorption were neglected in the model at both wavelengths involved.

One of the most interesting aspects of these results was the evolution dynamics involving the beam walk-off, which was studied for the first time in these experiments. When the nonlinearity is negligible, the expected linear behavior is observed: both extraordinary waves walk away in space from the ordinary fundamental field and diffract. A close inspection of the governing coupled nonlinear equations (1) shows that with no second-harmonic input, the second harmonic grows in the spatial region where both fundamentals overlap. Both on the self-focusing side of phase-matching and at phase-matching, the nonlinear interaction progressively changes the output beam profile as the peak intensity changes from 0.1 to 10 GW/cm^2 (see Figs. 7 and 8). In this regime, self-focusing dominates over diffraction. Down-conversion from the second harmonic to both fundamental fields then occurs, and the two fundamental beams start to be trapped by the second harmonic, preventing, at higher intensities, the mutual walk-off and diffraction. Beyond 5 GW/cm^2, self-focusing along the extraordinary axis takes place eventually

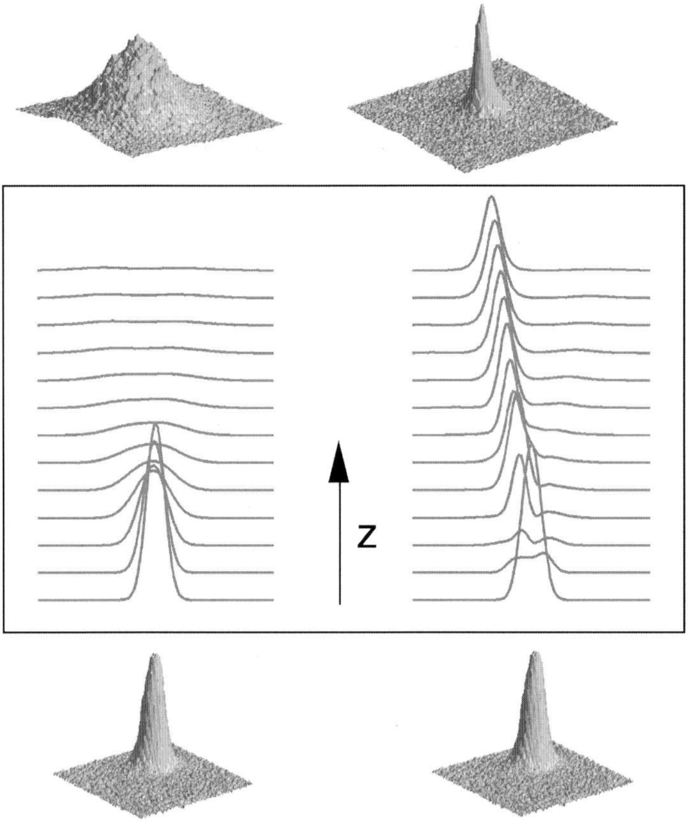

Fig. 6. Input and output profiles for low (100 MW/cm^2, *left*) and moderate (10 GW/cm^2, *right*) powers. The numerical simulations of the propagation of the beam in a 1 cm long KTP crystal are shown in the *middle*, for both cases. The effect of the spatial walk-off is clearly seen (from [134])

trapping the three beams into a cylindrically symmetric solitary wave above 10 GW/cm^2, defeating both diffraction and walk-off. This occurs at a location intermediate between the two original output beam centers formed by walk-off of the extraordinarily polarized beam. At high intensities, the instability causing the soliton splitting remains unexplained, and it may be due to the presence of a strong cubic nonlinearity (see Fig. 8).

Figure 9 shows that at phase-matching, the output fundamental beam remains locked to a stable waist of 12.5 μm, almost half of the input waist, clearly showing that diffraction is defeated (see also Fig. 7). A similar evolution into solitary waves is observed on *the self-focusing side of phase-matching* ($\Delta k\, L > 0$, with the convention $\Delta k = k_1(\omega) + k_2(\omega) - k_3(2\omega)$), with a lowering of the threshold for solitary-wave formation. On the other hand, a very different behavior is observed on *the negative side of phase-matching*, where self-defocusing occurs first, at powers below a phase-mismatch-dependent threshold. For example, at a phase mismatch of -5π, a threshold intensity of 30 GW/cm^2 is needed for solitons of 15 to 20 μm diameter to be formed (see Fig. 10). This negative phase mismatch behavior is attributed to wave-guiding in the parametric gain-guiding regime. Indeed, at these large input powers, during propagation the second-harmonic field strongly depletes the fundamental components; when it down-converts, parametric gain dominates and guides the two fundamental fields, which then trap the harmonic field.

The vectorial nature of the type II SHG interaction in KTP can be used for the purpose of all-optical switching [54–56]. Dragging of the mutually trapped waves at input intensities beyond the soliton threshold was achieved by control of the imbalance, i.e. intensity difference, between the input ordinary and extraordinary fundamental polarizations. For example, if the ordinary beam is more intense than the extraordinary, the soliton propagation direction is "dragged" towards the propagation direction of the isolated ordinary beam. Figure 11 shows numerical predictions of the transmission through an

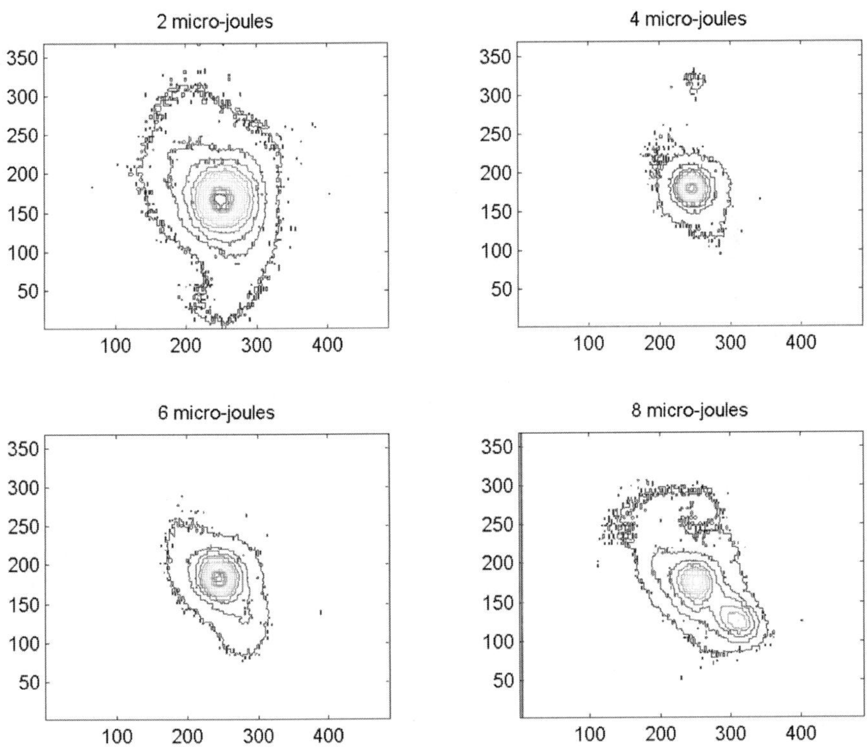

Fig. 7. Gray-scaled output profiles recorded for input fluences ranging from 2 to 8 μJ at phase-matching. The pulse duration was 35 ps. Note the narrowing of the beam profile associated with mutual trapping of the fields. Both the ordinary and the extraordinary input fundamental polarizations had exactly the same magnitude

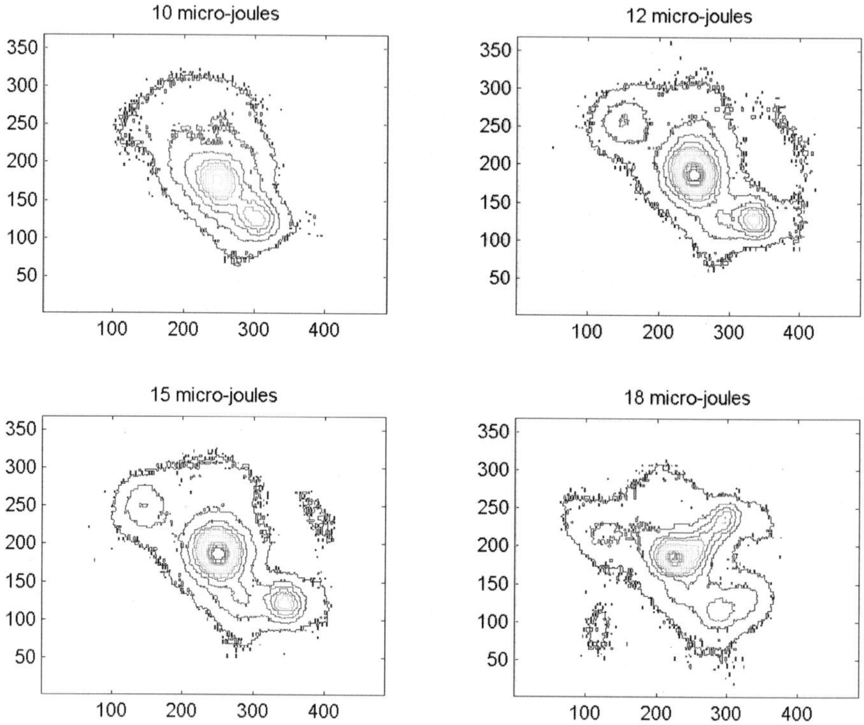

Fig. 8. As in Fig. 7 but for input fluences ranging from 10 to 18 μJ. Note that at these high intensities the soliton radiates in both the ordinary and the extraordinary channels, and splitting in the direction transverse to the walk-off direction is also observed

aperture placed at the back of a 1 cm long KTP crystal. A sharp steering is expected as a function of the imbalance between the two input fundamental polarizations. This behavior was experimentally observed by introducing a half-wave plate into the path of the input fundamental beam. When the input was biased towards the ordinary polarization, the solitary wave propagated along a direction perpendicular to the input face of the crystal, while a polarization imbalance towards the extraordinary side steered the solitonic beam to an angle close to the walk-off angle. As expected, for a critical position of the half-wave plate and enough input intensity, two solitons were observed simultaneously at the output of the crystal.

It is important to note that previous experiments showed that in the case of a type II interaction, the two fundamental beams do not need to be mutually coherent to achieve the described trapping and steering effects. In other words, two separate laser sources can be used at different wavelengths. However, if the second harmonic is used to control the evolution of the solitary wave, accurate phase control of the coherently generated input fundamen-

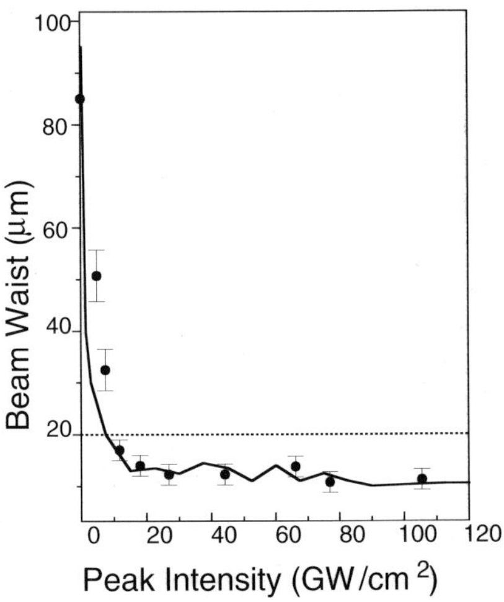

Fig. 9. Output beam width fitted to a Gaussian profile near the point of phase-matching. The input beam waist was 20 µm. Above 10 GW/cm^2 a narrow beam was observed at the output-associated with spatial-soliton formation (from [134])

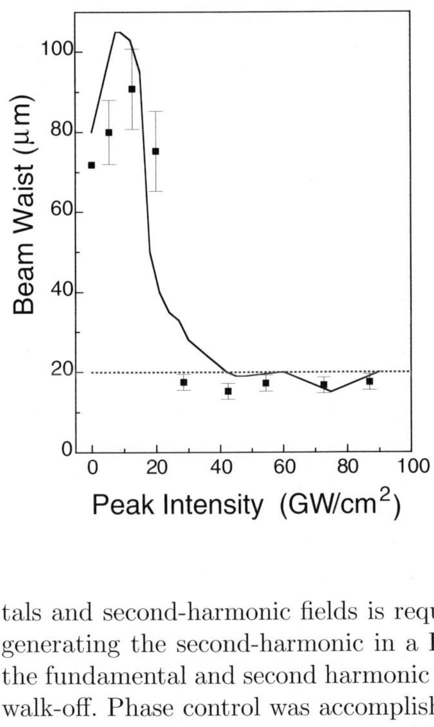

Fig. 10. Output beam waist in the negative phase-matching region associated with self-defocusing at low input powers. Above 30 GW/cm^2 a spatial soliton is recovered. The angular detuning from phase-matching corresponds to $\Delta k\, L = -5\pi$ (from [134])

tals and second-harmonic fields is required [135]. This was achieved by first generating the second-harmonic in a KDP crystal which laterally displaced the fundamental and second harmonic owing to birefringence-induced spatial walk-off. Phase control was accomplished with nitrogen-filled cell. Changing the pressure in the cell modifies the difference in the refractive index between the fundamental and second harmonic with a very good accuracy. When the second harmonic was co-launched and the appropriate relative phase was chosen, soliton formation was achieved. Figure 12 shows the steering coherently induced by a pressure difference, or, in other words, the relative phase difference between the input fundamental and second-harmonic fields. Note

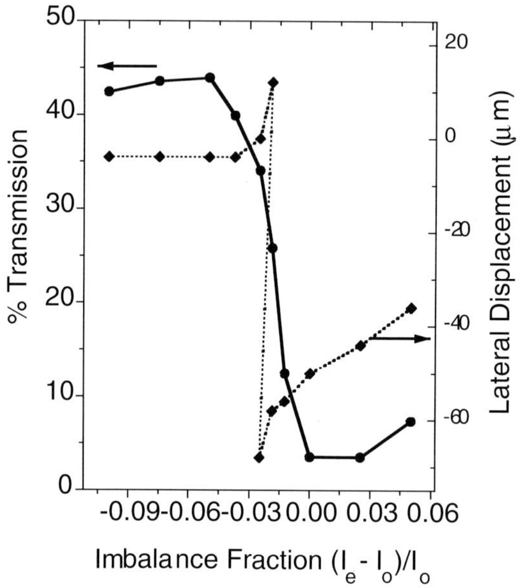

Fig. 11. Calculated transmission and lateral displacement as a function of the input imbalance between the two fundamental polarizations. Notice the critical behavior near equal magnitude, or zero imbalance (from [136])

that in this case, the steering direction is different from that of the natural walk-off. In other words, a soliton channel is induced by the second-harmonic seeding process.

All of the quadratic-soliton experiments described above required input intensities of the order of 10 GW/cm^2 because the nonlinearities are a few pm/V. The development of the QPM technique for parametric processes in $LiNbO_3$ has resulted in bulk samples with effective nonlinearities of the order of 17 pm/V. Bourliaguet et al. [137] were able to generate, for the first time, type I quadratic solitons in such samples at threshold intensities of the order of 1 GW/cm^2, working at 1064 nm. This geometry has the added advantage that there exists *no walk-off* between the fundamental and harmonic beams.

In recent work, Di Trapani and coworkers [138] extended the generation of quadratic solitons from second-harmonic generation to more general parametric processes. They reported the mutual trapping of multidimensional spatial solitary waves in a parametric amplifier based on a 15 mm long lithium triborate (LBO) crystal, where the signal and idler waves grew from quantum noise [138]. Using 1.5 ps pulses at 527 nm, they showed that beyond a threshold of 1 mJ and an input pump beam diameter of 57 μm, the output beam reduces to a spot size of 12 μm, while the conversion efficiency saturates. On the basis of modeling of the propagation in an LBO crystal, soliton formation and mutual trapping were inferred.

At the same time, Canva et al. [139], stimulated by the theoretical work of Leo and Assanto [56], reported similar results in a case of degenerate optical parametric amplification, where a weak seeding beam was used at half the

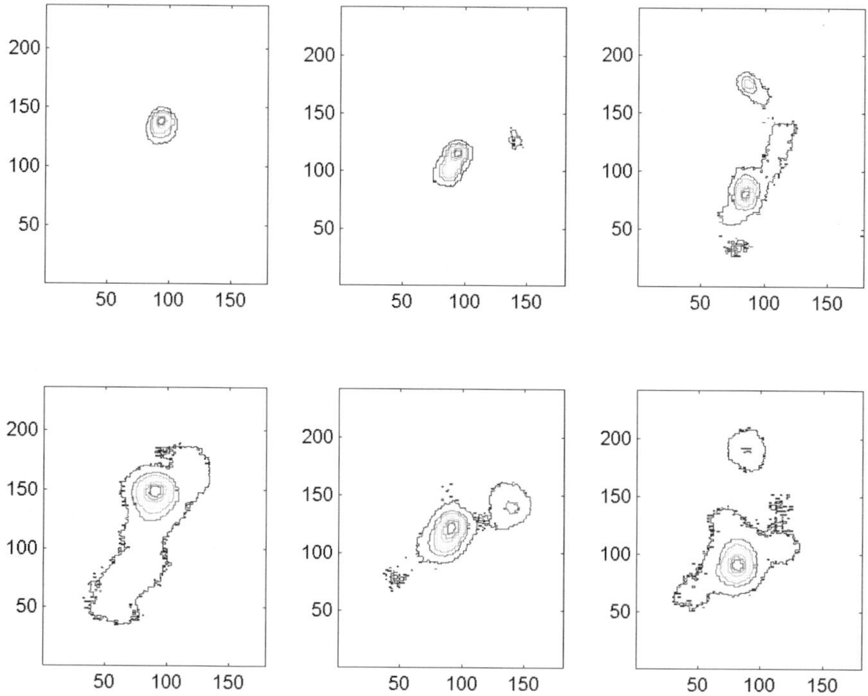

Fig. 12. Observation of mutual trapping and coherent steering of the fundamental and second-harmonic fields as a function of the relative phase of the input fundamental and second-harmonic fields. Note that the steering is achieved in a direction different from the natural spatial walk-off direction (from [135])

pump frequency in a KTP crystal. In this case, the effects of detuning of the phase mismatching of the pump and seeding beams were investigated in detail. One of the interesting aspects of this work was that the soliton's output fundamental field was constant over five orders of magnitude of the input seed, a very effective saturating amplifier. This result is a consequence of the high gain of the parametric amplification process combined with the eigenmode properties of the solitons, which in this case were determined by the magnitude of the pump beam.

The advantages (and disadvantages) of quadratic-soliton generation for efficient SHG were also considered, primarily numerically. Although standard weakly focused SHG geometries are much more efficient in terms of conversion efficiency, the experimental bandwidth over which SHG is obtained with quadratic-soliton generation can be an order of magnitude, or more, larger than for the weakly focused case.

As was mentioned above, the theory predicts the existence of stable (2+1)-dimensional two-wave parametric vortices on a nonvanishing two-frequency background, provided a contribution from a defocusing cubic nonlinearity is taken into account. However, in realistic $\chi^{(2)}$ materials this higher-order cubic nonlinearity is usually self-focusing, and therefore it cannot stop the development of parametric modulational instability of the background beam, making it hard to observe the vortices in an experiment. Nevertheless, the first successful experimental demonstration of a stable, localized two-wave optical vortex was reported by Di Trapani et al. [140], who employed a combined effect of transverse walk-off and finite beam size in order to eliminate the dangerous development of parametric modulational instability and to observe, finally, stable structures in parametric wave mixing that carried phase singularities in both of the harmonics.

3.2 Temporal Solitons and Light Bullets

The demonstration of temporal solitons in quadratic optical media has been hampered by the lack of materials with large enough nonlinearity and large GVD in their multifrequency transparency regions. In addition, the correct sign of the dispersion, required to achieve trapping of at least one of the waves involved in the interaction, is difficult to achieve because negative or anomalous dispersion occurs far away from the absorption edges of materials. Optical fibers with a quadratic response and a long enough path length are in their infancy but may prove in the future to be the ideal setting to observe temporal-soliton formation due to quadratic interactions, and perhaps to observe combined quadratic and cubic temporal nonlinear phase modulation effects. Note, however, that in the temporal domain, compensation of the walk-off of optical fields at different carrier frequencies is difficult to achieve.

Quite recently, a clever approach involving spatially tilting the phase front in a parametric amplification scheme has been used to demonstrate pulse compression and, later, spatiotemporal solitons in one time and one space dimension [141]. In other words, dispersing a pulse spatially in one of the transverse spatial dimensions of a beam allows the cancellation of the temporal walk-off in a well-controlled fashion. In addition, this approach can induce large effective (positive or negative) second-order dispersion in the temporal evolution of ultrafast pulses. We believe this approach is a promising alternative for achieving mutual trapping of waves involved in a parametric three-wave interaction. Note, however, that even in this case, very long crystals (or pulses under 50 fs in duration) are required to demonstrate mutual trapping via a three-wave interaction, and both higher-order dispersion and third-order nonlinear refractive and absorptive effects are needed to properly account for the evolution of the ultrafast pulses.

Di Trapani et al. [141] demonstrated effective pulse stabilization over one dispersion length, indicating that temporal solitons can really be generated in this way. A key to the success of the work was that they operated in the "cascading limit", a region of large positive phase mismatch, so that the harmonic component of the quadratic soliton was very small, and it was necessary to control GVD of the fundamental component only.

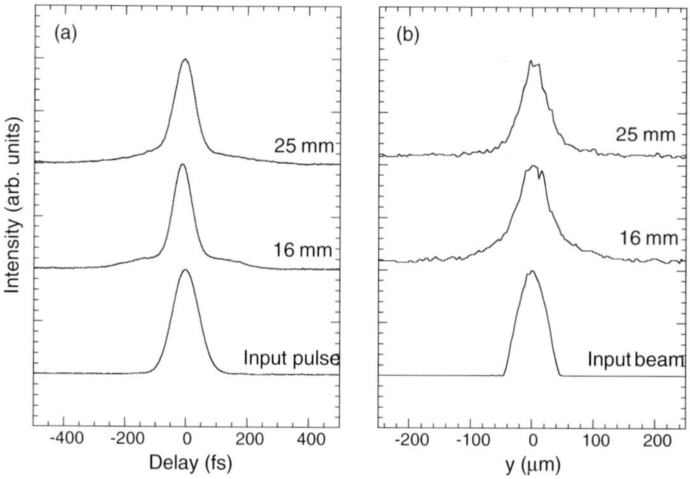

Fig. 13. Experimental temporal (**a**) and spatial (**b**) profiles of a spatiotemporal soliton generated experimentally in BBO for different propagation lengths, with $I = 8\mathrm{GW/cm}^2$ and $\Delta k = -60\pi/25$ mm (from [143])

This idea was then beautifully extended by Wise and coworkers [143,142], who achieved self-trapping in both space and time in a 1 cm long $LiIO_3$ crystal cut for type I phase matching [142], and later in a 25 mm long barium metaborate (Ba_2BO_4, BBO.) crystal [143] Wavefront tilting was used in the cascading limit to control the GVD in one spatial dimension, and the diffraction length in that dimension was set equal to the dispersion and nonlinear lengths by adjusting the input beam diameter and intensity. However, the second transverse beam dimension was made large so that there was no diffraction with propagation distance in that dimension. The evolution of the temporal and spatial profiles (measured also with an intermediate BBO crystal length of 17 mm when repeating the experiment with another crystal) is shown in Fig. 13; where a slight asymmetry is also observed. Thus, the results reported by Wise and coauthors provide the first example of a spatiotemporal quadratic soliton, or $(1+1)$-dimensional *parametric light bullet.*

3.3 Quadratic Solitons in Cavities

Work on optical parametric oscillators (OPOs) in the femtosecond regime has shown that for a singly resonant signal wave, enough dispersion could be accumulated over many intracavity round trips to achieve soliton-like action [144]. In their work, Reid et al. [144] compare the behavior of temporal pulses with that of solitons in fibers near the zero-dispersion point. A splitting of the optical spectrum, accompanied by the frequency shift characteristic of the temporal evolution of solitons in this regime, gave clear evidence of the solitonic nature of the waves involved in the process. One should note, however, that the quantitative origin of the mechanisms involved in the interaction was not discussed. In particular, the intricate interplay between intracavity loss, parametric gain and its saturation, and second- and third-order nonlinearity require someone to revisit this problem in greater detail. In particular, the realization of ultrafast pulses 10–20 fs in duration indicates that this approach to soliton formation has considerable practical potential [145,146]. One can imagine both the parametric gain and the source of self-phase modulation being produced by one or several intracavity elements with specific, engineered characteristics, for instance chosen to take advantage of resonance with other quadratic interactions such as second-harmonic generation in the ultraviolet [147].

3.4 Soliton Interaction

Following the demonstration of single and multidimensional spatial solitons, a few natural questions arise about their similarity to other types of solitons. For example, how will quadratic solitons interact, and will they undergo modulational instability under appropriate conditions? Both questions have been answered experimentally.

The Florida group has demonstrated that, in the cascaded regime, spatial solitons acted in accordance with the NLS limit and could interact in an almost elastic fashion [132]. In order to demonstrate the collision of quadratic solitons, they launched two beams in a $LiNbO_3$ waveguide in both parallel and crossed configurations. Figure 14 demonstrates the experimental results for such an interaction between parallel input solitons when the relative phase between the two launched beams was controlled by two glass wedges. For *in-phase solitons* the force is attractive, and it can lead to soliton fusion, as shown in Fig. 14a. At large angles of incidence the solitons pass right through each other, and *out-of-phase solitons* repel (Fig. 14c). Other relative phases lead to inelastic collisions and energy exchange between the solitons (Figs. 14b, d), in accordance with the interactions of solitons in nonintegrable nonlinear models. The overall behavior resembles closely that of "saturable" nonlinear media.

Collisions of (2+1)-dimensional spatial-soliton beams in quadratic optical media were investigated experimentally by Costantini et al. [148]. As expected from the theory of non-integrable systems, these authors were able to demonstrate both quasi-elastic and inelastic (fusion) interactions by controlling the angle at which the beams interact. The experiments were conducted in a 2 cm long KTP crystal, generating two quadratic solitons with similar thresholds for formation to those observed by Torruellas et al. [134].

3.5 Transverse Effects

The existence of modulational instability is well known in nonlinear optics. For example, in fibers it has been shown that pulses evolve into temporal

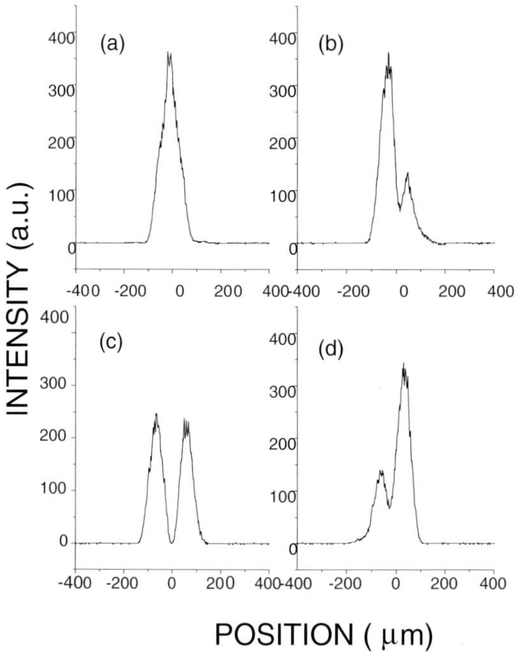

Fig. 14. Measured output beam profiles after the interaction of two solitary waves in the case of parallel launching at a temperature of 335.05 °C. The relative phase difference between the two beams is (**a**) 0, (**b**) $\pi/2$, (**c**) π, and (**d**) $3\pi/2$ (from [132])

solitons, and a CW beam breaks up into a recurring periodic train of pulses in the presence of self-modulation. This connection between modulational instability and solitons is quite general in nonlinear wave science (see, e.g., the recent review paper [47]).

In the spatial domain, modulational instability is associated with a spatial breakup of an elongated beam via the so-called *soliton transverse instability*. For quadratic solitons, such an effect was first predicted and studied theoretically [66,149–152], and then demonstrated experimentally in a 1 cm long KTP doubling crystal by Fuerst et al. [153]. Above a critical input intensity the elongated beam evolved into a single soliton beam [154]. Beyond an input intensity corresponding to the formation of several solitonic beams, a line of stable solitons emanated from the 1 cm long KTP crystal [153]. The generation of such a train of solitonic beams can be understood from simple physics, since the instability breaks the beam up into "filaments", which in turn undergo self-focusing and attempt to form solitons. Part of energy is radiated away, leading to the formation of a train of $(2+1)$-dimensional stable solitons, as demonstrated experimentally in Fig. 15.

The situation is subtly different in a slab waveguide and is more similar to a fiber than to a bulk medium. In this case, a finite beam excitation leads to an initial beam breakup via modulational instability. However, for a finite-sized beam, the excess energy can eventually leak out through the sides of the waveguide. For wide beams and samples of practical length (a few diffraction

Fig. 15. Experimental results demonstrating transverse modulational instability in a quadratic medium. (**a**) Elliptical input beam, (**b**) output beam at 48 GW/cm^2, and (**c**) output beam at 57 GW/cm^2 (from [153])

lengths), no solitons are formed, and strong interactions between the pulse maxima complicate the observed output considerably.

More recently, Wise and coworkers [155] reported the observation of *spatiotemporal instability* of quadratic solitons and their intensity-dependent filamentation. They performed experiments with 17 and 25 mm BBO crystals cut for type I phase matching, in a setting similar to that reported earlier by Liu et al. [143]. The resulting picture of soliton formation is somewhat similar to that observed by Fuerst et al. [153], but the structures observed were also localized in time instead of being localized only in one of the spatial dimensions, effectively resembling light bullets.

Spatial solitons can also be generated because of other types of instabilities (see the general discussion and examples in [47]). In particular, Torner and coworkers have predicted and verified experimentally that vortices embedded in an input fundamental beam with sufficient intensity to produce quadratic solitons can undergo a different form of transverse instability. An azimuthal instability leads to the generation of multiple quadratic solitons. Experiments were performed by Petrov et al. [156] using 2 cm long KTP crystals, and a typical result observed was the formation of three quadratic solitons generated as a result of the decay of a ring-like vortex beam carrying a phase singularity.

4 Concluding Remarks

We have presented a brief overview of the state of the art in the physics of parametric solitary waves in diffractive optical media with a nonlinear quadratic response. Most of our conclusions can be easily translated to systems with *dispersion* instead of *diffraction*, as well as to systems where dispersion and diffraction are both present (as in the case of light bullets). The most important achievement of both the early and the recent theoretical efforts has been to recognize that parametric wave interaction can support solitary waves in diffractive quadratic media, with properties somewhat similar to those known for solitary waves in Kerr-like optical media. In the latter case, solitons form owing to a nonlinear change of the refractive index and are stable only in systems with a single transverse dimension, requiring a saturable nonlinearity for their stabilization in higher dimensions. Quadratic solitons are remarkable in that the number of transverse dimensions seems to play less of a critical role. In that regard, quadratic materials seem ideal for the observation of multidimensional solitons involving both temporal and spatial dimensions.

Remarkably, a renaissance in the *theoretical* activity in this field has been triggered by the experimental discoveries of quadratic solitons in waveguides and bulk media, and the subsequent experimental studies of their generation, properties, and collisions. This has led to a much deeper understanding of many aspects and features of solitary waves in nonintegrable models not properly discussed in other branches of nonlinear physics nor in other branches of optics. As a simple consequence of the theoretical progress resulting from this effort, a series of novel fundamental theoretical results have been obtained, such as a generalized criterion for the stability of multiparameter solitary waves in nonintegrable models, the theory of soliton internal modes, the concept of walking solitons, a general asymptotic approach to soliton stability and to the instability-induced dynamics of solitary waves, and the theory of soliton interactions in a bulk medium, including soliton spiralling.

A number of important *experimental* results in this field have revealed the deep richness of parametric self-focusing and parametric multiwave spatial solitons. Here, we have presented only some aspects of the physics of spatial solitons in a quadratic medium, including analytical, numerical, and experimental results. We believe they demonstrate a number of very interesting features of this type of nonlinear waves. Additional efforts are, however, still required to achieve a complete understanding. We believe this is an exciting forefront area of modern nonlinear science that is very closely related to technological applications. We expect that quadratic solitons and cascading effects will soon play a role in modern laser designs involving new and improved nonlinear optical materials. The practicality of the concepts uncovered in this chapter, as discussed in Prof. Chiao's introduction to this book, may also end up having a profound impact in other areas of nonlinear science, such as in the physics of atom–molecular Bose–Einstein condensates, nonlinear atom optics, etc.

Acknowledgments

It is a great pleasure for us to thank and acknowledge the many colleagues with whom we had discussions and ongoing collaborations on this subject for the last few years. We would like to name especially Gaetano Assanto, Ole Bang, Alan Boardman, Alexander Buryak, Peter Drummond, Falk Lederer, Dmitry Pelinovsky, Rowland Sammut, Roland Schiek, Anatoly Sukhorukov, Andrey Sukhorukov, Lluis Torner, Stefano Trillo, and Frank Wise.

References

1. G.A. Askar'yan, Zh. Eksp. Teor. Fiz. **42**, 1567 (1962) [Sov. Phys. JETP **15**, 1088 (1962)].
2. M. Segev, G. Stegeman Phys. Today **51**, 42 (1998).
3. Yu.S. Kivshar, Opt. Quantum Electron. **30**, 571 (1998).
4. L.A. Ostrovskii, Pisma Zh. Eksp. Teor. Fiz. **5**, 331 (1967).
5. Yu.N., Karamzin, A.P. Sukhorukov, Pisma Zh. Exp. Teor. Fiz. **20**, 734 (1974) [JETP Lett. **20**, 339 (1974)].
6. Yu.N., Karamzin, A.P. Sukhorukov, Zh. Eksp. Teor. Fiz. **68**, 834 (1975) [Sov. Phys.-JETP. **41**, 414 (1975)].
7. Yu.N., Karamzin, A.P. Sukhorukov, T.S. Filipchuk, Vest. Mosk. Univ. Ser. Fiz. **19**(4), 91 (1978) [Moscow Univ. Phys. Bull. **33**, 73 (1978)].
8. A.A. Kanashov, A.M. Rubenchik, Physica D **4**, 122 (1981).
9. A.P. Sukhorukov, *Nonlinear Wave Interactions in Optics and Radiophysics* (Nauka, Moscow, 1988) p. 232 [in Russian].
10. G.I. Stegeman, D.J. Hagan, L. Torner, Opt. Quantum Electron. **28**, 1691 (1996).
11. L. Torner, in *Beam Shaping and Control with Nonlinear Optics*, ed. by F. Kajzer, R. Reinish (Plenum, New York, 1998), p. 229.
12. Yu.S. Kivshar, in: *Advanced Photonics with Second-Order Optically Nonlinear Processes*, ed. by A. Boardman, L. Pavlov, and S. Tanev (Kluwer, Dordrecht, 1999), p. 441.
13. D.L. Mills, *Nonlinear Optics: Basic Concepts* (Springer, Berlin, Heidelberg, 1991).
14. R.W. Boyd, *Nonlinear Optics* (Academic Press, San Diego, 1992).
15. A.V. Buryak, Yu.S. Kivshar, Phys. Lett. A **197**, 407 (1995).
16. A.V. Buryak, Yu.S. Kivshar, Phys. Rev. A **51**, R41 (1995).
17. A.V. Buryak, Yu.S. Kivshar, Opt. Lett. **20**, 834 (1995).
18. A.V. Buryak, Yu.S. Kivshar, Opt. Lett. **20**, 1080 (1995).
19. O. Bang, J. Opt. Soc. Am. B. **14**, 51 (1997).
20. K. Hayata, M. Koshiba, Phys. Rev. Lett. **71**, 3275 (1993).
21. R. Schiek, J. Opt. Soc. Am. B. **10**, 1848 (1993).
22. M.J. Werner, P.D. Drummond, J. Opt. Soc. Am. B. **10**, 2390 (1993).
23. M.J. Werner, P.D. Drummond, Opt. Lett. **19**, 613 (1994).
24. A.V. Buryak, Yu.S. Kivshar, Opt. Lett. **19**, 1612 (1994).
25. L. Torner, C.R. Menyuk, G.I. Stegeman, Opt. Lett. **19**, 1615 (1994).
26. D. Ferro, S. Trillo, Phys. Rev. E **51**, 4994 (1995).

27. L. Torner, Opt. Commun. **114**, 136 (1995).
28. A.D. Boardman, K. Xie, A. Sangarpaul, Phys. Rev. A **52**, 4099 (1995).
29. D. Mihalache, F. Lederer, D. Mazilu, L.-C. Crasovan, Opt. Eng. **35**, 1616 (1996).
30. H. He, M.J. Werner, P.D. Drummond, Phys. Rev. E **54**, 896 (1996).
31. M. Haelterman, S. Trillo, P. Ferro Opt. Lett. **22**, 84 (1997).
32. L. Torner, C.R. Menyuk, W.E. Torruellas, G.I. Stegeman, Opt. Lett. **20**, 13 (1995).
33. A.V. Buryak, Yu.S Kivshar, V.V. Steblina, Phys. Rev. A **52**, 1670 (1995).
34. A.V. Buryak, Yu.S. Kivshar, S. Trillo, Phys. Rev. Lett. **77**, 5210 (1996).
35. D. Mihalache, D. Mazilu, L.-C. Crasovan, L. Torner, Phys. Rev. E, **56**, R6294 (1997).
36. V.V. Steblina, Yu.S. Kivshar, M. Lisak, B.A. Malomed, Opt. Commun. **118**, 345 (1995).
37. S. Trillo, P. Ferro, Opt. Lett. **20**, 438 (1995).
38. H. He, P.D. Drummond, B.A. Malomed, Opt. Commun. **123**, 394 (1996).
39. V.M. Agranovich, S.A. Darmanyan, O.A. Dubovsky, A.M. Kamchatnov, E.I. Ogievetsky, T. Neidlinger, P. Reineker, Phys. Rev. B **53**, 15451 (1996).
40. A.A. Sukhorukov, Phys. Rev. E **61**, 4530 (2000).
41. D.E. Pelinovsky, A.V. Buryak, Y.S. Kivshar, Phys. Rev. Lett. **75**, 591 (1995).
42. L. Torner, D. Mihalache, D. Mazilu, N.N. Akhmediev, Opt. Lett. **20**, 2183 (1995).
43. A.C. Yew, A.R. Champneys, P.J. McKenna, J. Nonlinear Sci. **9**, 33 (1999).
44. C. Etrich, U. Peschel, F. Lederer, D. Mihalache, D. Mazilu, Opt. Quantum Electron. **30**, 881 (1998).
45. A.C. Yew, B. Sandstede, C.K.R.T. Jones, Phys. Rev. E **61**, 5886 (2000).
46. K. Hayata, M. Koshiba, Phys. Rev. A. **50**, 675 (1994).
47. Yu.S. Kivshar, D.E. Pelinovsky, Phys. Rep. **331**, 117 (2000).
48. R.Y. Chiao, E. Garmire, C.H. Townes, Phys. Rev. Lett. **13**, 479 (1964).
49. P.L. Kelley, Phys. Rev. Lett. **15**, 1005 (1965).
50. J.J. Rasmussen, K. Rypdal, Phys. Scr. **33**, 481 (1986).
51. L. Bergé, V.K. Mezentsev, J.J. Rasmussen, J. Wyller, Phys. Rev. A. **52**, R28 (1995).
52. L. Torner, W.E. Torruellas, G.I. Stegeman, C.R. Menyuk, Opt. Lett. **20**, 1952 (1995).
53. L. Torner, E.M. Wright, J. Opt. Soc. Am. B. **13**, 864 (1996).
54. G. Leo, G. Assanto, W.E. Torruellas, Opt. Lett. **22**, 7 (1997).
55. G. Leo, G. Assanto, W.E. Torruellas, Opt. Commun. **134**, 223 (1997).
56. G. Leo, G. Assanto, Opt. Lett. **22**, 1391 (1997).
57. L. Torner, D. Mihalache, D. Mazilu, E.M. Wright, W.E. Torruellas, G.I. Stegeman, Opt. Commun. **121**, 149 (1995).
58. L. Bergé, O. Bang, J.J. Rasmussen, V.K. Mezentsev, Phys. Rev. E. **55**, 3555 (1997).
59. O. Bang, Yu.S. Kivshar, A.V. Buryak, Opt. Lett. **22**, 1680 (1997).
60. O. Bang, Yu.S. Kivshar, A.V. Buryak, A. De Rossi, S. Trillo, Phys. Rev. E **58**, 5057. (1998).
61. D. Mihalache, D. Mazilu, L.-C. Crasovan, L. Torner, Opt. Commun. **137**, 113 (1997).
62. D. Mihalache, D. Mazilu, J. Dörring, L. Torner, Opt. Commun. **159**, 129 (1999).

63. D. Mihalache, D. Mazilu, B.A. Malomed, L. Torner, Opt. Commun. **169**, 341 (1999).
64. B.A. Malomed, P. Drummond, H. He, A. Berntson, D. Anderson, M. Lisak, Phys. Rev. E **56**, 4725 (1997).
65. D.V. Skryabin, W.J. Firth, Opt. Commun. **148**, 79 (1998).
66. D.V. Skryabin, W.J. Firth, Phys. Rev. Lett. **81**, 3379 (1998).
67. D.M. Baboiu, G.I. Stegeman, L. Torner, Opt. Lett. **20**, 2282 (1995).
68. C. Etrich, U. Peschel, F. Lederer, B.A. Malomed, Phys. Rev. A **52**, 3444 (1995).
69. C.B. Clausen, P.L. Christiansen, L. Torner, Opt. Commun. **136**, 185 (1997).
70. D.M. Baboiu, G.I. Stegeman J. Opt. Soc. Am. B. **14**, 3143 (1997).
71. C. Etrich, U. Peschel, F. Lederer, B.A. Malomed, Yu.S. Kivshar, Phys. Rev. E **54**, 4321 (1996).
72. V.V. Steblina, Yu.S. Kivshar, A.V. Buryak Opt. Lett. **23**, 156 (1997).
73. L. Torner, D. Mazilu, D. Mihalache Phys. Rev. Lett. **77**, 2455 (1996).
74. C. Etrich, U. Peschel, F. Lederer, B.A. Malomed, Phys. Rev. E **55**, 6155 (1997).
75. L. Torner, Opt. Lett. **23**, 1256 (1998).
76. B.A. Malomed, D. Anderson, M. Lisak, Opt. Commun. **126**, 251 (1996).
77. H.T. Tran, Opt. Commun. **118**, 581 (1995).
78. U. Peschel, C. Etrich, F. Lederer, B.A. Malomed, Phys. Rev. E **55**, 7704 (1997).
79. B.S. Azimov, A.P. Sukhorukov, D.V. Trukhov Izv. Akad. Nauk USSR Ser. Fiz. **51**, 229 (1987) [Bull. Acad. Sci. USSR Phys. **51**(2), 19 (1987)].
80. A.D. Capobianco, B. Costantini, C. De Angelis, D. Modotto, A. Lauteri Palma, G.F. Nalesso, C.G. Someda, Opt. Quantum Electron. **30**, 483 (1998).
81. A.V. Buryak, Yu.S. Kivshar, S. Trillo, J. Opt. Soc. Am. B **14**, 3110 (1997).
82. L.S. Telegin, A.S. Chirkin, Sov. J. Quantum Electron. **12**, 1354 (1982).
83. S. Trillo, S. Wabnitz, Opt. Lett. **17**, 1572 (1992).
84. A. Kobyakov, F. Lederer, O. Bang, Yu.S. Kivshar Opt. Lett. **23**, 506 (1998).
85. C.B. Clausen, O. Bang, Yu.S. Kivshar, Phys. Rev. Lett. **78**, 4749 (1997).
86. Yu.S. Kivshar, T. Alexander, S. Saltiel, Opt. Lett. **24**, 759 (1999).
87. Yu.S. Kivshar, A.A. Sukhorukov, S.M. Saltiel, Phys. Rev. E **60**, R5056 (1999).
88. M.V. Komissarova, A.P. Sukhorukov Bull. Russ. Acad. Sci. Phys. **56**, 1995 (1992).
89. M.A. Karpierz, Opt. Lett. **20**, 1677 (1995).
90. A.V. Buryak, Yu.S. Kivshar, S. Trillo, Opt. Lett. **20**, 1961 (1995).
91. S. Trillo, A.V. Buryak, Yu.S. Kivshar, Opt. Commun. **122**, 200 (1996).
92. O. Bang, L. Bergé, J.J. Rasmussen, Opt. Commun. **146**, 231 (1998).
93. A. De Rossi, G. Assanto, S. Trillo, W.E. Torruellas, Opt. Commun. **150**, 390 (1998).
94. T.J., Alexander, A.V. Buryak, Yu.S. Kivshar, Opt. Lett. **28**, 670 (1998).
95. T.J. Alexander, Yu.S. Kivshar, A.V. Buryak, R.A. Sammut, Phys. Rev. E **61**, 2042 (2000).
96. Z.K. Yankauskas, Izv. VUZov Radiofiz. **9**, 412 (1996) [Sov. Radiophys. **9**, 261 (1966)].
97. V.I. Kruglov, R.A. Vlasov, Phys. Lett. A **111**, 401 (1985).
98. W.J. Firth, D.V. Skryabin, Phys. Rev. Lett. **79**, 2450 (1997).
99. V. Tikhonenko, J. Christou, B. Luther-Davies, J. Opt. Soc. Am. B **12**, 2046 (1995).

100. V. Tikhonenko, J. Christou, B. Luther-Davies, Phys. Rev. Lett. **76**, 2698 (1996).
101. L. Torner, D.V. Petrov, Electron. Lett. **33**, 608 (1997).
102. L. Torner, D.V. Petrov, J. Opt. Soc. Am. B **14**, 2017 (1997).
103. J.P. Torres, J.M. Soto-Crespo, L. Torner, D.V. Petrov, J. Opt. Soc. Am. B **15**, 625 (1998).
104. J.P. Torres, Soto-Crespo, L. Torner, D.V. Petrov, Opt. Commun. **149**, 77 (1998).
105. L. Torner, J.P. Torres, D.V. Petrov, J.M. Soto-Crespo, Opt. Quantum Electron. **30**, 809 (1998).
106. Y. Silberberg, Opt. Lett. **15**, 1282 (1990).
107. D.E. Edmundson, R.H. Enns, Opt. Lett. **17**, 596 (1992).
108. D.E. Edmundson, R.H. Enns, Opt. Lett. **18**, 1609 (1993).
109. D. Mihalache, D. Mazilu, B. Maloned, L. Torner, Opt. Commun. **152**, 365 (1998).
110. F.F. So, S.R. Forrest, Y.Q. Shi, W.H. Steier, Appl. Phys. Lett. **56**, 674 (1990).
111. Y. Imanishi, S. Hattori, A. Kakuta, S. Numata, Phys. Rev. Lett. **71**, 2098 (1993).
112. T. Nanaka, Y. Mori, N. Nagai, Y. Nakagawa, N. Saeda, T. Nakahaki, A. Ishitani, Thin Solid Films **239**, 214 (1994).
113. V. Bulovic, S.R. Forrest, Chem. Phys. Lett. **238**, 88 (1995).
114. V.M. Agranovich, O.A. Dubovsky, Chem. Phys. Lett. **210**, 458 (1993).
115. V.M. Agranovich, O.A. Dubovsky, A.M. Kamchatnov, J. Chem. Phys. **28**, 13607 (1994).
116. V.M. Agranovich, P. Reineker, V.I. Yudson, Synth. Met. **64**, 147 (1994).
117. V.M. Agranovich, O.A. Dubovsky, A.M. Kamchatnov, Chem. Phys. **198**, 245 (1995).
118. V.M. Agranovich, A.M. Kamchatnov, Pis'ma Zh. Eksp. Teor. Fiz. **59**, 397 (1994) [JETP Lett. **59**, 424 (1994)].
119. V.M. Agranovich, S.A. Darmanyan, A.M. Kamchatnov, T.A. Leskova, A.D. Boardman, Phys. Rev. E. **55**, 1894 (1997).
120. S.A. Dubovsky, A.V. Orlov, Phys. Solid State **38**, 675 (1996) [Fiz. Tverd. Tela (St. Petersburg) **38**, 1221 (1996)].
121. S.A. Dubovsky, A.V. Orlov, Phys. Solid State **38**, 1067 (1996) [Fiz. Tverd. Tela (St. Petersburg) **38**, 1931 (1996)].
122. O. Bang, P.L. Christiansen, C.B. Clausen, Phys. Rev. E **56**, 7257 (1997).
123. S. Darmanyan, A. Kobyakov, F. Lederer, Phys. Rev. E **57**, 2344 (1998).
124. S. Darmanyan, A. Kamchatnov, F. Lederer, Phys. Rev. E **58**, R4120 (1998).
125. T. Peschel, U. Peschel, F. Lederer, Phys. Rev. E **57**, 1127 (1998).
126. A. Kobyakov, S. Darmanyan, T. Pertsch, F. Lederer, J. Opt. Soc. Am. B **16**, 1737 (1999).
127. A.B. Aceves, C. De Angelis, T. Peschel, R. Muschall, F. Lederer, S. Trillo, S. Wabnitz, Phys. Rev. E **53**, 1172 (1996).
128. S.M. Jensen, IEEE J. Quantum Electron. **18**, 1580 (1982).
129. A.A. Sukhorukov, Yu.S. Kivshar, O. Bang, C.M. Soukoulis, Phys. Rev. E **63**, 16615 (2001).
130. A.A. Sukhorukov, Yu.S. Kivshar, O. Bang, Phys. Rev. E **59**, R41 (1999).
131. R. Schiek, Y. Baek, G.I. Stegeman, Phys. Rev. E **53**, 1138 (1996).

132. Y. Baek, *Cascaded Second-Order Nonlinearities in Lithium Niobate Waveguides*, Ph.D. thesis, Department of Physics, University of Central Florida (1997).
133. R. Schiek, Y. Baek, G.I. Stegeman, W. Sohler, Opt. Lett. **24**, 83 (1999).
134. W.E. Torruellas, Z. Wang, D.J. Hagan, E.W. Van Stryland, G.I. Stegemanm, Phys. Rev. Lett. **74**, 5036 (1995).
135. W.E. Torruellas, Z. Wang, L. Torner, G.I. Stegeman, Opt. Lett. **20**, 1949 (1995).
136. W.E. Torruellas, G. Assanto, B.L. Lawrence, R.A. Fuerst, G.I. Stegeman, Appl. Phys. Lett. **68**, 1449 (1996).
137. B. Bourliaguet, V. Couderc, A. Barthélémy, G.W. Ross, P.G.R. Smith, D.C. Hanna, C. De Angelis, Opt. Lett. **24**, 1410 (1999).
138. P. Di Trapani, G. Valiulis, W. Chinaglia, A. Andreoni, Phys. Rev. Lett. **80**, 265 (1998).
139. M.T.G. Canva, R.A. Fuerst, D. Baboiu, G.I. Stegeman, G. Assanto, Opt. Lett. **22**, 1683 (1997).
140. P., Di Trapani, W. Chinaglia, S. Minardi, A. Piskarskas, G. Valiulis, Phys. Rev. Lett. **84**, 3843 (2000).
141. P. Di Trapani, D. Caironi, G. Valiulis, A. Dubietis, R. Danielius, A. Piskarskas, Phys. Rev. Lett. **81**, 570 (1998).
142. X. Liu, L.J. Qian, F.W. Wise, Phys. Rev. Lett. **82**, 4631 (1999).
143. X. Liu, K. Beckwitt, F. Wise, Phys. Rev. E **62**, 1328 (2000).
144. D.T. Reid, J.M. Dudley, M. Ebrahimzadeh, W. Sibbett, Opt. Lett. **19**, 825 (1994).
145. G. Cerullo, M. Nisoli, S. De Silvestri, Opt. Lett. **71**, 3616 (1997).
146. G.M. Gale, M. Cavallari, T.J. Driscoll, F. Hache, Opt. Lett. **20**, 1562 (1995).
147. F. Hache, A. Zeboulon, G. Gallot, G.M. Gale, Opt. Lett. **20**, 1556 (1995).
148. B. Costantini, C. De Angelis, A. Barthelemy, B. Bourliaguet, V. Kermene, Opt. Lett. **22**, 19 (1998).
149. A. De Rossi, S. Trillo, A.V. Buryak, Yu.S. Kivshar Opt. Lett. **22**, 868 (1997).
150. A. De Rossi, S. Trillo, A.V. Buryak, Yu.S. Kivshar, Phys. Rev. E **56**, R4959 (1997).
151. D.M. Baboiu, G.I. Stegeman, Opt. Lett. **23**, 31 (1998).
152. D.V. Skryabin, Phys. Rev. E **60**, 7511 (1999).
153. R.A. Fuerst, D.M. Baboiu, B. Lawrence, W.E. Torruellas, G.I. Stegeman, S. Trillo, S. Wabnitz, Phys. Rev. Lett. **78**, 2756 (1997).
154. R.A. Fuerst, B.L. Lawrence, W.E. Torruellas, G.I. Stegeman, Opt. Lett. **22**, 19 (1997).
155. X. Liu, K. Beckwitt, F. Wise, Phys. Rev. Lett. **85**, 1871 (2000).
156. D.V. Petrov, L. Torner, J. Martorell, R. Vilaseca, J.P. Torres, C. Cojocaru, Opt. Lett. **23**, 1444 (1998).

Bragg Solitons: Theory and Experiments

C. Martijn de Sterke, Benjamin J. Eggleton, and John E. Sipe

Summary. We discuss the propagation of Bragg solitons in one dimensional nonlinear periodic structures, such fibre Bragg gratings and corrugated semiconductor waveguide structures. Such solitons exhibit a wealth of exciting phenomena, such as slow propagation, pulse compression, and optical switching.

1 Introduction

Nonlinear periodic media have a number of intriguing properties. In this chapter we discuss the best known of these, Bragg solitons. The required refractive index, which can be achieved in a grating written in the core of an optical fiber (see Fig. 1), gives Bragg grating solitons their unique properties: arbitrary propagation velocities between 0 and the speed of light in the uniform medium (i.e. the medium without a grating), short formation and interaction lengths of around 1 cm, and a considerable degree to which variations of the grating properties can be used to vary the soliton parameters.

In this section we give a qualitative description of Bragg grating solitons. We first review the general properties of linear periodic media, and then discuss the features added by a Kerr nonlinearity. We do not give proofs here; for more rigorous derivations we refer to Sect. 2.

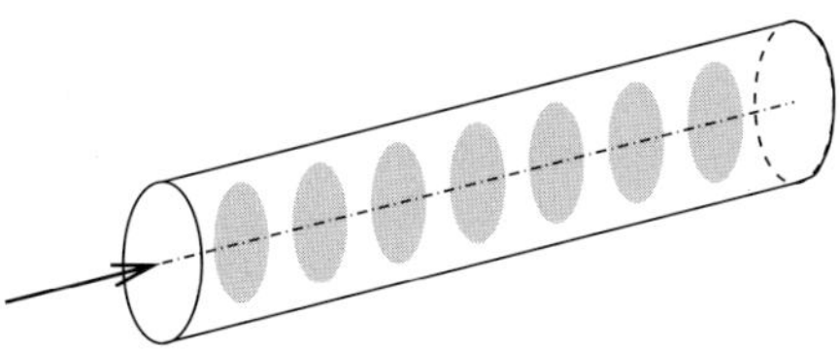

Fig. 1. Schematic of a grating written in the core of an optical fiber. The gray areas indicate regions of elevated refractive index. The *arrow* indicates the direction of light propagation

The best known linear property of periodic media is that they exhibit Bragg reflection. Though Bragg reflection is perhaps better known in the context of the scattering of X-rays or neutrons off crystals, the concept applies to any wave propagating through a periodic medium. In fact, the optical

Bragg reflection that we are concerned with here is more straightforward than that of other problems, since it is one-dimensional to a good approximation. A one-dimensional geometry is best illustrated by a thin-film stack, with a plane wave, incident normal to the layers. However, in Sect. 2 we argue that gratings in optical fibers and other guided-wave structures, such as that shown in Fig. 1, can be considered one-dimensional to a good approximation.

Bragg reflection can most easily be understood using Fig. 2, which shows schematically the refractive index n versus position in a "phase grating". In practice, the refractive-index distribution is usually smoother than that in the figure. However, the precise shape of the distribution does not matter for the argument here, as only the periodicity is important. In addition, we take the modulation depth of the grating to be small. This "shallow grating" assumption is not essential, but it simplifies the argument.

Since the grating is shallow, one might ask whether one ever notices the weak periodic refractive-index modulation. In fact, in general one does not. This can most easily be seen from the following argument. The reflection in Fig. 2 is due to Fresnel reflections off the various interfaces, half of which are indicated by arrows. These reflections are very weak, because the grating is shallow. For an arbitrary wavelength the weak reflections have no particular phase relation to each other, and therefore the *intensities* associated with each can be added, leading to a small reflectivity for the grating as a whole. In this case the periodicity, in effect, is thus not noticed.

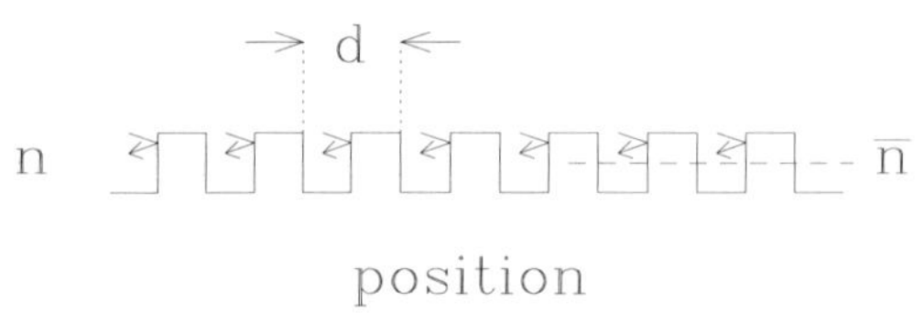

Fig. 2. Schematic of reflection off a periodic refractive-index structure. The average refractive index is $\bar{n}$; the period is d

But consider now the situation in which the waves reflected off two interfaces that are a period d apart are in phase. The condition for this is

$$M\lambda = 2\bar{n}d\ , \tag{1}$$

where λ is the vacuum wavelength of the light, M is a positive integer, and the other symbols are defined in Fig. 2. If the reflections off two interfaces that are a period apart are in phase, then, because of the periodicity, they are in phase with those off all other interfaces that are an integer number of periods away. One must now thus add the *amplitudes* of the Fresnel-reflected light. The total reflectivity is now a factor N larger than in the situation discussed in the previous paragraph, where N is the number of grating periods. Since in a typical fiber grating $N \approx 10^4$–10^5, the grating's Bragg reflectivity can be substantial, even though the reflectivity of each interface is small. In the experiments discussed here one is interested in the shortest Bragg wavelength; we thus set $M = 1$ in (1), and write the Bragg wavelength λ_B as

$$\lambda_B = 2\bar{n}d \; . \tag{2}$$

The above argument neglects the reflections off half the interfaces (see Fig. 2). However, at λ_B, the reflections off these interfaces do not completely cancel the reflections off the interfaces considered above. In fact, in shallow gratings with a duty cycle of 50%, the reflections of the two sets of interfaces add in phase for all odd reflection orders (M odd in (1)), though not for the even orders.

The calculated reflection spectrum of a typical uniform grating, showing the reflectivity versus wavelength, is given in Fig. 3. It shows that the grating reflects not only at λ_B, but also in a range of wavelengths around this value. The extent of this range is tedious to obtain from the simple arguments given above. However, according to coupled-mode theory (see Sect. 2), the range of wavelengths $\Delta\lambda$ where the grating reflects is given by

$$\frac{\Delta\lambda}{\lambda_B} \approx \frac{\Delta n}{\bar{n}} \; , \tag{3}$$

where Δn is the magnitude of the lowest Fourier component of the refractive-index distribution. For typical fiber gratings $\Delta\lambda/\lambda_B$ is in the range 10^{-4}–10^{-3}. Thus, since λ_B is typically around 1 μm, the grating reflects over a fraction of a nanometer. It is this property that makes fiber gratings useful in optical-fiber telecommunications networks. However, such applications [1] are beyond the scope of this review.

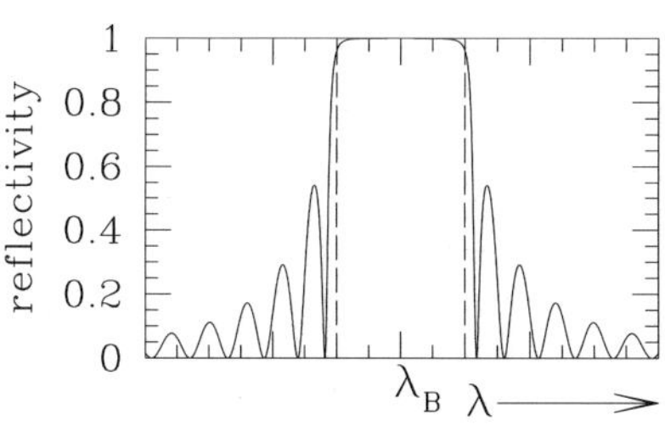

Fig. 3. Reflectivity versus wavelength for a typical Bragg grating. The *vertical dashed lines* indicate the edges of the photonic bandgap

The reflection spectrum in Fig. 3 consists of two parts. In the central region between the dashed lines, the "photonic bandgap", the large reflectivity is an intrinsic property of the grating, and is associated with an evanescently decaying electric field (see Sect. 2).[1] In contrast, outside the photonic

[1] The statement above refers strictly to the field envelope, rather than to the field itself; we neglect this distinction until Sect. 2.

bandgap the field is a propagating one, and the reflectivity can be considered to be due to Fabry–Perot-like effects in which the two edges of the grating act as mirrors [2]. Many of the experiments discussed in this review take place at frequencies just outside the photonic bandgap, where the residual grating reflectivity is undesirable. For this reason one uses *apodized* gratings, rather than uniform ones [3]. In an apodized grating the grating strength does not jump at the edges but, rather, increases smoothly until the maximum value is reached. This reduces the reflectivity associated with the grating's edges for frequencies outside the photonic bandgap, thus reducing the total grating reflectivity at these frequencies. Of course, the reflectivity remains high for frequencies inside the photonic bandgap (see Fig. 8).

In Sect. 2 we show that the group-velocity of the light vanishes inside the photonic bandgap; this is perhaps not surprising, given the evanescent decay of the field. For frequencies well away from λ_B the influence of the grating is negligible, and the group-velocity of the light is that of the uniform fiber $c/\bar{n}$, where c is the speed of light in vacuum. In between these two extremes, the light's group-velocity varies monotonically between 0 and $c/\bar{n}$. The low group-velocity is due to multiple reflections of the light off the grating, causing it to travel backwards for part of the time, leading to an increased path length for a given length of fiber.

For a typical fiber grating, the group-velocity varies between 0 and $c/\bar{n}$ within a fraction of a nanometer. Since the group-velocity is such a strong function of wavelength, the dispersion of a grating is huge [4]. Indeed, it has been shown that this dispersion is five or six orders of magnitude larger than the dispersion of typical fiber.[2] Thus, even though pulses with a frequency just outside the photonic bandgap of a grating travel slowly, the dispersion is so large that initially transform-limited pulses broaden rapidly. Note that on the short-wavelength side of the photonic bandgap the group-velocity increases with decreasing wavelength; here the dispersion is thus anomalous. In contrast, on the long-wavelength side of the gap the dispersion is normal.

The low group-velocity and strong dispersion of light propagating through a grating are illustrated in Fig. 4 [6], which shows experimental results for pulses transmitted by an apodized grating as a function of time, for various positive values of a detuning parameter Δ, where

$$\Delta = \frac{\omega - \omega_B}{c/\bar{n}} = -2\pi\bar{n}\,\frac{(\lambda - \lambda_B)}{\lambda_B^2}\ . \tag{4}$$

Here ω_B is the (angular) frequency associated with λ_B. Since $\Delta > 0$, and thus $\lambda < \lambda_B$, the dispersion is anomalous. The actual values of the detuning given in the caption do not matter at this point. Of interest here is only the fact that Δ is largest for the left-hand curve in Fig. 4, and decreases for the curves towards the right.

[2] This has allowed a fiber grating with a length of 10 cm to be used to compensate the dispersion of 100 km of standard telecommunication fiber [5]!

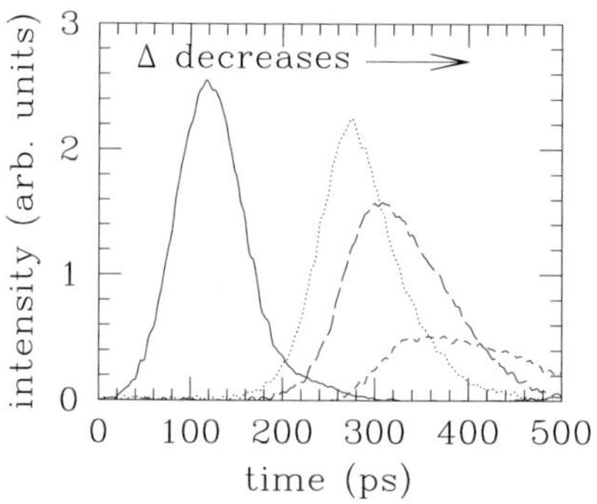

Fig. 4. Measured transmitted intensity versus time after propagation through a 55 mm long apodized fiber grating in the linear regime, for four values of the detuning, all of which are outside the photonic bandgap: $\Delta = 3612\ \mathrm{m}^{-1}$ (*solid curve*), $\Delta = 965\ \mathrm{m}^{-1}$ (*dotted curve*), $\Delta = 906\ \mathrm{m}^{-1}$ (*long-dashed curve*), $\Delta = 818\ \mathrm{m}^{-1}$ (short-dashed curve)

The experimental results in Fig. 4 confirm our discussion. For the solid curve the detuning is so large that the effect of the grating is negligible. It shows that the incoming pulses have a full width at half maximum (FWHM) of 80 ps. The group-velocity decreases with decreasing detuning, and thus the arrival time increases with decreasing detuning, as is confirmed in Fig. 4; the additional delay of over 200 ps for the short-dashed curve is substantial for propagation over only 55 mm. Note also that at the smallest detunings the transmitted pulse is broadened owing to the grating dispersion. Other aspects of these experiments are discussed in Sect. 4.2.

Let us now consider the effects of an optical nonlinearity. For our purposes it is sufficient to take a Kerr nonlinearity, according to which the refractive index n depends on the intensity I as

$$n(I) = n_l + n_2 I \ , \tag{5}$$

where n_l is the low-intensity value of the refractive index, and the magnitude of n_2 describes the strength of the nonlinearity. Here we take $n_2 > 0$, as in glass and GaAs at frequencies below their bandgaps; effects qualitatively similar to those described below, albeit in different frequency regimes, are observed for $n_2 < 0$.

The effect of a Kerr nonlinearity can be understood in different ways. In one of these ways, one observes that, with the inclusion of the nonlinearity, a system with strong dispersion and weak nonlinearity is obtained. It is well known that under very general conditions, the dynamics of such systems are described by the nonlinear Schrödinger equation (NLSE) to some approximation. The NLSE has a number of well-known properties [7], one of which is that it supports soliton solutions if, for positive nonlinearity, the dispersion is anomalous. The lowest-order soliton propagates without changing shape because the effect of the nonlinearity balances that of the dispersion. Applying this to the grating geometry under discussion, we would expect to find slowly propagating pulse-like solutions on the short-wavelength side of the photonic bandgap, which do not broaden, or at least broaden to a lesser degree than in the linear regime shown in Fig. 4 (these are "Bragg solitons") .

Figure 5 [6] shows results obtained under conditions similar to those of Fig. 4, except that the peak input intensity was about 10 GW/cm^2, where the effect of the Kerr nonlinearity is substantial. As in Fig. 4, the detuning is largest for the left-hand curve and decreases towards the right, indicating that the group-velocity again decreases with decreasing detuning. However, in contrast to Fig. 4, for the long-dashed and dotted curves, corresponding to intermediate detunings, the transmitted pulse is narrower than the incoming pulse [8]. This clearly demonstrates the effect of the nonlinearity and is associated with the formation of a soliton (for details, see Sect. 4). At the smallest detuning (short-dashed line), the dispersion is too large to be balanced by the nonlinearity, and the system behaves roughly as if it were linear, whereas at the largest detuning (solid line), the dispersion is too small for any effect to occur within the length of the grating.

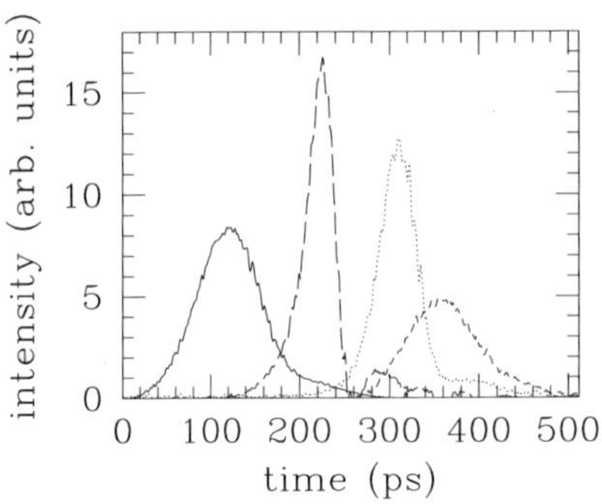

Fig. 5. Transmitted intensity versus time after high-intensity propagation through a 55 mm apodized fiber grating for four detunings, all of which are outside the photonic bandgap: $\Delta = 3612\ \mathrm{m}^{-1}$ (*solid curve*), $\Delta = 1053\ \mathrm{m}^{-1}$ (*long-dashed curve*), $\Delta = 847\ \mathrm{m}^{-1}$ (*dotted curve*), $\Delta = 788\ \mathrm{m}^{-1}$ (*short-dashed curve*)

An alternative way to see the effect of a Kerr nonlinearity was given in an earlier review article by de Sterke and Sipe [9]. The arguments given there lead to the conclusion that nonlinear periodic media support localized solutions; as mentioned above, these are referred to Bragg solitons. In Sect. 2 we present a more rigorous discussion of the properties for nonlinear periodic media.

2 Theory of Periodic Structures

Here we discuss the properties of periodic media using coupled-mode theory. Though this theory can be derived in different ways, in many of these the geometry is treated as if it were one-dimensional [2,10–12]. This can be justified as follows. Quite generally, the electric field can be written as a superposition over all the modes of the structure, only a few of which are relevant. Now if the perturbation associated with the grating is sufficiently weak, then the modal profiles are essentially unaffected by the perturbation. The calculation can then be cast in terms of the amplitudes of the modes of the unperturbed structure, leading to a calculation that is one-dimensional. This approximation is certainly valid for the shallow gratings considered here.

Coupled-mode theory can be developed at a number of different levels. The simplest level, and the level at which we shall be working here, is valid only for shallow gratings. At the start of the calculation one requires the Fourier coefficients of the refractive-index distribution; in the calculation it is then found that the size of the nth gap is proportional to the magnitude of the nth Fourier component (see, e.g., (3)). More accurate treatments have been developed according to two different "philosophies", as follows.

- The first requires the photonic band structure and associated Bloch functions of the one-dimensional periodic structure. These parameters, the calculation of which is straightforward by means of the transfer matrix method (e.g. [13]), for example, are then used as input to the coupled-mode theory [10].
- The second method does not require this elaborate input, as it is determined to some approximation as part of the procedure. This calculation is thus somewhat complicated, and gives order-by-order results, with the lowest order corresponding to standard coupled-mode theory [2,14].

As mentioned above, however, here we use standard coupled-mode theory.

2.1 Coupled-Mode Theory

Though the usual coupled-mode equations can be derived rigorously, usually a heuristic approach is taken. For brevity, we adopt the latter approach here, refering the reader to the literature for more rigorous methods [2,14]. We consider a refractive-index distribution of the form

$$n(z) = \bar{n} + \delta n(z) + \Delta n(z) \cos\left(\frac{2\pi}{d} z + \varphi(z)\right) + n_2 I \ . \tag{6}$$

Here z is the propagation direction, and, as indicated above, we neglect transverse dimensions; $\bar{n}$ is the nominal fiber refractive index; Δn is the amplitude of the grating and d is its nominal period; φ is the phase of the grating; δn represents (small) variations in the background refractive index; and the nonlinear n_2 term was discussed in Sect. 1. As we shall see below, for coupled-mode theory to be applicable we must assume that

- $\delta n, \Delta n, n_2 I \ll \bar{n}$, i.e. these all represent weak perturbations,
- δn, Δn, and φ are slowly varying on the scale of d.

These assumptions are satisfied for typical fiber gratings.

In coupled-mode theory, the linearly polarized electric field $E(z,t)$ is taken to be of the form [11]

$$E(z,t) = \left[E_+(z,t)\mathrm{e}^{+\mathrm{i}k_0 z} + E_-(z,t)\mathrm{e}^{-\mathrm{i}k_0 z}\right]\mathrm{e}^{-\mathrm{i}\omega_0 t} + \text{c.c.}\,, \tag{7}$$

where the $E_\pm$ are the amplitudes of the forward- and backward-propagating waves, and k_0 and ω_0, which are unspecified for now, are the nominal values of their wavenumber and frequency, respectively. ansatz (7) can be understood as follows: in the absence of the grating, the moduli of the mode amplitudes are of course constant. The presence of a weak grating can then be accounted for by taking these amplitudes to be slowly varying.

We now require the wave equation for the electric field,

$$\frac{\partial^2 E}{\partial z^2} - \frac{n^2(z)}{c^2}\frac{\partial^2 E}{\partial t^2} = 0\,, \tag{8}$$

where $n(z)$ describes the variations in the refractive index, and also the nonlinear effect of the form (5). To substitute (7) into (8), we require the second derivatives of the field with respect to space and time. Taking the space derivative as an example, we obtain

$$\left(-k_0^2 E_+ + 2\mathrm{i}k_0 E_{+,z} + E_{+,zz}\right)\mathrm{e}^{+\mathrm{i}(k_0 z - \omega_0 t)}\,, \tag{9}$$

with a similar expression for the E_- term, and where the subscript z indicates differentiation with respect to z. Now, in the usual heuristic derivation, it is argued that, since the $E_\pm$ vary slowly, the first term in parentheses in (9) dominates, with the second and third terms being progressively smaller. The third term is therefore dropped. It has been pointed out that this argument is erroneous, though the correct result is obtained [2]. If the same argument is applied to the temporal derivative, the wave equation takes the form

$$\begin{aligned}
&\left(-k_0^2 E_+ + 2\mathrm{i}k_0 E_{+,z}\right)\mathrm{e}^{+\mathrm{i}k_0 z} + \left(-k_0^2 E_- - 2\mathrm{i}k_0 E_{-,z}\right)\mathrm{e}^{-\mathrm{i}k_0 z} \\
&- \frac{1}{c^2}\left[\bar{n} + \delta n + n_2 I + \frac{\Delta n}{2}\left(\mathrm{e}^{\mathrm{i}(2\pi z/d+\varphi)} + \mathrm{e}^{-\mathrm{i}(2\pi z/d+\varphi)}\right)\right]^2 \\
&\times \left[\left(-\omega_0^2 E_+ - 2\mathrm{i}\omega_0 E_{+,t}\right)\mathrm{e}^{+\mathrm{i}k_0 z} + \left(-\omega_0^2 E_- - 2\mathrm{i}\omega_0 E_{-,t}\right)\mathrm{e}^{-\mathrm{i}k_0 z}\right] = 0\,,
\end{aligned} \tag{10}$$

where the intensity I has not yet been written in terms of the field. Note first that the dominant terms in (10) can be eliminated if we choose

$$\omega_0 = \frac{k_0 c}{\bar{n}}\,, \tag{11}$$

as expected. We also drop the smallest terms that appear when working out the products in (10) and use the shallowness of the grating and the weakness of the nonlinearity to write $(\bar{n}+\mathrm{d}n)^2 \approx \bar{n}^2 + 2\bar{n}\,\mathrm{d}n$, where $\mathrm{d}n$ stands for any of δn, Δn or $n_2 I$.

Since δn, Δn, and φ are slowly varying, the terms in the first and second lines of (10) have (spatial) frequency components around $\pm k_0$. In contrast, the terms in the last line of (10) have frequency components around the four (spatial) frequencies $\pm k_0 \pm 2\pi/d$. One therefore argues that to have sensible solutions one must require that $k_0 \approx \pi/d$. This implies that terms at the frequencies $\pm(k_0 + 2\pi/d)$ are ignored. Since k_0 is as yet undefined, we may choose it to be

$$k_0 = \pi/d\,, \tag{12}$$

though other choices would be equally valid, provided they were close to π/d. Note from (2) that the choice in (12) corresponds to choosing the nominal vacuum wavelength of the field to equal λ_B. Considering then the terms that appear with $\exp(+\mathrm{i}k_0 z)$ separately, we obtain

$$+2\mathrm{i}k_0 E_{+,z} + \frac{2\mathrm{i}\omega_0\bar{n}^2}{c^2}E_{+,t} + \frac{2\omega_0^2\bar{n}\,\delta n}{c^2}E_+ + \frac{2\omega_0^2\bar{n}\,n_2 I}{c^2}E_+ + \frac{2\omega_0^2\bar{n}\Delta n}{c^2}E_-\mathrm{e}^{+\mathrm{i}\varphi} = 0\,, \tag{13}$$

and a similar equation for the $\exp(-\mathrm{i}k_0 z)$ terms. These equations are essentially the result of interest. To get them into their final form we divide both equations by $2k_0$, and we set $F_\pm = E_\pm \exp(\pm\mathrm{i}\varphi/2)$. The final step involves the intensity I in the nonlinear terms. Expressing the Poynting vector in terms of the $E_\pm$ using (7), and ignoring terms associated with third-harmonic generation owing to the absence of phase-matching, we obtain the coupled-mode equations

$$\begin{aligned}
+\mathrm{i}\frac{\partial F_+}{\partial z} + \frac{\mathrm{i}\bar{n}}{c}\frac{\partial F_+}{\partial t} + \delta F_+ + \kappa F_- + \Gamma|F_+|^2 F_+ + 2\Gamma|F_-|^2 F_+ &= 0\,,\\
-\mathrm{i}\frac{\partial F_-}{\partial z} + \frac{\mathrm{i}\bar{n}}{c}\frac{\partial F_-}{\partial t} + \delta F_- + \kappa F_+ + \Gamma|F_-|^2 F_- + 2\Gamma|F_+|^2 F_- &= 0\,.
\end{aligned} \tag{14}$$

Here the first and second nonlinear terms in each equation describe self- and cross-phase modulation , and

$$\begin{aligned}
\kappa(z) &= \frac{\pi}{\lambda_\mathrm{B}}\Delta n(z)\,,\\
\delta(z) &= \frac{2\pi}{\lambda_\mathrm{B}}\delta n(z) - \frac{1}{2}\frac{\mathrm{d}\varphi}{\mathrm{d}z}\,,\\
\Gamma &= \frac{2k_0}{Z_0}n_2\,,
\end{aligned} \tag{15}$$

where Z_0 is the vacuum impedance. Thus, in the linear limit, the grating is fully specified by the functions κ and δ; as shown below, these correspond to spatial variations in the grating strength, and the Bragg frequency, respectively. Note that by changing the definition of the $F_\pm$, which can be achieved by including a prefactor in (7), the value of Γ can be adjusted arbitrarily. Often, this freedom is used to define the $F_\pm$ such that $|F_\pm|^2$ are the energy flows in the $\pm z$ directions. However, we do not do this here.

Equations (14) are the coupled-mode equations that form the basis of the theoretical and numerical work described below. Recall that our derivation is open to suspicion; for example, assuming that the envelopes are slowly varying is not justified in the light of the terms varying as $\exp[\pm\mathrm{i}(2\pi/d+k_0)z]$ that were dropped. Nonetheless, the type of argument used is sound. Rigorous derivations have been given [2,10,14] and also lead to (14). Thus, even though our derivation is somewhat cavalier, the resulting equations are correct.

2.1.1 Dispersion Relation and Eigenfunctions

The dispersion relation of the coupled-mode equations indicates the properties of low-amplitude plane waves in a uniform grating. To obtain these, we set $E_\pm(z,t) = a_\pm \exp[\mathrm{i}(Qz - c\Delta t/\bar{n})]$, where the constants $a_\pm$ are the amplitudes of the forward- and backward-propagating components, and Δ [cf. (4)] and Q signify small deviations of the frequency and wavenumber from ω_0 and k_0, respectively. We also drop the nonlinear terms in (14) and we take δ and κ to have the desired (constant) values. This leads to the set of coupled (algebraic) equations

$$\begin{aligned}(\Delta - Q + \delta)a_+ + \kappa a_- &= 0\,,\\ \kappa a_+ + (\Delta + Q + \delta)a_- &= 0\,.\end{aligned} \tag{16}$$

Eliminating the amplitudes $a_\pm$, we find that

$$\Delta = -\delta \pm \sqrt{Q^2 + \kappa^2}\,, \tag{17}$$

which is the dispersion relation for the coupled-mode equations, and which is illustrated in Fig. 6.

The dispersion relation exhibits a bandgap for frequencies Δ such that $-\delta-\kappa < \Delta < -\delta+\kappa$. It thus has a width of 2κ, centered about $\Delta = -\delta$. Thus, as mentioned, the Bragg frequency, which is at the center of the photonic bandgap, is determined by δ. Using (11) and (12), it can be seen that this corresponds to an actual frequency

$$\omega = \omega_\mathrm{B} - \frac{2\pi}{\lambda}\frac{c}{\bar{n}}\delta n + \frac{1}{2}\frac{c}{\bar{n}}\frac{\mathrm{d}\varphi}{\mathrm{d}z}\,, \tag{18}$$

where the last two terms on the right-hand side are small corrections to the nominal Bragg frequency ω_B due to deviations of the background refractive index from the nominal refractive index, and of the period of the grating from its nominal value. The (photonic) bandgap discussed here was discussed earlier in Sect. 1. Light with a frequency inside the photonic bandgap cannot propagate through the structure and is reflected, whereas light with a frequency outside the gap is reflected to a lesser degree. The width 2κ of the

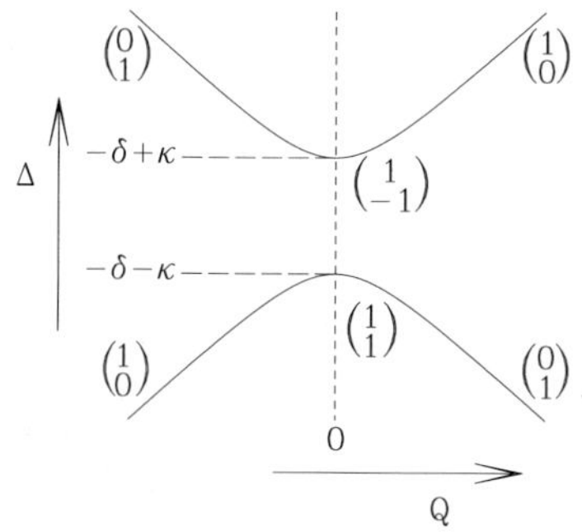

Fig. 6. Dispersion relation associated with the coupled-mode equations (14), showing the detuning $\varDelta$ versus the local wavenumber Q. Also indicated are the (unnormalized) eigenstates (see text)

gap, which is consistent with (3), is determined by the grating strength via the first equation of (15).

It was discussed in Sect. 1 why the group-velocity outside the photonic bandgap can differ substantially from that in the uniform fiber. This is confirmed by a calculation based on the dispersion relation, according to which the group-velocity v_{g} is given by

$$v_{\mathrm{g}} = \frac{\mathrm{d}\omega}{\mathrm{d}k} = \frac{c}{\bar{n}}\frac{\mathrm{d}\varDelta}{\mathrm{d}Q} = \pm\frac{c}{\bar{n}}\frac{Q}{\sqrt{\kappa^2+Q^2}} = \pm\frac{c}{\bar{n}}\sqrt{1-\frac{\kappa^2}{(\varDelta+\delta)^2}}\,, \tag{19}$$

and is thus proportional to the slope of the dispersion relation. Thus, consistently with Fig. 6, the magnitude of the group-velocity vanishes at the bandgap edges where $\varDelta+\delta=\pm\kappa$, whereas the group-velocity is largest far from the Bragg resonance, where it equals that in the uniform fiber.

In a similar way, one may calculate the group-velocity dispersion $\omega_2 = \mathrm{d}^2\omega/\mathrm{d}k^2$ of the grating, corresponding to the curvature of the dispersion relation, which is of the form

$$\omega_2 = \pm\frac{c}{\bar{n}}\frac{\kappa^2}{(\kappa^2+Q^2)^{3/2}} = \pm\frac{c}{\bar{n}}\frac{\kappa^2}{(\varDelta+\delta)^{3/2}}\,. \tag{20}$$

Thus the magnitude of the group-velocity dispersion (GVD) peaks at the edges of the photonic bandgap, and decreases further away from the Bragg resonance. Below we shall also use $k_2 = \mathrm{d}^2k/\mathrm{d}\omega^2$, which is found to be

$$k_2 = -\left(\frac{\bar{n}}{c}\right)^2\frac{\kappa^2}{[(\varDelta+\delta)^2-\kappa^2]^{3/2}}\,. \tag{21}$$

Note that k_2 diverges at the band edges, where $\varDelta+\delta=\pm\kappa$. This is because at these frequencies the inverse group-velocity also diverges (see (19)). One therefore often prefers to use ω_2 when working with gratings. From (20) and (21) it is easy to confirm that the grating dispersion is much larger than that of the uniform fiber (see also Sect. 1) [5]. The argument outlined here can be extended to include dispersion of increasingly higher orders. It is then found that the grating's cubic dispersion, for example, can also be much larger than that of the uniform fiber [4].

The discussion thus far has focused on the *eigenvalues* of the system (16). Here we briefly consider the *eigenvectors*, or "Bloch functions", giving the amplitudes of the forward- and backward-propagating modes $a_{\pm}$. Some of the eigenvectors are indicated in Fig. 6: far from the Bragg resonance the eigenvectors $\psi \equiv (a_+, a_-)^{\mathrm{T}}$ are $\psi = (1,0)^{\mathrm{T}}$ and $\psi = (0,1)^{\mathrm{T}}$. Thus, from (7), far from the Bragg resonance, the grating's eigenstates are purely forward or backward propagating, just as in a uniform medium. This is not surprising, as we would expect the effect of the grating to be small for frequencies far from the Bragg resonance. At frequencies closer to the Bragg resonance, the eigenvectors of the grating exhibit mixing, with the maximum degree of mixing occurring at the photonic-bandgap edges. According to Fig. 6, at the bottom and top of the photonic bandgap the eigenvectors are $\psi = (1, \pm 1)^{\mathrm{T}}$. Since $Q = 0$, we find from (7) that the spatial part of the electric field can be written as $\cos(k_0 z)$ and $\sin(k_0 z)$, respectively. Thus the eigenfunctions vary between forward- and backward-propagating plane waves, at frequencies far from the Bragg resonance, to standing waves at the edges. It is not surprising to find standing waves at the bandgap edges, since the energy flow vanishes.

2.2 CW Solutions to the Coupled-Mode Equations

Although the main emphasis in this review is on the properties of Bragg solitons, we briefly discuss some of the CW solutions to the nonlinear coupled-mode equations (14). These were first discussed by Winful et al. in 1979 [12], who showed that they can be written in terms of Jacobian elliptic functions. These authors showed that, at high intensities, a grating can have perfect transmission for frequencies inside the linear photonic bandgap. The reason for this is that, at high intensities, the envelope functions $E_{\pm}$ are periodic functions of position.[3] If the length of the grating equals this period, then the electric fields at the front and the back of the grating are identical and the transmissivity is then thus unity.

An associated feature is that the transmissivity of a grating is multistable, leading to possibilities of all-optical switching . Some of these schemes are reviewed in Sect. 4.1. By considering the fully time-dependent coupled-mode equations, it was shown later [15–17] that many of the higher-order branches of the multistability curves are unstable. The instability, in fact, leads to the formation of Bragg solitons, to be discussed in Sect. 2.4.

[3] Strictly, the envelopes are always periodic, but the period diverges in the linear limit.

2.3 Nonlinear-Schrödinger-Equation Limit

It is well known that (almost) all systems with a weak nonlinearity and a strong dispersion, can, to some approximation, be described by the NLSE. Here we discuss the NLSE limit of (14). This limit can be derived by writing the electric field as the product of a single Bloch function and a slowly varying envelope function. This is justified if (i) the dispersion (see (20) and (21)) does not vary appreciably over the spectrum of the electric field, and (ii) the intensity is not too high, viz. $n_2 I \ll \Delta n$. In the experiments performed to date, these conditions are mostly well satisfied.

Following the discussion above, the envelope functions $E_\pm$ are written as

$$\begin{pmatrix} E_+ \\ E_- \end{pmatrix} = (a\psi_\oplus + b\psi_\ominus + \ldots)\, \mathrm{e}^{\mathrm{i}(Qz - \Delta_\oplus t)} \, . \tag{22}$$

Here $\psi_\oplus$ and $\psi_\ominus$ are the two eigenvectors of (16), corresponding to the two Bloch functions at Q, and $\Delta_\oplus$ is the eigenfrequency associated with $\psi_\oplus$ (see (17)); the $\oplus$ refers to the upper branch in Fig. 6, while the $\ominus$ refers to the lower branch. Further, a and $b \ll a$ are slowly varying envelope functions; since $a \gg b$, the first term in brackets dominates. In applying (22), we assume that the spectrum of the electric field is predominantly above the Bragg resonance since the high-frequency Bloch function dominates (recall that $a \gg b$ in (22)). A suitable ansatz for fields with frequencies mostly below the Bragg resonance is obtained by exchanging $\oplus$ and $\ominus$ in (22).

Equation (22) is now substituted into the coupled-mode equations (14), and terms of similar size are collected using the multi-scale (or multiple scales) method. The lowest-order result of interest is then that a satisfies [18]

$$\frac{\partial a}{\partial z} + \frac{1}{v_\mathrm{g}}\frac{\partial a}{\partial t} = 0 \, , \tag{23}$$

where v_g is given by (19). Thus, to this order, the envelope a travels at its group-velocity. If terms at the next order are collected, the NLSE is obtained [18],

$$\left(\frac{\partial a}{\partial z} + \frac{1}{v_\mathrm{g}}\frac{\partial a}{\partial t}\right) + \frac{k_2}{2}\frac{\partial^2 a}{\partial t^2} + \alpha |a|^2 a = 0 \, , \tag{24}$$

Here k_2 is as given in (21), and the nonlinear coefficient α is

$$\alpha = \frac{3 - v^2}{2v}\Gamma = \left(\frac{4\pi}{\lambda_\mathrm{B} Z_0}\right)\frac{3 - v^2}{2v} n_2 \, , \tag{25}$$

where $Z_0 = 377\ \Omega$ is the vacuum impedance, and $v = \bar{n} v_\mathrm{g}/c$ is the group-velocity in units of $c/\bar{n}$. Note that $\alpha \geq \Gamma$ in general. In particular, the factor $(3 - v^2)/2 \geq 1$ is due to the fact that, unlike the case for plane waves, the amplitudes of the Bloch functions vary with position, the more so with decreasing v. The extra factor v^{-1} is a trivial geometrical factor associated with the reduced group-velocity. On the basis of the fact that a grating structure can lead to $\alpha > \Gamma$, it has been argued that the performance of nonlinear devices can be improved by including appropriate gratings to enhance the effective nonlinearity [19].

The use of the NLSE rather than the coupled-mode equations has a number of advantages. An important advantage is that it is integrable , and has soliton solutions. The best known of these, the fundamental soliton, can be written as

$$a = \sqrt{\frac{k_2}{\alpha}} \frac{1}{t_0} \operatorname{sech}(t/t_0)\, \mathrm{e}^{\mathrm{i}k_2 z/(2t_0^2)} , \tag{26}$$

where t_0 is an arbitrary constant representing the soliton's width. Pulses of this shape propagate through the grating without changing shape, in spite of the strong dispersion and nonlinearity. Another important parameter of the NLSE is the soliton period z_0, the length scale over which fundamental soliton can change their shape, and the oscillation period of higher-order solitons [7]. Since $z_0 \propto 1/\beta_2$ [7], it is much smaller than in a uniform fiber, and, in fact, is typically a few centimeters, rather than hundreds of meters [6].

As a final comment, it should be noted that the NLSE can also be derived directly from the wave equation for the electric field, and the intermediate step of deriving the coupled-mode equations is thus not required; the NLSE can thus be applied to both shallow and deep gratings [10].

2.4 Bragg Grating Solitons and Gap Solitons

Any solution to the NLSE must correspond, via (22) and (25), to an approximate solution to the coupled-mode equations. Indeed, the one-soliton solution (26) approximates a two-parameter family of solutions to the coupled-mode equations [18], first discussed by Christodoulides and Joseph [20] and first given in their most general form by Aceves and Wabnitz [21]. The two free parameters are the pulse's velocity, the magnitude of which can be anywhere between 0 and $c/\bar{n}$, and a detuning parameter. The first experimental observation of Bragg solitons was reported by Eggleton et al. [22]. Their stability was discussed by Barashenkov et al. [23] and Rossi et al. [24]. In the context of high-intensity pulse propagation through a grating, we refer to these solutions as *Bragg grating solitons*, or Bragg solitons for short. As a matter of convention, if the spectrum of these solitons overlaps the photonic bandgap considerably, they are referred to as *gap solitons*, first observed by Taverner et al. [25]. This term was coined by Chen and Mills on the basis of a numerical analysis of the CW solutions for a nonlinear grating [26].

Gap solitons can be considered to be one of the most striking effects in Bragg gratings, in that the contrast between the linear and nonlinear regimes is very strong: in the linear regime most of the light is reflected, but at high intensities low-velocity, narrow pulses emerge. While here we discuss only bright solitons, dark solitons and solitons on a finite background have also been found as solutions to the coupled-mode equations [27]. As a final comment, though the work discussed here refers to temporal solitons, a geometry in which spatial Bragg solitons can be observed was proposed by Feng [28].

3 Experimental Considerations: Grating Fabrication

3.1 Optical-Fiber Bragg Gratings

Here we provide a brief overview of the different holographic methods used to fabricate optical-fiber gratings. The formation of an optical fiber grating relies on the mechanism of photosensitivity in germanosilicate glass. [29] Briefly, when the fiber core, which contains germanium, is exposed to ultraviolet light with wavelengths around $\lambda = 193$ nm or $\lambda = 240$ nm, the refractive index increases permanently. The dynamics of the refractive-index change with exposure and the details of the chemistry of the photosensitivity are not completely understood and are beyond the scope of this discussion. For excellent overviews of this topic, we refer to the book by Kashyap [1] and the review article by Campbell and Kashyap [30].

The first observation of refractive-index changes in germanosilicate glass was reported by Hill and coworkers [29] in 1978. They described a permanent grating written in the core of an optical fiber by an argon laser line at 488 nm launched into the fiber. It was later understood that the formation of this grating relied on a two-photon process [31]. It was not until many years later that researchers at United Technologies demonstrated a versatile method for writing gratings. This method uses two ultraviolet beams, which are generated outside the fiber and then made to interfere in the core [32], with the period of the interference made to match the period of the desired grating. This allowed the writing of gratings to reflect light in the fiber at the very important telecommunications wavelength (≈ 1.5 μm) and paved the way for manufacture of gratings. Several different interferometric arrangements have been developed, the most generic being the amplitude interferometer illustrated schematically in Fig. 7, in which the UV beam is split into two beams of equal intensity that are later recombined after traversing different paths. The period of the grating d is then related to the irradiation wavelength λ_W and the half-angle between the intersecting beams θ via $d = \lambda_W/(2\sin\theta)$. Bulk interferometers such as those described above are susceptible to vibration and require sources with good temporal and spatial coherence. These factors limit the length of the gratings that can be fabricated with them.

A major step forward occurred when researchers at CRC Laboratories in Canada [33] and Bell Laboratories in the USA [34] demonstrated the writing of optical-fiber gratings using a phase mask. This phase mask consists of

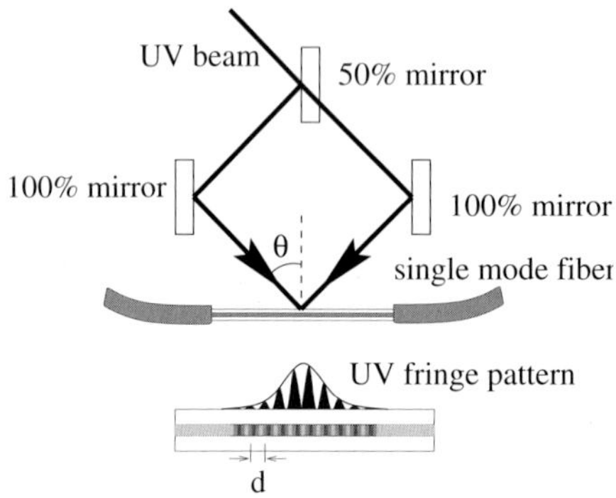

Fig. 7. Experimental setup for the production of Bragg gratings in a photosensitive optical fiber by UV irradiation interferometry

a relief grating, etched in a silica plate, created either holographically or by electron beam lithography. The principle of the phase mask is based on the diffraction of an incident ultraviolet beam into several orders. Grating fabrication relies on interference between the +1 and −1 orders in the fiber core. The fringe period is one-half the period of the mask. The other diffracted orders are regarded as a nuisance, and several methods for reducing them have been proposed [33]. Typically, the zeroth order contains less than 5% of the total power illuminating the optical fiber.

There are many advantages to using phase masks: the requirements on the laser beam's temporal and spatial coherence are reduced; the sensitivity to environmental fluctuations is reduced as the fiber is placed in close proximity to the mask; and the ultraviolet beam can be scanned along the length of the mask [35], allowing the fabrication of very long gratings, the length of which is limited by the mask length.

Other developments include the demonstration of enhanced photosensitivity as a result of loading the fiber with hydrogen [36]. Hydrogenation is typically carried out by diffusing hydrogen molecules into the fiber core at high pressure and temperature. Permanent refractive-index changes as high as 0.01 – over an order of magnitude higher than could be achieved without hydrogen loading – have been demonstrated using this technique [36]. Williams et al. demonstrated that boron could play a similar role to hydrogen when included as a codopant in the core [37]. Typically, a combination of hydrogen loading and boron doping is used to achieve efficient grating fabrication.

As mentioned in Sect. 1, apodization reduces the sidelobes in the reflection spectrum of the grating. This is critical for efficient coupling of pulses into the grating and is also essential for implementation in telecommunication systems [38,39]. Several different schemes for apodizing gratings have been proposed. One method for producing apodized fiber Bragg gratings, demonstrated by Strasser et al. [39], involves two steps. First, an ultraviolet beam is scanned along the length of the phase mask. The intensity of the UV beam is varied as the beam is scanned, creating the modulated component of the grating. Then the unmodulated component is imprinted in the same manner, without the phase mask. The postprocessing of the grating ensures that the average refractive index of the grating remains constant along the length of the grating, even though its strength varies. Typically, the apodization profile is a raised cosine function (Fig. 8).

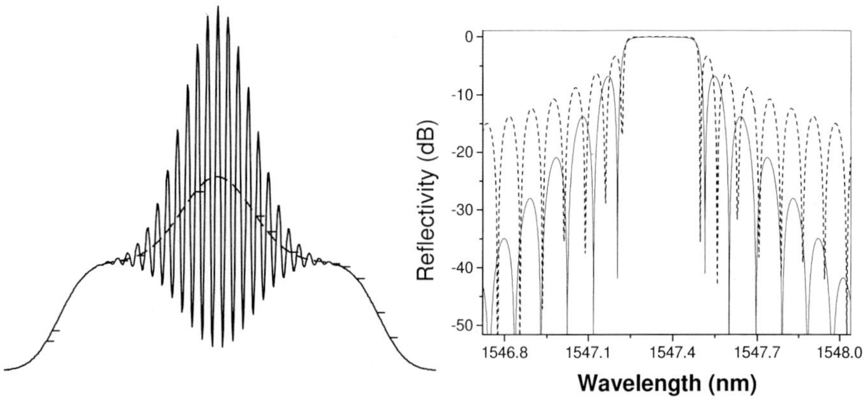

Fig. 8. Schematic of an apodized fiber grating, showing the refractive index versus position (*left*) and the reflection spectrum (*right*); the *dashed line* shows the spectrum without apodization, and the *dotted line* the spectrum with apodization

Complex, chirped gratings have been fabricated using a variety of techniques. There are two distinct methods for obtaining complex grating structures: (i) the introduction of nonuniform perturbations into a uniform grating by, for example, temperature or strain tuning, and (ii) by complex interferometry. Schemes of type (i) have been used to realize a range of complex grating devices [40,41]. The advantage of this method is that the grating chirp can be adjusted, for example by varying the temperature gradient [42]. Slusher et al. [43] performed an experiment in which a linear temperature gradient was imposed along the length of an optical-fiber grating, allowing them to study the dynamics of complex nonlinear pulses on chirped fiber gratings (see Sect. 5.3). The ability to adjust the chirp gradient allowed pulse propagation to be studied in different regimes. In schemes of type (ii) complex chirps are introduced into the grating by interfering two nonplanar UV beams in the core of the optical fiber. Sugden et al. [44] demonstrated that complex, chirped gratings can be written using such a technique. Alternatively, by varying the velocity of the UV beam when fabricating a grating by the scanning method [35], one can also generate nonuniform grating profiles [45].

3.2 Semiconductors

Semiconductor waveguides have many desirable features for applications in nonlinear optics, including high optical nonlinearity, large refractive-index modulation, and good optical confinement. Though the interaction lengths can be quite long, and the losses can be low, these two latter quantities are typically not as good as in optical-fiber gratings. Fabrication of integrated semiconductor waveguide filters has commonly been achieved with a two-step process in which the waveguides are defined prior to the formation of the grating. Here we briefly review the fabrication of semiconductor grating filters, focusing on geometries that were used in nonlinear-switching experiments.

Sankey et al. [46] used silicon-on-silicon structures for their nonlinear experiments. These structures consisted of a thin, monocrystalline silicon layer separated from a silicon substrate by an insulating layer, which was usually an oxide or nitride of silicon. The waveguides were fabricated using a bond-and-etch-back technique. The gratings were then written using a high-resolution electron beam lithography system. The optical nonlinearity exploited here is due to the photogeneration of a population of real carriers through the absorption of a light pulse, and hence is only approximately of the Kerr type used in most theoretical analyzes.

More recently, in experiments performed by Millar et al. [47], the grating and waveguide were defined simultaneously, using a simple one-step electron-beam-based approach. These devices used a weak grating on a strip-loaded waveguide to minimize losses. This allowed the fabrication of 4–6 mm long high-quality grating filters with scattering losses as low as 2 dB/cm. These authors used molecular-beam epitaxy to grow the underlying AlGaAs wafer. The structure comprised a lower cladding layer 4 μm thick, containing 24% Al; a waveguide layer 1.5 μm thick, containing 18% Al; and an upper cladding layer 1 μm thick, containing 24% Al. The waveguides were then etched using reactive-ion etching, in order to achieve the desired grating strength.

Coriasso et al. [48] reported all-optical switching and pulse routing in a distributed-feedback multiple-quantum-well (MQW) nonlinear waveguide. MQWs exhibit large, efficient nonlinearities at photon energies slightly below the fundamental exciton resonance, when the material is sufficiently transparent. As in the experiments by Sankey et al. [46], the optical nonlinearity exploited was due to photogeneration of a population of real carriers. The semiconductor heterostructure was grown by chemical-beam epitaxy and consisted of 30 InGaAs quantum wells with a width of 3.7 nm, separated by 10.0 nm unstrained InAlAs barriers. This later was surrounded by a 0.25 μm InP buffer layer and a 0.3 μm InP capping layer. A single-mode ridge waveguide was then fabricated by reactive-ion etching into the capping layer, leaving stripes 2 μm wide and 0.1 μm thick. A grating was then formed with a period of 0.240 μm; the grating was defined by electron beam lithography and etched into the capping region to a depth of 0.2 μm by reactive-ion etching.

3.3 Chalcogenide Glasses

Chalcogenide glasses are promising materials for use as linear and nonlinear integrated optical elements. This is mainly due to their high optical transparency in the near and far infrared, and their extremely high optical nonlinearity. Nonlinearities as high as 100 times that of silica have been demonstrated in fiber-based chalcogenides [49], and recent measurements with bulk samples have revealed nonlinearities as high as 1000 times that of silica [50]. Such high nonlinearities are required if these nonlinear effects are ever to be exploited in actual telecommunications applications.

Spälter et al. [51] recently demonstrated record nonlinear phase shifts in planar chalcogenide glass waveguides. These guides were written in glass films using photodarkening. The nonlinearity of the waveguide, inferred from the nonlinear pulse propagation experiments, was determined to be 100 times higher than in silica, with an acceptable level of two-photon absorption, having an associated figure of merit [52] of $n_2/(\beta\lambda) \approx 17$, where β is the two-photon absorption coefficient [51]. Further work by Lenz et al. demonstrated that nonlinearities as high as 500 times the value for silica are possible, and even higher values are possible with some further material optimization [50]. These authors investigated a variety of Se-based chalcogenides using Z-scan techniques to determine both the nonlinearity and the nonlinear absorption accurately.

It has also been shown that chalcogenide glasses can be photosensitive when exposed to near-bandgap illumination. For example, As_2S_3, a common chalcogenide, exhibits strong photosensitivity at wavelengths close to its bandgap, around 514 nm [53]. The index changes in thin films of chalcogenides are due to both photodarkening and photoexpansion of the glass [54]. Saliminia et al. [55] recently demonstrated first- and second-order Bragg reflectors fabricated at telecommunication wavelengths in single-mode monolayer and multilayer chalcogenide glass planar waveguides. They employed near-bandgap illumination and used an interferometric technique. The main mechanism responsible for the index changes in these planar samples was photodarkening [55].

3.4 Experimental Apparatus

In Sect. 4.2 we present a summary of the experimental results obtained by Eggleton et al. using optical-fiber Bragg gratings [6]. To lay a foundation for describing these results, we describe here the experimental apparatus used. More generally, this experiment is typical of the high-power experiments that have been performed by others, for example Taverner et al. [25,56].

The experimental setup is shown in Fig. 9 [57]. The light source was a mode-locked Nd:YLF laser generating pulses at a wavelength of 1053.2 nm. The laser was mode-locked using an active intracavity modulator, generating a train of pulses with a duration of approximately 80 ps. To achieve even

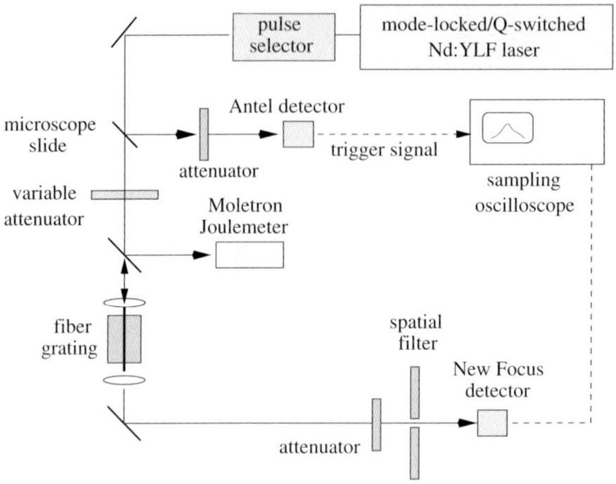

Fig. 9. Schematic of our experimental setup. Note that the figure is not drawn to scale, and that the path length between the grating and the detector is about 5 m

higher peak intensities required simultaneous Q-switching of the laser to obtain repetitive Q-switched bursts of short, intense, mode-locked pulses, at a repetition rate of 500 Hz. Under these operating conditions, every 2 ms the laser emitted a train of pulses, each with a duration of approximately 100 ps. Because the individual pulses in the mode locked Q-switched train had different peak intensities, a single pulse was selected using an electro-optic modulator. This allowed a single, unique pulse, repeated at a rate of 500 Hz, to be studied.

By operating at low repetition rate, we maintained the average power below 1 mW, minimizing the possibility of optical damage resulting from thermal effects. Several forms of photoinduced optical damage were observed in these fiber experiments (both at the fiber input facet and in the grating itself) and were attributed to multi-photon processes.

Key to these experiments was the fact that the pulse incident upon the grating was very close to transform limited. Owing to the very high intensities achieved in the YLF cavity, self-phase modulation can lead to spectral broadening [6,58,59]. This was particularly true near the peak of the Q-switched train, where the nonlinear phase shift could exceed 2π. By choosing a pulse near the leading edge of the Q-switched train, we obtained a pulse with high peak power and small "prechirp", corresponding to a nonlinear phase shift of 0.13π.

Because the central wavelength of the laser was fixed, it was necessary to tune the grating instead. Fortunately, this was quite straightforward and could be achieved by strain-tuning the grating. The fiber grating was mounted onto a precision mechanical translation stage. The length of untreated fiber before the grating was minimized to less than 5 mm. This ensured minimal spectral broadening in the pigtail of the fiber. This broadening was undesirable, as it affected propagation through the fiber and complicated the interpretation of results. The length of fiber after the grating was 20 cm and was used to strip the cladding modes that can "leak" through the grating. The transmitted light was then detected using a fast photodiode with a response time of 9 ps. The net time resolution, including trigger jitter, was approximately 20 ps.

A reflection spectrum of the 55 mm apodized Bragg grating, shown in Fig. 10, was obtained by strain tuning; the resolution of about 0.01 nm was limited by the precision mechanical translation stage. The steep edges on the short- and long-wavelength sides of the photonic bandgap are characteristic of apodized gratings, and differ from those for uniform gratings (see Fig. 8). The grating bandwidth is approximately 0.17 nm, corresponding to a maximum coupling strength of $\kappa = 700\ \mathrm{m}^{-1}$ in the uniform section of the grating.

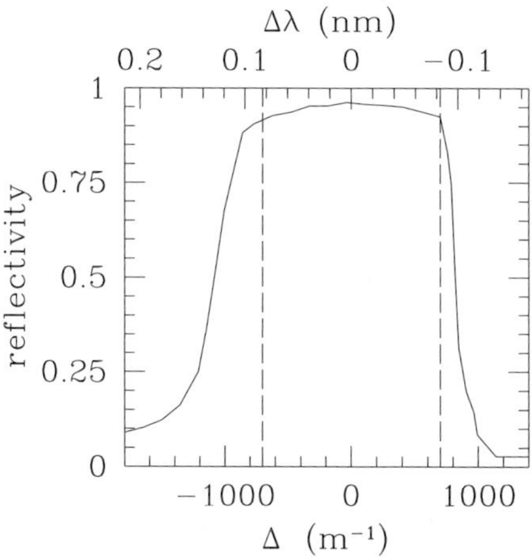

Fig. 10. Measured grating reflection spectrum, showing the reflectivity versus the wavelength (*top scale*) and versus the detuning (*bottom scale*). The origin is at the inferred bandgap center. The *vertical dashed lines* are estimates of the bandgap edges

4 Summary of Experimental Results

4.1 Continuous-Wave Experiments

The first experimental investigation of nonlinear optics in periodic structures was performed by La Rochelle et al. [60]. In this experiment, cross-phase modulation [7] was employed in an optical-fiber Bragg grating to achieve switching of a probe beam by a control beam. The control pulse was detuned from the photonic bandgap, so soliton effects were not observed. Furthermore, the control pulse used in these experiments was longer than the length of the grating, and thus only CW dynamics were observed. A subsequent paper by Lauzon et al. [42] investigated numerically the full time dependence of cross-phase modulation in the grating.

The first detailed experimental study of all-optical switching in a nonlinear periodic structure was reported by Sankey et al. [46]. The structure used in these experiments was a corrugated, silicon-on-insulator optical waveguide. As discussed in Sect. 3.2, the nonlinearities of the samples were due to free-carrier effects and were estimated to be such that $n_2 \approx 10^{-10}$ cm^2/W, which is more than five orders of magnitude larger than the nonlinearity of SiO_2. In contrast to the experiments of La Rochelle et al., here the incident pulses were coincident with the photonic bandgap and were reflected at low intensities. To achieve the high peak intensities necessary for switching of the bandgap, a Q-switched Nd:YAG laser, emitting pulses at 1.064 μm, was used. By operating at this wavelength, they could exploit the semiconductor nonlinearities (see Sect. 3.2). This high nonlinearity came at the expense of a long response time, determined to be approximately 1–5 ns [46]. The pulses had a width of 25–50 ns, and the pulse energy was between 1 and 20 μJ. These experiments demonstrated optical switching, true optical hysteresis, and a tendency toward unstable behavior at high intensities. In particular, these experiments demonstrated self-switching of a Bragg grating, whereby the grating acts as a mirror at low intensities, while at high intensities the nonlinearity shifts the photonic bandgap, resulting in an increase in transmission. These experiments were later extended to optically resonant periodic-electrode structures, and out-of-plane coupling into higher-order Bragg resonances [61].

Others have demonstrated a range of novel effects in nonlinear semiconductor periodic structures. For example, He and Cada theoretically and experimentally investigated switching in GaAs/AlAs uniform distributed-feedback gratings [62] and in combined distributed-feedback/Fabry–Perot structures. In contrast to the experiments of Sankey et al., these authors used the Fabry–Perot fringes that exist outside the photonic bandgap, rather than the photonic bandgap itself. These "out of band" resonances can be very narrow, providing the advantage of reducing the switching power. The reduced switching power does, however, come at the expense of speed. In these experiments, He and Cada also observed optical bistability and signal regeneration.

In a later experiment by Janz et al. [63], low-threshold optical bistability was observed in an asymmetric $\lambda/4$ distributed-feedback GaAs/AlAs heterostructure. In these experiments the nonlinearity was due to carrier generation, as the incident light wavelength was in the tail of the interband absorption. The $\pi/2$ phase shift creates a resonance in the photonic bandgap, the position of which depends on the incident intensity [64]. Such resonances can be very narrow, resulting in low-power optical switching, but at the expense of reduced switching speed. Janz et al. observed bistability thresholds that were 60 times lower than in a comparable uniform grating.

Other experiments have exploited thermo-optic nonlinearities in semiconductors. Fernando et al. [65] demonstrated thermo-optic switching in a 92 layer $Si/Si_{0.7}Ge_{0.3}$ distributed Bragg reflector. They observed switching times of less than 20 ns, with contrast ratios of better than 50%.

Herbert and Malcuit [66] demonstrated optical limiting in a three-dimensional periodic structure. They employed an ordered colloidal suspension immersed in a Kerr medium. At high peak intensities, they observed a dynamic shift in the Bragg resonance. In these experiments, the incident light was detuned to be just outside the photonic bandgap at low intensities. As the input intensity was increased, thermal nonlinearities resulted in an increase in the refractive index, shifting the photonic bandgap to longer wavelengths. This led to an increase in reflection and a decrease in the corresponding transmission.

4.2 Pulse Experiments

High-intensity grating experiments using (almost) uniform gratings were performed by Eggleton et al. [6]. The experimental setup was described in Sect. 3. Some of the results have already been shown in Fig. 5, which shows compression of input pulses due to the combined effects of the dispersion and the nonlinearity. The grating length of 55 mm is sufficient to see these effects, since the soliton period is of the order of a few centimeters [6].

Figure 11 shows the width versus the detuning at an estimated pulse energy of 0.20 μJ, corresponding to a peak power of roughly 11 GW/cm^2. The filled circles indicate experimental results; some of these are derived from the results in Fig. 5. The trends in Fig. 11 are well understood: at large detunings, the effect of the grating is negligible and the pulse is unaffected by it. At small

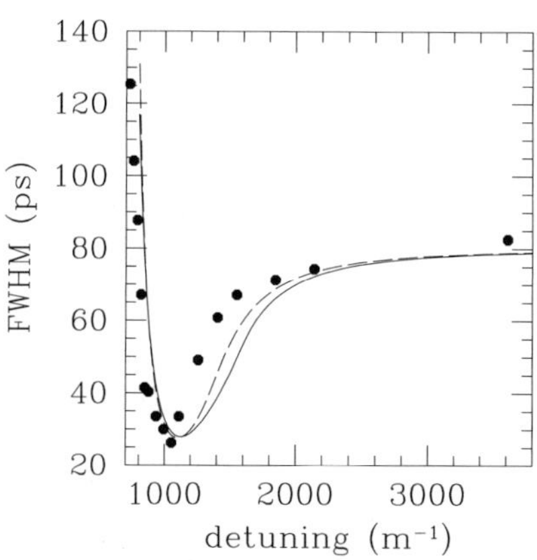

Fig. 11. Experimental results derived from the transmitted intensity versus time at at a pulse energy of 0.20 μJ. The *filled circles* indicate experimental results. Also shown are numerical results based on (14) (*solid line*) and (24) (*dashed line*)

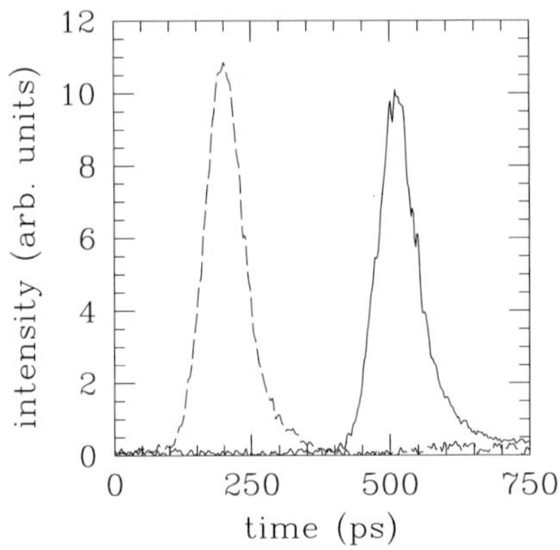

Fig. 12. Transmitted pulse at large detuning (*dashed curve*) and small detuning (*solid curve*). Note that the pulses have similar peak powers, but that the latter is delayed by about 300 ps

detunings, the grating dispersion dominates and the effect of the nonlinearity is small. The main difference of the results in Fig. 11 from the linear results (see Sect. 1) is that soliton compression occurs at intermediate values of the detuning [8]. The lines in Fig. 11 give the results of two calculations. The solid line follows from the coupled-mode equations (14), whereas the dashed line follows from the NLSE (24). The agreement between the theoretical and the experimental results is comparable for the two calculations.

Note from Fig. 11 that the pulse width is unaffected at $\Delta \approx 850\ \mathrm{m}^{-1}$. This does not mean that the pulse is not affected, as the delay at this detuning is substantial. Rather, at this detuning the input pulse is very close to being a fundamental Bragg grating soliton of the periodic structure, and the pulse should then thus propagate without changing shape, but with reduced group-velocity. This is illustrated in Fig. 12, which shows the transmitted pulse for a large detuning where the grating does not affect the pulse width (dashed line), and for a small detuning (solid line). The two traces clearly have roughly the same shape, but the solid curve is delayed by roughly 300 ps. Thus, by the argument above, the solid curve shows a Bragg grating soliton after transmission through the grating. The delay of 300 ps, over a grating of 55 mm length, implies that the average group-velocity of the pulse in the uniform section of the grating is $0.5c/\bar{n}$, the lowest group-velocity reported.

The results are similar to those in Fig. 11 for a wide range of pulse energies. However, when the pulse energy reaches roughly 0.40 μJ, higher-order soliton effects become apparent for some detunings. To illustrate this we refer to Fig. 13, in which we present the results from a set of measurements [67] different from that shown in Figs. 4, 5, 11, and 12. In these measurements a 60 mm apodized grating was used; since $\kappa = 205\ \mathrm{m}^{-1}$, it was substantially weaker than the grating used for the earlier results. The incoming pulse had a

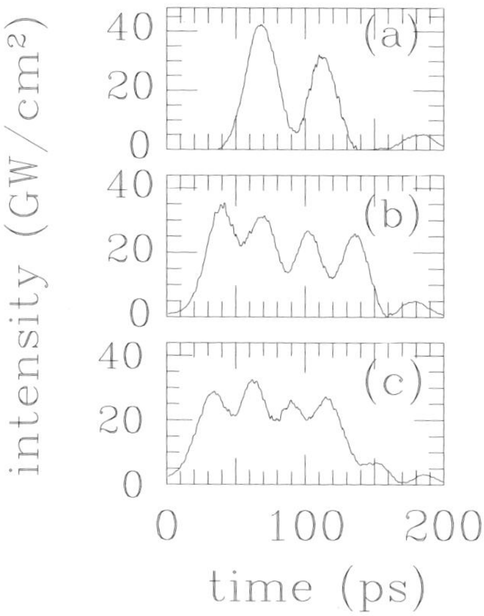

Fig. 13. Measured transmitted intensity versus time for three different detunings, showing the tunability of the instability. The detunings are (**a**) 250 mm^{-1}, (**b**) 269 mm^{-1}, and (**c**) 277 mm^{-1}. The bandgap edge is at $\Delta = 205$ mm^{-1}

width of 150 ps, and a peak power of 30 GW/cm^2. The three results shown in Fig. 13 were taken at different detunings, increasing from Fig. 13a to Fig. 13c; the actual values are given in the caption. Note that the repetition rate of the transmitted pulses depends strongly on the detuning; the decrease of the pulse spacing with increasing detuning is consistent with analytic results based on modulational instability in the NLSE [67]. Briefly, modulational instability is the process whereby a CW signal becomes unstable to amplitude fluctuations [7]. From studies of the NLSE it is known that, for positive nonlinearities, modulational instability occurs for anomalous dispersion, and that it leads to a train of solitons. Returning to Fig. 13, note that at the small detuning in Fig. 13a the instability is better developed than at the larger detunings in Fig. 13b, while it is least well developed in Fig. 13c. This is also consistent with results based on the NLSE, which show that the growth rate of modulational instability increases with decreasing detuning [67].

The results in Fig. 13 indicate that a grating can be used as a tunable pulse generator; the inverse repetition rate of the pulse train in this figure varies by almost a factor of two, between 42 ps in Fig. 13a and 27 ps in Fig 13c. Such an application cannot be implemented in a uniform fibre, since there the repetition rate is essentially independent of frequency.

The experiments discussed thus far were performed by Eggleton et al. at Bell Labs, Lucent Technologies, in the USA. Experiments were also performed by Broderick et al. at the Optoelectronic Research Centre at the University of Southampton in the UK. We now discuss some of their work.

The Southampton group used a directly modulated diode laser that was amplified by a series of fiber amplifiers [68]. The pulse width could be varied but was around 1 ns, and the center wavelength was around $\lambda = 1530$ nm [25]. This longer pulse length had an important consequence for the experiments, since clear grating soliton effects are only observable if the pulse spectrum is narrower than the width of the photonic bandgap. Because of the narrower pulse spectrum, the photonic bandgap could be narrower and so, from (15) and (17), the grating did not need as strong as that used by the Bell Labs group. The grating used in the experiments was apodized, with a sinusoidal envelope. The maximum value of κ for this grating was roughy 60 m^{-1}, which is roughly an order of magnitude smaller than that used by Eggleton et al.; the grating had a length of 80 mm. Figure 14 shows the time-resolved transmission through this grating as a function of time at various values of the input power, but at a fixed, positive detuning [25]. The leftmost feature in each trace is a part of the pulse that is detuned far from the grating and is thus not affected by it. Its transmission properties therefore also do not depend on power; however, this feature acts as a useful marker when comparing the traces. Apart from this feature, at low intensities the pulse is mostly reflected. At higher intensities, an increasing number of Bragg grating solitons can be identified, similarly to the results shown in Fig. 13; at a peak power of 8 kW, corresponding to a pulse energy of about 16 μJ, four solitons are observed. A simultaneous measurement shows that the reflection decreases with increasing power. It should be noted that a key difference between the data in Figs. 14 and 13 is that in the former, the frequency spectrum of the input pulse overlaps the local photonic bandgap in the region near the center of the grating (recall that the grating strength has a sine profile, and is thus strongest in the center). This is the reason that the pulse is mostly reflected at low intensities. The authors therefore point out that the solitons shown in

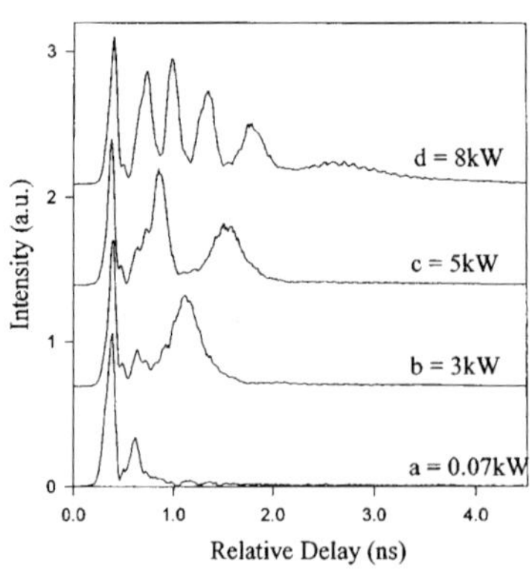

Fig. 14. Transmitted intensity versus time for fixed detuning at four different powers. Each trace is normalized to the incident pulse [25]

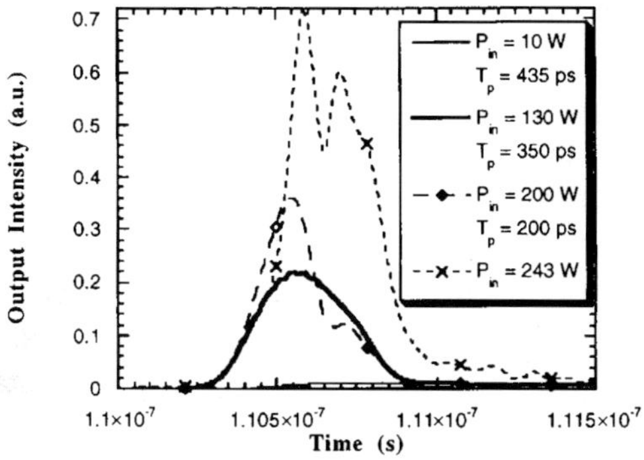

Fig. 15. Transmitted intensity versus time at fixed detuning at four different powers and pulse lengths, at a fixed detuning in a GaAs sample [47]

Fig. 14 are associated with the bandgap of the grating, and thus assert the observation of gap solitons rather than Bragg grating solitons. Recall that, in contrast, the results of the Bell Labs group are for frequencies outside the photonic bandgap of the grating and all power is transmitted, independent of input power.

In further work, in collaboration with the University of Glasgow, a number of experiments were performed with GaAs waveguide structures [47]. The important features of this geometry were discussed in Sect. 3.2. Figure 15 is similar to Fig. 14, but the data were taken with a GaAs sample containing a 4 mm grating with a Bragg wavelength around 1532 nm (the pulse widths vary between the traces). Clearly, qualitatively similar behavior to that in Figs. 13 and 14 is observed, in that the pulse develops multiple peaks. The refractive-index contrast in the grating used to obtain the results shown in Fig. 15 is similar to that in fibers, even though in principle it could be made much larger. Note also that the required power is over an order of magnitude smaller than that used to obtain the data in Fig. 14.

5 Further Developments

In this section we review some recent theoretical and experimental work that illustrates more general properties of nonlinear periodic media.

5.1 Optical Push Broom

We now discuss the optical push broom [69,70], which can be used for pulse compression. Though this effect is not directly related to Bragg grating solitons, it is a powerful illustration of the properties of nonlinear periodic media. The optical-push-broom scheme requires a strong pump pulse and a weak probe pulse. The probe frequency is on the long-wavelength side of the photonic bandgap, whereas the pump is detuned far from the Bragg wavelength.

In the push broom scheme the probe is initially ahead of the pump, as shown schematically in Fig. 16. Since the probe is tuned close to the Bragg frequency, it travels slowly (see (19)), and the pump, the velocity of which is not affected by the grating, thus catches up with it. It is well known that in a uniform medium with a positive Kerr nonlinearity, the leading edge of a strong pulse shifts to longer wavelengths because of self-phase modulation [7]. Since the pump is detuned far from the grating, the periodicity does not matter, and thus the pump's leading edge is redshifted by the same mechanism. When the pump catches up with the probe, then, by cross-phase modulation [7], it shifts the probe to longer wavelengths as well, starting with the probe's trailing edge, as illustrated in Fig. 16.

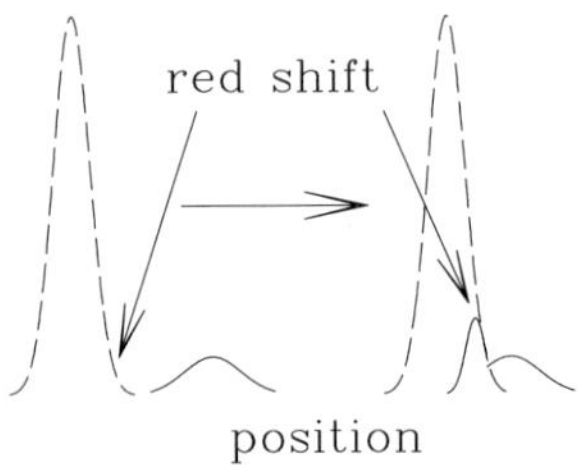

Fig. 16. Schematic of the optical-push-broom geometry. The strong pump pulse catches up with the probe pulse and shifts it to the red

Now recall that the probe is tuned to the long-wavelength side of the photonic bandgap, and is shifted to even longer wavelengths by its interaction with the pump. Therefore, according to (19), the probe speeds up. However, since this occurs first at the probe's trailing edge, the trailing edge catches up with the leading edge, and *the probe is thus compressed* by the interaction with the pump. Since the probe's asymptotic velocity equals that of the pump, in the final stages of the process the probe "rides" on the leading edge of the pump [69].

The compression of a probe pulse by a strong pump pulse was verified experimentally by Broderick et al. in an 80 mm fiber grating [70]. In their experiment the probe was CW, rather than pulsed. However, this does not affect the description given above. One of their main results is shown in Fig. 17, which shows the transmitted pump and probe intensities (dashed and solid lines, respectively) versus time. It is clear that the probe "rides" on the leading edge of the probe, as required, and has been compressed to a width of about 100 ps. It should be noted that the experimental results here are in excellent agreement with numerical simulations of the coupled-mode equations [70]. Approximate schemes for understanding the push broom have also been developed [71,72]. Pulse compression in a fiber grating in reflection has also been demonstrated [73].

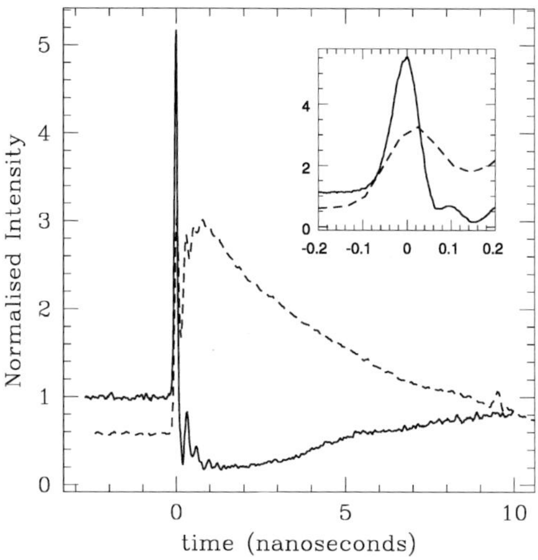

Fig. 17. Demonstration of the optical push broom by Broderick et al. [70], showing the intensity of the pump (*dashed line*) and probe (*solid line*) as a function of time. The inset *shows* more detail around the arrival time of the pulses. The pump and probe are shown on different vertical scales

5.2 Frequency Conversion and Generation

In Sect. 5.1 it was mentioned that the optical push broom has been simulated using the coupled-mode equations (14). In the simplest of these simulations, the pump is simply modeled as a refractive index variation that propagates through the grating at uniform velocity and without changing shape. In more realistic analyzes, the pump is modeled as an actual electric-field envelope that is detuned far from the Bragg resonance. These two types of simulations give very similar results.

Now consider again the simulations in which the pump is modeled as an electric-field envelope. If the detuning from the Bragg resonance is decreased, then effects other than the optical push broom are observed, particularly the generation of new frequencies. Frequency generation in gratings has been studied for many years, particularly second-harmonic generation [74–78].[4] The essence of this phenomenon, which was studied experimentally in 1976 [76], is that the grating provides a mechanism for phase matching, because close to the Bragg resonance the dispersion curve of a periodic structure is shifted with respect to that of the uniform medium. Gratings can also provide resonant enhancement, particularly around reflection zeros (see Fig. 3).

Similar, though richer effects have been predicted to occur in the presence of a grating in a *cubically* nonlinear medium [79,80]. In the presence of a strong

[4] This work should be distinguished from quasi-phase matching, in which the second-order susceptibility is varied periodically.

pump and a weak probe, a number of new frequencies can then be generated via four-wave mixing [7], both in the CW and in the pulsed regimes. A very rich dynamics has been found which depends on the detuning from the grating and on the wavevector mismatch of the underlying medium [79,80].

5.3 Nonuniform Gratings

The experiments discussed in Sect. 4 were all performed with uniform or apodized gratings. An obvious generalization is to consider nonuniform gratings. Nonlinear experiments with linearly chirped gratings were performed by Slusher et al. [43]. The chirp was obtained by applying a temperature gradient along the length of the fiber. Thus the position at which a frequency was reflected was approximately a linear function of position, and this effect could thus, in principle, be used for compensating the quadratic dispersion of an optical fiber [1,81]. For this reason, chirped gratings have been studied extensively in the linear regime.

In the experiments of Slusher et al. [43], high-intensity pulses were incident on the grating. Because of the nonlinear effects, the broadening of the pulses was much smaller than that expected at low intensities [82]. However, over a wide range of parameters, the pulses appeared to split inside the grating, resulting in the reflection of two pulses. An analysis of the pulse splitting was recently carried out [83]. According to this work, the pulse splitting occurs because the grating nonuniformity causes higher-order solitons to break up during propagation. Experimental work with other types of nonuniform gratings was reported by Eggleton et al. [84] and Broderick et al. [73].

5.4 Novel Nonlinear Grating-Based Compressors

Short optical pulses are desirable in a large variety of applications where time resolution, a large bandwidth, or a high peak power is necessary. Moreover, in telecommunication applications pulses with shorter duration will be needed as the per-channel capacity of communication systems increases beyond 10 Gb/s. In many cases nonlinear pulse compression is required to produce such pulses; both soliton and nonsoliton techniques have been demonstrated [7]. Typically, the Kerr nonlinearity of an optical fiber is used in conjunction with dispersion supplied by bulk devices, such as diffraction gratings, prisms, or an additional piece of fiber. In the case of optical solitons, the dispersion and nonlinearity occur simultaneously in the optical fiber [7]. The intrinsic dispersion of standard optical fibers is small (typically $\beta_2 \approx -20\ \mathrm{ps}^2/\mathrm{km}$ at communications wavelengths), which leads to long dispersion lengths. Thus compact pulse compressors are difficult to realize. In contrast, periodic structures, such as fiber Bragg gratings and photonic crystals, have large dispersion (see Sect. 2.1), which allows scaling down of the compressor to centimeter lengths. Various pulse compression schemes based on fiber Bragg gratings have been proposed to achieve small, versatile, all-fiber pulse compressors [85–87], some of which are discussed below.

The first grating-based compressor [85] is based on a two-section device, similar in principle to the Tomlinson compressor [88]. In this scheme, first proposed by Tomlinson et al., a fiber with positive dispersion broadens the spectrum and imposes a nearly linear, positive chirp on a high-intensity pulse through the combined effects of self-phase modulation and dispersion. The linear chirp is then compensated for by another dispersive element, such as a diffraction grating. By the use of a Bragg grating, the size of the compressor can be reduced substantially [85]. In this case the pulse is tuned to the long-wavelength side of the photonic bandgap and exploits the grating's strong positive dispersion. Numerical simulations of (14) have shown that efficient pulse compression can be achieved [86].

A second grating-based scheme, referred to as the adiabatic Bragg soliton compressor, was proposed by Lenz and Eggleton [86]. This scheme is based on the principle of adiabatic soliton compression [89] in a nonuniform grating in which the dispersion decreases along the grating [86]. This is achieved, for example, in a grating in which, apart from possible tapers at the edges, the strength κ decreases monotonically with position (see (21)). The adiabatic soliton compressor has the advantage that it only requires one grating, and produces clean, nearly transform-limited pulses. The NLSE model for nonlinear pulse propagation in a Bragg grating, discussed in Sect. 2.3, can be extended to treat nonlinear pulse propagation in a nonuniform Bragg grating and was used to design a compact adiabatic Bragg grating compressor [86]. A further extension is a proposal for a high-repetition-rate source of optical pulse trains based on the beating of two CW sources in a nonuniform Bragg grating [90]. Sending the waves through such a grating transforms the beat signal into a train of fundamental solitons, with the repetition rate determined by the frequency difference of the CW sources [90].

5.5 Polarization Effects

Our discussion of Bragg soliton effects to this point has neglected the fact that light has two polarizations. Two consequences of polarization for nonlinear propagation in gratings have been studied, both of which are discussed below.

5.5.1 Optical AND Gate

In 1993 Lee and Ho [91] proposed an all-optical AND gate based on a birefringent grating structure. Its operation relies on the fact that the bandgaps associated with the two polarizations are displaced from each other in frequency owing to the birefringence [91]. The frequancy of a pulse containing both polarizations is chosen such that, in the linear limit, both polarizations are within their bandgaps, but one more deeply than the other. At low intensities the pulse is reflected, while at high intensities the cross-phase modulation

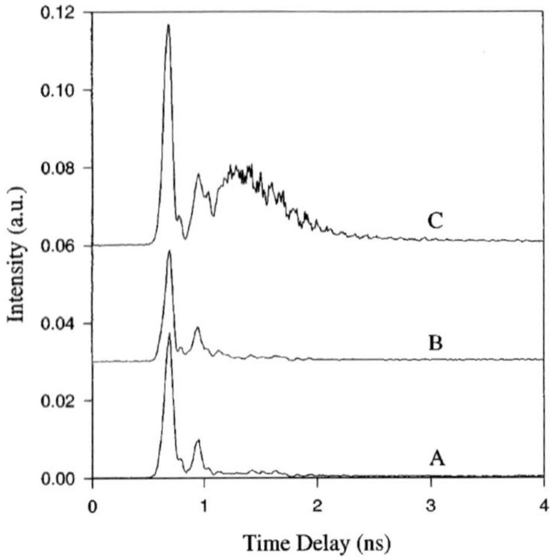

Fig. 18. Experimental results showing a grating acting as an AND gate, by Taverner et al. [56]. Two individual, orthogonally polarized pulses are reflected by the grating and the transmission is low (traces A and B). However, when they are incident simultaneously the transmission is substantial (trace C)

between the polarizations can switch one or both polarization out of the gap. This switching operation was observed by Taverner et al. [56] in 1998; their results are shown in Fig. 18. Traces A and B show that the transmittance through the grating is low when a pulse with either of the two orthogonal polarizations is incident on the grating. However, when both pulses are incident simultaneously, the transmission is high (trace C). A switching contrast of 17 dB was achieved, and the switching power was 2.5 kW. The high-intensity feature is tuned well away from the Bragg resonance and is not affected by it (see Sect. 4.2).

5.5.2 Polarization Rotation and Polarization Instability

It is well known that when intense light propagates through a uniform medium, the major axis of a beam of elliptically polarized light rotates at a rate that depends on the intensity, as first observed by Maker and Terhune [92]. In 1995 Barad and Silberberg [93] proposed a nonlinear switch based on the soliton analogue of this effect; the soliton nature of the pulse guarantees that the azimuthal angle of the polarization is constant across the pulse. They observed a rotation of 0.5 radians over 2 m of fiber. If the medium is birefringent, even more interesting effects can appear. For example, the propagation of linearly polarized light along the fast axis becomes unstable at high intensities, as the nonlinearity makes the medium, in effect, optically isotropic ("polarization instability" [7,94]).

Both these effects were observed experimentally in a fiber grating geometry [95]. Here we discuss the results showing the polarization instability. In the experiments it was found that for input polarizations near the fast axis, and input frequencies just above the photonic bandgap ($\Delta = 1.3\kappa$), the threshold for the onset of the instability is decreased by almost an order of magnitude compared with frequencies far away from the bandgap. A result obtained from these experiments is shown in Fig. 19, in which the dotted lines indicate the transmitted energy along the slow axis, and the solid lines the transmitted along the fast axis. The lower curves show the linear results; the higher curves show results for a peak input power of 2 $\mathrm{GW/cm^2}$. In the linear regime there is no exchange between the two polarizations. However, at high intensities the energy along the slow axis is larger than that along the fast axis.

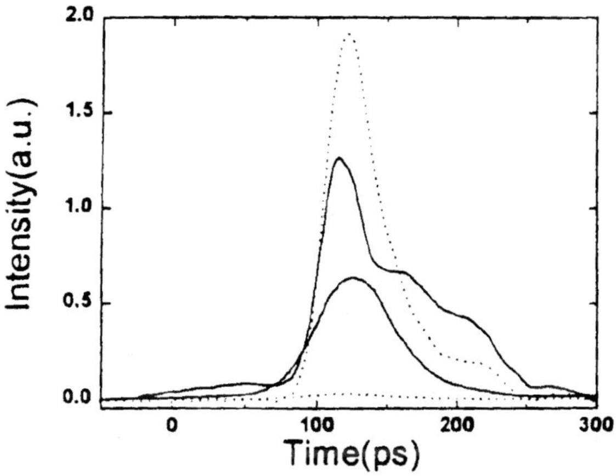

Fig. 19. Polarization instability in a fiber grating. Shown is the transmitted energy along the slow axis (*solid lines*) and fast axis (*dotted lines*), at a low intensity (*lower curves*) and high intensity (*upper curves*). The light is incident close to the fast axis

The analysis [96,97] follows the approach outlined in this review. In the nonlinear Schrödinger limit there are two coupled equations, one for each polarization, and in the coupled-mode description there four amplitude equations, describing the forward- and backward-propagating waves of each polarization. In optical fibers the growing of the grating itself leads to some birefringence, and of course polarization preserving fibers can be used if a larger birefringence is desired. Fiber grating structures offer the possibility of not only birefringence, which can be studied in virgin fibers, but significant differences in the group velocities of the two polarizations near the gap.

The dynamics of pulse propagation in the experiments discussed here involves all the complicated interplay between dispersion and nonlinearity that one can observe with a single polarization, together with the richness that comes from the variation of these properties with polarization, a variation that is determined self-consistently by the light pulse itself. Despite a number of exact and approximate solutions for special cases, much remains to be explored here both theoretically and experimentally.

5.6 Copropagating Geometries

Nonlinear periodic structures have also been studied extensively in geometries where they provide phase matching to a forward-propagating mode. The corresponding grating period is typically anywhere from 100 μm to 1 cm, as opposed to roughly the wavelength of light. The associated bandwidth of the resonance can be as high as hundreds of nanometers. In contrast to Bragg gratings, these long-period gratings are not inherently dispersive, and only under a few circumstances have soliton-like phenomena been described [98]. Well-known examples include rocking filters [99] that couple two copropagating polarization modes in a birefringent fiber, and long-period gratings that couple a core mode to a cladding mode [100].

In long-period fiber gratings, a periodic perturbation in the refractive index of the core provides phase matching between a core mode and a cladding mode propagating in the same direction [101]. Strong coupling occurs over a range of wavelengths, manifesting itself in a transmission resonance, where core light is coupled into the cladding. Often the cladding light is subsequently lost owing to scattering and loss associated with the polymer jacket. The wavelength-dependent coupling to cladding modes can be exploited as the basis of an all-fiber optical switch [101]. The typical operation of such a device is that at low intensities, the long-period grating resonantly couples the core mode to the copropagating cladding mode, whereas at high intensities the transmission resonance shifts, thus effectively decoupling the modes. This leads to a power dependence of the coupling ratio between the modes. Nonlinear pulse propagation in long-period fiber gratings was studied by Eggleton et al. [101,102]. They observed optical switching, pulse reshaping, and optical limiting at intensities in the range of 1–20 GW/cm^2. Figure 20 shows a typical result from these experiments illustrating both superlinear and sublinear transmission as a function of incident power. Similar results have been obtained using rocking filters [103–105]

5.7 Gratings with Quadratic Nonlinearities

In the geometries that have been discussed until now, the source of the nonlinearity is the optical Kerr effect, a cubically nonlinear effect. However, it has been shown that a quadratic nonlinearity can have qualitatively similar effects, but can be considerably stronger [106] (see also the chapters by Torruellas et al. and by Trillo and Haelterman in this volume). Briefly, in

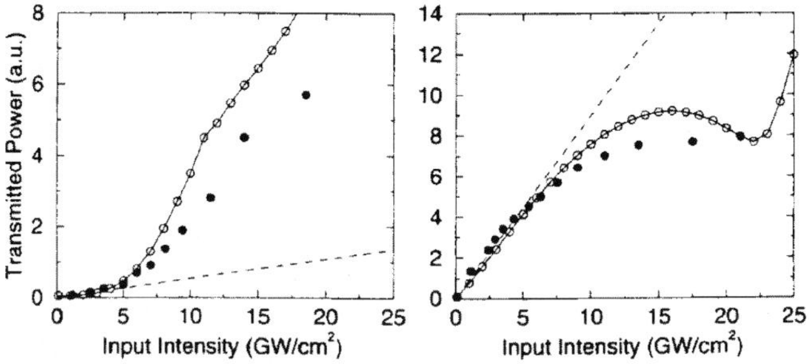

Fig. 20. Power transmitted through a grating versus incident power, when the incident pulses are on (*left*) and off (*right*) resonance. The *filled circles* show the experimental results, while the *open circles* show results of numerical simulations. The *dashed lines* show the linear results

a quadratically nonlinear medium, some of the energy is carried in the second harmonic of the driving frequency. Since (a) the refractive indices at the fundamental and second-harmonic frequencies differ, and (b) the fraction of energy in the second harmonic increases with intensity, there is a contribution to the phase of the field that depends on the intensity.

The first to point out that a quadratic effect could be used to generate Bragg grating solitons was Kivshar, who considered the limit in which the refractive-index difference between the fundamental and second-harmonic frequencies is large (the "cascading limit") [107]. He showed that in this limit the relevant equations are (14), but without the cross-phase modulation terms. This was generalized in subsequent work [108–110], in which soliton solutions were found under more general conditions.

This work was followed by more general theoretical investigations, including discoveries of novel types of solutions, for example dark solitons and bright–dark solitons [111,112], investigation of the excitation, manipulation, and stability of such solitons [113–115], a derivation for deep gratings [116,117], and investigations of applications [117] and modulational instability [118]. Associated experiments have not yet been reported, and for this reason we keep the discussion here brief. More details can be found in Sect. 6 of the chapter by Trillo and Haelterman in this volume.

6 Outlook

The experiments reported thus far have been preformed with uniform gratings or in gratings that behave in a similar way, for example apodized gratings. An exception is the work by Slusher et al., discussed in Sect. 5.3, in which a linearly chirped grating was used. Work with uniform gratings is continuing, and, with the use of new geometries, such as those formed in chalcogenide glasses [51], semiconductors [47], or very long gratings [1,119,120], the area will remain fruitful for many years. One of the main results that may be expected is the generation of grating solitons at much lower intensities than is possible at present.

Nonuniform gratings form another new challenge. It has been shown that, in such gratings, at low velocities, Bragg solitons can be described to some approximation by Newtonian-like equations in which the potential is determined by the grating [121]. The linearly chirped grating used by Slusher et al., for example, leads to a linear potential ramp [82]. A class of structures that has been considered theoretically is that of gratings with a localized defect [82]. In such gratings the light can, in principle, be trapped in or around the defect in a similar way to that in which charged particles can be trapped by electric fields. This possibility is unique to gratings and is not practical in uniform media.

Other generalizations that are being investigated are higher-dimensional periodic structures, consisting, for example, of arrays of microspheres of a polymer [122] which form hexagonal closed-packed or face-centerd cubic arrays, or combinations of these. The refractive-index variations between the polymer and air in these structures are large, and a deep-grating analysis is thus required. More importantly, coupled-mode theory as discussed in Sect. 2 is a one-dimensional theory, and needs to be suitably generalized to describe the geometries discussed here. Some aspects of higher-dimensional nonlinear media have been discussed by John et al. [123].

A final generalization is the inclusion of variations not only in the real part of the refractive index but also in the imaginary part, i.e. gain and loss. Such effects have been included theoretically by Poladian [124] and by Maywar and Agrawal [125], and have been considered experimentally by Bieber and Brown in a semiconductor geometry [61]. They can be included in the coupled-mode equations (14) by taking κ to be complex. The novel element that is added by including these effects is that the gain dynamics needs to be included. Since the gain depends on the electric field in the structure, we find that not only does the grating determine the fields, but the fields also determine the grating, which must now be treated as a dynamical variable. Hence the character of the problem changes completely.

In conclusion, the future for the study of periodic media with a nonlinearity is bright. Now that the properties of (almost) uniform gratings have been clarified, the next set of challenges involves applying this understanding to materials and geometries that make applications such as logic gates and pulse compressors more realistic, and also to the exploration of richer geometries as discussed in Sects. 5 and 6.

Acknowledgments

Over the years we have benefited from interactions with many excellent collaborators, including Alejandro Aceves, Awdah Arraf, Neil Broderick, Tom Brown, Hao He, Takeshi Iizuka, Peter Krug, Gadi Lenz, Natasha Litchinitser,

François Ouellette, Suresh Pereira, Leon Poladian, David Psaila, Dick Slusher, Stefan Spälter, Mike Steel, Tom Strasser, and Eduard Tsoy. C.M. de Sterke is also with the Australian Photonics Cooperative Research Centre, 101 National Innovation Centre, Australian Technology Park, Eveleigh, NSW 1430, Australia. This work was supported, in part, by the Australian Research Council, the National Science & Engineering Research Council of Canada, and Photonics Research Ontario.

References

1. R. Kashyap, *Fiber Bragg gratings*, 1st ed. (Academic Press, San Diego, 1999).
2. J.E. Sipe, L. Poladian, C.M. de Sterke, J. Opt. Soc. Am. A **11**, 1307–1320 (1994).
3. P.S. Cross, H. Kogelnik, Opt. Lett. **1**, 43–45 (1977).
4. P.St.J. Russell, J. Mod. Opt. **38**, 1599–1619 (1991).
5. B.J. Eggleton, T. Stephens, P.A. Krug, G. Dhosi, Z. Brodzeli, F. Ouellette, Electron. Lett. **32**, 1610–1611 (1996).
6. B.J. Eggleton, C.M. de Sterke, R.E. Slusher, J. Opt. Soc. Am. B **16**, 587–599 (1999).
7. G.P. Agrawal, *Nonlinear Fiber Optics*, 2nd ed. (Academic Press, San Diego, 1995).
8. H.G. Winful, Appl. Phys. Lett. **46**, 527–529 (1985).
9. C.M. de Sterke, J.E. Sipe, Gap solitons, in *Progress in Optics XXXIII*, ed. by E. Wolf, (Elsevier, Amsterdam, 1994), pp. 203–260.
10. C.M. de Sterke, D.G. Salinas, J.E. Sipe, Phys. Rev. E **54**, 1969–1989 (1996).
11. H. Kogelnik, C.V. Shank, Appl. Phys. Lett. **18**, 152–154 (1971).
12. H.G. Winful, J.H. Marburger, E. Garmire, Appl. Phys. Lett. **35**, 379–381 (1979).
13. M. Braun, *Differential Equations and Their Applications* (Springer, New York, 1977), Chap. 3, 447–458.
14. T. Iizuka, C.M. de Sterke, Phys. Rev. E **61**, 4491–4499 (2000).
15. H.G. Winful, G.D. Cooperman, Appl. Phys. Lett. **40**, 298–300 (1982).
16. C.M. de Sterke, J.E. Sipe, Phys. Rev. A **42**, 2858–2869 (1990).
17. C.M. de Sterke, Phys. Rev. A **45**, 8252–8258 (1992).
18. C.M. de Sterke, B.J. Eggleton, Phys. Rev. E **59**, 1267–1269 (1999).
19. H.-M. Keller, S. Pereira, J.E. Sipe, Opt. Commun. **170**, 35–40 (1999).
20. D.N. Christodoulides, R.I. Joseph, Phys. Rev. Lett. **62**, 1746–1749 (1989).
21. A.B. Aceves, S. Wabnitz, Phys. Lett. A **141**, 37–42 (1989).
22. B.J. Eggleton, R.E. Slusher, C.M. de Sterke, P.A. Krug, J.E. Sipe, Phys. Rev. Lett. **76**, 1627–1630 (1996).
23. I.V. Barashenkov, D.E. Pelinovsky, E.V. Zemlyanaya, Phys. Rev. Lett. **80**, 5117 (1998).
24. A.D. Rossi, C. Conti, S. Trillo, Phys. Rev. Lett. **81**, 85–88 (1998).
25. D. Taverner, N.G.R. Broderick, D.J. Richardson, R.I. Laming, M. Ibsen, Opt. Lett. **15**, 328–330 (1998).
26. W. Chen, D.L. Mills, Phys. Rev. Lett. **58**, 160–163 (1987).
27. J. Feng, F.K. Kneubühl, J. Quantum Electron. **29**, 590–597 (1993).

28. J. Feng, Opt. Lett. **18**, 1302–1304 (1993).
29. K.O. Hill, Y. Fujii, D.C. Johnson, B.S. Kawasaki, Appl. Phys. Lett. **32**, 647–649 (1978).
30. R.J. Campbell, R. Kashyap Int. J. Optoelectron. **9**, 33–57 (1994).
31. D.K. Lam, B.K. Garside, Appl. Opt. **20**, 440–445 (1981).
32. G. Meltz, W.W. Morey, W.H. Glenn, Opt. Lett. **14**, 283–285 (1989).
33. K.O. Hill, B. Malo, F. Bilodeau, D.C. Johnson, J. Albert, Appl. Phys. Lett. **62**, 1035–1037 (1993).
34. D.Z. Anderson, V. Mizrahi, T. Erdogan, A.E. White, Electron. Lett. **29**, 566-568 (1993).
35. J. Martin, F. Ouellette, Electron. Lett. **30**, 812–813 (1994).
36. P.J. Lemaire, R.M. Atkins, V. Mizrahi, W.A. Reed, Electron. Lett. **29**, 1191 (1993).
37. D.L. Williams, B.J. Ainslie, R. Armitage, R. Kashyap, R. Campbell, Electron. Lett. **29**, 45 (1993).
38. B. Malo, D.C. Johnson, F. Bilodeau, J. Albert, K.O. Hill, Electron. Lett. **31**, 223-225, (1995).
39. T. A. Strasser, P.J. Chandonnet, J. DeMarko, C.E. Soccolich, J.R. Pedrazzani, D.J. Digiovanni, M.J. Andrejco, D.S. Shenk, presented at *Optical Fiber Communications Conference*, San Diego, USA, postdeadline paper PD8 (1996).
40. B.J. Eggleton, J.A. Rogers, P.S. Westbrook, T.A. Strasser, Photon. Technol. Lett. **11**, 854–856 (1999).
41. J.A. Rogers, B.J. Eggleton, J.R. Pedrazzani, T.A. Strasser, Appl. Phys. Lett. **74**, 3131–3133 (1999).
42. J. Lauzon, S. Thibault, J. Martin, F. Ouellette, Opt. Lett. **19**, 2027–2029 (1994).
43. R.E. Slusher, B.J. Eggleton, T.A. Strasser, C.M. de Sterke, Opt. Express **3**, 465–475 (1998).
44. K. Sugden, I. Bennion, A. Molony, N.J. Copner, Electron. Lett. **30**, 440–442 (1994).
45. W.H. Loh, M.J. Cole, M.N. Zervas, S. Barcelos, R.I. Laming, Opt. Lett **20**, 2051–2053 (1995).
46. N.D. Sankey, D.F. Prelewitz, T.G. Brown, Appl. Phys. Lett. **60**, 1427-1429 (1992).
47. P. Millar, R.M. De La Rue, T.F. Krauss, J.S. Aitchison, Opt. Lett. **24**, 685–687 (1999).
48. C. Coriasso, D. Campi, C. Cacciatore, L. Faustini, C. Rigo, A Stano, Opt. Lett. **23**, 183–185 (1998).
49. M. Asobe, Opt. Fiber Technol. **3**, 142-145 (1997).
50. G. Lenz, J. Zimmermann, T. Katsufuji, M.E. Lines, H.Y. Hwang, S. Spälter, R.E. Slusher S.-W. Cheong, J.S. Sanghera, I. D. Aggarwal, Opt. Lett. **25**, 254–256 (2000).
51. S. Spälter, G. Lenz, H.Y. Hwang, J. Zimmermann, S-W. Cheong, T. Katsufuji, R.E. Slusher, Nonlinear pulse dynamics in planar chalcogenide glass waveguides, *Nonlinear Guided Waves Topical Meeting*, Dijon, France, postdeadline paper PD-6 (1999).
52. V. Mizrahi, K. W. DeLong, G. Stegeman, M.A. Saifi, M.J. Andrejco, Opt. Lett. **14**, 1140–1142 (1989).

53. M. Asobe, T. Ohara, I. Yokohama, T. Kaino, Electron. Lett. **32**, 1611–1613 (1996).
54. S. Ramachandran and S.G. Bishop, Appl. Phys. Lett. **74**, 13–15 (1999).
55. A. Saliminia, A. Villeneuve, T. Galstyan, S. La Rochelle, K. Richardson, J. Lightwave Technol. **17**, 837–842 (1999).
56. D. Taverner, N.G.R. Broderick, D.J. Richardson, M. Ibsen, R.I. Laming, Opt. Lett. **23**, 259–261 (1998).
57. B.J. Eggleton, C.M. de Sterke, R.E. Slusher, J. Opt. Soc. Am. B **14**, 2980–2993 (1997).
58. D. Von der Linde, J. Quantum Electron. **QE-8**, 328-338 (1972).
59. B.J. Eggleton, G. Lenz, R.E. Slusher, N.M. Litchinitser, Appl. Opt. **37**, 7055-7061 (1998).
60. S. La Rochelle, V. Mizrahi, G. Stegeman, Electron. Lett. **26** 1459–1460 (1990).
61. A.E. Bieber, T.G. Brown, Appl. Phys. Lett. **71**, 861 (1997).
62. J. He, M. Cada, J. Quantum Electron. **27**, 1182–1188 (1991).
63. S. Janz, J. He, Z.R. Wasilewski, Appl. Phys. Lett. **67**, 1051–1053 (1995).
64. S. Radic, N. George, G.P. Agrawal, Opt. Lett. **19**, 1789–1791 (1994).
65. C. Fernando, S. Janz, J.M. Baribeau, R. Normandin, J. S. Wright, Electron. Lett. **30**, 901-903 (1994).
66. C.J. Herbert, M.S. Malcuit, Opt. Lett. **18**, 1783–1785 (1993).
67. B.J. Eggleton, C.M. de Sterke, A.B. Aceves, J.E. Sipe, T.A. Strasser, R.E. Slusher, Opt. Commun. **149**, 267–271 (1998).
68. D. Taverner, D.J. Richardson, L. Dong, J.E. Caplen, K. Williams, and R.V. Penny, Opt. Lett. **22**, 378-380 (1997).
69. C.M. de Sterke, Opt. Lett. **17**, 914–916 (1992).
70. N.G.R. Broderick, D. Taverner, D.J. Richardson, M. Ibsen, R.I. Laming, Phys. Rev. Lett. **79**, 4566–4569 (1997).
71. M.J. Steel, C.M. Sterke, Phys. Rev. A **49**, 5048–5055 (1994).
72. M.J. Steel, D.G.A. Jackson, C.M. de Sterke, Phys. Rev. A. **50**, 3447–3452 (1994).
73. N.G.R. Broderick, D. Taverner, D.J. Richardson, M. Ibsen, R. I. Laming, Opt. Lett. **22**, 1837–1839 (1997).
74. N. Bloembergen and A.J. Sievers, Appl. Phys. Lett. **17**, 483–485 (1970).
75. C.L. Tang and P.P. Bey, J. Quantum Electron. **QE-9**, 9–17 (1973).
76. J.P. van der Ziel, M. Ilegems, Appl. Phys. Lett. **28**, 437–439 (1976).
77. A. Yariv, P. Yeh, J. Opt. Soc. Am. **67**, 438–448 1977).
78. M. Scalora, M.J. Bloemer, A.S. Manka, J.P. Dowling, C.M. Bowden, R. Viswanathan, J.W. Haus, Phys. Rev. A **56**, 3166–3174 (1997).
79. M.J. Steel, C.M. de Sterke, J. Opt. Soc. Am. B **12**, 2445–2452 (1995).
80. M.J. Steel, C.M. de Sterke, Opt. Lett. **21**, 420–422 (1996).
81. F. Ouellette, Opt. Lett. **12**, 847–849 (1987).
82. N.G.R. Broderick, C.M. de Sterke, Phys. Rev. E **58**, 7941–7951 (1998).
83. E.N. Tsoy, C.M. de Sterke, Phys. Rev. E. **62**, 2882–2890 (2000).
84. B.J. Eggleton, C.M. de Sterke, and R.E. Slusher, Opt. Lett. **21**, 1223–1225 (1996).
85. G. Lenz, B.J. Eggleton, N.M. Litchinitster, J. Opt. Soc. Am. B **15**, 715–721 (1998).
86. G. Lenz, B. J. Eggleton, J. Opt. Soc. Am. B **15**, 2979–2985 (1998).

87. B.J. Eggleton, G. Lenz, and N.M. Litchinitser, J. Fiber Integrated Opt., in press.
88. W.J. Tomlinson, R.H. Stolen, C.V. Shank, J. Opt. Soc. Am. B **1**, 139–143, (1984).
89. S V. Chernikov, P.V. Mamyshev, J. Opt. Soc. Am. B **8**, 1633–1641 (1991).
90. N.M. Litchinitser, G.P. Agrawal, B.J. Eggleton, G. Lenz, Opt. Express **3**, 411-417, (1998).
91. S. Lee, S.-T. Ho, Opt. Lett. **18**, 962–964 (1993).
92. P.D. Maker, R.W. Terhune, Phys. Rev. **137**, A801–A818 (1965).
93. Y. Barad, Y. Silberberg, Opt. Lett. **20**, 246–248 (1995).
94. H.G. Winful, Opt. Lett. **11**, 33–35 (1986).
95. R.E. Slusher, S. Spälter, B.J. Eggleton, S. Pereira, J.E. Sipe, Bragg grating enhanced nonlinear polarization mode conversion, *Bragg Gratings, Photosensitivity & Poling in Glass Waveguides*, postdeadline paper PD-2, Stuart, Florida (1999).
96. S. Pereira, J.E. Sipe, Opt. Express **3**, 418–432 (1998).
97. S. Pereira, J.E. Sipe, Phys. Rev. E., submitted.
98. S. Wabnitz, Opt. Lett. **14**, 1071–1073 (1989).
99. R.H. Stolen, A. Ashkin, W. Pleibel, J.M. Dziedzic, Opt. Lett. **9**, 300–302 (1984).
100. A.M. Vengsarkar, P.J. Lemaire, J.B. Judkins, V. Bhatia, T. Erdogan, J.E. Sipe, J. Lightwave Technol. **14**, 58–65, (1996).
101. B.J. Eggleton, R.E. Slusher, J.B. Judkins, J.B. Stark, A.M. Vengsarkar, Opt. Lett. **22**, 883–885 (1997).
102. J.N. Kutz, B.J. Eggleton, J.B. Stark, R.E. Slusher, IEEE J. Select. Topics Quantum Electron. **3**, 1232–1245 (1997).
103. S. Trillo, S. Wabnitz, N. Finlayson, W.C. Banyai, C.T. Seaton, G.I. Stegeman, R.H. Stolen, Appl. Phys. Lett. **53**, 837–840 (1998).
104. C.G. Krautchik, G.I. Stegeman, R.H. Stolen, Appl. Phys. Lett. **61**, 1751–1754 (1992).
105. D.C. Psaila, C.M. de Sterke, F. Ouellette, Opt. Commun. **153**, 111–117 (1998).
106. G.I. Stegeman, D.J. Hagan, L. Torner, Opt. Quantum Electron. **28**, 1691–1740 (1996).
107. Y.S. Kivshar, Phys. Rev. E **51**, 1613–1615 (1995).
108. T. Peschel, U. Peschel, F. Lederer, B.A. Malomed, Phys. Rev. E **55**, 4730–4739 (1997).
109. C. Conti, S. Trillo, G. Assanto, Phys. Rev. Lett. **78**, 2341–2344 (1997).
110. H. He, P. D. Drummond, Phys. Rev. Lett. **78**, 4311–4314 (1997).
111. S. Trillo, Opt. Lett. **21**, 1111–1113 (1996).
112. C. Conti, S. Trillo, G. Assanto, Phys. Rev. E **57**, R1251–R1254 (1998).
113. C. Conti, G. Assanto, S. Trillo, Opt. Lett. **22**, 1350–1352 (1997).
114. C. Conti, S. Trillo, G. Assanto, Phys. Rev. E **59**, 2467–2470 (1999).
115. C. Conti, A. De Rossi, S. Trillo, Opt. Lett. **23**, 1265–1267 (1998).
116. A. Arraf, C.M. de Sterke, Phys. Rev. E **58**, 7951–7958
117. C. Conti, S. Trillo, G. Assanto, Opt. Express **3**, 389–404 (1998).
118. H. He, A. Arraf, C.M. de Sterke, P.D. Drummond, B.A. Malomed, Phys. Rev. E **59**, 6064–6078 (1999).
119. R. Kashyap, A. Swanton, R.P. Smith, Electron. Lett. **35**, 1871–1872 (1999).

120. J.F. Brennan III and D.L. La Brake, Realization of >10 m long chirped fiber Bragg gratings, *Bragg Gratings, Photosensitivity & Poling in Glass Waveguides*, paper ThD2, Stuart, Florida (1999).
121. N.G.R. Broderick, C.M. de Sterke, Phys. Rev. E **52**, 4458–4464 (1995).
122. D.M. Mittleman, J.F. Bertone, P. Jiang, K.S. Hwang, V.L. Colvin, J. Chem. Phys. **111**, 345–354 (1999).
123. N. Aközbek, S. John, Phys. Rev. E **57**, 2287–2319 (1998).
124. L. Poladian, Phys. Rev. E **54**, 2963–2975 (1996).
125. D.N. Maywar, G.P. Agrawal, Opt. Express **3**, 440–446 (1998).

Stability of Spatial Optical Solitons

Yuri S. Kivshar and Andrey A. Sukhorukov

Summary. We present a brief overview of the basic concepts of the soliton stability theory and discuss some characteristic examples of the instability-induced soliton dynamics, in application to spatial optical solitons described by the NLS-type nonlinear models and their generalizations. In particular, we demonstrate that the soliton internal modes are responsible for the appearance of the soliton instability, and outline an analytical approach based an a multi-scale asymptotic technique that allows to analyze the soliton dynamics near the marginal stability point. We also discuss some results of the rigorous linear stability analysis of fundamental solitary waves and nonlinear impurity modes. Finally, we demonstrate that multi-hump vector solitary waves may become stable in some nonlinear models, and discuss the examples of stable (1+1)-dimensional composite solitons and (2+1)-dimensional dipole-mode solitons in a model of two incoherently interacting optical beams.

1 Introduction

Spatial optical solitons are known to originate from the nonlinearity-induced diffraction suppression and beam self-trapping in a bulk dielectric medium [1]. Since, generally speaking, the effect of diffraction is strong, considerable optical nonlinearities are required in order to compensate for the diffraction-induced beam spreading. As a result, in the vicinity of a self-trapped beam the refractive index experiences large deviations from a Kerr-type dependence, and the models of generalized nonlinearities that describe beam self-trapping phenomena and spatial solitons become *nonintegrable*.

For solitary waves of nonintegrable nonlinear models, *linear stability* is one of the crucial issues, since only stable (or weakly unstable) self-trapped beams can be observed in an experiment. The study of soliton stability in nonlinear optical materials has a long history, also associated with the study of nonlinear waves in other media such as plasmas and fluids. The stability of one-parameter solitary waves is already well understood for both fundamental (single-hump and nodeless) solitons [2–5] and solitons with nodes and multiple humps [6,7]. The pioneering results of Vakhitov and Kolokolov [2], known these days as the *Vakhitov–Kolokolov stability criterion*, found their rigorous justification in a general mathematical theory developed by Grillakis et al. [8]. Although the corresponding stability and instability theorems for the scalar nonlinear Schrödinger (NLS) models formally extend to the case of multiparameter solitons [8], most of the examples analyzed so far correspond to solitary waves with a *single parameter*. Only very recently has a systematic analysis of more general cases has been carried out, in connection with the study of three-wave parametric solitons in quadratic (or $\chi^{(2)}$) optical media (for a review, see [9]).

Recent progress in the study of soliton instabilities can be associated with the application of a *multiscale asymptotic bifurcation theory* developed for *weakly unstable* stationary, nonlinear localized waves. In the framework of this theory, the unstable eigenvalue of the associated linear spectral problem is treated as a *small parameter* of the asymptotic expansions [10], and the resulting equation for the slowly varying soliton propagation constant may also account for weakly nonlinear effects that describe the long-time evolution of linearly unstable solitary waves, and for an effective saturation of the soliton instability due to higher-order nonlinear effects. In the case of multiparameter solitary waves, a simplified version of this asymptotic method applied near a marginal stability point (or, in general, a surface) is reduced to finding certain determinants constructed from the derivatives of the system invariants [11–14]. However, the validity of this bifurcation theory has no rigorous proof, and it can only be used to *estimate* the domains of soliton stability and instability. Since more general *oscillatory instabilities* may also occur [7,15–18], numerical simulations are often required in order to verify the predictions of the asymptotic theory (see, e.g., [13] for an example).

Recently, a general matrix criterion for the stability and instability of *multicomponent solitary waves* was derived [19] for a system of N incoherently coupled NLS equations. In this general approach, soliton stability is studied as a constrained variational problem reduced to finite-dimensional linear algebra. Unstable eigenvalues of the linear stability problem for multicomponent solitary waves were shown to be connected with negative eigenvalues of the Hessian matrix constructed for the energetic surface of N-component spatially localized stationary solutions.

The analysis and, correspondingly, the stability criteria obtained in [19] can be extended, at least in principle, to other types of solitary waves, such as incoherent solitons in non-Kerr media, parametric solitary waves in $\chi^{(2)}$ optical media, etc. In all such cases, the results on the stability and instability of solitons can be readily obtained, with a *rigorous generalization* of some of the previously known results of the multiscale asymptotic theory. However, in each of those cases some additional analysis is required, in order to clarify whether the results of the asymptotic theory completely define the stability properties of multicomponent solitary waves. Beyond the validity of the multiscale analysis, oscillatory instabilities may occur, and studies should rely solely on the numerical analysis of the corresponding eigenvalue problems.

In this chapter, we present a brief overview of the basic concepts of soliton stability theory and discuss some characteristic examples of the instability-induced soliton dynamics, in application to spatial optical solitons described by NLS-type models and their generalizations. First of all, we demonstrate the crucial role played in the stability theory by the internal modes of solitons, which are an important characteristic of solitary waves in nonintegrable nonlinear models. In particular, we study an example model of the NLS equation with higher-order nonlinearity, in order to show that an internal mode gives birth to a soliton instability. Near the marginal stability point, soliton stability can be analyzed by a multiscale asymptotic technique, where the growth rate of the instability is treated as a small perturbation parameter. We also discuss some results of the rigorous linear stability analysis of fundamental solitary waves and nonlinear impurity modes. More recent studies of higher-order solitons have revealed that the so-called *multi-hump vector solitary waves* may become stable in some nonlinear models, and we discuss the stability of $(1+1)$-dimensional two- and three-hump composite solitons created by incoherent interaction of two optical beams in a photorefractive nonlinear medium. The final part of this chapter is devoted to the stability of solitons in higher dimensions and, in particular, it discusses very recent results on the symmetry-breaking instability of a $(2+1)$-dimensional vortex-mode [39] composite soliton and the formation of a rotating dipole-like structure associated with a robust radially asymmetric *dipole-mode vector soliton*, a new composite object resembling a "molecule of light".

2 Linear Eigenvalue Problem

To discuss the stability properties of spatial optical solitons, we consider the nonintegrable dimensionless generalized NLS equation that describes the self-focusing of a $(1+1)$-dimensional beam in a waveguide geometry,

$$\mathrm{i}\frac{\partial\psi}{\partial z} + \frac{\partial^2\psi}{\partial x^2} + \mathcal{F}(I;x)\psi = 0\,, \tag{1}$$

where $\psi(x,z)$ is the dimensionless complex envelope of the electric field, x is the transverse spatial coordinate, z is the propagation distance, $I=|\psi(x,z)|^2$ is the beam intensity, and the real function $\mathcal{F}(I;x)$ characterizes both the linear and the nonlinear properties of the dielectric medium, for which we assume that $\mathcal{F}(0;\pm\infty)=0$. Stationary spatially localized solutions of the model (1) have the standard form $\psi(x,z)=\Phi(x;\beta)\mathrm{e}^{\mathrm{i}\beta z}$, where β is the soliton propagation constant ($\beta>0$) and the real function $\Phi(x;\beta)$ vanishes for $|x|\to\infty$. An important conserved quantity of the soliton in the model (1) is its *power*, defined as

$$P(\beta) = \int_{-\infty}^{+\infty} |\psi(x,z)|^2\,\mathrm{d}x = \int_{-\infty}^{+\infty} \Phi^2(x;\beta)\,\mathrm{d}x\,. \tag{2}$$

To find the linear stability conditions, we consider the evolution of a small-amplitude perturbation of the soliton representing the solution in the form

$$\psi(x,z) = \left\{\Phi(x;\beta) + [v(x) - w(x)]\,\mathrm{e}^{\mathrm{i}\lambda z} + [v^*(x) + w^*(x)]\mathrm{e}^{-\mathrm{i}\lambda^* z}\right\}\mathrm{e}^{\mathrm{i}\beta z}\;, \quad (3)$$

where the star stands for complex conjugation, and obtain the linear eigenvalue problem for $v(x)$ and $w(x)$,

$$L_0 w = \lambda v\;, \quad L_1 v = \lambda w\;,$$

$$L_j = -\frac{\mathrm{d}^2}{\mathrm{d}x^2} + \beta - U_j\;, \quad (4)$$

where $U_0 = \mathcal{F}(I;x)$ and $U_1 = \mathcal{F}(I;x) + 2I[\partial\mathcal{F}(I;x)/\partial I]$.

A stationary solution of the model (1) is stable if none of the eigenmodes of the corresponding linear problem (4) have exponentially growing amplitudes, i.e. $\mathrm{Im}(\lambda) = 0$. It can be demonstrated that the continuum part of the linear spectrum of the problem (4) consists of *two symmetric branches* corresponding to real eigenvalues with absolute values $|\lambda| > \beta$, and therefore only discrete eigenstates are responsible for the stability properties of localized waves. Then, the corresponding eigenmode solutions fall into the following categories:

- *internal modes* with real eigenvalues describe periodic oscillations;
- *instability modes* correspond to purely imaginary eigenvalues;
- *oscillatory instabilities* can occur when the eigenvalues are complex.

In what follows, we present several approaches that allow us to study analytically and numerically the structure of the discrete spectrum in order to determine the linear stability properties of solitary waves. We also analyze the nonlinear evolution of unstable solitons.

3 Soliton Internal Modes and Stability

Since soliton instabilities always occur in nonintegrable models, it is interesting to know what kind of distinct features of solitary waves in nonintegrable models might be responsible for their instabilities. It is commonly believed that the solitary waves of nonintegrable nonlinear models differ from the solitons of integrable models only in the character of the soliton interactions: unlike the case for "proper" solitons, interaction of solitary waves is accompanied by radiation (see, e.g., [20] and references therein). However, the soliton instabilities are associated with nontrivial effects of a different nature that are generic for localized waves of nearly integrable and nonintegrable models. In particular, a small perturbation to an integrable model may create an *internal mode* of a solitary wave [21]. This effect is beyond anything that can be treated by a regular perturbation theory, because solitons of integrable models do not possess internal modes. But in nonintegrable models such modes

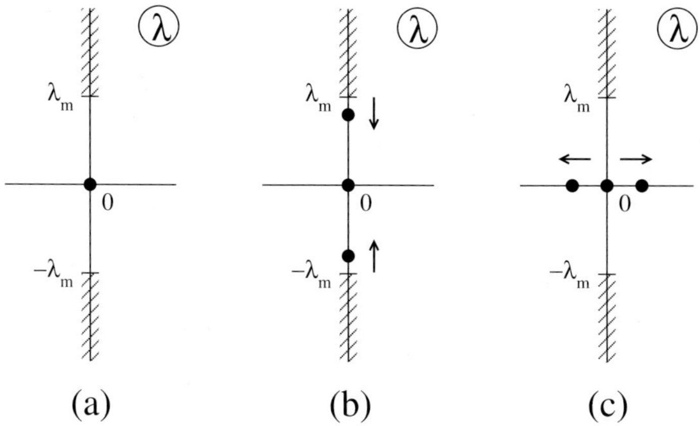

Fig. 1. Schematic representation of the origin of bifurcation-induced soliton instabilities: (**a**) spectrum of the integrable cubic NLS model, (**b**) bifurcation of the soliton internal mode, (**c**) collision of the internal mode with the neutral mode, resulting in the Vakhitov–Kolokolov-type soliton instability

may introduce *qualitatively new features* into the system dynamics and, in particular, lead to the appearance of soliton instabilities.

To demonstrate that internal modes are generic for nonintegrable models, we consider a weakly perturbed cubic NLS equation with the nonlinear term

$$\mathcal{F}(I;x) = I + \epsilon f(I) , \tag{5}$$

where $f(I)$ describes a deviation from the Kerr nonlinear response, and ϵ is a small parameter. Then, the stationary solution can be expressed asymptotically as $\Phi(x;\beta) = \Phi_0(x) + \epsilon\Phi_1(x) + O(\epsilon^2)$, where $\Phi_0(x) = \sqrt{2\beta}\,\mathrm{sech}\,(\sqrt{\beta}x)$ is the soliton of the cubic NLS equation, and $\Phi_1(x)$ is a localized correction derived from (1) and (5). Neglecting the second-order corrections, we find the following results for the effective potentials of the linearized eigenvalue problem (4): $U_0 = \Phi_0^2 + \epsilon\widetilde{U}_0$ and $U_1 = 3\Phi_0^2 + \epsilon\widetilde{U}_1$, where $\widetilde{U}_0 = f(\Phi_0^2) + 2\Phi_0\Phi_1$ and $\widetilde{U}_1 = f(\Phi_0^2) + 2\Phi_0^2 f'(\Phi_0^2) + 6\Phi_0\Phi_1$, and the prime denotes differentiation with respect to the argument.

The linear eigenvalue problem (4) and (5) can be solved exactly at $\epsilon = 0$ (see, e.g., [22]). Its discrete spectrum contains only a degenerate eigenvalue at the origin, $\lambda = 0$, corresponding to the so-called *soliton neutral mode* (see Fig. 1a). We can show that a small perturbation can lead to the creation of an internal mode (which corresponds to two symmetric discrete eigenvalues), which bifurcates from the continuous spectrum band, as shown in Fig. 1b. To be definite, we consider the upper branch of the spectrum and suppose that the cut-off frequencies $\lambda_{\mathrm{m}} = \pm\beta$ are not shifted by the perturbation. Then, the frequency of the internal mode can be represented in the form

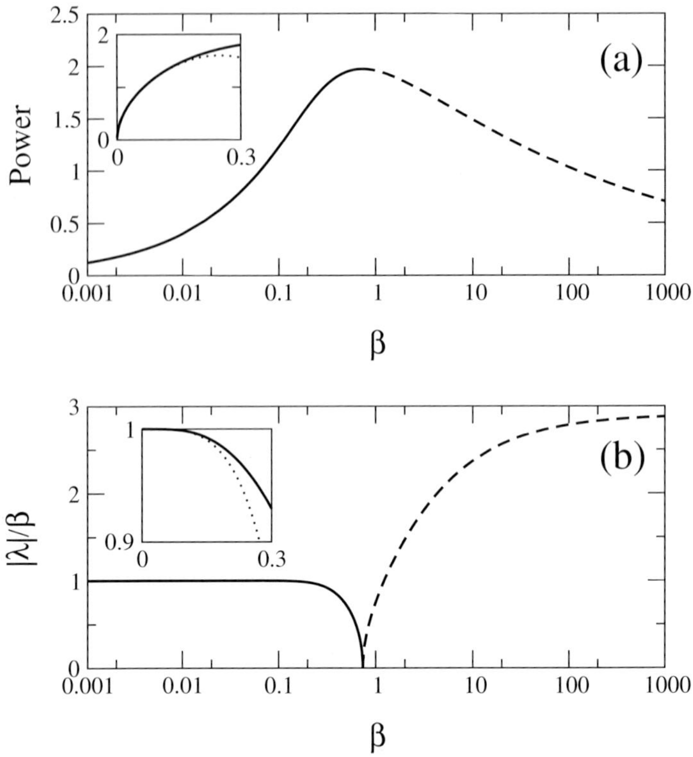

Fig. 2. Soliton instability in the model (1), (5) and (7) for $\epsilon = 1$, represented through (**a**) the power dependence $P(\beta)$ and (**b**) the evolution of the discrete eigenvalue of the problem (3), which defines a soliton internal mode (*solid lines*) and an instability mode (*dashed lines*). Solitons for $\beta > \beta_{\rm cr}$, i.e. for $\mathrm{d}P/\mathrm{d}\beta < 0$ (dashed curves in (a) and (b)), are linearly unstable. The *dotted lines* show the asymptotic dependences calculated analytically

$\lambda = \beta - \epsilon^2\kappa^2$, where κ is defined by the following [21]:

$$|\kappa| = \frac{1}{4}\mathrm{sign}(\epsilon)\int_{-\infty}^{\infty}\left\{V(x,\beta)\widetilde{U}_1V(x;\beta) + W(x,\beta)\widetilde{U}_0W(x;\beta)\right\}dx\,. \qquad (6)$$

Here $\{V(x;\beta), W(x;\beta)\}$ are the eigenfunctions of the cubic NLS equation calculated at the edge of the continuous spectrum, $V(x;\beta) = 1 - 2\,\mathrm{sech}^2(\sqrt{\beta}x)$, and $W(x;\beta) = 1$. A soliton internal mode appears if the right-hand side of (6) is positive.

As an important example, we consider the case of the NLS equation (1), where (5) is perturbed by a higher-order power-law nonlinear term,

$$f(I) = I^3\,. \qquad (7)$$

The first-order correction to the soliton profile can be found in the form

$$\Phi_1(x) = -\frac{\sqrt{2}\beta^{5/2}\left[2\cosh(2\sqrt{\beta}x) + \cosh(4\sqrt{\beta}x)\right]}{3\cosh^5(\sqrt{\beta}x)} .$$

With the help of (6), it is easy to show that for $\epsilon > 0$ a perturbed NLS soliton possesses an internal mode, which can be found analytically near the edge of the continuum spectrum, with

$$\lambda = \beta\left[1 - \left(\frac{64\epsilon}{15}\right)^2 \beta^4 + O(\beta^6)\right] . \tag{8}$$

For high intensities, the additional nonlinear term (7) is no longer small, and the soliton solutions, together with the associated linear spectrum, must be calculated numerically. The power dependence $P(\beta)$, calculated with the help of (2) for the soliton of the model (1), (5), and (7), is presented in Fig. 2a, together with the discrete eigenvalue of the linearized problem (4), shown in Fig. 2b. First of all, we notice that the asymptotic theory provides accurate results for small-intensity solitons, i.e. for $\beta < 0.1$ (shown by the dotted curves in the insets). Secondly, the slope of the power dependence changes its sign at the point $\beta = \beta_{\mathrm{cr}}$, where the soliton internal mode vanishes when it collides with the neutral mode, as depicted in Figs. 1b,c. At that point, the soliton stability changes owing to the appearance of a pair of unstable (purely imaginary) eigenvalues (dashed curves in Fig. 2b). In Sect. 4 below, we prove rigorously a link between the stability of the soliton and the slope of the dependence $P(\beta)$.

For $\beta \approx \beta_{\mathrm{cr}}$, the instability-induced dynamics of an unstable soliton can be described by approximate equations for the soliton parameters derived by means of the multiscale asymptotic technique (see Sect. 5 below) but, in

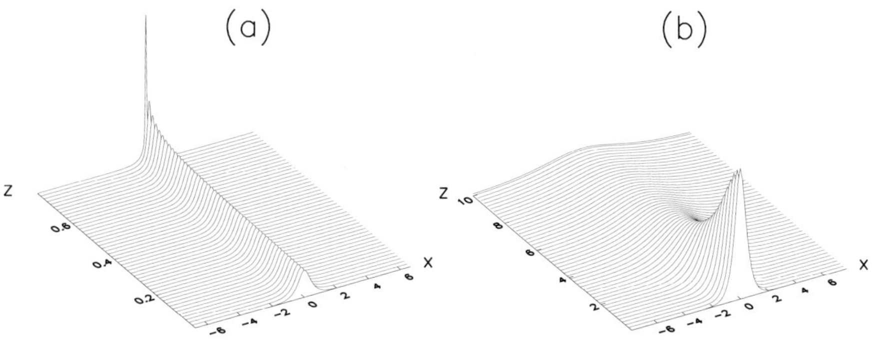

Fig. 3. Evolution of a perturbed unstable NLS soliton in the model (1), (5), and (7) for $\epsilon = 1$, and $\beta = 2$, in the case of (**a**) increased power (collapse) and (**b**) decreased power (switching to a low-amplitude stable state). The initial power was changed by 1% compared with the exact soliton solution

general, we need to perform numerical simulations in order to study the evolution of linearly unstable solitons. In Fig. 3, we show two different types of instability-induced soliton evolution in our model. In the first case, a small perturbation that effectively *increases* the soliton power results in an unbounded growth of the soliton amplitude and subsequent beam collapse (Fig. 3a). In the second case, a small *decrease* of the soliton power leads to the switching of a soliton from an unstable branch (dashed line in Fig. 2a) to a stable branch (solid line in Fig. 2a), as is shown in Fig. 3b. This latter scenario becomes possible because, in the model under consideration, all small-amplitude solitons are stable. However, if the small-amplitude solitons are unstable, the soliton beam does not converge to a stable state but, instead, diffracts. Thus, in NLS-type nonlinear models there exist *three distinct types* of instability-induced soliton dynamics [10].

4 Stability Criterion for Fundamental Solitons

Direct investigation of the eigenvalue problem (4) is a complicated task which, in general, does not yield a complete analytical solution. However, for a class of "fundamental" solitary waves (i.e. solitons with no nodes), the analysis can be greatly simplified. First, we reduce the system (4) to a single equation,

$$L_0 L_1 v = \lambda^2 v \,, \tag{9}$$

for which the stability condition requires all eigenvalues λ^2 to be positive. It is straightforward to show that $L_0 = L^+ L^-$, where $L^\pm = \pm \mathrm{d}/\mathrm{d}x + \Phi^{-1}(\mathrm{d}\Phi/\mathrm{d}x)$, and thus, instead of (9), one can consider an auxiliary eigenvalue problem (see, e.g., [23]),

$$L^- L_1 L^+ \tilde{v} = \lambda^2 \tilde{v} \,, \tag{10}$$

which reduces to (9) after the substitution $v = L^+ \tilde{v}$. Since the operator $L^- L_1 L^+$ is Hermitian, all eigenvalues λ^2 of (9) and (10) are real, and in this case *oscillatory instabilities do not occur.*

The properties of the operators L_j $(j = 0, 1)$ have been well studied in the literature, particularly, as a characteristic example of the spectral theory of second-order differential operators (see, e.g., [24]). For our problem, we use two general mathematical results about the spectrum of the linear eigenvalue problem $L_j \varphi_n^{(j)} = \lambda_n^{(j)} \varphi_n^{(j)}$:

- the eigenvalues can be ordered as $\lambda_{n+1}^{(j)} > \lambda_n^{(j)}$, where $n \geq 0$ defines the number of zeros in the corresponding eigenfunction $\varphi_n^{(j)}$;
- for a "deeper" potential well, $\widetilde{U}_j(x) \geq U_j(x)$, the corresponding set of the eigenvalues is shifted "down", i.e. $\widetilde{\lambda}_n^{(j)} \leq \lambda_n^{(j)}$.

Let us first discuss the properties of the operator L_0, for which the soliton neutral mode is an eigenstate, i.e. $L_0\Phi(x;\beta) = 0$. As we have assumed earlier, $\Phi(x;\beta) > 0$ is the ground state solution with no nodes and, therefore, $\lambda_n^{(0)} > \lambda_0^{(0)} = 0$ for $n > 0$. This means that the operator L_0 is positive definite on the subspace of the functions orthogonal to $\Phi(x;\beta)$, which allows us to use several general theorems [2,4,6–8] in order to link the stability properties of solitons to the number of negative eigenvalues of the operator L_1. Specifically,

- soliton instability appears if there are two (or more) negative eigenvalues, i.e. $\lambda_1^{(1)} < 0$;
- solitons are always stable if the operator L_1 is positive definite;
- in the intermediate case, the soliton stability depends on the slope of the power dependence $P(\beta)$, according to the Vakhitov–Kolokolov stability criterion, [2] i.e. the soliton is *stable* if $\partial P/\partial\beta > 0$, and it is *unstable* otherwise.

Thus, to distinguish between these cases, it is sufficient to determine the signs of the zeroth and first eigenvalues of the operator L_1.

To demonstrate the validity of the Vakhitov–Kolokolov criterion, we follow the standard procedure [2]. First, we note that, for fundamental solitons with no nodes, both the direct, L_0, and inverse, L_0^{-1}, operators exist, and they are positive definite for any function orthogonal to $\Phi(x;\beta)$, which can only be a neutral eigenmode of (9) and therefore can be ignored. Thus, by applying the inverse operator L_0^{-1} to (9), we obtain another linear problem with the same spectrum:

$$L_1 v = \lambda^2 L_0^{-1} v\,, \tag{11}$$

where $v(x)$ now satisfies the orthogonality condition

$$\langle v|\Phi\rangle \equiv \int_{-\infty}^{+\infty} v^*(x)\Phi(x;\beta)\,\mathrm{d}x = 0\,. \tag{12}$$

Second, we multiply both sides of (11) by the function $v^*(x)$, integrate over x, and obtain the following result:

$$\lambda^2 = \frac{\langle v|L_1 v\rangle}{\langle v|L_0^{-1} v\rangle}\,. \tag{13}$$

Because the denominator in (13) is positive definite if v satisfies (12), the sign of this ratio depends only on the numerator. An instability will appear if there exists an eigenvalue $\lambda^2 < 0$ (so that λ is imaginary), and this is possible only if

$$\min\,\langle v|L_1 v\rangle\ < 0\,, \tag{14}$$

where we normalize the function $v(x)$ to make the expression in (14) finite, as follows:

$$\langle v|v\rangle = 1 \,. \tag{15}$$

In order to find the minimum in (14) under the constraints given by (12) and (15), we use the method of Lagrange multipliers, and look for a minimum of the following functional:

$$\mathcal{L} = \langle v|L_1 v\rangle - \nu \langle v|v\rangle - \mu \langle v|\Phi\rangle \,,$$

where the real parameters ν and μ are unknown. With no lack of generality, we assume that $\mu \geq 0$, as otherwise the sign of the function $v(x)$ can be inverted. The extremal point of the functional $\mathcal{L}$ can be then found from the condition $\delta\mathcal{L}/\delta v = 0$, where δ denotes the variational derivative. As a result, we obtain the following equation:

$$L_1 v = \nu v + \mu\Phi \,, \tag{16}$$

where the values of ν and μ must be chosen in such a way that the conditions (12) and (15) are satisfied. Then, it follows from above that $\langle v|L_1 v\rangle = \nu \langle v|v\rangle$ and, according to (14), the stationary state is unstable if and only if there exists a solution with $\nu < 0$.

The operator L_1 has a full set of orthogonal eigenfunctions $\varphi_n^{(1)}$ [24], i.e. $\left\langle \varphi_n^{(1)}|\varphi_m^{(1)}\right\rangle = 0$ if $n \neq m$. The spectrum of L_1 consists of discrete $(\lambda_n^{(1)} < \beta)$ and continuous $(\lambda_n^{(1)} \geq \beta)$ parts, and the norms of the eigenmodes, $\left\langle \varphi_n^{(1)}|\varphi_n^{(1)}\right\rangle$, can be scaled to unity and a delta function, respectively. Then, we can decompose the function $v(x)$ in the following way:

$$v(x) = \sum_n D_n \varphi_n^{(1)}(x) + \int_\beta^{+\infty} D_n \varphi_n^{(1)}(x)\, \mathrm{d}\lambda_n^{(1)} \,, \tag{17}$$

where the sum is over the eigenvalues of the discrete spectrum of L_1 only. The coefficients in (17) can be found as $D_n = \left\langle \varphi_n^{(1)}|v\right\rangle$. The function $\Phi(x;\beta)$ can be decomposed in a similar way, with the coefficients $C_n = \left\langle \varphi_n^{(1)}|\Phi\right\rangle$. Then, (16) can be reduced to

$$D_n = \begin{cases} \mu C_n/(\lambda_n^{(1)} - \nu), \text{ if } C_n \neq 0 \text{ and } \mu > 0 \,, \\ 1, \text{ if } C_n = 0,\ \mu = 0, \text{ and } \nu = \lambda_n^{(1)} \,, \\ 0, \text{ otherwise} \,. \end{cases} \tag{18}$$

Note that we should not consider degenerate solutions $v \equiv 0$, and therefore $\mu = 0$ is only possible if $\nu = \lambda_n^{(1)}$ and $C_n = 0$. In order to find the Lagrange multiplier ν, we substitute (17) and (18) into the orthogonality condition (12), and obtain the following equation for the parameter ν:

$$Q(\nu) \equiv \langle v|\Phi\rangle = \sum_n C_n D_n^* + \int_\beta^{+\infty} C_n D_n^* \, \mathrm{d}\lambda_n^{(1)} = 0 \,. \tag{19}$$

As has been mentioned above, instability appears if there exists a root $\nu < 0$, and thus we need to determine the sign of the minimal ν that solves (19). Because the lowest-order modes of the operators L_0 and L_1, $\Phi(x;\beta)$ and $\varphi_0^{(1)}$, respectively, do not contain zeros, the coefficient $C_0 \neq 0$. Then, from the structure of (19), it follows that $Q(\nu < \lambda_0^{(1)}) > 0$, and thus solutions are only possible for $\nu > \lambda_0^{(1)}$, meaning that if $\lambda_0^{(1)} \geq 0$ the stationary state $\Phi(x;\beta)$ is *stable*.

We notice that the function $Q(\nu)$ is monotonic in the interval $(-\infty, +\infty)$ for $\mu > 0$ and $\lambda_0^{(1)} < \nu < \lambda_n^{(1)}$, where $n \geq 1$ corresponds to the smallest eigenvalue with $C_n \neq 0$. Then it immediately follows that instability appears if $\lambda_1^{(1)} < 0$ and $C_1 \neq 0$. On the other hand, if $C_1 = 0$, the corresponding eigenmode $\varphi_1^{(1)}$ satisfies (16) and the constraints (12) and (15) with $\nu = \lambda_1^{(1)}$ and $\mu = 0$ and, therefore, an instability is always present if $\lambda_1^{(1)} < 0$.

The last possible scenario is when $\lambda_0^{(1)} < 0$ and $\lambda_1^{(1)} \geq 0$, so that the modes with $\nu = \lambda_n^{(1)}$ do not give rise to instability; we search for the solutions with $\mu > 0$. Because $\lambda_n^{(1)} > \lambda_1^{(1)} \geq 0$, the sign of the solution ν is determined by the value of $Q(0)$. Indeed, if $Q(0) > 0$, the function $Q(\nu)$ vanishes at some $\nu < 0$ which indicates *instability*, and vice versa. From (16) and (19), it follows that $Q(\nu = 0) = \langle L_1^{-1}\mu\Phi|\Phi\rangle$. To calculate this value, we differentiate the equality $L_0\Phi = 0$ with respect to the propagation constant and obtain

$$L_1 \frac{\partial \Phi}{\partial \beta} = -\Phi \,, \tag{20}$$

which finally gives $\mathrm{sign}[Q(0)] = -\mathrm{sign}(\mathrm{d}P/\mathrm{d}\beta)$. These results provide a proof of the stability conditions outlined above.

An important case is the stability of a soliton in a homogeneous medium, when $\mathcal{F}(I)$ does not depend on x. Then, a fundamental soliton has a symmetric profile with a single maximum, and $\mathrm{d}\Phi/\mathrm{d}x$ is the first-order neutral mode of the operator L_1, i.e. $\lambda_1^{(1)} = 0$. Therefore, in such a case the stability of the soliton follows directly from the slope of the dependence $P(\beta)$.

The linear stability discussed above should be compared with the more general *Lyapunov stability theorem*, which states that in a conservative system a stable solution (in the Lyapunov sense) corresponds to an extremal point of an invariant such as the system Hamiltonian, provided it is bounded from below (or above). For the NLS equation, this means that a soliton solution is a stationary point of the Hamiltonian H for a fixed power P, and is found from the variational problem $\delta(H + \beta P) = 0$. In order to prove Lyapunov stability, one needs to demonstrate that, for a class of spatially localized solutions, the soliton corresponds to a minimum of the Hamiltonian when P is fixed. This can be shown rigorously for a homogeneous medium with cubic nonlinearity [4], i.e. for $\mathcal{F}(I;x) = I$, and follows from the integral inequality

$$H > H_s + (P^{1/2} - P_s^{1/2}) , \tag{21}$$

where the subscript "s" indicates the values of the corresponding integral calculated for the NLS soliton. Condition (21) proves the stability of the soliton for both small and finite-amplitude perturbations; and a similar relation can be derived for a generalized nonlinearity and is consistent with the Vakhitov–Kolokolov criterion (see [4] for details).

5 Marginal Stability Point: Asymptotic Analysis

As follows from the linear stability analysis, solitons in a homogeneous medium are unstable when the slope of the power dependence is negative, i.e. for $\mathrm{d}P/\mathrm{d}\beta < 0$. Near the marginal stability point $\beta = \beta_{\mathrm{cr}}$ defined by the condition $(\mathrm{d}P/\mathrm{d}\beta)_{\beta=\beta_{\mathrm{cr}}} = 0$, where the growth rate of the instability is small, we can derive a *general analytical asymptotic model* which describes not only linear instabilities but also the nonlinear long-term evolution of unstable solitons. Such an approach must be based on a nontrivial modification of soliton perturbation theory [20] . Indeed, the standard soliton perturbation theory is usually applied to analyze the dynamics of solitons under the action of external perturbations. Here we are dealing with a *qualitatively different physical problem*, where an unstable bright soliton evolves under the action of its "own" perturbations. As a result of the development of such an instability, the soliton propagation constant varies slowly along the propagation direction, i.e. $\beta = \beta(z)$. As the growth rate of the instability is small near the threshold $\beta = \beta_{\mathrm{cr}}$, we can assume that the profile of the perturbed soliton varies slowly with z and that the soliton evolves almost adiabatically (i.e. it remains self-similar). Therefore, we can develop an *asymptotic theory* representing the solution to the original model (1) in the form $\psi = \phi(x;\beta;Z)\exp[\mathrm{i}\beta_0 z + \mathrm{i}\epsilon \int_0^Z \beta(Z')\,\mathrm{d}Z']$, where $\beta = \beta_0 + \epsilon^2 \Omega(Z)$, $Z = \epsilon z$, and $\epsilon \ll 1$. Here the constant value β_0 is chosen to be in the vicinity of the critical point β_{cr}. Then, using an asymptotic multiscale expansion in the form $\phi = \Phi(x;\beta) + \epsilon^3\phi_3(x;\beta;Z) + O(\epsilon^4)$, we obtain the following equation for the soliton propagation constant β (details can be found in [10,25]):

$$M(\beta_{\mathrm{cr}})\frac{\mathrm{d}^2\Omega}{\mathrm{d}Z^2} + \frac{1}{\epsilon^2}\frac{\mathrm{d}P}{\mathrm{d}\beta}\bigg|_{\beta=\beta_0}\Omega + \frac{1}{2}\frac{\mathrm{d}^2P}{\mathrm{d}\beta^2}\bigg|_{\beta=\beta_{\mathrm{cr}}}\Omega^2 = 0 . \tag{22}$$

Here $P(\beta)$ and $M(\beta)$ are calculated through the stationary soliton solution, and

$$M(\beta) = \int_{-\infty}^{+\infty}\left(\frac{1}{\Phi(x;\beta)}\int_0^x \Phi(x';\beta)\,\frac{\partial\Phi(x';\beta)}{\partial\beta}dx'\right)^2 dx > 0 .$$

A remarkable result which follows from this asymptotic analysis is the following. In the generalized NLS equation (1), the dynamics of solitary waves near the marginal stability point $\beta = \beta_{\rm cr}$ can be described by a simple collective-coordinate model (22) which is equivalent to the equation of motion of an effective (inertial and conservative) particle of mass $M(\beta_{\rm cr})$ with a coordinate Ω, moving under the action of a potential force proportional to the difference $P_0 - P(\beta)$, where $P_0 = P(\beta_0)$.

The first two terms in (22) give the result of the linear stability theory, according to which the soliton is *linearly unstable* if $\mathrm{d}P/\mathrm{d}\beta < 0$. The nonlinear term in (22) allows us to describe not only the linear but also the long-term nonlinear dynamics of an unstable soliton and, moreover, to identify qualitatively several different scenarios of instability-induced soliton dynamics near the marginal stability point, as discussed in [10,25].

6 Nonlinear Guided Waves and Impurity Modes

6.1 Model and General Remarks

In this section, we demonstrate how the basic concepts and results of soliton stability theory can be employed to study the stability of nonlinear guided waves in inhomogeneous media. In particular, we consider the model (1) with a real function $\mathcal{F}(I;x)$ that describes both *nonlinear* and *inhomogeneous* properties of an optical medium. This kind of problem has a number of important physical applications, ranging from the nonlinear dynamics of solids to the theory of nonlinear photonic crystals and waveguide arrays, in optics. In its application to the theory of guided electromagnetic waves, the model we discuss below describes a special case of a stratified (or layered) dielectric medium for which nonlinear guided waves and their stability have been analyzed over the last 20 years [26,27].

Following the original study presented in [28], we consider a simplified case where the inhomogeneity is localized in a small region and is created by a thin layer embedded in a nonlinear medium. Then, if the corresponding wavelength is much larger than the layer thickness, the response of the layer in the continuum-limit approximation can be described by a delta function and, therefore, we can write

$$\mathcal{F}(I;x) = F(I) + \delta(x)G(I)\,, \tag{23}$$

where the functions $F(I)$ and $G(I)$ characterize the properties of the bulk medium and the layer, respectively. The model (1), (23) describes, a special case of a more general problem of the existence and stability of nonlinear guided waves in a stratified medium with a Kerr or non-Kerr nonlinear response (see, e.g., [26]).

When a thin layer is introduced into the system, the translational invariance of the model is broken at the location of the layer, i.e. at $x = 0$. The nonlinear modes of such an inhomogeneous model can be found as spatially localized solutions of the following equation:

$$-\beta\Phi + \frac{\mathrm{d}^2\Phi}{\mathrm{d}x^2} + \mathcal{F}(\Phi^2; x)\Phi = 0 . \tag{24}$$

We notice that, for the problem at hand, the profiles of localized modes should satisfy a *homogeneous equation* on both sides of the layer and, therefore, that they can be constructed by using the solutions of (24) with $\mathcal{F}(I;x) = F(I)$. Owing to the translational symmetry, such a localized solution can be represented in the form $\Phi_0(x - x_0)$, where x_0 is an arbitrary shift. We also note that the profile is symmetric, i.e. $\Phi_0(x) = \Phi_0(-x)$, it does not contain zeros, i.e. $\Phi_0(x) > 0$, and it has a single hump, since $\mathrm{d}\Phi_0/\mathrm{d}|x| < 0$ at $x \neq 0$. Therefore, the general solution of the inhomogeneous equation (24) for a spatially localized normal mode can be represented in the following form:

$$\Phi(x) = \begin{cases} \Phi_0(x - x_0), \; x \geq 0 , \\ \Phi_0(x - s x_0), \; x \leq 0 , \end{cases} \tag{25}$$

where x_0 and s are defined through the layer parameters. From the continuity condition at $x = 0$, it follows that $s = \pm 1$. Thus, the parameter s defines the type of symmetry of the localized mode, i.e. the mode is *symmetric* for $s = -1$, and *asymmetric* for $s = +1$. The parameter x_0 is defined by the following matching condition:

$$(1 - s)\frac{\mathrm{d}\Phi_0}{\mathrm{d}x}(x_0) = G(I_0)\Phi_0(x_0) , \tag{26}$$

where $I_0 = \Phi_0^2(x_0)$. We note that the power of the asymmetric modes is the same as that for solitons in a homogeneous medium, $P_0 = \int_{-\infty}^{+\infty} \Phi_0^2(x)\,\mathrm{d}x$, while the power of the symmetric modes (hereafter denoted as P) is smaller than P_0 if $x_0 < 0$, and larger otherwise.

As follows from (25), the function $\Phi(x)$ is sign definite. In order to apply the general stability results derived for the fundamental solitons of NLS-type equations in Sect. 4, we need to study the spectral properties of the linear operator L_1. Owing to the presence of the layer, the number of negative eigenvalues of the operator L_1 depends on the mode parameter β. As has been demonstrated in [27], if an eigenvalue passes though zero and changes its sign, a so-called *critical point* appears in the power dependence $P(\beta)$ calculated for a localized mode. One such type of critical point is a *bifurcation*, where the mode belongs simultaneously to two families of localized solutions, symmetric and asymmetric ones. Thus, the stability properties of these distinct types of nonlinear localized modes can be expected to be different, and we discuss these two cases separately.

6.2 Stability of Symmetric and Asymmetric Modes

6.2.1 Symmetric Modes

In this case it can be verified that, similarly to the solitary waves of homogeneous models (see Sect. 4 above), the operator L_1 has a neutral mode $\mathrm{d}\Phi/\mathrm{d}|x|$ with a zero eigenvalue, but only for a special value of the layer response,

$$G_1^{\mathrm{cr}} = 2\left(\frac{\mathrm{d}\Phi_0(x)}{\mathrm{d}x}\right)^{-1} \frac{\mathrm{d}^2\Phi_0(x)}{\mathrm{d}x^2}\bigg|_{x=x_0}, \tag{27}$$

where $G_1 = G(I_0) + 2I_0[\mathrm{d}G/\mathrm{d}I]_{I=I_0}$. For $x_0 > 0$ the function $\Phi(x)$ has two humps, and its derivative $\mathrm{d}\Phi/\mathrm{d}|x|$ corresponds to the second-order mode with $\lambda_2^{(1)} = 0$. On the other hand, for $x_0 < 0$, the neutral mode describes a ground-state eigenfunction with $\lambda_0^{(1)} = 0$. From the spectral theorem [24], it follows that this eigenvalue increases for $G_1 < G_1^{\mathrm{cr}}$, and decreases otherwise. Additionally, we note that owing to the symmetry of the potentials in the linear eigenvalue problem, i.e. $U_j(x) = U_j(-x)$, the amplitude of the first-order eigenmode vanishes at the layer location ($x = 0$), and thus its eigenvalue $\lambda_1^{(1)}$ does not depend on G_1. However, the inequality $\lambda_0^{(1)} < \lambda_1^{(1)} < \lambda_2^{(1)}$ must be fulfilled for any G_1, and we immediately come to the conclusion that $\lambda_1^{(1)} < 0$ if $x_0 > 0$, and $\lambda_1^{(1)} > 0$ if $x_0 < 0$. It is also straightforward to verify that for $x_0 = 0$ we must have $G(I_0) = 0$, and that the derivative $\mathrm{d}\Phi(x)/\mathrm{d}x$, which has a single zero at $x = 0$, is a neutral mode of the linear operator L_1, i.e. $\lambda_1^{(1)} = 0$. In Table 1, we summarize the properties of the operator L_1 and the corresponding general conditions for the stability of *symmetric nonlinear localized modes*.

Table 1. Conditions for stability of symmetric modes

Condition	L_1 spectrum	Stability
$x_0 > 0$	$\lambda_0^{(1)} < \lambda_1^{(1)} < 0$	Unstable
$x_0 \le 0$, $G_1 \le G_1^{\mathrm{cr}}$	$0 \le \lambda_0^{(1)} < \lambda_1^{(1)}$	Stable
$x_0 \le 0$, $G_1 > G_1^{\mathrm{cr}}$, $\partial P/\partial\beta > 0$	$\lambda_0^{(1)} < 0 \le \lambda_1^{(1)}$	Stable
$x_0 \le 0$, $G_1 > G_1^{\mathrm{cr}}$, $\partial P/\partial\beta \le 0$	$\lambda_0^{(1)} < 0 \le \lambda_1^{(1)}$	Unstable

The next important step is to connect the spectral characteristics of the linear operator L_1 (and thus the stability of nonlinear localized modes) with the power dependence $P(\beta)$. First, we notice that at the *bifurcation point*, defined as the point at which $P(\beta_b) = P_0(\beta_b)$, the parameter x_0 passes through zero and the eigenvalue $\lambda_1^{(1)}$ changes its sign (see Table 1). In particular, for $P(\beta) > P_0(\beta)$ such a mode should have a *double-hump profile* and, therefore, it is *always unstable*. In contrast to these unstable double-hump modes, a family of asymmetric modes emerges at the bifurcation point.

6.2.2 Asymmetric Modes

Although asymmetric modes have the profiles of solitons, the corresponding spectrum of the operator L_1 can be different. Only in the special case $G_1 = 0$ does there exist a first-order neutral mode with $\lambda_1^{(1)} = 0$, i.e. $L_1(\mathrm{d}\Phi/\mathrm{d}x) = 0$.

There can exist several families of asymmetric modes, each of them characterized by a constant intensity at the layer I_0, so that $\mathrm{sign}(G_1) = \mathrm{sign}[\mathrm{d}G/\mathrm{d}I]_{I=I_0}$ at the point where the layer response vanishes, i.e.$G(I_0) = 0$. If G_1 is negative, the stability of the localized mode is the same as the stability of a soliton in the bulk medium. On the other hand, the modes corresponding to positive G_1 are unstable, and they can be attracted to or repelled from the nonlinear layer. By performing an analysis similar to that in [27], we find that for the asymmetric modes, $G_1 > 0$ if after the bifurcation point the branch of the symmetric modes is lower than that of the asymmetric modes, i.e. $P(\beta) < P_0(\beta)$. In the opposite case, we have $G_1 < 0$. In Table 2 we summarize the stability conditions for *asymmetric nonlinear localized modes*.

Table 2. Conditions for stability of asymmetric modes

Condition	L_1 spectrum	Stability
$G_1 > 0$	$\lambda_0^{(1)} < \lambda_1^{(1)} < 0$	Unstable
$G_1 \leq 0$, $\partial P_0/\partial\beta > 0$	$\lambda_0^{(1)} < 0 \leq \lambda_1^{(1)}$	Stable
$G_1 \leq 0$, $\partial P_0/\partial\beta \leq 0$	$\lambda_0^{(1)} < 0 \leq \lambda_1^{(1)}$	Unstable

6.3 Example: Competing Power-Law Nonlinearities

To discuss a characteristic example of the stability theory briefly outlined above, we consider a power-law nonlinear response of both the bulk medium and the layer,

$$F(I) = I^{\sigma} , \quad G(I) = a + bI^{\gamma} , \tag{28}$$

where the nonlinearities are characterized by the positive powers σ and γ, and the coefficients a and b account for the linear and nonlinear properties, respectively, of the layer.

We consider the case where the layer itself possesses *repulsive nonlinearity* (i.e. $b < 0$). Owing to a competition of localizing and delocalizing nonlinearities, the power dependence of the family of localized *symmetric* modes can become *multivalued*, i.e. two or three states may correspond to the same value

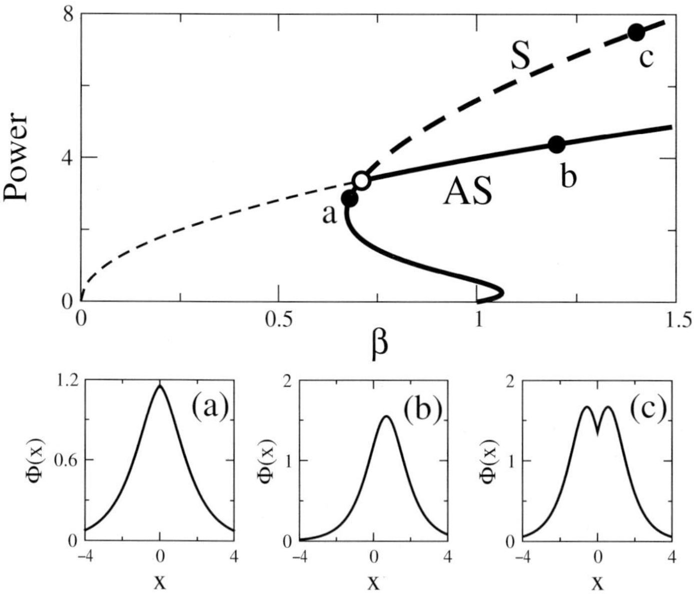

Fig. 4. Power vs. propagation constant for $\sigma = 1$, $\gamma = 2$, $a = 2$, and $b = -1$. The *solid* and *dashed* parts of the main curves correspond to stable and unstable localized modes, respectively. *Open circle*: the bifurcation point from the symmetric (S) to asymmetric (AS) modes. *Thin dashed line*: the soliton power $P_0(\beta)$. The modes corresponding to the points marked "a", "b", and "c" in the main part of the figure are shown in the diagram in the lower part of the figure

of the mode propagation constant (see Fig. 4). Moreover, such states can exist both below and above the linear cutoff.

The branch of the asymmetric solutions emerges at the bifurcation point, β_b, where the power of the symmetric localized modes, $P(\beta_\mathrm{b})$, coincides with the power of solitons in the homogeneous medium, $P_0(\beta_\mathrm{b})$ (open circle in Fig. 4). The stability of the asymmetric modes for $b < 0$ is the same as that of the solitons in the bulk medium, i.e. the modes are stable for $\sigma < 2$ and unstable otherwise. On the other hand, after the bifurcation point, i.e. for $\beta > \beta_\mathrm{b}$, the symmetric localized mode becomes two-humped and, therefore, is unstable to asymmetric perturbations. In Fig. 5, we illustrate the evolution of a two-hump symmetric mode when the power of the bulk nonlinearity is below the instability threshold ($\sigma < 2$). In this case, a symmetry-breaking instability is observed, and the mode is transferred to one side of the layer. Such an excitation of a stable asymmetric state is accompanied by slowly decaying quasi-periodic oscillation of the mode amplitude.

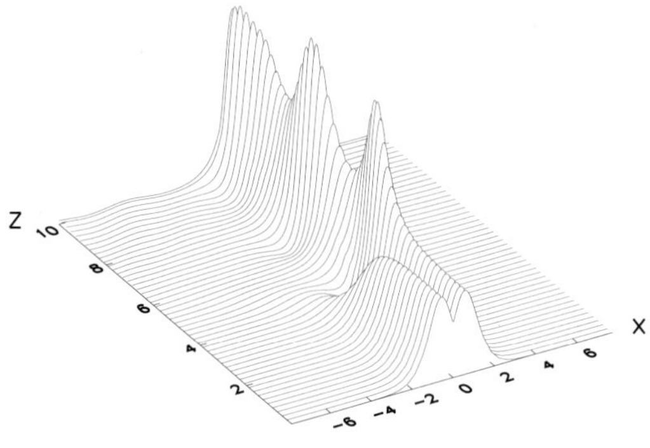

Fig. 5. Evolution of a slightly perturbed symmetric localized mode of the kind shown in Fig. 4c. An unstable two-hump mode eventually transforms into a stable asymmetric mode

7 Multicomponent Solitary Waves

7.1 General Theory: *N* Coupled NLS Equations

As has been already mentioned in the introduction to this chapter, the Vakhitov–Kolokolov stability criterion is valid for one-parameter solitary waves, such as scalar non-Kerr solitons, two-wave parametric solitary waves in a quadratic medium, etc. As we demonstrated in Sect. 6, the Vakhitov–Kolokolov criterion can also describe, in some cases, the stability of nonlinear guided waves in a scalar model. However, when solitary waves described by nonintegrable nonlinear models possess many parameters, the stability analysis becomes more involved, and in many cases a direct numerical study of the corresponding eigenvalue problem is required. Nevertheless, in some cases it is possible to extend the stability analysis to multiparameter solitary waves. Below, we follow the recent original work by Pelinovsky and Kivshar [19] and demonstrate how a multiparameter generalization of the Vakhitov–Kolokolov stability criterion can be derived for a system of N coupled NLS equations. Such a theory includes, as a limiting case, an analysis of bifurcations near the marginal stability curve.

We consider a system of N incoherently coupled NLS equations,

$$\mathrm{i}\frac{\partial \psi_n}{\partial z} + d_n \nabla^2{}_{\boldsymbol{x}} \psi_n + \left(\sum_{m=1}^{N} \gamma_{nm} |\psi_m|^2 \right) \psi_n = 0 \;, \tag{29}$$

where ${\nabla_{\boldsymbol{x}}}^2$ stands for the Laplacian in the D-dimensional space $\boldsymbol{x}{=}(x_1,\ldots,x_D)$, and all the coefficients d_n are assumed to be positive. When one of the variables of the vector $\boldsymbol{x}$ stands for time, (29) describes the spatiotemporal dynamics of self-focused and self-modulated light in the form of so-called *light bullets* .

Provided the symmetry conditions $\gamma_{nm} = \gamma_{mn}$ are satisfied, the system (29) conserves the Hamiltonian

$$H = \int_{-\infty}^{\infty} \mathrm{d}\boldsymbol{x} \left(\sum_{n=1}^{N} \mathrm{d}_n \left| \nabla_{\boldsymbol{x}} \psi_n \right|^2 - \frac{1}{2} \sum_{n=1}^{N} \sum_{m=1}^{N} \gamma_{nm} |\psi_n|^2 |\psi_m|^2 \right) ,$$

the individual mode powers $P_n = (1/2)\int |\psi_n|^2\,\mathrm{d}\boldsymbol{x}$, and the total field momentum. Localized solutions of (29) for the multicomponent fundamental solitary waves are defined in a standard form as $\psi_n = \Phi_n(\boldsymbol{x})\mathrm{e}^{\mathrm{i}\beta_n z}$, where the $\Phi_n(\boldsymbol{x})$ are real functions *with no nodes*, and β_n are the *positive* propagation constants. Such localized solutions can be also obtained as stationary points of the Lyapunov functional,

$$\Lambda[\boldsymbol{\psi}] = H[\boldsymbol{\psi}] + \sum_{n=1}^{N} \beta_n P_n[\boldsymbol{\psi}] \,, \tag{30}$$

and, therefore, the first variation of Λ vanishes, whereas the second variation $\delta^2\Lambda[\psi]$ defines the stability properties: the negative directions of the second variation correspond to unstable eigenvalues in the soliton stability problem.

The stability problem is defined by minimizing the second variation of the Lyapunov functional $\Lambda[\boldsymbol{\psi}]$,

$$\delta^2\Lambda = \int_{-\infty}^{\infty} \mathrm{d}\boldsymbol{x}\,\left[\langle \boldsymbol{u}|\mathbf{L}_1\boldsymbol{u}\rangle + \langle \boldsymbol{w}|\mathbf{L}_0\boldsymbol{w}\rangle\right] , \tag{31}$$

where $\boldsymbol{u}(\boldsymbol{x})$ and $\boldsymbol{w}(\boldsymbol{x})$ are perturbations of the multicomponent solitary wave taken in the form $\boldsymbol{\psi} = \boldsymbol{\Phi}(\boldsymbol{x}) + [\boldsymbol{u} + \mathrm{i}\boldsymbol{w}]\,(\boldsymbol{x})\mathrm{e}^{\lambda z}$, and the scalar product is defined as $\langle \boldsymbol{f}|\boldsymbol{g}\rangle = \sum_{n=1}^{N} f_n^* g_n$. The matrix Sturm–Liouville operator $\mathbf{L}_0$ has a diagonal form with elements

$$(L_0)_{nn} = -d_n \nabla^2{}_{\boldsymbol{x}} + \beta_n - \sum_{m=1}^{N} \gamma_{nm}\Phi_m^2 \,,$$

and the matrix operator $\mathbf{L}_1$ has elements

$$(L_1)_{nn} = -d_n \nabla^2{}_{\boldsymbol{x}} + \beta_n - \sum_{m=1}^{N} \gamma_{nm}\Phi_m^2 - 2\gamma_{nn}\Phi_n^2 \,,$$

on the diagonal, and $(L_1)_{nm} = -2\gamma_{nm}\Phi_n\Phi_m$ off the diagonal. Similarly to the stability problem for one-component solitary waves discussed in Sects. 2 and 4, the matrix operators $\mathbf{L}_0$ and $\mathbf{L}_1$ define the linear eigenvalue problem for the stability of multicomponent solitary waves,

$$\mathbf{L}_1\boldsymbol{u} = -\lambda\boldsymbol{w} \,, \quad \mathbf{L}_0\boldsymbol{w} = \lambda\boldsymbol{u} \,. \tag{32}$$

Both the linear problem (32) and minimization problem (31) must satisfy a set of N additional constraints,

$$F_n = \int_{-\infty}^{\infty} \mathrm{d}\boldsymbol{x}\,\langle \Phi_n \boldsymbol{e}_n|\boldsymbol{u}\rangle = 0 \,, \tag{33}$$

where $\boldsymbol{e}_n$ is the nth unit vector, which correspond to the conservation of the individual powers P_n under the action of a vector perturbation $(\boldsymbol{u}, \boldsymbol{w})$.

First of all, we recall that the one-parameter solitary waves with no nodes ($N = 1$) are stable in the framework of the constrained variational problem (31)–(33) provided the energetic surface $\Lambda_s(\boldsymbol{\beta}) = \Lambda[\boldsymbol{\Phi}]$ is concave upwards, i.e.

$$\frac{d^2 \Lambda_s}{d\beta_1^2} = \frac{dP_1}{d\beta_1} > 0\,. \tag{34}$$

Under this condition, the linear eigenvalue problem (32) has *no unstable eigenvalues*, i.e. eigenvalues with a positive real part λ. Otherwise, the second variation (31), constrained by the N conditions (33), has a *single negative direction* that corresponds to a single positive eigenvalue λ in the linear eigenvalue problem (32) (see [2,4] and the discussion above). The stability criterion for scalar (or one-component) NLS solitons holds when the self-adjoint operator $\mathbf{L}_1$ has a single negative eigenvalue, i.e. when the second variation (31), without the constraint (33) imposed, has a single negative direction. If the latter condition is not satisfied, as happens for solitary waves with nodes, the fundamental criterion for soliton instability can be extended only in special cases, while more generic mechanisms of oscillatory instabilities, associated with complex eigenvalues of the linear eigenvalue problem, may appear, beyond the predictions of the fundamental criterion.

In some particular cases, the soliton stability analysis can be extended to multicomponent solitary waves. In particular, this is possible for a system of incoherently coupled NLS equations (29) . We assume that the number of negative directions (eigenvalues) of the second variation $\delta^2 \Lambda$ is fixed, and we denote this number as $n(\Lambda)$. The unstable eigenvalues λ of the linear problem (32) are connected with some negative eigenvalues of the matrix $\mathbf{U}$ defined by the elements

$$U_{nm} = \frac{\partial^2 \Lambda_s}{\partial\beta_n \partial\beta_m} = \frac{\partial P_n}{\partial\beta_m} = \frac{\partial P_m}{\partial\beta_n}\,. \tag{35}$$

The matrix $\mathbf{U}$ is the *Hessian matrix* of the soliton energy surface $\Lambda_s(\boldsymbol{\beta})$. We denote the number of positive eigenvalues of the matrix $\mathbf{U}$ as $p(U)$, and the number of its negative eigenvalues as $n(U)$, so that $p(U) + n(U) \le N$, since some eigenvalues may be zeros in a degenerate (bifurcation) case. As is shown in [19], both $p(U)$ and $n(U)$ satisfy some additional constraints,

$$p(U) \le \min\{N, n(\Lambda)\}\,, \quad n(U) \ge \max\{0, N - n(\Lambda)\}\,. \tag{36}$$

With these notations, the following results on the stability and instability of multicomponent fundamental solitary waves of the coupled NLS equations (29) can be formulated and proved [19]:

- the linear problem (32) may have a maximum of $n(\Lambda)$ unstable eigenvalues λ, all *real* and *positive*;

- a multicomponent fundamental soliton is *linearly unstable* if $p(U) < n(\Lambda)$; then the linear problem (32) has $n(\Lambda) - p(U)$ real (positive or zero-becoming-positive) eigenvalues λ;
- a multicomponent fundamental soliton is *linearly stable* if $p(U) = n(\Lambda)$ ($\leq N$); in the case $n(\Lambda) = N$ this criterion implies that the energetic surface $\Lambda_s(\boldsymbol{\beta})$ is concave upwards in the $\boldsymbol{\beta}$ space;
- a single eigenvalue λ crosses a *marginal stability curve* when the matrix $\mathbf{U}$ possesses a zero-becoming-negative eigenvalue; the normal form of the instability-induced dynamics of multicomponent solitary waves resembles the equation of motion of an effective classical particle subjected to an N-dimensional potential field,

$$E = \frac{1}{2} \sum_{n=1}^{N} \sum_{m=1}^{N} M_{nm} \frac{\mathrm{d}\nu_n}{\mathrm{d}z} \frac{\mathrm{d}\nu_m}{\mathrm{d}z} + W(\boldsymbol{\beta}, \boldsymbol{\nu}) \,. \tag{37}$$

Here M_{nm} are the elements of the positive-definite "mass matrix",

$$M_{nm} = \int_{-\infty}^{\infty} \mathrm{d}\boldsymbol{x} \sum_{k=1}^{N} G_{kn}(\boldsymbol{x}) G_{km}(\boldsymbol{x}) \,,$$

where

$$G_{kn}(\boldsymbol{x}) = \frac{1}{\Phi_k(\boldsymbol{x})} \left(\int_0^{\boldsymbol{x}} \mathrm{d}\boldsymbol{x}' \, \Phi_k(\boldsymbol{x}') \frac{\partial \Phi_k(\boldsymbol{x}')}{\partial \beta_n} \right) ,$$

$\boldsymbol{\nu}$ is a vector describing a perturbation to the soliton parameters $\boldsymbol{\beta}$, and $W(\boldsymbol{\beta}, \boldsymbol{\nu})$ is an effective potential energy, defined as

$$W(\boldsymbol{\beta}, \boldsymbol{\nu}) = \Delta H(\beta) + \sum_{n=1}^{N} (\beta_n + \nu_n) \, \Delta P_n(\beta) \,,$$

where $\Delta H(\beta) = H_\mathrm{s}(\boldsymbol{\beta} + \boldsymbol{\nu}) - H_\mathrm{s}(\boldsymbol{\beta})$, and $\Delta P_n(\beta) = P_{\mathrm{s}n}(\boldsymbol{\beta} + \boldsymbol{\nu}) - P_{\mathrm{s}n}(\boldsymbol{\beta})$, where the index "s" indicates values calculated on the energy surface. A proof of these results and a comparison with the theorems of Grillakis et al. [7,8] are presented in [19].

7.2 Example: Two Coupled NLS Equations

In order to demonstrate how this general theory can be applied to a particular physical problem, we consider the case of *two incoherently coupled NLS equations*,

$$\begin{aligned} \mathrm{i}\frac{\partial \psi_1}{\partial z} + \frac{\partial^2 \psi_1}{\partial x^2} + \left(|\psi_1|^2 + \gamma |\psi_2|^2\right) \psi_1 = 0 \,, \\ \mathrm{i}\frac{\partial \psi_2}{\partial z} + \frac{\partial^2 \psi_2}{\partial x^2} + \left(|\psi_2|^2 + \gamma |\psi_1|^2\right) \psi_2 = 0 \,, \end{aligned} \tag{38}$$

where γ is a coupling parameter. The system (38) is a two-component reduction of the general N-component system (29) for $d_1 = d_2 = 1$, $\gamma_{11} = \gamma_{22} = 1$, and $\gamma_{12} = \gamma_{21} = \gamma$. An explicit soliton solution of (38) can easily be found for $\beta_1 = \beta_2 = \beta$ and $\gamma > -1$, in the form

$$\Phi_1(x) = \Phi_2(x) = \sqrt{\frac{2\beta}{1+\gamma}}\, \mathrm{sech}\left(\sqrt{\beta}x\right) . \tag{39}$$

This solution describes a two-component solitary wave with components of equal amplitudes, and it corresponds to the straight line $\beta_1 = \beta_2$ in the parameter plane (β_1, β_2) of a general *two-parameter family of solitary waves* of the model (38). For $-1 < \gamma \leq 0$, two-parameter solitons exist everywhere in the plane (β_1, β_2), while for $\gamma > 0$, the soliton existence region is restricted by two bifurcation curves, $\beta_2 = \omega_\pm(\gamma)\beta_1$, where

$$\omega_\pm(\gamma) = \left(\frac{\sqrt{1+8\gamma}-1}{2}\right)^{\pm 2} . \tag{40}$$

Approximate analytical expressions for a two-component solitary wave can also be obtained in the vicinity of one of the bifurcation curves (40), when one of the components of the composite solitary wave becomes small, while the other component is described by a scalar NLS equation. From a physical point of view, this corresponds to a situation where one of the component creates an effective waveguide that guides the other component, and this is known to describe the so-called *pulse-shepherding effect* in optical fibers, where the large-amplitude component plays the role of a shepherding pulse [29]. The composite soliton, which describes, for example, a large pulse ψ_1 guiding a small pulse ψ_2, can be found in the form [19]

$$\Phi_1 = R_0(x) + \epsilon^2 R_2(x) + O(\epsilon^4) , \quad \Phi_2 = \epsilon S_1(x) + O(\epsilon^3) , \tag{41}$$

and this solution is valid in the vicinity of the bifurcation curve,

$$\beta_2 = \omega_+(\gamma)\beta_1 + \epsilon^2 \omega_{2+}(\gamma)\beta_1 + O(\epsilon^4) . \tag{42}$$

The main terms of the asymptotic series (41), (42) are given by

$$R_0 = \sqrt{2\beta_1}\, \mathrm{sech}\left(\sqrt{\beta_1}x\right), \quad S_1 = \sqrt{\beta_1}\, \mathrm{sech}^{\sqrt{\omega_+}}\left(\sqrt{\beta_1}x\right),$$

and

$$\omega_{2+} = \frac{\int_{-\infty}^{\infty} \mathrm{d}x\, (S_1^4 + 2\gamma R_0 R_2 S_1^2)}{\int_{-\infty}^{\infty} \mathrm{d}x\, S_1^2} ,$$

and the second-order correction $R_2(x)$ is a solution of the equation

$$\left[-\partial_x^2 + \beta_1 - 6\beta_1\, \mathrm{sech}^2\left(\sqrt{\beta_1}x\right)\right] R_2 = \gamma R_0 S_1^2 .$$

From the existence domain of this two-component soliton, it follows that $\omega_{2+}(\gamma) > 0$ for $0 < \gamma < 1$, and $\omega_{2+}(\gamma) < 0$ for $\gamma > 1$. For $\gamma = 1$ (the so-called *Manakov model*), the family of two-parameter composite solitons becomes degenerate: it exists on the line $\beta_1 = \beta_2$ but, generally, it is different from the one-parameter solution (39). The coupled solitons are known to be stable for the integrable case $\gamma = 1$. Here we apply the stability theory developed above and prove that the $(1+1)$-dimensional two-parameter solitons, including the solitons of equal amplitude (39), are *stable* for $\gamma \geq 0$ and *unstable* for $\gamma < 0$.

Following [19] and Sect. 7.1, we evaluate the indices $p(U)$ and $n(\Lambda)$ for the explicit solution (39). As follows from (38) and (39), the Hessian matrix $\mathbf{U}$ with elements (35) can be found in the form

$$\frac{\partial P_1}{\partial \beta_1} = \frac{\partial P_2}{\partial \beta_2} = \frac{1}{\sqrt{\beta}(1+\gamma)} \quad \text{and} \quad \frac{\partial P_1}{\partial \beta_2} = \frac{\partial P_2}{\partial \beta_1} = -\frac{\gamma}{\sqrt{\beta}(1+\gamma)} .$$

Therefore, the Hessian matrix has $p(U) = 2$ positive eigenvalues for $-1 < \gamma < 1$, and $p(U) = 1$ positive eigenvalue for $\gamma > 1$. On the other hand, the linear matrix operator $\mathbf{L}_1$ given after (31) can be diagonalized for linear combinations of the eigenfunctions, $v_1 = u_1 + u_2$ and $v_2 = u_1 - u_2$, such that

$$\left[-\partial_x^2 + \beta - 6\beta \operatorname{sech}^2\left(\sqrt{\beta}x\right)\right] v_1 = \mu v_1 ,$$

$$\left[-\partial_x^2 + \beta - 2\beta\frac{(3-\gamma)}{(1+\gamma)} \operatorname{sech}^2\left(\sqrt{\beta}x\right)\right] v_2 = \mu v_2 . \tag{43}$$

Both of the operators in (43) are linear Schrödinger operators with solvable sech-type potentials, and the corresponding eigenvalue spectra are well studied. The first operator always has a single negative eigenvalue for $\mu = -3\beta$, whereas the second operator has no negative eigenvalues for $\gamma > 1$, has a single negative eigenvalue for $0 < \gamma < 1$, and has two negative eigenvalues for $-1 < \gamma < 0$. Thus, in total, there exist $n(\Lambda) = 3$ negative eigenvalues for $-1 < \gamma < 0$, $n(\Lambda) = 2$ negative eigenvalues for $0 < \gamma < 1$, and $n(\Lambda) = 1$ negative eigenvalue for $\gamma > 1$.

Applying the stability and instability results of [19] summarized in Sect. 7.1, we come to the conclusion that the soliton solution (39) with equal amplitudes is *linearly stable* for $\gamma > 0$, since in this domain $p(U) = n(\Lambda) = \{1, 2\}$, and it is *linearly unstable* for $-1 < \gamma < 0$, since in this domain $p(U) = 2 < n(\Lambda) = 3$.

Finally, we mention the other limiting case that describes the shepherding effect (see (41) and (42)). In this case, the elements (35) of the Hessian matrix $\mathbf{U}$ can also be calculated in an explicit analytical form, and it can be shown that the Hessian matrix $\mathbf{U}$ has $p(U) = 2$ positive eigenvalues for $0 < \gamma < 1$, and $p(U) = 1$ positive eigenvalue for $\gamma > 1$.

On the other hand, the linear matrix operator $\mathbf{L}_1$ cannot be diagonalized for a soliton in the shepherding regime (41), unless $\epsilon = 0$. In the latter, decoupled, case, it has a single negative eigenvalue at $\mu = -3\beta_1$ and a double, degenerate zero eigenvalue. When $\epsilon \neq 0$, the zero eigenvalue shifts to become $\mu = -2\omega_{2+}(\gamma)\beta_1\epsilon^2 + O(\epsilon^4)$. Therefore, the matrix operator $\mathbf{L}_1$ for the soliton (41) has $n(\Lambda) = 2$ negative eigenvalues for $0 < \gamma < 1$, and $n(\Lambda) = 1$ negative eigenvalue for $\gamma > 1$. Thus, we come to the conclusion that the soliton in the shepherding regime is stable for $\gamma > 0$ since $p(U) = n(\Lambda) = \{1, 2\}$.

8 Multihump Solitary Waves

In the simplest case, a spatial optical soliton is created by one beam of a certain polarization and frequency, and it can be viewed as a self-trapped mode of an effective waveguide that it induces in the bulk medium [30]. When a spatial soliton is composed of two (or more) modes of the induced waveguide [31], its structure becomes rather complicated, and the soliton intensity profile may display several peaks. This happens when one of the components creates an effective waveguide that guides higher-order modes of the other component. The resulting solitary waves are usually referred to as *multihump solitary waves*. Their modal structure is more complex than that of the multicomponent fundamental solitons considered in Sect. 7.

For a long time, it was believed that *all types* of multihump solitary waves were *linearly unstable*, except for the special case of neutrally stable solitons in the integrable Manakov model. In contrast, the experimental work of Mitchell et al. [32] reported the observation of stationary structures resembling multihump solitary waves. Moreover, a recent analysis [13] has confirmed that multihump solitons supported by incoherent interaction of two optical beams in a photorefractive medium can indeed be stable. Below, we address this issue, following the original results of Ostrovskaya et al. [13].

In the experiments [32], spatial multihump solitary waves were generated by incoherent interaction of two optical beams in a biased photorefractive medium. The corresponding model is described by a system of two coupled nonlinear equations for the normalized beam envelopes, $u(x, z)$ and $w(x, z)$, as follows:

$$\mathrm{i}\frac{\partial u}{\partial z} + \frac{1}{2}\frac{\partial^2 u}{\partial x^2} + \frac{u(|u|^2 + |w|^2)}{1 + s(|u|^2 + |w|^2)} - u = 0 \, ,$$

$$\mathrm{i}\frac{\partial w}{\partial z} + \frac{1}{2}\frac{\partial^2 w}{\partial x^2} + \frac{w(|u|^2 + |w|^2)}{1 + s(|u|^2 + |w|^2)} - \lambda w = 0, \tag{44}$$

where the transverse (x) and propagation (z) coordinates are measured in units of $(L_\mathrm{d}/k)^{1/2}$ and L_d, respectively; here L_d is a diffraction length and k is the wavevector. The parameter λ has the meaning of the ratio of the nonlinear propagation constants β_1 and β_2 of the two components, and s is an effective *saturation parameter* (see also Sect. 10 below).

We look for stationary, z-independent, solutions of (44) in which both components $u(x)$ and $w(x)$ are real and vanish as $|x| \to \infty$. Different types of such two-component localized solutions, existing for $0 < \{\lambda, s\} < 1$, can be characterized by the total power, $P(\lambda, s) = P_u + P_w$, where $P_u = \int_{-\infty}^{\infty} |u|^2 \, dx$ and $P_w = \int_{-\infty}^{\infty} |w|^2 \, dx$. If one of the components is small (e.g. $w/u \sim \varepsilon$), (44) becomes decoupled and, in the leading order, the equation for the u component has a solution $u_0(x)$ in the form of a fundamental, sech-like, soliton with no nodes. The second equation can then be considered as an eigenvalue problem for the "modes" $w_n(x)$ of a waveguide created by the soliton $u_0(x)$ with an effective refractive-index profile $u_0^2(x)/[1 + s u_0^2(x)]$. The parameter s determines the total number of guided modes and the cutoff value for each mode, $\lambda_n(s)$. Therefore, a two-component vector soliton (u_0, w_n) consists of a fundamental soliton and an n-order mode of the waveguide that it induces in the medium, and it can be characterized by its "state vector" $|0, n\rangle$.

On the $P(\lambda)$ diagram, continuous branches representing $|0, n\rangle$ solitons emerge at the points of bifurcation $\lambda_n(s)$ of one-component solitons. Families of two-component composite solitons can be found by a numerical relaxation technique [13], and some results of these calculations are presented in Fig. 6, for $|0, 1\rangle$ and $|0, 2\rangle$ solitons at $s = 0.8$. Observing the modification of the soliton profiles with changing λ (see inset in Fig. 6), one can see that the modal description of two-component solitons is valid only near the points of bifurcation. For $\lambda \gg \lambda_n$, the amplitude of an initially small w component *grows* and the soliton-induced waveguide deforms. The two- and three-hump solitons are members of the soliton families $|0, 1\rangle$ (branch ABC) and $|0, 2\rangle$

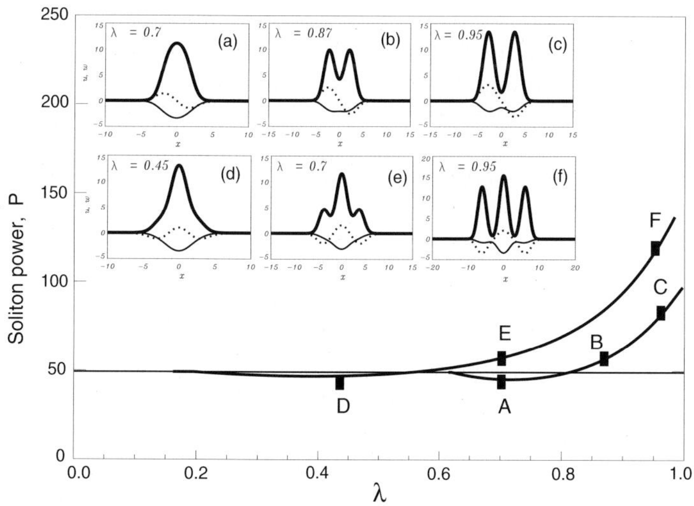

Fig. 6. Soliton bifurcation diagram at $s = 0.8$. *Horizontal line*: fundamental u soliton. *Curve ABC*: branch of $|0, 1\rangle$ solitons. *Curve DEF*: branch of $|0, 2\rangle$ solitons. *Inset*: Transverse profiles of u (*thin lines*) and w (*dashed lines*) fields, and total intensity (*thick lines*), shown for the marked points [13]

(branch DEF) originating at different bifurcation points. At $\lambda \sim \lambda_n(s)$, while the w component remains small, all $|0,n\rangle$ solitons are *single-humped*, as shown in Figs. 6a,d. As the amplitude of the component w grows with increasing λ, the total intensity profile, $I(x) = u^2(x) + w^2(x)$, develops $(n+1)$ humps (see Figs. 6b,e), and at sufficiently large λ the u component itself becomes *multihumped* (Figs. 6c,f). The separation distance between the soliton humps tends to infinity as $\lambda \to 1$.

To analyze the linear stability of these multihump solitons, we seek solutions of (44) in the form of weakly perturbed solitary waves, $u(x,z) = u_0(x) + \varepsilon[F_u(x,z) + \mathrm{i}G_u(x,z)]$ and $w(x,z) = w_n(x) + \varepsilon[F_w(x,z) + \mathrm{i}G_w(x,z)]$, where $\varepsilon \ll 1$. Setting $F_{u,w} \sim f_{u,w}(x)\mathrm{e}^{\mu z}$, $G_{u,w} \sim g_{u,w}(x)\mathrm{e}^{\mu z}$, one can obtain the following eigenvalue problems:

$$\begin{aligned} \hat{\mathcal{L}}_1\hat{\mathcal{L}}_0\boldsymbol{g} &= -\Lambda\boldsymbol{g}\,, \\ \hat{\mathcal{L}}_0\hat{\mathcal{L}}_1\boldsymbol{f} &= -\Lambda\boldsymbol{f}\,. \end{aligned} \tag{45}$$

Here $\boldsymbol{g} \equiv (g_u, g_w)^{\mathrm{T}}$, $\boldsymbol{f} \equiv (f_u, f_w)^{\mathrm{T}}$, $\Lambda = \mu^2$, and

$$\hat{\mathcal{L}}_{0,1} = \begin{pmatrix} -(1/2)(\mathrm{d}^2/\mathrm{d}x^2) + 1 - a_{0,1} & b_{0,1} \\ b_{0,1} & -(1/2)(\mathrm{d}^2/\mathrm{d}x^2) + \lambda - c_{0,1} \end{pmatrix},$$

where $a_0 = c_0 = I/(1+sI)$, $b_0 = 0$, $a_1 = a_0 + 2u_0^2/(1+sI)^2$, $c_1 = c_0 + 2w_n^2/(1+sI)^2$, and $b_1 = -2u_0w_n/(1+sI)^2$.

Note that $\hat{\mathcal{L}}_1\hat{\mathcal{L}}_0$ and $\hat{\mathcal{L}}_0\hat{\mathcal{L}}_1$ are adjoint operators with identical spectra. Therefore, we can consider the spectrum of only one of these operators, e.g. $\hat{\mathcal{L}}_1\hat{\mathcal{L}}_0$. Considering the complex Λ plane, it is straightforward to show that $\Lambda \in (-\infty, -\lambda^2)$ is a continuum part of the spectrum with unbounded eigenfunctions. Stable bounded eigenmodes of the discrete spectrum, i.e. the *soliton internal modes*, can have eigenvalues only inside the gap, $-\lambda^2 < \Lambda < 0$. According to the general theory outlined in Sect. 2, the presence of either a positive Λ or a pair of complex conjugated Λs implies soliton instability.

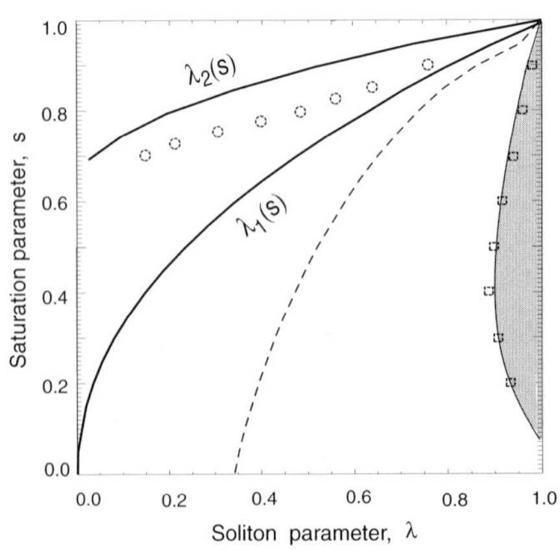

Fig. 7. Existence and stability domains for two- and three-hump solitons. Shown are the existence thresholds $\lambda_1(s)$ and $\lambda_2(s)$ for the $|0,1\rangle$ and $|0,2\rangle$ soliton families. *Dashed line*: the line where the total intensity of the $|0,1\rangle$ soliton develops humps. *Shaded region*: instability domain for two-hump solitons determined by generalized Vakhitov–Kolokolov criterion. *Squares* and *circles*: instability threshold for two- and three-hump solitons, respectively, obtained by numerical solution of the eigenvalue problem [13]

Numerical solution of the eigenvalue problem (45) shows that both types of solitary-wave solutions can be stable in a certain region of their existence domain, see Fig. 7. In the case of two-hump solitons, the appearance of the instability is related to the fact that close to the curve where the total intensity of the soliton becomes two-humped (see Fig. 7, dashed line), a pair of soliton internal modes splits from the continuum spectrum into the gap.

With the aid of an analytical asymptotic technique [10], it is possible to show that a perturbation mode with a small but positive eigenvalue, and therefore a linear instability of the general localized solution (u, w), appears if the determinant $J\{u, w\}$, defined as

$$J = \frac{\partial(P_u, P_w)}{\partial(\beta_1, \beta_2)} = \frac{P_u}{2s}\frac{\partial P_w}{\partial \lambda} - \frac{P_w}{2s}\frac{\partial P_u}{\partial \lambda} + \frac{\partial P_u}{\partial s}\frac{\partial P_w}{\partial \lambda} - \frac{\partial P_w}{\partial s}\frac{\partial P_u}{\partial \lambda},$$

changes its sign. The instability threshold given by the condition $J = 0$ is, in fact, the Vakhitov–Kolokolov stability criterion [2] generalized to the case of two-parameter composite solitons, as described in Sect. 7. However, such a generalization is rigorous for fundamental solitons only, whereas here we are dealing with higher-order solitary waves. As a result, in this case the determinant condition does not necessarily give the threshold of the leading instability and, therefore, the presence of other instabilities (which are not associated with the bifurcation condition $J = 0$ and could even have larger growth rates) is still possible. As was shown above, this is indeed true for the three-hump solitons, whereas the stability condition for two-hump solitons is well defined by the determinant condition $J = 0$, as verified by numerical solution of the linear eigenvalue problem (see Fig. 7).

9 Oscillatory Instabilities

When the conditions of the Vakhitov–Kolokolov criterion [2] or the theorem of Grillakis et al. [7,8], and their generalizations discussed in the preceding sections, are not satisfied, for example when there exists more than one negative eigenvalue of the linear eigenvalue problem in the scalar case, the study of soliton stability must be based on the results of numerical simulations, and no simple stability criterion that involves the system invariants can be constructed. In such a case the unstable eigenvalues are *complex* and, therefore, the corresponding soliton instability is called *oscillatory*. Oscillatory instabilities can usually be associated with resonances between two (or more) soliton internal modes or with resonances between a soliton internal mode and the edge of the continuous radiation spectrum. The former case is schematically represented in Fig. 8.

One of the first examples of oscillatory instabilities of solitary waves was found and analyzed by Barashenkov et al. [15] for a case of optical gap solitons described, in the framework of the coupled-mode theory, by two nonlinear coupled equations for the dimensionless amplitudes of the forward- and

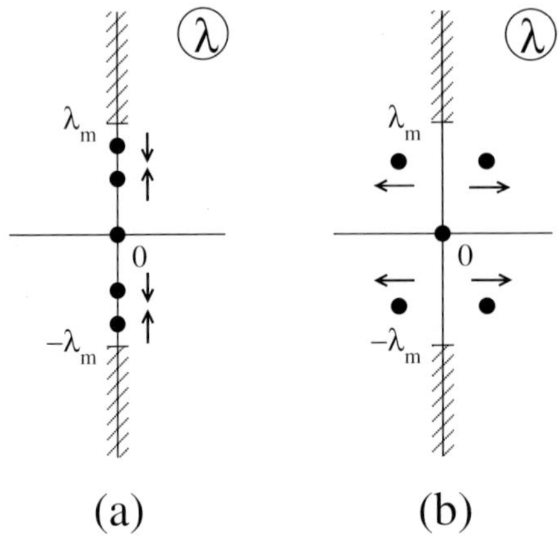

Fig. 8. Schematic representation of one of the scenarios for the soliton oscillatory instability that occurs via collision of two soliton internal modes

backward-propagating waves,

$$\begin{aligned} &\mathrm{i}\left(\frac{\partial u}{\partial z}+\frac{\partial u}{\partial x}\right)+\left(|u|^2+\sigma|v|^2\right)u+v=0\,,\\ &\mathrm{i}\left(\frac{\partial v}{\partial z}-\frac{\partial v}{\partial x}\right)+\left(|v|^2+\sigma|u|^2\right)v+u=0\,. \end{aligned} \tag{46}$$

Here σ is the cross-phase-modulation parameter, and the case $\sigma = 0$ corresponds to the exactly integrable massive Thirring model of field theory.

The fact that the system (46) is integrable at $\sigma = 0$ can be employed [15] to study bifurcations of new eigenvalues from the edge of the continuous spectrum in the case of small but nonzero σ, similarly to the case of a scalar weakly perturbed, cubic NLS equation (see Sect. 3). The additional eigenvalues correspond to internal modes of a gap soliton, and collision of two such eigenvalues results in a soliton oscillatory instability, characterized by a pair of complex eigenvalues with a positive real part.

The existence of soliton oscillatory instabilities is usually associated with multiparameter solitary waves such as optical gap solitons (in the latter case, the two independent parameters are the soliton frequency and velocity). Another example of a soliton oscillatory instability was recently discussed by Mihalache et al. [17] for soliton propagation in birefringent optical fibers in the presence of walk-off, self-phase modulation, cross-phase modulation, and parametric four-wave-mixing effects. It was found that the condition for linear marginal stability of the two-parameter solitary waves of that model is not necessarily given by an explicit determinant criterion, because oscillatory instabilities associated with complex eigenvalues were also found to exist in that model.

One more interesting example of oscillatory instabilities of solitary waves has been found for a *dark soliton* for the generalized Ablowitz–Ladik equation and a discrete NLS model of the form [18]

$$\mathrm{i}\frac{\partial \psi_n}{\partial z} + C(\phi_{n+1} + \psi_{n-1}) + |\psi_n|^2\psi_n = 0 \, . \tag{47}$$

It was revealed that even weak inherent discreteness of a nonlinear model can lead to instabilities of solitary waves and, in particular, in the model (47), an instability of a dark soliton on a fixed background, $|\psi_\infty|^2 = 1$, appears for $C > C_{\rm cr} \approx 0.076$ and is of the oscillatory type. Additionally, it was found that oscillatory instabilities may appear in *two different scenarios*, and they may occur owing to either a resonance between radiation modes and a soliton internal mode, or a resonance between two soliton internal modes (as shown in Fig. 8).

10 Symmetry-Breaking Instabilities

All the problems discussed above involve perturbations acting on a soliton in a space of the same dimension. However, a solitary wave may also become unstable when it is subjected to the action of perturbations of higher dimensions. We define these cases as *symmetry-breaking soliton instabilities*, which can be further classified into two categories: *transverse instabilities* and *modulational instabilities* . These two categories are characterized by the type of soliton evolution, which may be affected either by higher spatial dimensions (such as in the breakup of a plane soliton) or by temporal modulation (such as in the temporal instability of spatial solitons and the formation of light bullets).

A comprehensive overview of the soliton self-focusing and transverse instabilities of solitons has recently been presented by Kivshar and Pelinovsky [33], and we refer to that review paper for detailed discussion of a general theory of symmetry-breaking (both transverse and modulational) soliton instabilities and of different examples of instability-induced soliton dynamics, in the case when a plane solitary wave is subjected to perturbations of higher dimensions. A less studied case is the so-called *azimuthal instability* of radially symmetric solitary waves, recently observed experimentally for a number of physical systems. Below we present an example of such a symmetry-breaking instability.

First of all, we recall that in higher dimensions there exist different types of radially symmetric ring-like solitary waves of higher order, even in a scalar NLS equation with generalized nonlinearity [34]. Such modes consist of a bright central spot surrounded by one or more rings, and all those structures are known to be unstable, evolving into a fundamental radially symmetric soliton and radiation, or displaying more complicated decay-induced collapse-free evolution in a saturable nonlinear medium [35].

More interesting types of radially symmetric localized modes are associated with a nonzero angular momentum of the solitary wave, carrying a topological charge (or "spin") m, such that $u(r,\phi) = u(r)\mathrm{e}^{\mathrm{i}m\phi}$, where $m = \pm 1, \pm 2, \ldots$ Such solitary waves are known as *vortex solitons* (see, e.g., [36]) and have been found for various types of nonlinear models (see, e.g., [37]). All types of these radially symmetric ring-like solitary waves are also *unstable* with respect to azimuthally dependent perturbations, and the decay scenario is determined by the value of the angular momentum. In particular, owing to modulational instability, rings with a charge $m = \pm 1$ decay into pairs of fundamental bright solitons ("filaments") of opposite phase, which strongly repel each other and always move apart along straight trajectories satisfying momentum conservation [37]. Various filamentation scenarios have been observed in experiments on solitons in Rb vapors (see, e.g., [38]).

In scalar beam propagation, symmetry-breaking instabilities of higher-order radially symmetric solitary waves (with or without angular momentum) usually lead to decay or filamentation of solitons. However, for vectorial (or multicomponent) nonlinear modes, the decay scenarios of ring-like solitary waves become more complicated and lead to novel phenomena. One recent example of this kind is the prediction of the existence and stability of a novel type of robust vector soliton – a *dipole-mode vector soliton*, which is a radially *asymmetric* two-component self-trapped beam in a bulk medium that can be created via the symmetry-breaking instability of a vortex-mode composite soliton, predicted earlier in [39].

To discuss this phenomenon, we follow the original work [40] and consider two incoherently interacting beams that propagate along the z direction in a bulk saturable nonlinear medium. This model corresponds, in the so-called isotropic approximation, to the experimentally realized self-trapped beams in photorefractive crystals, and, in dimensionless form, it can be described by a system of two coupled NLS equations for the slowly varying beam envelopes E_1 and E_2,

$$\mathrm{i}\frac{\partial E_{1,2}}{\partial z} + \Delta_\perp E_{1,2} - \frac{E_{1,2}}{1 + |E_1|^2 + |E_2|^2} = 0\,, \tag{48}$$

where $\Delta_\perp$ is the transverse Laplacian. Stationary solutions of (48) can be found in the form

$$E_1 = \sqrt{\beta_1}u(x,y)\mathrm{e}^{\mathrm{i}\beta_1 z}\,, \quad E_2 = \sqrt{\beta_1}v(x,y)\mathrm{e}^{\mathrm{i}\beta_2 z}\,,$$

where β_1 and β_2 are two independent propagation constants. Measuring the transverse coordinates in units of $\sqrt{\beta_1}$ and introducing the ratio of the propagation constants, $\lambda = (1-\beta_1)/(1-\beta_2)$, we obtain from (48) the normalized stationary equations for u and v,

$$\begin{aligned} &\Delta_\perp u - u + uf(I) = 0\,, \\ &\Delta_\perp v - \lambda v + vf(I) = 0\,, \end{aligned} \tag{49}$$

where $f(I) = I(1+sI)^{-1}$, $I = u^2 + v^2$, and the parameter $s = 1 - \beta_1$ plays the role of an effective saturation parameter, as in (44) (Sect. 8).

Radially symmetric vortex-mode vector solitons of the model (49) can be found in the form [39]

$$u(x,y) = u(r) , \quad v(x,y) = v(r)e^{im\phi} ,$$

where m is an integer (the topological charge). From the physical point of view, such self-trapped states can be regarded as composite solitons consisting of a self-induced waveguide created in the field u, and the first-order doughnut-type guided mode in the field v, the latter resembling the structure of a Gauss–Laguerre (LG_0^1) mode of a cylindrically symmetric optical waveguide. Importantly, when such a vortex-mode composite soliton propagates in a nonlinear medium, it undergoes a symmetry-breaking oscillatory-type instability, as shown in Fig. 9. However, in sharp contrast to the vortex decay scenario in the scalar NLS model, the two fundamental solitons created as the result of this breakup do not move apart but, instead, remain trapped in a rotating dipole-like structure, as clearly seen in Fig. 9b. This dipole structure corresponds to an asymmetric vector solitary wave of a different type, which can be thought of as a composite self-trapped state of a soliton-induced waveguide that guides a dipole mode with the shape of a Gauss–Hermite (HG_{10}) laser mode. Recent studies revealed that such dipole-mode solitary waves are indeed extremely robust objects that survive moderate and large perturbations [40]. Dipole-mode composite solitons have been recently created experimentally in strontium barium niobate (SBN) photorefractive crystals, both by a phase-imprinting technique and as a product of the symmetry-breaking instability of radially symmetric vortex-mode solitons [41,42].

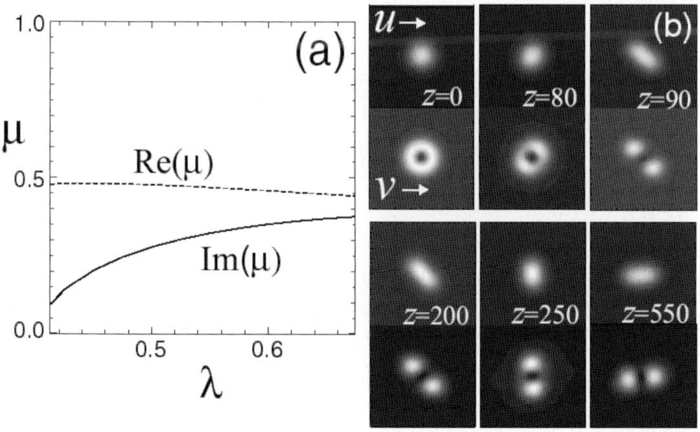

Fig. 9. (**a**) Eigenvalue of the leading unstable mode for the vortex-mode vector soliton ($s = 0.5$), and (**b**) typical decay scenario of the vortex-mode soliton near cutoff ($s = 0.65$, $\lambda = 0.6$). Shown are the intensity distributions of the u and v components in the (x, y) plane [40]

More recently, the unique robustness of the dipole-mode soliton was demonstrated, theoretically and experimentally, in a study of their collisions with scalar solitons and other dipoles [43].

11 Concluding Remarks

We have presented a brief overview of the key concepts and basic results of soliton stability theory applied to spatial optical solitons, emphasizing the most recent results on the stability of multicomponent solitary waves, the study of oscillatory instabilities of solitary waves, and symmetry-breaking instabilities of solitons in higher dimensions. We have demonstrated that the theory of soliton instabilities is closely related to the concept of soliton internal modes, an important feature of the solitary waves of many nonintegrable nonlinear models.

The simplest and most familiar class of soliton instabilities in NLS-type nonlinear models is described by the Vakhitov–Kolokolov criterion, $\mathrm{d}P/\mathrm{d}\beta < 0$, where P is the soliton power and β is its propagation constant. This criterion is valid for one-component fundamental solitary waves provided the L_1 operator of the linear eigenvalue problem possesses exactly one negative eigenvalue. We have shown that a nontrivial generalization of this criterion to the case of multicomponent solitary waves is also possible, but requires more restrictive conditions.

When rigorous results from linear stability theory are not available, a simple asymptotic theory allows one to find a condition for the marginal stability point (or, in general, a marginal stability surface) that may be further verified numerically. One such case, where the result of the asymptotic theory has no rigorous proof yet, even though it has been verified numerically, is the stability of two-hump solitary waves created by incoherent interaction of two optical beams in a photorefractive nonlinear medium. The multiscale asymptotic analysis is valid in the region of marginal stability, and is a powerful tool that also allows one to study weakly nonlinear effects in the instability-induced evolution of solitons and to describe different scenarios of the dynamics of unstable solitons analytically.

In many other cases, the eigenvalues of the corresponding linear problem are complex, and the study of soliton stability must be carried out by means of a numerical analysis of the corresponding linear eigenvalue problem. This is the case for the so-called oscillatory instabilities of solitary waves, for instance those known to exist for gap solitons described by a coupled-mode theory. Such instabilities are not predicted by simple integral criteria; they appear because of a resonance between two soliton internal modes or between a soliton internal mode and the edge of the continuous spectrum.

Many interesting problems of soliton stability theory and of the dynamics of unstable optical solitons can be found in higher dimensions, where the symmetry-breaking instability of plane or radially symmetric solitary waves brings new features into the soliton dynamics. In particular, we have demonstrated that a vortex-like solitonic mode that always breaks up into fundamental bright solitons in a scalar model may evolve, in a vector model, into a rotating dipole-like structure associated with a robust dipole-mode vector soliton.

The study of instabilities is an important direction in the research on spatial optical solitons that links the theoretical concepts of self-trapping of a beam and stationary self-guided beam propagation with the possibility of experimental verification. After all, wave instabilities are probably the most remarkable physical phenomena that can occur in nature, and their study allows researchers to reveal many important and fascinating properties of nonlinear physical systems.

Acknowledgments

Many of our colleagues contributed to the topics briefly discussed in this chapter. We are especially indebted to Dmitry Pelinovsky and Elena Ostrovskaya for a fruitful collaboration, and for their valuable contribution to the results presented here and to our understanding of soliton stability.

References

1. M. Segev, G. Stegeman, Phys. Today **51**, No. 8, 42 (1998); G.I. Stegeman, M. Segev, Science **286**, 1518 (1998); Yu.S. Kivshar, Opt. Quantum Electron. **30**, 571 (1998).
2. N.G. Vakhitov, A.A. Kolokolov, Izv. Vyssh. Uchebn. Zaved. Radiofiz. **16**, 1020 (1973) [Radiophys. Quantum Electron. **16**, 783 (1973)]; A.A. Kolokolov, Izv. Vyssh. Uchebn. Zaved. Radiofiz. **17**, 1332 (1974) [Radiophys. Quantum Electron. **17**, 1016 (1974)].
3. J. Shatah, W. Strauss, Commun. Math. Phys. **100**, 173 (1985); M.I. Weinstein, Commun. Pure Appl. Math. **39**, 5 (1986).
4. E.A. Kuznetsov, A.M. Rubenchik, V.E. Zakharov, Phys. Rep. **142**, 103 (1986).
5. V.G. Makhankov, Yu.P. Rybakov, V.I. Sanyuk, Usp. Fiz. Nauk **164**, 121 (1994) [Phys.–Uspekhi **62**, 113 (1994)].
6. C.K.R.T. Jones, J. Diff. **71**, 34 (1988); Ergod. Theory Dyn. Syst. **8***, 119 (1988).
7. M. Grillakis, Commun. Pure Appl. Math. **41**, 747 (1988); Commun. Pure Appl. Math. **43**, 299 (1990).
8. M. Grillakis, J. Shatah, W. Strauss, J. Funct. Anal. **74**, 160 (1987); J. Funct. Anal. **94**, 308 (1990).
9. L. Torner, in *Beam Shaping and Control with Nonlinear Optics*, ed. by F. Kajzer, R. Reinisch (Plenum, New York, 1998), p. 229; Yu.S. Kivshar, in: *Advanced Photonics with Second-Order Optically Nonlinear Processes*, ed, by A.D. Boardman, K. Pavlov, S. Tanev (Kluwer, Dordrecht, 1999), p. 451.
10. D.E. Pelinovsky, A.V. Buryak, Yu.S. Kivshar, Phys. Rev. Lett. **75**, 591 (1995); D.E. Pelinovsky, V.V. Afanasjev, and Yu.S. Kivshar, Phys. Rev. E **53**, 1940 (1996).

11. A.V. Buryak, Yu.S. Kivshar, S. Trillo, Phys. Rev. Lett. **77**, 5210 (1996); C. Etrich, U. Peschel, F. Lederer, B.A. Malomed, Phys. Rev. E **55**, 6155 (1997); A. De Rossi, C. Conti, S. Trillo, Phys. Rev. Lett. **81**, 85 (1998).
12. D.E. Pelinovsky, R.H.J. Grimshaw, in: *Advances in Fluid Mechanics*, Vol. 12, ed. by L. Debnath, S.R. Choudhury (Computational Mechanics Publications, Southampton, 1997), p. 246.
13. E.A. Ostrovskaya, Yu.S. Kivshar, D.V. Skryabin, W.J. Firth, Phys. Rev. Lett. **83**, 296 (1999).
14. D.V. Skryabin, Physica D **139**, 186 (2000).
15. I.V. Barashenkov, D.E. Pelinovsky, E.V. Zemlyanaya, Phys. Rev. Lett. **80**, 5117 (1998).
16. Y.A. Li, K. Promislow, Physica D **124**, 137 (1998).
17. D. Mihalache, D. Mazilu, L. Torner, Phys. Rev. Lett. **81**, 4353 (1998).
18. M. Johansson, Yu.S. Kivshar, Phys. Rev. Lett. **82**, 85 (1999).
19. D.E. Pelinovsky, Yu.S. Kivshar, Phys. Rev. E **62**, 8668 (2000).
20. Yu.S. Kivshar, B.A. Malomed, Rev. Mod. Phys. **63**, 761 (1989).
21. Yu.S. Kivshar, D.E. Pelinovsky, T. Cretegny, M. Peyrard, Phys. Rev. Lett. **80**, 5032 (1998).
22. D.J. Kaup, Phys. Rev. A **42**, 5689 (1990).
23. M.M. Bogdan, A.S. Kovalev, I.V. Gerasimchuk, Fiz. Nizk. Temp. **23**, 197 (1997) [Low Temp. Phys. **23**, 145 (1997)].
24. E.C. Titchmarsh, *Eigenfunction Expansions Associated with Second-Order Differential Equations* (Oxford University Press, London, 1958).
25. Yu.S. Kivshar, V.V. Afanasjev, A.V. Buryak, D.E. Pelinovsky, in: *Physics and Applications of Optical Solitons in Fibers*, ed. by A. Hasegawa (Kluwer, Dordrecht, 1996), pp. 75–88.
26. G.I. Stegeman, C.T. Seaton, W.H. Hetherington, A.D. Boardman, P. Egan, in: *Electromagnetic Surface Excitations*, ed. by R.F. Wallis, G.I. Stegeman (Springer-Verlag, Berlin, Heidelberg, 1986);
R.W. Micallef, Yu.S. Kivshar, J.D. Love, D. Burak, R. Binder, Opt. Quantum Electron. **30**, 751 (1998).
27. D.J. Mitchell, A.W. Snyder, J. Opt. Soc. Am. B **10**, 1572 (1993).
28. A.A. Sukhorukov, Yu.S. Kivshar, O. Bang, J.J. Rasmussen, P.L. Christiansen, Phys. Rev. E **63**, 036601 (2001).
29. C. Yeh, L. Bergman, Phys. Rev. E **57**, 2398 (1998); Phys Rev. E **60**, 2306 (1999); Yu.S. Kivshar, E.A. Ostrovskaya, Pulse shepperding and multi-channel soliton transmission in bit-parallel-wavelength optical fiber links, arXiv: patt-sol/9912008 (1999).
30. R.Y. Chiao, E. Garmire, C.H. Townes, Phys. Rev. Lett. **13**, 479 (1964).
31. A.W. Snyder, S.J. Hewlett, D.J. Mitchell, Phys. Rev. Lett. **72**, 1012 (1994).
32. M. Mitchell, M. Segev, D.N. Christodoulides, Phys. Rev. Lett. **80**, 4657 (1998).
33. Yu.S. Kivshar, D.E. Pelinovsky, Phys. Rep. **331**, 117 (2000).
34. Z.K. Yankauskas, Izv. Vyssh. Uchebn. Zaved. Radiofiz. **9**, 412 (1966) [Sov. Radiophys. **9**, 261 (1966)]; H.A. Haus, Appl. Phys. Lett. **8**, 128 (1966).
35. D.E. Edmundson, Phys. Rev. E **55**, 7636 (1997).
36. V.I. Kruglov, R.A. Vlasov, Phys. Lett. A **111**, 401 (1985).
37. W.J. Firth, D.V. Skryabin, Phys. Rev. Lett. **79**, 2450 (1997); J.P. Torres, J.M. Soto-Crespo, L. Torner, D.V. Petrov, J. Opt. Soc. Am. B **15**, 625 (1998).
38. V. Tikhonenko, J. Christou, B. Luther-Davies, J. Opt. Soc. Am. B **12**, 2046 (1995).

39. Z.H. Musslimani, M. Segev, D.N. Christodoulides, M. Soljačić, Phys. Rev. Lett. **84**, 1164 (2000);
J.N. Malmberg, A.H. Carlsson, D. Anderson, M. Lisak, E.A. Ostrovskaya, Yu.S. Kivshar, Opt. Lett. **25**, 643 (2000).
40. J.J. García-Ripoll, V.M. Perez-García, E.A. Ostrovskaya, Yu.S. Kivshar, Phys. Rev. Lett. **85**, 82 (2000).
41. W. Krolikowski, E.A. Ostrovskaya, C. Weilnau, M. Geisser, G. McCarthy, Yu.S. Kivshar, C. Denz, B. Luther-Davies, Phys. Rev. Lett. **85**, 1424 (2000).
42. T. Carmon, C. Anastassiou, S. Lan, D. Kip, Z.H. Musslimani, M. Segev, D. Christodoulides, Opt. Lett. **25**, 1113 (2000).
43. J.J. García-Ripoll, V.M. Perez-García, W. Krolikowski, Yu.S. Kivshar, Scattering of light by molecules of light, arXiv: nlin.PS/0006047 (2000); J.J. García-Ripoll, V.M. Perez-García, W. Krolikowski, G. McCarthy, B. Luther-Davies, D. Neshev, E.A. Ostrowskaya, and Yu.S. Kivshar, Nonlinear Guided Waves and Their Applications, Clearwater, Florida, USA, 25–28 March 2001, OSA Technical Digest. pp. 455–457.

Nonlinear Wave Collapse

Luc Bergé

Summary. We review the general properties of optical pulses that produce wave collapse in bulk Kerr media. First, analytical tools emphasizing the threshold power for collapse are recalled for the case in which one wave component is described by a single nonlinear Schrödinger equation. Next, the case of several light waves is investigated using a system of coupled nonlinear Schrödinger equations. Depending on their initial separation distance and power, two waves are shown either to disperse or to collapse individually, or to attract each other to form a central lobe, which may blow up at a finite propagation distance. Furthermore, the influence of four-wave mixing and walk-off between two components is detailed. It is shown that near phase matching, four-wave mixing can reinforce the collapse by lowering the self-focusing power threshold, whereas walk-off inhibits the collapse by detrapping the waves. Finally, collapse in media that promote an interplay between cubic and quadratic nonlinearities is discussed.

1 Introduction

For several decades, the question of optimizing the propagation of light beams in nonlinear bulk media has generated increasing interest in connection with the development of high-power laser sources [1,2]. In particular, the self-trapping and self-focusing of intense beams have been extensively investigated since the 1960s. Self-trapped beams evolving in three spatial dimensions are known to be unstable in an ideal Kerr material [3], and to self-focus until a catastrophic collapse occurs when the input power exceeds a critical value. This value is usually computed in the framework of the nonlinear Schrödinger (NLS) equation, with two transverse dimensions in the diffraction plane and one longitudinal dimension for the propagation axis [4]. In a physical medium, once self-focusing is initiated, the local wave intensity starts to increase and collapse is arrested by saturation of the Kerr response, which may promote the formation of spatial solitons. In the present investigation, however, we shall ignore these saturation effects and allow the waves to collapse freely in self-focusing regimes. Although "unphysical", the collapse singularity emerging in the medium will enable us to stress more the localization mechanism of the waves during their growth stage, before saturation becomes efficient.

Recently, the topic of interacting optical components has opened up a novel area of investigation. For instance, numerous works [5–8] display evidence that in a Kerr (cubic) medium, a beam with arbitrary polarization decomposes into two components with opposite circular polarizations, so that the slowly varying complex envelopes of the resulting scalar waves obey two distinct, nonlinearly coupled NLS equations, which can be written in the generic form [7–14]

$$i(\partial_z + \boldsymbol{\delta}_j \cdot \boldsymbol{\nabla})E_j - \beta_j E_j + \boldsymbol{\nabla}^2 E_j + (|E_j|^2 + A|E_{3-j}|^2)E_j + BE_{3-j}^2 E_j^* = 0 \,, \quad (1)$$

where $j = 1, 2$. In (1), the asterisk denotes the complex conjugate, and the first term accounts for the propagation of the electric field components $E_1(\boldsymbol{r}, z)$ and $E_2(\boldsymbol{r}, z)$ along the z axis, expressed in a frame moving with the group velocity of the beam. The Laplacian describes the transverse diffraction effects and is equal to $\boldsymbol{\nabla}^2 \equiv \partial_x^2 + \partial_y^2$ for a diffraction plane spanned by the radius vector $\boldsymbol{r} = (x, y)$. For technical convenience, this diffraction plane may formally be extended to $D \leq 3$ transverse dimensions, including a temporal dimension to account for anomalous group velocity dispersion (GVD). The second term, where $\boldsymbol{\delta}_1 = -\boldsymbol{\delta}_2 = \boldsymbol{\delta}$, describes so-called "walk-off" effects, such as linear spatial convection or a group velocity difference due to GVD. The third term, where $\beta_1 = -\beta_2 = \beta$, stands for the mismatch in wavenumbers between the components, and the last term represents the contribution of four-wave mixing (FWM) processes, which parametrically mix the components. The constants A and B are positive and measure the strength of the nonlinear coupling between the two waves. Their values are determined by the symmetry of the medium and also by the interaction geometry. For example, in isotropic media, they satisfy $A + B = 1$, as E_1 and E_2 represent two orthogonal polarizations of a vector field.

When the FWM and walk-off are zero, (1) applies basically to waves that are incoherently coupled. In the one-dimensional case ($D = 1$) the equations admit soliton solutions, which were discovered by Manakov [6]. The existence of these solitons was shown by a direct application of inverse scattering transform (IST) techniques, which can be employed under specific conditions that involve a low dimensionality and severe constraints on the nonlinearity coefficients. From a physical viewpoint, these constraints impose the condition that the ratio between the self-phase and cross-phase modulations in the coupling constants must be equal to unity, while the self-phase modulation coefficients need to be identical for the two polarizations. Partly because of the difficulty in overcoming FWM processes, Manakov spatial solitons have only been recently detected in experiments [9]; here the interaction material allowed the two components to be incoherently coupled so as to average the FWM terms to zero. Moreover, coupled solitons undergo linear walk-off when $\boldsymbol{\delta} \neq \boldsymbol{0}$, which can be associated with either the difference between their group velocities in a birefringent fiber [7,8,10–12] or the angle between the transverse and carrier wavevectors, for example in biased photorefractive crystals [13,14].

In previous investigations, finding exact soliton solutions was possible because the propagation equations involved one transverse dimension. In contrast, for media that allow diffraction in the transverse plane and/or have anomalous GVD, the total number of transverse dimensions, D, is increased and coupled waves can self-focus until a collapse singularity is formed, which forbids the formation of stable solitons. In this respect, McKinstrie and Russel [15] showed that two light waves described by (1) with $B = \delta = 0$ could be mutually entrained while promoting collapse events, if they are initially located around different centroids. Self-focusing conditions for two superimposed waves were detailed in [15,16], while for initially separated waves, a critical distance between two Gaussian pulses was identified as the minimal distance beyond which they may behave independently of one another. From these results, it therefore appears necessary to establish classification criteria, which discriminate between the different interactions that coupled waves may undergo.

Besides, an important challenge in optics concerns second-harmonic generation in asymmetric media. Harmonic generation occurs when spectral components from two (or more) waves with frequencies ω_1 and ω_2 are mixed such that the generated frequency is the sum of the input frequencies. Second harmonics have already been produced experimentally with media that do not have inversion symmetry and have the periodicity required for phase matching. Mathematically, we can assume that a pump wave at a frequency ω_1 is incident on such a material. The resulting equations satisfied by the slowly varying envelopes of the pump field E_1 and its second harmonic E_2 then take the form of coupled-amplitude equations [17–19]. If the fundamental wave has only one polarization component, these equations basically resemble (1) with neither walk-off ($\boldsymbol{\delta} = \mathbf{0}$) nor FWM ($B = 0$). The first equation is, nevertheless, supplemented by a quadratic contribution $E_1^* E_2 \mathrm{e}^{-\mathrm{i}\Delta k\, z}$, while the second one possesses a quadratic nonlinearity $E_1^2 \mathrm{e}^{+\mathrm{i}\Delta k\, z}$, both of which originate from $\chi^{(2)} : \chi^{(2)}$ cascading processes and depend on the residual wavevector mismatch $\Delta k = 2k_1(\omega_1) - k_2(2\omega_1)$. Solitons created by such mutually trapped components have been observed in a recent experiment [20], where the phase-matchable quadratic material, provided by a KTP crystal, generated solitary waves when the intensity in the input beam was high enough. From a theoretical point of view, solitary waves formed by the fundamental and second-harmonic fields at a power above a threshold have been shown to exist and to be stable for a large class of physical parameters in purely quadratic media [21]. However, even in materials where the $\chi^{(2)}$ nonlinearity is initially dominant, cubic $\left(\chi^{(3)}\right)$ nonlinearities can lead to quite a different evolution, by promoting collapse at a finite propagation distance. Thus, a crucial issue for all-optical switching technologies based on materials with both $\chi^{(2)}$ and $\chi^{(3)}$ nonlinearities is to determine whether the coupled components are capable of self-focusing before a destructive collapse.

This contribution reviews the above topics in the light of wave collapse. We first recall in Sect. 2 the main properties of collapsing solutions for an NLS equation describing a single component. In Sect. 3, we detail the dynamics of multiple components in purely cubic media. We derive some criteria for blowup and coalescence between wave envelopes in terms of a critical separation distance that depends on the powers of the individual waves. In Sect. 4, the inclusion of four-wave mixing and walk-off is discussed for coherently/incoherently coupled waves in Kerr media, and in Sect. 5, a similar analysis is applied to a medium that promotes the interplay between cubic and quadratic nonlinearities.

2 General Results on the Collapse of a Single Component

We first recall some basic properties of the NLS equation for one wave, whose electric-field envelope, $E(\boldsymbol{r}, z)$, is governed by

$$\mathrm{i}\partial_z E + \boldsymbol{\nabla}^2 E + |E|^2 E = 0 \, . \tag{2}$$

For technical convenience, we shall employ some standard notations for denoting L^p norms, i.e., for any function f,

$$\|f\|_p \equiv \left(\int |f|^p \, \mathrm{d}\boldsymbol{r} \right)^{1/p} \quad \Rightarrow \quad \|f\|_p^p \equiv \int |f|^p \, \mathrm{d}\boldsymbol{r} \, . \tag{3}$$

We shall henceforth suppose that the wavefunction $E(\boldsymbol{r}, z)$, which evolves from the incident (initial) datum $E(\boldsymbol{r}, 0) \equiv E_0(\boldsymbol{r})$, is localized in the transverse plane and decays to zero at infinity. Two main integrals of motion are associated with $E(\boldsymbol{r}, z)$, namely, the dimensionless power

$$P \equiv \|E\|_2^2 = \int |E|^2 \, \mathrm{d}\boldsymbol{r} \tag{4}$$

and Hamiltonian

$$H \equiv \|\boldsymbol{\nabla} E\|_2^2 - \frac{1}{2}\|E\|_4^4 = \int |\boldsymbol{\nabla} E|^2 \, \mathrm{d}\boldsymbol{r} - \frac{1}{2}\int |E|^4 \, \mathrm{d}\boldsymbol{r} \, , \tag{5}$$

from which (2) can be derived through the Hamilton formulation $\mathrm{i}\partial_z E = \delta H / \delta E^*$. In addition, a fundamental relation governing the mean square radius of the solutions E can be established. This mean square radius, often called the "virial integral" , reads

$$I(z) \equiv \frac{1}{P} \int |\boldsymbol{r} - \langle \boldsymbol{r} \rangle|^2 |E|^2 \, \mathrm{d}\boldsymbol{r} \, . \tag{6}$$

Its evolution equation is obtained by multiplying (2) by $(\boldsymbol{r} \cdot \boldsymbol{\nabla} E^*)$ and $|\boldsymbol{r}|^2 E^*$. Combining the real and imaginary parts of the results thus yields [4]

$$\partial_z^2 I(z) = \frac{4}{P} \left(2H_0 + \frac{1}{2}(2 - D)\|E\|_4^4 \right) , \quad H_0 \equiv H - \frac{P}{4} |\partial_z \langle \boldsymbol{r} \rangle|^2 \, . \tag{7}$$

Here, $\partial_z \langle \boldsymbol{r} \rangle$ is the velocity of the center of mass of the wavepacket; governed by the relation $\partial_z \langle \boldsymbol{r} \rangle = \boldsymbol{M}/P$, where $\boldsymbol{M} \equiv 2\,\mathrm{Im} \int (E^* \boldsymbol{\nabla} E)\, \mathrm{d}\boldsymbol{r}$ is the conserved wave momentum, it satisfies $\partial_z^2 \langle \boldsymbol{r} \rangle = \boldsymbol{0}$. From the identity (7), one can then infer that the mean square radius of a given localized wave tends to zero at a finite propagation distance, $z_c < +\infty$, under some specific conditions, among which are the condition of multidimensionality $D \geq 2$ and the requirement that $H_0 < 0$. Although other conditions involving prefocusing ($\partial_z I(0) < 0$) are available in the current literature [4], the condition $H_0 < 0$, meaning that nonlinearities continuously dominate over wave dispersion, applies to any initial shape of the waves and we shall retain it for the sake of simplicity. So, waveforms with $H_0 < 0$ have a typical transverse scale that tends to zero as $z \to z_c$, provided that $D \geq 2$. This property characterizes self-focusing dynamics that achieve a collapse. For beams centered on the origin with zero velocity, H_0 reduces to H, and the virial identity simplifies to

$$\partial_z^2 I(z) \leq \frac{8H}{P}\,, \quad I(z) \equiv \frac{\|\boldsymbol{r}E\|_2^2}{P}\,, \tag{8}$$

where the strict equality applies only to the so-called "critical" case $D = 2$. Because of the vanishing of $I(z)$, i.e. $I(z) \to 0$, the L^2 norm of the gradient of E must, moreover, diverge by virtue of the obvious inequality

$$\|E\|_2^2 \leq \frac{2}{D} \|\boldsymbol{\nabla} E\|_2 \|\boldsymbol{r}E\|_2\,, \tag{9}$$

and the wavefunction can be shown to blow up, with its maximum amplitude growing to infinity at the center [22]. Let us recall that this blowup generally takes place before the complete vanishing of the virial integral, which only yields a *maximum* collapse distance [21]. Indeed, a wave blowup occurs before the total vanishing of $I(z)$, because only a finite amount of power is captured in the collapse process. For instance, at the critical dimension $D = 2$, this finite amount corresponds to the smallest L^2 norm obtainable from the radially symmetric stationary solution $E(\boldsymbol{r}, z) = R_0(|\boldsymbol{r}|)\mathrm{e}^{\mathrm{i}z}$ of (2). The solution R_0, often termed the "soliton" or the "ground state", is even, real, positive and unique. Its associated power, denoted by P_c, enters the following lower bound on the Hamiltonian:

$$\|\boldsymbol{\nabla} E\|_2^2 \left(1 - \frac{P}{P_c}\right) \leq H < +\infty\,, \tag{10}$$

which can be established after using the Sobolev inequality $\|E\|_4^4 \leq C \|\boldsymbol{\nabla} E\|_2^2 \|E\|_2^2$ with the best constant $C_{\mathrm{best}} = 2/P_c$ [23]. As a wave collapse is characterized by the divergence of the gradient norm of E, the requirement $P > P_c$ is thus a *necessary condition* for initiating the collapse, and P_c yields a critical power for self-focusing or a threshold power for collapse. At higher dimensions ($D > 2$), the Sobolev inequality can be generalized as follows:

$$\|E\|_4^4 \leq \frac{4}{DP_c} \left(\frac{4}{D} - 1\right)^{D/2-1} \|\boldsymbol{\nabla} E\|_2^D \|E\|_2^{4-D}\,, \tag{11}$$

and for $D = 3$ a criterion for collapse, sharper than $H < 0$, arises from (7) as $H < P_c^2/P$ when the gradient norms are initially above $3P_c^2/P$ [24].

In connection with this property, an important point is to know whether NLS systems can support the formation of stable stationary solutions of the form $E(\boldsymbol{r}, z) = R(|\boldsymbol{r}|, \lambda)\mathrm{e}^{\mathrm{i}\lambda z}$, where R obeys the differential equation

$$-\lambda R + \boldsymbol{\nabla}^2 R + R^3 = 0 \,. \tag{12}$$

In a focusing cubic medium, $\lambda > 0$ guarantees the existence of such localized modes whenever $D < 4$. This parameter λ can be normalized to unity through the rescaling transformation $R(\boldsymbol{r}, \lambda) = \sqrt{\lambda} R_0(\sqrt{\lambda}\boldsymbol{r})$, and the power associated with the ground states R_0 takes the values $P_\mathrm{c} \equiv \int |R_0|^2 \,\mathrm{d}\boldsymbol{r} = 11.68$ for $D = 2$ and $P_\mathrm{c} = 18.94$ for $D = 3$ [23,24]. For radially symmetric solutions, the problem of soliton stability is usually solved by perturbing R as

$$E(r, z) = [R(r, \lambda) + v(r, z) + \mathrm{i}w(r, z)]\mathrm{e}^{\mathrm{i}\lambda z} \,, \tag{13}$$

where $v, w \ll R$ are real functions. Standard perturbation techniques [25–27] then enable one to show that a sufficient condition for the stability of the ground states is given by

$$\frac{\mathrm{d}P_s}{\mathrm{d}\lambda} > 0 \,, \tag{14}$$

where $P_\mathrm{s} \equiv \|R\|_2^2$. Conversely, $\mathrm{d}P_s/\mathrm{d}\lambda \leq 0$ leads to instability. Using the rescaling transformation given above, it is easy to see that the condition (14) can only be realized for $D = 1$, from which sech-shaped solitons are stable. Instead, at higher dimensions, all NLS ground states are unstable, in accordance with the conditions on the number of dimensions required for collapse.

To end this section, we recall that an easy way to describe the dynamics of collapsing beams is provided by the so-called variational (or average-Lagrangian) approach. This method assumes that the wave dynamics may be described through a finite number of z-dependent parameters in a trial function that approximately fits the true NLS solutions. This trial function is constructed from a test function, ϕ, chosen from Gaussians or sech functions in the standard approach [28,29]. For waveforms centered at the origin, the trial function usually depends on the rescaled spatial variables $\boldsymbol{\xi} \equiv \boldsymbol{r}/a(z)$, where $a(z)$ represents the transverse radius of the wave, and can be expressed as

$$E(\boldsymbol{r}, z) = \frac{1}{\sqrt{J(z)}}\phi\left(\frac{\boldsymbol{r}}{a(z)}\right)\exp\left[\mathrm{i}\theta(z)|\boldsymbol{r}|^2 + \mathrm{i}\zeta(z)\right] . \tag{15}$$

Here, $\theta(z)$ denotes a quadratic phase chirp parameter and $\zeta(z)$ is an arbitrary function of z. $J(z)$ is a real amplitude factor scaling like $a^D(z)$, such that $J(z)$ vanishes as $z \to z_\mathrm{c}$. Inserting (15) into the Lagrangian $L = (\mathrm{i}/2)\int(E^*\partial_z E -$

$E\partial_z E^*)\mathrm{d}\boldsymbol{r} - H$ of (2) and functionally differentiating L with respect to all the z-dependent parameters provides the dynamical equation

$$\frac{1}{4}\mathrm{d}_z^2 a = \frac{K_1}{a^3} - \frac{P}{P_\mathrm{c}}\frac{K_2}{a^{D+1}}\,, \tag{16}$$

where K_1 and K_2 are integral coefficients computed from the test function ϕ. From this relation, it is readily seen that unlike the case for $D = 1$, any solution with $D \geq 2$ satisfying $P > K_1 P_\mathrm{c}/K_2$ collapses, with $|E| \to +\infty$ and $a(z) \to 0$ at a finite propagation distance. This distance, denoted by $z_\mathrm{c}^{\max}$, is the same as that given by the virial identity and it generally overestimates the distance of wave blowup, z_c, as revealed by numerical simulations. At the critical dimension $D = 2$, a more correct estimate of the blowup distance can be given for Gaussian beams by an empirical formula numerically established by Marburger [30]: $z_\mathrm{c} \simeq 0.367/[(\sqrt{P/P_\mathrm{c}} - 0.852)^2 - 0.0219]^{1/2}$.

3 Interaction Regimes of Several Collapsing Components

We now investigate the dynamical properties of $n \geq 2$ copropagating waves. With this aim, we first ignore FWM and walk-off processes and extend the equation set (1) into the general form

$$\mathrm{i}\partial_z E_j + \boldsymbol{\nabla}^2 E_j + \sum_k \Lambda_{jk}|E_k|^2 E_j = 0\,. \tag{17}$$

Here, the summation symbol with indices (j,k) refers to a summation over $1 \leq j,k \leq n$, and the coupling coefficients Λ_{jk} are positive. The system (17) conserves the individual powers $P_j \equiv \|E_j\|_2^2$ separately, and also the total Hamiltonian

$$H = \sum_j \|\boldsymbol{\nabla} E_j\|_2^2 - \sum_{j,k}\frac{\Lambda_{jk}}{2}\|E_j E_k\|_2^2\,, \quad \sum_{j,k} \equiv \sum_{j=1}^{n}\sum_{k=1}^{n}\,. \tag{18}$$

Conservation of H requires symmetric nonlinearity coefficients satisfying $\Lambda_{jk} = \Lambda_{kj}$ for $j \neq k$, which is assumed here. In this case, we can derive a dynamical relation for the total center of mass, $\langle \boldsymbol{r}\rangle = P^{-1}\int \boldsymbol{r}\sum_j |E_j|^2\,\mathrm{d}\boldsymbol{r}$, where $P \equiv \sum_j P_j$. From straightforward algebra, the center of mass (centroid) of the jth wave, $\langle \boldsymbol{r}_j\rangle \equiv P_j^{-1}\int \boldsymbol{r}|E_j|^2\,\mathrm{d}\boldsymbol{r}$, is found to be governed by

$$\partial_z^2\langle \boldsymbol{r}_j\rangle = \frac{2}{P_j}\int |E_j|^2\boldsymbol{\nabla}\sum_{k\neq j}\Lambda_{jk}|E_k|^2\,\mathrm{d}\boldsymbol{r}\,, \tag{19}$$

so that the total center of mass again obeys the identity $\partial_z^2\langle \boldsymbol{r}\rangle = \boldsymbol{0}$. Repeating the computational stages of Sect. 2, we can, moreover, establish a virial-type identity for the total mean square radius

$$I(z) \equiv \frac{1}{P} \int |\boldsymbol{r} - \langle \boldsymbol{r} \rangle|^2 \sum_j |E_j|^2 \, \mathrm{d}\boldsymbol{r}. \tag{20}$$

This relation consists of the generalized form of (7):

$$\partial_z^2 I(z) = \frac{4}{P} \left[2H_0 + \left(\frac{2-D}{2} \right) \sum_{j,k} \Lambda_{jk} \|E_j E_k\|_2^2 \right] , \; H_0 \equiv H - \frac{P}{4} |\partial_z \langle \boldsymbol{r} \rangle|^2 , \tag{21}$$

which reduces to $\partial_z^2 I(z) = 8H/P$ for waves with zero velocity at $D = 2$ [15]. Hence, in this critical case, we can use both the Cauchy–Schwarz and the Sobolev inequalities to bound the nonlinear contributions in H as follows:

$$\|E_j E_k\|_2^2 \le \frac{1}{2} \|E_j\|_4^4 + \frac{1}{2} \|E_k\|_4^4 \le \|\boldsymbol{\nabla} E_j\|_2^2 \frac{P_j}{P_\mathrm{c}} + \|\boldsymbol{\nabla} E_k\|_2^2 \frac{P_k}{P_\mathrm{c}} . \tag{22}$$

This yields, for $\Lambda_{jk} = \Lambda_{kj}$,

$$H \ge \sum_{j=1}^{n} \|\boldsymbol{\nabla} E_j\|_2^2 \left(1 - \sum_{k=1}^{n} \Lambda_{jk} \frac{P_j}{P_c} \right) . \tag{23}$$

From this estimate, initial states with $H < 0$ promote the collapse of *all* the wave components, because the total virial integral is the direct sum of positive integrals $I_j(z) \equiv P^{-1} \int |\boldsymbol{r} - \langle \boldsymbol{r} \rangle|^2 |E_j|^2 \, \mathrm{d}\boldsymbol{r}$, implying $I_j(z) \to 0$ for every $j = 1, ..., n$ as $I(z) \to 0$. This property implies, in turn, that each wave component blows up such that $\|\boldsymbol{\nabla} E_j\|_2 \to +\infty$, since inequality (9) still applies in the form $P_j^2/P \le (2/D)^2 I_j \|\boldsymbol{\nabla} E_j\|_2^2$. Therefore, preventing the collapse requires that the partial norms P_j must all be *below* a certain threshold, defined by

$$P_j < \frac{P_\mathrm{c}}{\sum_k \Lambda_{jk}} , \quad k = 1, ..., n . \tag{24}$$

The gradient norms remain bounded by H, which is then positive. Conversely, choosing P_j *above* this bound may initiate the collapse. In the case of two identical coupled waves, triggering the collapse thus needs $P_1 = P_2 > P_\mathrm{c}/(\Lambda_{11} + \Lambda_{12})$, and special interaction regimes can be expected for which waves amalgamate into one collapsing lobe as their total mean square radius, including their mutual separation distance, decreases with z.

To specify these interaction regimes, we consider in-phase Gaussian waves with incident shapes

$$E_j(\boldsymbol{r}, 0) = \sqrt{\frac{P_j}{\pi \rho^2}} \exp\left(- \frac{|\boldsymbol{r} - \langle \boldsymbol{r}_j(0) \rangle|^2}{2\rho^2} \right) \tag{25}$$

and equal transverse radii ρ. For such initial envelopes, H can be expanded as

$$H = \sum_{j=1}^{n} \frac{P_j}{\rho^2} \left(1 - \sum_{k=1}^{n} \frac{\Lambda_{jk} P_k}{4\pi} \mathrm{e}^{-\Delta_{jk}^2(0)/2\rho^2} \right) , \tag{26}$$

where $\Delta_{jk}(0) \equiv |\langle \boldsymbol{r}_j(0) \rangle - \langle \boldsymbol{r}_k(0) \rangle|$. By using $(P_j - P_k)^2 \geq 0$, we deduce that each wave contributing to $H < 0$ will self-focus provided that its power exceeds the threshold

$$P_j > \frac{4\pi}{\sum_k \Lambda_{jk} \mathrm{e}^{-\Delta_{jk}^2(0)/2\rho^2}} , \tag{27}$$

which decreases faster as the separation distance $\Delta_{jk}(0)$ becomes small. Thus, superimposed waves with $\Delta_{jk}(0) = 0$ will collapse together if their partial powers exceed the critical value

$$P_j > P_{\mathrm{c}}^0 \equiv \frac{4\pi}{\sum_k \Lambda_{jk}} > \frac{P_{\mathrm{c}}}{\sum_k \Lambda_{jk}} , \tag{28}$$

whereas in the limit of well-separated structures ($\Delta_{jk}(0) \to +\infty$ for $j \neq k$), we recover the threshold for free collapse events, $P_j > P_{\mathrm{c}}^{\mathrm{f}} \equiv 4\pi/\Lambda_{jj} > P_{\mathrm{c}}/\Lambda_{jj}$. Here, 4π is the self-focusing power of isolated Gaussians with $\Lambda_{jj} = 1$.

Now, we denote by $\Delta_0 \equiv \Delta_{jk}(0)$ the initial distance separating the centroids of neighboring waves with $j \neq k$. We search for a critical distance Δ_{c} such that, for $\Delta_0 > \Delta_{\mathrm{c}}$, the waves evolve independently of each other, owing to the exponential decrease of the interaction term in H. To do so, we formally rewrite H as $H_{\mathrm{free}} + H_{\mathrm{int}}(\Delta_0)$, where $H_{\mathrm{free}} \equiv \sum_j (P_j/\rho^2)(1 - \Lambda_{jj} P_j/4\pi)$, and conjecture that the interaction will be strong if the initial separation Δ_0 satisfies $|\sum_j H_j| \ll |H_{\mathrm{int}}(\Delta_0)|$. The critical distance below which the waves interact significantly is then given by the zeros of H and is determined by

$$\sum_{j \neq k} \Lambda_{jk} P_j P_k \mathrm{e}^{-\Delta_{\mathrm{c}}^2/2\rho^2} = \sum_j P_j |4\pi - \Lambda_{jj} P_j| . \tag{29}$$

For two equal waves with $\Lambda_{11} = \Lambda_{22}$, the expression for Δ_{c} then reads

$$\Delta_{\mathrm{c}} = \rho \left(2 \ln \left| \frac{P \Lambda_{12}}{8\pi - \Lambda_{11} P} \right| \right)^{1/2} , \tag{30}$$

where $P/2 = P_1 = P_2$. By employing the virial properties of the NLS equation, we can then deduce different regimes of interaction. For instance, if $P_{\mathrm{c}}^0 < P_1 < P_{\mathrm{c}}^{\mathrm{f}}$, both waves disperse if $H > 0$ and $\Delta_0 > \Delta_{\mathrm{c}}$, but they can fuse into a self-focusing central lobe with $H < 0$, if they satisfy $\Delta_0 < \Delta_{\mathrm{c}}$ initially. Conversely, if $P_1 > P_{\mathrm{c}}^{\mathrm{f}}$, the waves self-focus independently with $\Delta_0 > \Delta_{\mathrm{c}}$, whereas, if they start with $\Delta_0 < \Delta_{\mathrm{c}}$, they may still amalgamate into a collapsing central lobe.

These interaction regimes can be classified systematically by using a refinement emphasized in [32]. Let us denote by θ the ratio $\theta = |H_{\text{int}}(\Delta_c)/H_{\text{free}}|$, above which interaction between waves becomes significant. This ratio, which must be identified numerically according to the initial data, may a priori differ from unity, and it is clear that, owing to the self-localizing action of the cubic nonlinearity, it will be stronger in collapse regimes than in spreading regimes, i.e. $\theta_s \equiv \theta_{\text{spreading}} < \theta_c \equiv \theta_{\text{collapse}} \leq 1$. Introducing θ into Δ_c is straightforward and leads to the refined estimate $\Delta_c(\theta) = \rho[2\ln|P\Lambda_{12}/\theta(8\pi - \Lambda_{11}P)|]^{1/2}$. From this estimate, four distinct regimes of interaction clearly arise:

I $|H_{\text{int}}/H_{\text{free}}| < \theta_s$, $H > 0$: the two waves disperse independently, with, for example, $P_1 < P_c^f$ when $\Delta_0 > \Delta_c(\theta_s)$.
II $|H_{\text{int}}/H_{\text{free}}| > \theta_s$, $H > 0$: the two waves merge and then disperse, with, for example, $P_1 < P_c^f$ when $\Delta_c < \Delta_0 < \Delta_c(\theta_s)$.
III $|H_{\text{int}}/H_{\text{free}}| > \theta_c$, $H < 0$: the waves fuse and then collapse, with, for example, $P_c^0 < P_1 < P_c^f$ when $\Delta_0 < \Delta_c$, or $P_1 > P_c^f$ when $\Delta_0 < \Delta_c(\theta_c)$.
IV $|H_{\text{int}}/H_{\text{free}}| < \theta_c$, $H < 0$: the waves collapse independently, with, for example, $P_1 > P_c^f$ for $\Delta_0 > \Delta_c(\theta_c)$.

Figure 1 shows, by the dashed lines, the domains of the plane (Δ_0, P) that separate the different interactions expected from the virial results and (30). The improved estimate $\Delta_c(\theta)$, which divides the diffraction range into two subregimes of interaction, is also shown by the outermost dashed lines,

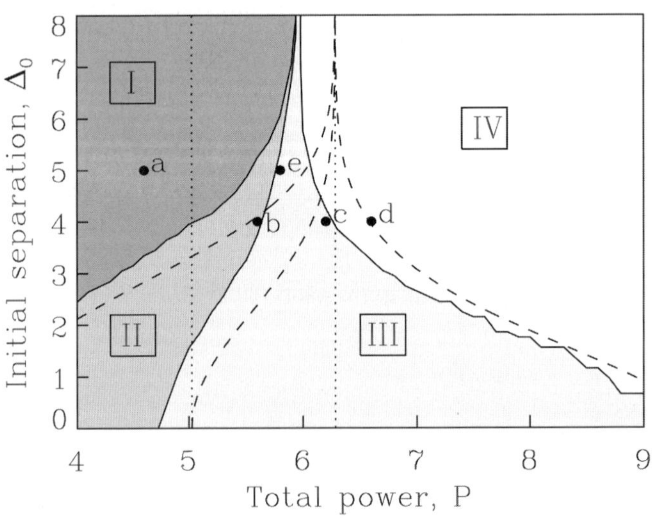

Fig. 1. Overview graph of (Δ_0, P) showing the regions of interaction between two Gaussian pulses with $P_1 = P_2$, $\Lambda_{11} = \Lambda_{22} = 4$, $\Lambda_{12} = \Lambda_{21} = 1$, and $\rho = 2$, identified numerically. The *dashed lines* represent $\Delta_c(\theta_s)$, $\Delta_c(\theta = 1)$, and $\Delta_c(\theta_c)$ for $\theta_s = 0.25$ and $\theta_c = 0.75$. The *dotted lines* indicate the bounds $2P_c^0 = 5.03$ and $2P_c^f = 6.28$

for $\theta_{\text{spread}} = 0.25$ and $\theta_{\text{col}} = 0.75$. The four distinct interaction regimes are then mapped as regions of independent spreading for weak-power waves or independent collapse for high-power waves when $\Delta_0 > \Delta_c(\theta)$, and regions of fusion and spreading or fusion and collapse when $\Delta_0 \leq \Delta_c(\theta)$. The solid lines delimit the same domains identified through numerical computations. These lines show good agreement with the boundaries established from theoretical arguments. Figure 2 details these interaction regimes in the transverse plane for the points a,b,c,d indicated in Fig. 1.

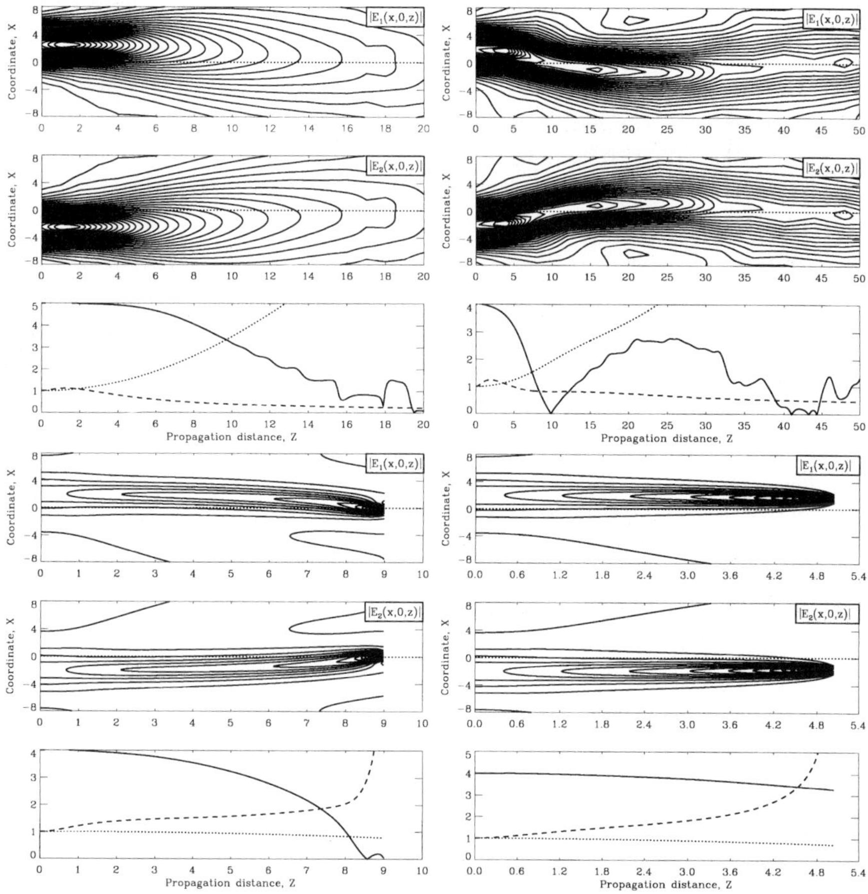

Fig. 2. Contour plots of $|E_1(x, 0, z)|$ (*top*) and $|E_2(x, 0, z)|$ (*middle*) for the four data points a–d of Fig. 1, illustrating the distinct types of interaction: independent spreading (*top left*), fusion and spreading (*top right*), fusion and collapse (*bottom left*), and independent collapse (*bottom right*). For each configuration, the *bottom figure* shows the evolution of the separation distance (*solid line*), the normalized virial $I(z)/I(0)$ (*dotted line*) and the wave amplitude $|E_1(z)/E_1(0)|$ (*dashed line*)

Interacting beams can also be described with the variational method briefly recalled in Sect. 2, by inserting

$$E_j(\boldsymbol{r},z)=\frac{\phi_j(\boldsymbol{\xi}_j)}{a_j^{D/2}(z)}\exp\left(\mathrm{i}\theta_j(z)|\boldsymbol{r}-\langle\boldsymbol{r}_j\rangle|^2+\frac{\mathrm{i}}{2}\boldsymbol{\psi}_j(z)\cdot(\boldsymbol{r}-\langle\boldsymbol{r}_j\rangle)+\mathrm{i}\zeta_j(z)\right)$$

into the Lagrangian $L=(\mathrm{i}/2)\sum_j\int(E_j^*\partial_z E_j-E_j\partial_z E_j^*)\,\mathrm{d}\boldsymbol{r}-H$, where H is defined by (18). Here, ϕ_j is a real, even test function depending on the rescaled spatial variables $\boldsymbol{\xi}_j\equiv(\boldsymbol{r}-\langle\boldsymbol{r}_j\rangle)/a_j(z)$, where $\langle\boldsymbol{r}_j\rangle\equiv\langle\boldsymbol{r}_j(z)\rangle$ denotes the centroid of the jth wave and $a_j(z)$ its typical, z-varying radius; $\theta_j(z)$ accounts for the quadratic chirp parameter, and the vector $\boldsymbol{\psi}_j(z)$ is related to the velocity of the centroid displacement $\langle\boldsymbol{r}_j\rangle$. By inserting the above trial function into L and taking functional derivatives, we obtain a dynamical system describing the global tendencies in the propagation of the coupled waves. For $D=2$, two symmetric Gaussian pulses with $P_1=P_2$, modeled around the test function $\phi_j(\boldsymbol{\xi}_j)=\sqrt{P_j/\pi}\mathrm{e}^{-\xi_j^2/2}$ $(j=1,2)$, are then found to have their separation vector and radial width $a_1=a_2$ governed by the dynamical relations [31]

$$d_z^2\boldsymbol{\Delta}=-\frac{2P_1}{\pi a_1^4}\Lambda_{12}\,\boldsymbol{\Delta}\,\exp\left(-\frac{\Delta^2}{2a_1^2}\right),\quad \boldsymbol{\Delta}(z)\equiv\langle\boldsymbol{r}_1(z)\rangle-\langle\boldsymbol{r}_2(z)\rangle\,, \tag{31}$$

$$\frac{1}{4}d_z^2a_1=\left(1-\frac{\Lambda_{11}P_1}{4\pi}\right)\frac{1}{a_1^3}-\frac{\Lambda_{12}P_1}{4\pi a_1^4}a_1\left(1-\frac{\Delta^2}{2a_1^2}\right)\exp\left(-\frac{\Delta^2}{2a_1^2}\right). \tag{32}$$

When the waves are initially at rest and superimposed, with $\Delta_0=0$, they stay mutually overlapped during propagation, and collapse occurs provided that $P_1>P_\mathrm{c}^0=4\pi/(\Lambda_{11}+\Lambda_{12})$, in accordance with (28). In the opposite limit $\Delta_0\to+\infty$, the equation for $a_1(z)$ restores the usual self-focusing threshold for one Gaussian wave: $P_1>P_\mathrm{c}^\mathrm{f}\equiv 4\pi/\Lambda_{11}$. Between these two extreme limits, equal waves with an initial separation $\Delta_0\leq\Delta_\mathrm{c}$ mutually attract and merge into one collapsing central lobe if $P_1=P_2$ satisfies $P_\mathrm{c}^0<P_1<P_\mathrm{c}^\mathrm{f}$, as expected.

4 Collapse with Four-Wave Mixing and Walk-Off

We now study the influence of walk-off and FWM using the starting model (1). Here, the total power $P=P_1+P_2=\|E_1\|_2^2+\|E_2\|_2^2$ is always conserved, and each individual power $P_j\equiv\|E_j\|_2^2$, where $j=1,2$, is invariant in the particular case $B=0$ only. Equations (1) also conserve the Hamiltonian

$$H=\sum_{j=1}^{2}\left(\|\boldsymbol{\nabla}E_j\|_2^2+\beta_jP_j-\frac{\chi_j}{2}-\frac{B}{2}\mathrm{Re}\int(E_{3-j}^2E_j^{*2})\,\mathrm{d}\boldsymbol{r}\right.$$
$$\left.+\int|E_j|^2\boldsymbol{\delta}_j\cdot\boldsymbol{\nabla}\arg(E_j)\,\mathrm{d}\boldsymbol{r}\right),$$

where $\chi_j \equiv \|E_j\|_4^4 + A\|E_j E_{3-j}\|_2^2$. Furthermore, a virial identity for the total mean square radius (20) can be derived as follows:

$$\partial_z^2 I = \frac{4DH}{P} + \frac{4}{P}\sum_{j=1}^{2}\left((2-D)\|\boldsymbol{\nabla} E_j\|_2^2 - D\beta_j P_j + \frac{\delta_j^2 P_j}{2} - B\boldsymbol{\delta}_j\cdot \mathrm{Im}\int \boldsymbol{r}(E_{3-j}^2 E_j^{*2})\,\mathrm{d}\boldsymbol{r}\right) - 2|\partial_z\langle \boldsymbol{r}\rangle|^2 - 2\langle \boldsymbol{r}\rangle\cdot\partial_z^2\langle \boldsymbol{r}\rangle\,, \tag{33}$$

where $\partial_z\langle \boldsymbol{r}\rangle = \boldsymbol{M}^\delta/P$ and $\boldsymbol{M}^\delta \equiv 2\,\mathrm{Im}\sum_{j=1}^2\int(E_j^*\boldsymbol{\nabla}E_j)\,\mathrm{d}\boldsymbol{r} + \sum_{j=1}^2 \boldsymbol{\delta}_j P_j$.

Considering first FWM without walk-off, we set $B \neq 0$ and $\boldsymbol{\delta} = \boldsymbol{0}$. The virial relation (33) can then be bounded from above for $D \geq 2$, in such a way that a collapse with $I(z) \to 0$ always occurs whenever $H + |\beta|P$ is negative. For $D = 2$, the threshold power for collapse can be estimated by bounding H by means of (22), and blowup never occurs when each power P_j is less than

$$\mathcal{P}_\mathrm{c}^B \equiv P_\mathrm{c}/(1+A+B)\,. \tag{34}$$

In terms of the conserved total power, collapse is certainly avoided when P is below $\mathcal{P}_\mathrm{c}^B$, but owing to power transfers between the two components, the bound $P < \mathcal{P}_\mathrm{c}^B$ can significantly underestimate the actual threshold for collapse if $B \neq 0$. In spite of this, when the two components are initially identical, (1) with $\beta = 0$ reduces by symmetry to the single-wave NLS equation $\mathrm{i}\partial_z E + \boldsymbol{\nabla}^2 E + (1+A+B)|E|^2 E = 0$, and power transfers are eliminated. The condition $P > 2\mathcal{P}_\mathrm{c}^B$ can thus yield a reasonably good estimate of the exact collapse threshold for small mismatch values, provided that $E_1(\boldsymbol{r},0) \simeq E_2(\boldsymbol{r},0)$. In this situation, (34) shows that FWM lowers the power threshold for collapse and thereby strengthens the self-focusing dynamics, compared with the case of incoherently coupled waves ($B = 0$). For $\beta = 0$, the threshold power $2\mathcal{P}_\mathrm{c}^B$ has been found to agree with its numerical counterpart [33]. For finite $\beta \neq 0$, however, this estimate is no longer valid, because the waves mutually exchange power, and this leads to sharp resonances at moderate mismatches ($\beta > 1$) for $P > 2\mathcal{P}_\mathrm{c}^0 \equiv 2P_\mathrm{c}/(1+A)$. At large mismatches ($\beta \gg 1$), the FWM terms are easily averaged out ($B \to 0$) after the phase transform $E_j \to E_j \exp(-\mathrm{i}\beta_j z)$ is applied, so that power transfers disappear and the collapse threshold is simply given by $2\mathcal{P}_\mathrm{c}^0$. On the whole, FWM enhances the self-focusing of waves when they are close to exact phase matching only. This enhancement may cease for waves with a finite phase mismatch.

We now consider the opposite situation of walk-off ($\boldsymbol{\delta} \neq \boldsymbol{0}$) without FWM ($B = 0$). As the nonlinearities depend only on the intensities, the phase mismatch plays no role in the collapse dynamics, which amounts to setting $\beta_j = 0$. In this case, the individual powers P_j are preserved and, for $D \geq 2$, the virial integral satisfies

$$\partial_z^2 I \leq \frac{8}{P}H + 2\delta^2 - \frac{2}{P^2}|\boldsymbol{M}^\delta|^2\,, \tag{35}$$

where the strict equality applies to $D = 2$ only. From (35), collapse thus inevitably develops when $H + \delta^2 P/4 \leq 0$. This requirement means that to promote collapse, the incident waves must have a high enough power, not only to ensure $H < 0$, but also to overcome the walk-off contribution in $\delta^2 P/4$. Therefore, the collapse dynamics should be significantly delayed when $\boldsymbol{\delta} \neq \mathbf{0}$.

Let us check this property for two-dimensional Gaussian beams (25) centered at the origin of the coordinates. Without walk-off, these beams have a power ratio $\alpha = P_1/P_2$ and a Hamiltonian $H = (P/\rho^2)(1 - P/\mathcal{P}_{\text{th}})$, where $\mathcal{P}_{\text{th}}$ denotes their collapse threshold $\mathcal{P}_{\text{th}} = 4\pi(1+\alpha)^2/(1+\alpha^2+2\alpha A)$. With walk-off, the virial estimate (35) then yields the effective critical power,

$$\mathcal{P}_{\text{c}}^{\delta} = \mathcal{P}_{\text{th}}\left[1 + \delta^2\rho^2\alpha/(1+\alpha)^2\right] , \tag{36}$$

which increases with the length δ. This threshold is associated with a total collapse at the center $\langle \boldsymbol{r} \rangle = \mathbf{0}$. For a given power, walk-off is thus able to delay and even arrest such a collapse, as can be guessed from the virial expression $I(z) = (4z^2/\rho^2\mathcal{P}_{\text{th}})(\mathcal{P}_{\text{c}}^{\delta} - P) + \rho^2$. In [33], the agreement between the theoretical estimate (36) and numerical integrations of (1) was rather good for equal waves with $P_1 = P_2$ at small δ values. Nonetheless, for large δ, walk-off significantly affects the wave motion. To clarify this point, Fig. 3 shows that walk-off competes with collapse by detrapping both components in the transverse plane. The centroids of the components are shifted from $\langle \boldsymbol{r} \rangle = \mathbf{0}$ and fuse again at the origin, where collapse occurs (Fig. 3a). Also, when the two waves contain enough power to produce individual collapses ($P_1 = P_2 > 4\pi$), walk-off can make the waves separate from each other and self-focus on their own centers of mass, far away from the origin (Fig. 3b). In this case, (35) emphasizes that walk-off becomes important whenever $|H| \ll \delta^2 P/4$, which yields the bound $\delta > (2/\rho)\sqrt{P/\mathcal{P}_{\text{th}} - 1}$, above which the two components can self-focus separately. In the diffraction regime, where collapse does not occur, the two components can either cross or go away from each other, while they both disperse. For equal components modeled by symmetric Gaussians $|E_j|^2 = (P/2\pi\rho^2)\mathrm{e}^{-|\boldsymbol{r}-\langle \boldsymbol{r}_j\rangle|^2/\rho^2}$, it is easily shown from the equation for $\langle \boldsymbol{r}_j \rangle$ that their individual centroids undergo a harmonic motion, which is damped in the collapse regimes and relaxes otherwise. The constraint that oscillations develop only if the maximal displacement of this centroid remains finite imposes, moreover, a critical value on the walk-off length: $\delta_j^2 \leq \delta_{\text{c}}^2 = AP/2\pi\rho^2$ [33]. Comparable oscillations in the self-attraction or separation of optical components were revealed earlier in [12] for one-dimensional birefringent media.

Finally, the above results can formally be extended to the nonlinear coupling of $n > 2$ waves described by

$$\mathrm{i}(\partial_z + \boldsymbol{\delta}_j \cdot \boldsymbol{\nabla})E_j + \boldsymbol{\nabla}^2 E_j + \sum_{k=1}^{n} A_{jk}|E_k|^2 E_j + \sum_{k=1}^{n} B_{jk} E_k^2 E_j^* = 0 , \tag{37}$$

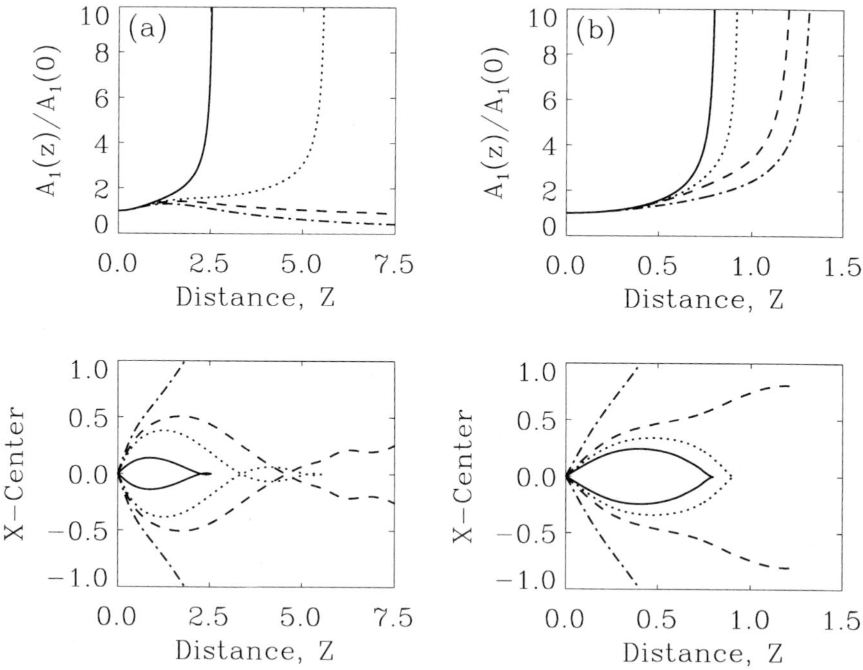

Fig. 3. Dynamics ($A_1 = |E_1|$, centroid along x) of 2D coupled waves undergoing walk-off for **(a)** $P = 16$ with $\delta = 0.3$ (*solid line*), $\delta = 0.7$ (*dotted line*), $\delta = 0.8$ (*dashed line*), and $\delta = 1.0$ (*dash–dotted line*), and **(b)** $P = 40$ with $\delta = 1.0$ (*solid line*), $\delta = 1.3$ (*dotted line*), $\delta = 1.6$ (*dashed line*), and $\delta = 3.0$ (*dash–dotted line*)

where δ_j denotes the convection length of the jth component; A_{jk} and B_{jk}, where $B_{jj} = 0$ $(1 \le j,k \le n)$, are positive, symmetric nonlinearity coefficients corresponding to the cubic and FWM terms, respectively. The total power $P = \sum_{j=1}^{n} P_j$ is conserved, and when walk-off is disregarded ($\boldsymbol{\delta}_j = \mathbf{0}$), equal waves with zero mismatch can collapse under the condition $P > nP_c/\sum_{j\neq k=1}^{n}(A_{jj} + A_{jk} + B_{jk})$, in which FWM diminishes the threshold power for self-focusing. On the other hand, by ignoring the FWM terms and setting $B_{jk} = 0$, the virial relation generalizes (35) and predicts wave collapse when $H + \sum_{j=1}^{n} \delta_j^2 P_j/4 \le 0$.

5 Interplay between Quadratic and Cubic Nonlinearities

We now apply the previous analysis to the competition between quadratic and cubic nonlinearities in anisotropic crystals. In the absence of FWM and walk-off, the equations (1) include quadratic contributions of the form $E_1^* E_2 \mathrm{e}^{-\mathrm{i}\Delta k\, z}$ and $E_1^2 \mathrm{e}^{+\mathrm{i}\Delta k\, z}$ for the fundamental and second-harmonic waves, respectively.

Simple rescaling and phase transformations like $E_2 \to E_2 e^{i\Delta k\, z}$ can be applied, which modify (1) into

$$i\partial_z E_1 + \nabla^2 E_1 + E_1^* E_2 + |E_1|^2 E_1 + A|E_2|^2 E_1 = 0 \,, \tag{38}$$

$$2i\partial_z E_2 + \nabla^2 E_2 - \beta E_2 + \frac{1}{2}E_1^2 + \Lambda|E_2|^2 E_2 + A|E_1|^2 E_2 = 0 \,, \tag{39}$$

where the factor 2 in the first term of (39) originates from the ratio $k_2/k_1 \simeq 2$ for components that are near phase matching. Written in a convenient form [19], this equation set is valid as long as the fundamental frequency ω_1 and its second harmonic $\omega_2 = 2\omega_1$ are far from any internal material resonance. The slowly varying complex envelopes of the fundamental field, $E_1(\boldsymbol{r}, z)$, and second harmonic, $E_2(\boldsymbol{r}, z)$, are assumed to propagate with a constant polarization along the z axis and to disperse within a general transverse plane of dimension $D \leq 3$. The quantity β is related to the normalized phase mismatch $(2k_1 - k_2)/k_1$, while the positive coupling coefficients Λ and A involve the Fourier components of the second- and third-order susceptibility tensors.

In the absence of cubic terms, the possibility of generating stable soliton-like beams in quadratic media offers important possible applications in all-optical switching [34]. In purely $\chi^{(2)}$ media, various types of one- and two-dimensional stationary waves, including bright and dark solitons, have been identified numerically by means of standard shooting techniques. Their transverse profiles, scaling properties, and stability were analyzed in terms of the total power required to exceed a threshold, and in terms of the mismatch β. In particular, close to phase resonance, all the stationary waves were shown to be stable [35–37]. However, stable solitons may not necessarily form in the presence of cubic nonlinearities, because of collapse-type instabilities. It is thus worth determining the conditions for which collapse occurs when both nonlinearities are mixed together.

For this purpose, we recall the main integrals and relations associated with (38) and (39) for localized solutions E_1 and E_2. These solutions preserve the total power $P = \|E_1\|_2^2 + 4\|E_2\|_2^2 \equiv P_1 + 4P_2$ and the Hamiltonian

$$H \equiv \|\nabla E_1\|_2^2 + \|\nabla E_2\|_2^2 + \beta P_2 - \mathrm{Re}\int (E_1^2 E_2^*)\,\mathrm{d}\boldsymbol{r} - \frac{\chi}{2} \,, \tag{40}$$

where $\chi \equiv \|E_1\|_4^4 + \Lambda\|E_2\|_4^4 + 2A\|E_1 E_2\|_2^2$. Moreover, for waves superimposed at the center, a virial relation governing their mean square radius

$$I(z) = \frac{1}{P}(\|\boldsymbol{r}E_1\|_2^2 + 4\|\boldsymbol{r}E_2\|_2^2) \equiv I_1(z) + 4I_2(z)$$

can be derived, by means of explicit calculations, and reads [16]

$$P\,\partial_z^2 I(z) = 2(4-D)(\|\nabla E_1\|_2^2 + \|\nabla E_2\|_2^2) + 2D(H - \beta P_2) - D\chi \,. \tag{41}$$

We now establish analytical criteria for the absence or occurrence of collapse. For the sake of clarity, we first discuss the simpler case of media that have a purely quadratic nonlinearity: we omit the cubic contributions, and bound the resulting virial relation (41) from below as follows:

$$\partial_z^2 I(z) \geq \frac{4-D}{2P}(\|\nabla E_1\|_2^2 + 4\|\nabla E_2\|_2^2) + \frac{2D}{P}(H - \beta P_2) \,. \tag{42}$$

Next, by using the inequality (9), we construct the estimate $P = P_1 + 4P_2 \leq (4/D^2)(\|\nabla E_1\|_2^2 + 4\|\nabla E_2\|_2^2)\, I(z)$ and insert it into (42). This leads to

$$\partial_z^2 I(z) \geq \frac{4-D}{8}\frac{D^2}{I(z)} + \frac{C}{P} \,, \tag{43}$$

where C denotes the constant $C = 2D(H - \beta P/4)$ if $\beta > 0$ or $C = 2DH$ if $\beta \leq 0$. Let us now suppose that a collapse occurs at a given distance z_c (or $z_c^{\max}$), in the usual sense that $I(z) \to 0$ as $z \to z_c$. Then, necessarily, $I(z)$ must decrease from a certain distance z_0 until it vanishes at $z_c > z_0$. Therefore, we integrate (43) from $z_0 \leq z$ after multiplying it by $\partial_z I(z) < 0$, which yields

$$E(z) \equiv [\partial_z I(z)]^2 + \frac{4-D}{4} D^2 \ln\left(\frac{1}{I(z)}\right) - \frac{2C}{P} I(z) \leq E(z_0) < +\infty \,. \tag{44}$$

Since the constant $E(z_0)$ is finite, (44) shows that passing to the limit $I(z) \to 0$ is impossible for $D \leq 3$, which proves that in media with a quadratic nonlinearity, a wave collapse cannot take place with a total mean square radius $I(z) = I_1(z) + 4I_2(z)$ tending to zero. Note, however, that proving the absence of collapse from the nonvanishing of the virial integral $I(z)$ constitutes a weaker result than proving that the gradient norms remain bounded from above. As inferred from (9), the boundedness of gradient norms indeed ensures that the corresponding virial integrals never vanish for a given power P, whereas the converse implication is not true in general.

Let us now introduce the cubic nonlinearities and restrict our analysis to the dimensional case $D = 2$, as three-dimensional collapses have already been treated in [16]. From the results in Sect. 3, it follows that for a purely cubic medium ($E_1^* E_2 = E_1^2/2 = \beta = 0$) each partial power is preserved during propagation, and collapse can develop only if at least one of the two constraints

$$P_1 > P_1^c \equiv \frac{P_c}{(1+A)} \quad \text{and/or} \quad P_2 > P_2^c \equiv \frac{P_c}{(\Lambda + A)} \tag{45}$$

is satisfied. Returning to the complete equations (38) and (39), none of the partial powers is individually conserved, and we can only say that we expect that self-focusing will be favored for initial conditions with sufficiently high powers. From the cubic contributions χ in (41), those powers are optimized when $P_1(0) > P_1^c$ and $P_2(0) > P_2^c$, yielding

$$P > \mathcal{P}_c \equiv \frac{P_c}{1+A} + \frac{4P_c}{\Lambda + A} \,. \tag{46}$$

This requirement is an *optimal* condition for self-focusing in $\chi^{(2)}$–$\chi^{(3)}$ media. It suggests that the quadratic nonlinearities should not efficiently hinder the collapse induced by the cubic nonlinearities at very high powers. Furthermore, the additional constraint $H - \beta P_2 < 0$ from (41) indicates that the power threshold for self-focusing can change according to the value and sign of the mismatch parameter β. In particular, from $H < 0$ alone, it appears that a positive mismatch may lead to an increase of the self-focusing threshold, which can alter the basic estimate $P > \mathcal{P}_c$.

Conversely, no collapse develops for $D = 2$ when the partial powers satisfy the conditions opposite to (45). Indeed, we can bound the virial identity (41) as follows:

$$P\,\partial_z^2 I(z) \geq 4\left[\|\nabla E_1\|_2^2\left(1-(1+A)\frac{P_1}{P_c}\right) + \|\nabla E_2\|_2^2\left(1-(\Lambda+A)\frac{P_2}{P_c}\right)\right] + C\,,$$

where C is the constant defined above. It is then easy to prove that if the constraints on the partial powers $P_1(z) < P_1^c$ and $P_2(z) < P_2^c$ are *always* satisfied, $I(z)$ can never reach zero in that case. Expressed in terms of the total power $P = P_1(0) + 4P_2(0)$, this estimate can be bounded once more to ensure the absence of collapse when

$$P < \mathcal{P}_{\text{low}} \equiv \min\left(P_1^c,\, 4P_2^c\right)\,. \tag{47}$$

Results of numerical integrations of (38) and (39) at the critical dimension $D = 2$ have been presented in [16,38]. Computations were performed with identical Gaussian waves and for different values of β. The numerically revealed value of the critical total power involved in a collapse event for $\beta = 0$ was found to lie rather close to $\mathcal{P}_c$, between the two theoretical limits $\mathcal{P}_c$ and $\mathcal{P}_{\text{low}}$ [16], while for moderate positive values of β, the collapse threshold increased, as expected [38]. For intermediate powers lying in the range $\mathcal{P}_{\text{low}} < P < \mathcal{P}_c$, nothing definite can be concluded. On the one hand, power was observed to be transferred from the second harmonic to the fundamental, which was able to promote a collapse in the case $\beta = 0$. The transfer of power between harmonics, introduced by the $\chi^{(2)}$ nonlinearity, can thus generate collapse in a case where collapse would be absent in a purely cubic medium, for which P_1 and P_2 are individually conserved. On the other hand, analytical arguments based on a Lyapunov stability approach [16] have emphasized the possibility of forming stable solitons, provided that the soliton power be less than the upper limit $\mathcal{P}_c$. Recent numerical computations [39] have indeed shown that stable solitons, detected for small effective mismatches $-1 < \beta < 2$, exist only at powers below $\mathcal{P}_c$. In this respect, spatiotemporal solitons that were one-dimensional in space and underwent GVD have been observed experimentally in $LiIO_3$ crystals with a large $\chi^{(2)}$ value [40], where they reached a constant beam size and pulse duration for small and large phase mismatches.

6 Conclusion

We have reviewed the basic properties of collapse of a single wave at high dimension numbers $D \geq 2$ and have recalled the various regimes of mutual interaction between nonlinear light beams described by coupled NLS equations. Four typical regimes of interaction between two symmetric Gaussian waves arise naturally in this analysis. When both pulses are initially separated by a distance Δ_0 larger than a critical value Δ_c, they may either spread out or collapse independently, depending on their incident powers. Conversely, for $\Delta_0 < \Delta_c$, they merge before ultimately dispersing or collapsing as one entity. This critical distance of separation, Δ_c, depends on the power contained in each wave. Criteria for coalescence have been developed on the basis of the vanishing of the virial integral $I(z)$, which includes the separation distance between the wave centroids, at a finite propagation distance. Finally, we have analyzed the influences of four-wave mixing (FWM), walk-off, and quadratic nonlinearities on the collapse of optical pulses in focusing Kerr media. Whereas quadratic nonlinearities tend to counteract the collapse, but do not stop it at high input powers, FWM enhances the self-focusing, but for small phase mismatches only. In contrast, walk-off inhibits the collapse to some extent, by separating the components from each other and by causing their centroids to undergo nonlinear oscillations.

Acknowledgments

The author thanks Dr. Ole Bang, who actively collaborated in most of the work presented in this contribution.

References

1. A.C. Newell, J.V. Moloney, *Nonlinear Optics* (Addison-Wesley, Redwood City, California, 1991).
2. G.P. Agrawal, *Nonlinear Fiber Optics* (Academic Press, New York, 1989).
3. R.Y. Chiao, E. Garmire, C.H. Townes, Self-trapping of optical beams, Phys. Rev. Lett. **13**, 479 (1964); P.L. Kelley, Self-focusing of optical beams, Phys. Rev. Lett. **15**, 1005 (1965).
4. J. Juul Rasmussen, K. Rypdal, Blow-up in nonlinear Schrödinger equations – I: a general review, Phys. Scr. **33**, 481 (1986).
5. A.L. Berkhoer, V.E. Zakharov, Self-excitation of waves with different polarizations in nonlinear media, Zh. Eksp. Teor. Fiz. **58**, 903 (1970) [Sov. Phys. JETP **31**, 486 (1970)].

6. S.V. Manakov, On the theory of two-dimensional stationary self-focusing of electromagnetic waves, Zh. Eksp. Teor. Fiz. **65**, 505 (1973) [Sov. Phys. JETP **38**, 248 (1974)].
7. C.R. Menyuk, Stability of solitons in birefringent optical fibers: I. Equal propagation amplitudes, Opt. Lett. **12**, 614 (1987).
8. C.R. Menyuk, Stability of solitons in birefringent optical fibers: II. Arbitrary amplitudes, J. Opt. Soc. Am. B **5**, 392 (1988).
9. J.U. Kang, G.I. Stegeman, J.S. Aitchison, N. Akhmediev, Observation of Manakov spatial solitons in AlGaAs planar waveguides, Phys. Rev. Lett. **76**, 3699 (1996).
10. N. Akhmediev, J.M. Soto-Crespo, Dynamics of solitonlike pulse propagation in birefringent optical fibers, Phys. Rev. E **49**, 5742 (1994).
11. C.R. Menyuk, Nonlinear pulse propagation in birefringent optical fibers, IEEE J. Quantum Electron. **QE-23**, 174 (1987); D.N. Christodoulides, R.I. Joseph, Vector solitons in birefringent nonlinear dispersive media, Opt. Lett. **13**, 53 (1988).
12. X.D. Cao, C.J. McKinstrie, Solitary-wave stability in birefringent optical fibers, J. Opt. Soc. Am. B **10**, 1202 (1993).
13. D. Mihalache, D. Mazilu, L. Torner, Stability of walking vector solitons, Phys. Rev. Lett. **81**, 4353 (1998).
14. D.N. Christodoulides, T.H. Coskun, M. Mitchell, M. Segev, Theory of incoherent self-focusing in biased photorefractive media, Phys. Rev. Lett. **78**, 646 (1997).
15. C.J. McKinstrie, D.A. Russel, Nonlinear focusing of coupled waves, Phys. Rev. Lett. **61**, 2929 (1988).
16. L. Bergé, O. Bang, J. Juul Rasmussen, V.K. Mezentsev, Self-focusing and solitonlike structures in materials with competing quadratic and cubic nonlinearities, Phys. Rev. E **55**, 3555 (1997).
17. A.V. Buryak, Yu.S. Kivshar, S. Trillo, Optical solitons supported by competing nonlinearities, Opt. Lett. **20**, 1961 (1995).
18. S. Trillo, A.V. Buryak, Yu.S. Kivshar, Modulational instabilities and optical solitons due to competition of $\chi^{(2)}$ and $\chi^{(3)}$ nonlinearities, Opt. Commun. **122**, 200 (1996).
19. O. Bang, Dynamical equations for wave packets in materials with both quadratic and cubic response, J. Opt. Soc. Am. B **14**, 51 (1997).
20. W.E. Torruellas, Z. Wang, D.J. Hagan, E.W. Van Stryland, G.I. Stegeman, L. Torner, C.R. Menyuk, Observation of two-dimensional spatial solitary waves in a quadratic medium. Phys. Rev. Lett. **74**, 5036 (1995).
21. L. Bergé, Wave collapse in physics: Principles and applications to light and plasma waves, Phys. Rep. **303**, 259 (1998).
22. R.T. Glassey, On the blowing up of solutions to the Cauchy problem for nonlinear Schrödinger equations, J. Math. Phys. **18**, 1794 (1977).
23. M.I. Weinstein, Nonlinear Schrödinger equations and sharp interpolation estimates, Commun. Math. Phys. **87**, 567 (1983)
24. E.A. Kuznetsov, J. Juul Rasmussen, K. Rypdal, S.K. Turitsyn, Sharper criteria for the wave collapse, Physica D **87**, 273 (1995).
25. N.G. Vakhitov, A.A. Kolokolov, Stationary solutions of the wave equation in a medium with nonlinearity saturation, Izv. Vuz. Radiofiz. **16**, 1020 (1973) [Radiophys. Quantum Electron. **16**, 783 (1975)].

26. A.A. Kolokolov, Stability of stationary solutions of nonlinear wave equations, Izv. Vuz. Radiofiz. **17**, 1332 (1974) [Radiophys. Quantum Electron. **17**, 1016 (1976)].
27. E.A. Kuznetsov, A.M. Rubenchik, V.E. Zakharov, Soliton stability in plasmas and hydrodynamics, Phys. Rep. **142**, 103 (1986).
28. D. Anderson, Variational approach to nonlinear pulse propagation in optical fibers, Phys. Rev. A **27**, 3135 (1983).
29. M. Desaix, D. Anderson, M. Lisak, Variational approach to collapse of optical pulses, J. Opt. Soc. Am. B **8**, 2082 (1991).
30. J.H. Marburger, Self-focusing: Theory, Prog. Quantum Electron. **4**, 35 (1975).
31. L. Bergé, Coalescence and instability of copropagating nonlinear waves, Phys. Rev. E **58**, 6606 (1998).
32. O. Bang, L. Bergé, J. Juul Rasmussen, Fusion, collapse and stationary bound states of incoherently-coupled waves in bulk cubic media, Phys. Rev. E **59**, 4600 (1999).
33. L. Bergé, O. Bang, W. Krolikowski, Influence of four-wave mixing and walk-off on the self-focusing of coupled waves, Phys. Rev. Lett. **84**, 3302 (2000).
34. G.I. Stegeman, D.J. Hagan, L. Torner, $\chi^{(2)}$ cascading phenomena and their applications to signal processing, mode-locking, pulse compression and solitons, Opt. Quantum Electron. **28**, 1691 (1996).
35. S.K. Turitsyn, Stability of two- and three-dimensional optical solitons in a media with quadratic nonlinearity, Pis'ma Zh. Eksp. Teor. Fiz. **61**, 458 (1995) [JETP Lett. **61**, 469 (1995)].
36. L. Bergé, V.K. Mezentsev, J. Juul Rasmussen, J. Wyller, Formation of stable solitons in quadratic nonlinear media, Phys. Rev. A **52**, R28 (1995).
37. D.E. Pelinovsky, A.V. Buryak, Yu.S. Kivshar, Instability of solitons governed by quadratic nonlinearities, Phys. Rev. Lett. **75**, 591 (1995).
38. O. Bang, L. Bergé, J. Juul Rasmussen, Wave collapse in bulk media with quadratic and cubic responses, Opt. Commun. **146**, 231 (1998).
39. O. Bang, Yu.S. Kivshar, A.V. Buryak, A. De Rossi, S. Trillo, Two-dimensional solitary waves in media with quadratic and cubic nonlinearity, Phys. Rev. E **58**, 5057 (1998).
40. X. Liu, L.J. Qian, F.W. Wise, Generation of optical spatiotemporal solitons, Phys. Rev. Lett. **82**, 4631 (1999).

Discrete Solitons

Falk Lederer, Sergey Darmanyan, and Andrey Kobyakov

Summary. In waveguide arrays the interplay of discrete diffraction and nonlinear phase modulation evoked by either quadratic or cubic on-site nonlinearities may lead to the formation of discrete solitons. Domains of existence as well as peculiarities of this particular type of spatial solitons will be identified. It will turn out that in addition to the canonical types of solitons there are novel solutions, peculiar for discrete systems, as twisted and flat-top bright solitons. Stability issues are addressed in detail.

1 Introduction

The year 1955 witnessed the origin of the study of nonlinear dynamics in discrete systems when Fermi, Pasta, and Ulam reported, in their seminal paper [1], the observation of a recurrence phenomenon in a nonlinear lattice. Because matter itself is discrete, this issue is currently attracting renewed interest in the field of basic nonlinear physics. In particular, if a spatial excitation involves only a few constituents the very discreteness of the nonlinear system matters, and a continuous approximation ceases to correctly predict the dynamical behavior of the excitation. In this context, questions that concern the existence and dynamics of intrinsically localized excitations, frequently referred to as *discrete solitons* or *localized modes*, are attracting steadily growing interest. These discrete solitons represent a particular type of spatial soliton and are the subject of this chapter.

Initially, discrete solitons have been studied in atomic chains (discrete lattices) with an on-site cubic nonlinearity (see, e.g., [2–7]). But very early on it turned out that this concept could be successfully extended to other fields, such as proton dynamics in hydrogen-bonded chains, the transport of excitation energy in biophysical systems and molecular crystals, the motion of localized waves on electrical lattices, and, last but not least, the propagation of light in arrays of coupled nonlinear optical waveguides [8,9]. Many of these systems can be described by versions of the discrete nonlinear Schrödinger equation (DNLSE).

Over the past few years, it has turned out that waveguide arrays represent a convenient, easily accessible laboratory on a macroscopic scale for the experimental verification of numerous theoretical predictions ([10–12]; see also the chapter by Silberberg and Stegeman in this book). Among these experimental achievements are the proof of the existence of and a study of the dynamics of discrete solitons [10–12], and a demonstration of optical Bloch oscillations [13,14]. Beyond their relevance to the study of fundamental physical issues, nonlinear waveguide arrays may have potential for future all-optical switching and routing schemes [9,12,15–17].

The primary aim of this contribution is to describe the basic effects behind the formation of discrete solitons in waveguide arrays. Thus we shall show how "discrete diffraction" acts and how the mutual interplay of this controllable kind of diffraction with nonlinearly induced phase modulation may lead to modulational instability (MI) of plane wave solutions, resulting eventually in intrinsic localization, i.e. soliton formation. For potential applications such as "discrete" steering and switching, as a consequence of both competition between soliton motion and self-trapping and various destabilization scenarios, the reader is referred to the literature.

Until recently, the main emphasis was on discrete-soliton formation (i) for fairly wide excitations (where many waveguides are involved) and (ii) in arrays with scalar cubic nonlinearities (self-phase modulation only); see the original paper [8]. For a survey of potential all-optical effects see [9], for the first experiments see [10,11] and for a review see [18]. Because wide discrete solitons have many properties in common with spatial solitons in the corresponding continuous system (a film waveguide), which are the subject of other chapters in this volume, we shall primarily focus our attention on the case of narrow excitations (strong localization), discussed more recently in the literature [12,16,19–24]. In this case the effects of discreteness appear in the most pronounced way. On the other hand vectorial nonlinear interactions, i.e. cubic nonlinearities with self- and cross-phase modulation [21,22] and quadratic nonlinearities [23–26], add another degree of freedom to the dynamical system, and thus the solutions exhibit a richer diversity than discrete solitons in scalar Kerr media do. As an example of this kind of interaction, we consider here the quadratic nonlinear case.

It is intuitively clear that the field dynamics in a nonlinear array may be considered as a general case (nonlinear localization, nonlinear trapping) that covers two limiting cases, viz. spatial-soliton formation in a film waveguide (nonlinear localization, no trapping) and nonlinear switching in a two-core coupler (linear localization, nonlinear trapping). From the mathematical point of view, the nonlinear evolution equations that describe the two latter cases are integrable, at least for a scalar cubic nonlinearity. A typical array of evanescently coupled waveguides is shown in Fig. 1.

We start by introducing the basic mathematical models describing field propagation in waveguide arrays with various nonlinearities and proceed to the discussion of the dispersion relations for linear and nonlinear plane wave solutions. Nonlinear plane waves are then investigated for modulational instability. In doing so, we can identify domains in parameter space where both canonical bright solitons with a vanishing background and more exotic local-

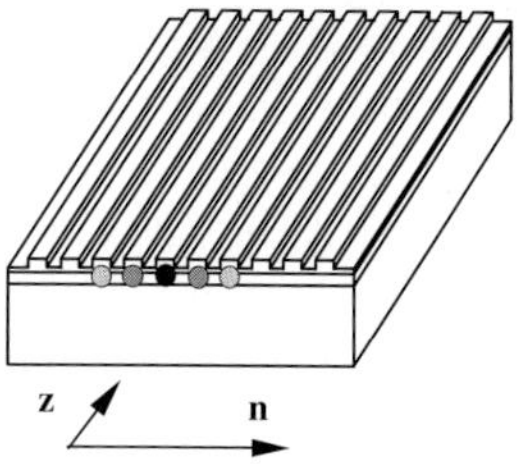

Fig. 1. Array of evanescently coupled waveguides

ized solutions with a finite background may potentially form. Here we are only concerned with these types of solitons, and refer to the literature as far as conventional dark solitons are concerned [22,27]. Moreover, we emphasize the peculiarities of "discrete diffraction", which is a synonym for coupling in an array, and its consequences for the formation of different types of discrete solitons. In particular, we shall concentrate primarily on the discussion of solutions which have no analogues in continuous systems and are thus a direct consequence of discreteness. The stability issues regarding strongly localized discrete solitons are reviewed because of their relevance to potential applications.

2 Basic Mathematical Models

The evolution of the slowly varying envelopes of the guided-mode fields in a homogeneous array consisting of infinitely many identical single-mode waveguides with nearest-neighbor interactions can be described by different types of discrete nonlinear Schrödinger equations; where details of the derivation can be found elsewhere [26]. For a Kerr nonlinearity we obtain

$$\mathrm{i}\frac{\mathrm{d}A_n}{\mathrm{d}Z} + C_1 A_n + C_2(A_{n+1} + A_{n-1}) + \Gamma_3 \left|A_n\right|^2 A_n = 0\,, \tag{1}$$

where A_n is the mode amplitude in the nth waveguide, C_1 and C_2 are corrections to the propagation constant and the coupling coefficient, respectively, both of which are associated with the linear overlap of the modes in adjacent guides, and Γ_3 is essentially the Kerr coefficient weighted with the effective mode area.

For a quadratic nonlinearity, we have to account for the field envelopes at the fundamental (A_n) and the second-harmonic (B_n) frequency leading to

$$\mathrm{i}\frac{\mathrm{d}A_n}{\mathrm{d}Z} + C_{\mathrm{a}1} A_n + C_{\mathrm{a}2}(A_{n+1} + A_{n-1}) + \Gamma_2 A_n^* B_n \exp(-\mathrm{i}\,\Delta k\, Z) = 0\,, \tag{2a}$$

$$\mathrm{i}\frac{\mathrm{d}B_n}{\mathrm{d}Z} + C_{\mathrm{b}1} B_{2n} + C_{\mathrm{b}2}(B_{n+1} + B_{n-1}) + \Gamma_2 A_n^2 \exp(\mathrm{i}\,\Delta k\, z) = 0\,. \tag{2b}$$

Here Γ_2 denotes the effective quadratic nonlinear coefficient and $\Delta k = 2k(\omega) - k(2\omega)$ is the propagation-constant mismatch. In deriving these equations we have assumed that the pulses are sufficiently long that we can neglect group-velocity dispersion in (1), (2a), and (2b) and temporal walk off in (2a) and (2b). A further simplification can be achieved by introducing a scaling length (e.g. the array length L_0), performing a gauge transformation, and normalizing the amplitudes by using the peak intensity (for details, see e.g., [9,24,26]). The dimensionless equations read now, for the cubic case,

$$\mathrm{i}\frac{\mathrm{d}a_n}{\mathrm{d}z} + c(a_{n+1} + a_{n-1}) + \gamma\left|a_n\right|^2 a_n = 0\,, \tag{3}$$

and for the quadratic case,

$$\mathrm{i}\frac{\mathrm{d}a_n}{\mathrm{d}z} + c_{\mathrm{a}}(a_{n+1} + a_{n-1}) + 2\bar{\gamma}a_n^* b_n = 0\,, \tag{4a}$$

$$\mathrm{i}\frac{\mathrm{d}b_n}{\mathrm{d}Z} + c_{\mathrm{b}}(b_{n+1} + b_{n-1}) + \beta b_n + \bar{\gamma}a_n^2 = 0\;, \tag{4b}$$

where the normalized mismatch $\beta = -\left(\Delta k + 2C_{\mathrm{a}1} - C_{\mathrm{b}1}\right)L_0$ includes also the corrections to the propagation constants. For the sake of flexibility we keep the nonlinear coefficients in both equations although they could be normalized to unity. Equations (3), (4a), and (4b) tell us that discrete-soliton formation will be the result of an interplay between coupling and nonlinearity in the cubic case, whereas in the quadratic case any deviation from phase-matching the mismatch β adds an additional parameter. Both in the cubic and in the quadratic case, the evolution equations yield only two conserved quantities, viz. the total power and the Hamiltonian, and are thus *nonintegrable*. These conserved quantities can be written as

$$P = \sum_{n=-\infty}^{\infty}\left|a_n\right|^2, \qquad H = -\sum_{n=-\infty}^{\infty}\left[c\left(a_n^* a_{n+1} + a_n a_{n+1}^*\right) + \gamma\left|a_n\right|^4\right] \tag{5}$$

in the cubic case and as

$$P = \sum_{n=-\infty}^{\infty}\left(\left|a_n\right|^2 + 2\left|b_n\right|^2\right),$$

$$H = -\sum_{n=-\infty}^{\infty}\left(c_{\mathrm{a}}a_n^* a_{n+1} + c_{\mathrm{b}}b_n b_{n-1}^* + \frac{1}{2}\beta\left|b_n\right|^2 + \bar{\gamma}a_n^2 b_n^* + \mathrm{c.c.}\right) \tag{6}$$

in the quadratic case. The lack of momentum conservation hints at soliton trapping rather than force-free motion across the array. These equations are particularly helpful in double-checking the accuracy of numerical solutions.

Moreover, frequently they can be used to evaluate the stability and mobility of soliton solutions. Discrete solitons centered either at a waveguide (*on-site* or *odd* solutions) or between two guides (*intersite* or *even* solutions) with the same power can have different Hamiltonians. Because of the periodicity of the array, the Hamiltonian is periodic with period $n = 1$, and is frequently termed the *Peierls–Nabarro potential.* The *Peierls-Nabarro barrier* reflects the difference between the minimum and the maximum Hamiltonian. This barrier increases with increasing localization and, accordingly, restricts the mobility of solitons. Moreover, conclusions concerning the stability of solitons can be drawn from the Peierls–Nabarro potential [7]. We shall return to this issue later.

3 Discrete Diffraction and Modulational Instability

3.1 Linear Plane Waves and Discrete Diffraction

From continuous systems it is well known that soliton formation requires a linear correlation effect such as dispersion, diffusion, or diffraction. In discrete systems this correlation is achieved by evanescent coupling between modes in adjacent waveguides. This coupling resembles diffraction in that it spreads an excitation of finite width across the array and can thus be termed *discrete diffraction* . Diffraction has its very origin in a varying z-dependent phase shift for different transverse wavevector components. This is a consequence of the dispersion relation of linear waves, which relates the longitudinal wavevector component (k_L) to the transverse component (q). The nature of the diffraction is controlled by the particular form of the dispersion relation. Obviously, in the linear case, any of the equations (3), (4a), or (4b) can be used to derive the dispersion relation. We insert the plane wave solution $a_n(z) = a\exp[\mathrm{i}(qn + k_\mathrm{L}z)]$ and obtain the dispersion relation $k_\mathrm{L} = 2c\cos q$. It is evident that linear coupling in an array leads to a very specific kind of diffraction ("discrete diffraction"). The continuum limit ($k_\mathrm{L} = c\left(2 - q^2\right)$) applies only for $q \ll 1$, i.e. for small transverse wavevector components (spatial frequencies), i.e. a wide initial excitation, and a small initial phase tilt. If we define the diffraction coefficient as $D = \partial^2 k_\mathrm{L}/\partial q^2$, we may straightforwardly conclude that it can be controlled with respect to size and sign in the discrete case by q, since $D = -2c\cos q$, whereas it is fixed in the continuum limit ($D = -2c$). This has important consequences for soliton formation. As for ordinary diffraction, the discrete analogue of diffraction can be conveniently described by solving the linear version of (4a) or (3) in Fourier space. Inserting $a_n(z) = (1/2\pi)\int_{-\pi}^{\pi} \tilde{a}(q,z)\exp(\mathrm{i}qn)\,\mathrm{d}q$, we obtain the Fourier amplitudes $\tilde{a}(q,z) = \tilde{a}(q,0)\exp(2\mathrm{i}cz\cos q)$, where $\tilde{a}(q,0)$ is the Fourier transform of the initial distribution $a_n(0)$. Returning to the original space, we obtain the "diffracted" field $a_n(z)$ for an arbitrary complex initial distribution $a_m(0)$,

$$a_n(z) \propto \sum_{m=-\infty}^{\infty} \mathrm{i}^{n-m} J_{n-m}(2cz)a_m(0) = \sum_{m=-\infty}^{\infty} G_{nm}(2cz)a_m(0) , \tag{7}$$

where J_n is the Bessel function of the first kind and G_{nm} can be considered the Green's function of the problem.

Although the diffraction has different signs for $q_0 = 0$ (unstaggered solution) and $q_0 = \pi$ (staggered solution), the beam spreading is identical. But, as in the temporal analogue the phase evolution is different, and it might be expected that the two cases would require different signs of nonlinearity for soliton formation. Discrete diffraction exhibits another remarkable property, namely that the beam with the largest possible initial phase tilt , which determines the transverse velocity of a finite excitation, $q_0 = \pi/2$, propagates in a diffractionless manner across the array. This situation resembles pulse propagation at zero dispersion in fibers. In Fig. 2 the discrete diffraction of an initial excitation $u_m(0) = \exp\left(-m^2/4\right) \exp \mathrm{i} q_0 m$ for $q_0 = 0$ and $q_0 = \pi/2$ is shown.

It is obvious that for localized solutions to exist, their wavevectors have to be situated outside the linear band, i.e. $|k_{\mathrm{sol}}| > 2c$ is required. In the quadratic nonlinear case, waves with two frequencies are involved, and this requirement becomes even more stringent. In this case the wavevector of the mutually locked localized solution must not be situated inside the linear band for either the fundamental harmonic (FH) or the second harmonic (SH), i.e. $|k_{\mathrm{sol}}| > \max\left(2c_{\mathrm{a}}, c_{\mathrm{b}} + \beta/2\right)$ and $|k_{\mathrm{sol}}| < \min\left(-2c_{\mathrm{a}}, -c_{\mathrm{b}} + \beta/2\right)$. The splitting of the permitted wavevector domain that appears for both types of nonlinearity is an effect of the discreteness itself. In the continuous case, this domain is connected and is either above or below the linear dispersion curve (for more details see [17,18]).

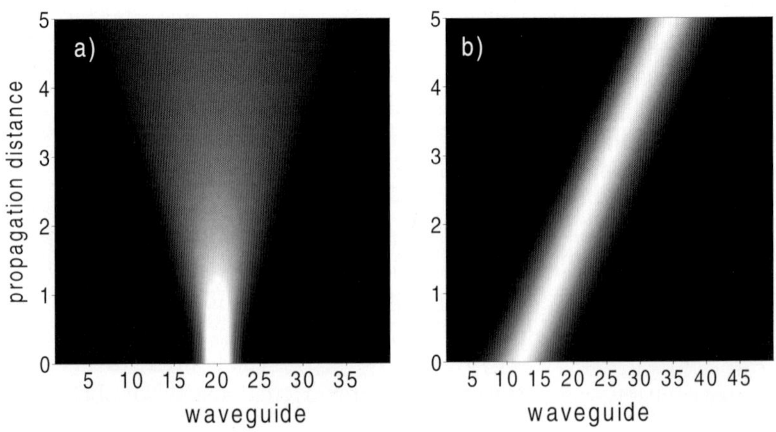

Fig. 2. Discrete diffraction of a Gausian input beam: (**a**) $q_0 = 0$, (**b**) $q_0 = \pi/2$

3.2 Nonlinear Plane Waves and Their Modulational Instability

Now we return to the nonlinear equations (3) and (4a), (4b). We seek the simplest solutions, i.e. nonlinear plane waves $a_n(z) = a_0 \exp[\mathrm{i}(qn + k_{\mathrm{pw}} z)]$, $b_n(z) = b_0 \exp[2\mathrm{i}(qn + k_{\mathrm{pw}} z)]$. These plane waves are only solutions if they obey the nonlinear dispersion relation

$$k_{\mathrm{pw}} = 2c\cos q + \gamma a_0^2 \tag{8}$$

for a cubic nonlinearity or

$$(k_{\mathrm{pw}} - 2c_{\mathrm{a}}\cos q)\,(k_{\mathrm{pw}} - c_{\mathrm{b}}\cos 2q - \beta/2) = \bar{\gamma} a_0^2 \tag{9}$$

for a quadratic nonlinearity. In what follows, we are concerned with stationary solitons only, i.e. we can restrict ourselves to so-called *unstaggered* ($q = 0 \curvearrowright \sigma = 1$) and *staggered* ($q = \pi \curvearrowright \sigma = -1$) solutions, where the parameter σ describes the topology of the soliton. Thus we obtain, in the cubic, case

$$k_{\mathrm{pw}} = 2c\sigma + \gamma a_0^2 \,. \tag{10}$$

In the quadratic case, (9) can be rewritten to obtain a dispersion relation that is controlled by the SH amplitude only and an additional equation that relates the FH and SH amplitudes, as follows:

$$\begin{aligned} k_{\mathrm{pw}} &= 2\,(\bar{\gamma} b_0 + \sigma c_{\mathrm{a}})\,, \\ a_0^2 &= 4b_0^2 + \frac{b_0}{\bar{\gamma}}\,(4\sigma c_{\mathrm{a}} - 2c_{\mathrm{b}} - \beta) > 0\,. \end{aligned} \tag{11}$$

We note, for further use, that strong localization requires the nonlinearity to prevail over linear coupling and thus that the dispersion will be essentially controlled by the nonlinearity.

Equations (10) and (11) give an initial hint of where bright localized nonlinear excitations are allowed to exist, because they require that k_{pw} is situated outside the linear bands. To be more specific, it is well known that the formation of conventional bright solitons requires the instability of plane wave solutions against weak transverse modulations (modulational instability, MI), whereas solitons with a nonzero background or with a flat-top shape can exist in domains of stable plane wave solutions.

3.2.1 Cubic Nonlinearity

To probe the stability of plane wave solutions, we insert

$$a_n(z) = [a_0 + \delta a_n(z)] \exp[\mathrm{i} k_{\mathrm{pw}} z] \tag{12}$$

into (3), and linearize in the small quantities δa_n and δa_n^* to obtain

$$\mathrm{i}\frac{\mathrm{d}}{\mathrm{d}z}\delta a_n + \sigma c\,(\delta a_{n+1} + \delta a_{n-1} - 2\delta a_n) + \gamma\,|a_0|^2\,(\delta a_n + \delta a_n^*) = 0 \tag{13}$$

and the corresponding complex conjugate equation. Now we insert the transverse modulations

$$\delta a_n(z) = \alpha\, \mathrm{e}^{\mathrm{i}Qn}\, \mathrm{e}^{gz}, \qquad \delta a_n^*(z) = \beta\, \mathrm{e}^{\mathrm{i}Qn}\, \mathrm{e}^{gz} \tag{14}$$

into (13), where Q is the wavevector of the modulation and g is the complex eigenvalue. If $\Re(g) > 0$ for any Q, the perturbation grows and the solution is modulationally unstable, whereas $\Re(g) < 0$ implies stability. $\Re(g) = G$ is frequently termed the *MI gain*. The eigenvalues are given by

$$g^2 = 8c \sin^2 \frac{Q}{2} \left(\sigma \frac{\gamma}{2} |a_0|^2 - c \sin^2 \frac{Q}{2} \right) . \tag{15}$$

So, we may conclude that for MI to emerge requires $\sigma\gamma > 0$, i.e. unstaggered solutions ($q = 0$) are modulationally unstable for a focusing nonlinearity ($\gamma > 0$), whereas MI of staggered solutions ($q = \pi$) requires a self-defocusing nonlinearity ($\gamma < 0$). A study of MI for arbitrary transverse wavevectors q can be found elsewhere [28].

3.2.2 Quadratic Nonlinearity

Because of the vectorial nature of the nonlinear interaction, the situation is more involved in this case. Details of the stability analysis can be found in the literature [23,24,29]. Here we briefly review the results. We insert

$$a_n(z) = [a_0 + \delta a_n(z)] \exp[\mathrm{i}k_{\mathrm{pw}} z] , \qquad b_n(z) = [b_0 + \delta b_n(z)] \exp[2\mathrm{i}k_{\mathrm{pw}} z] \tag{16}$$

into (4a) and (4b), respectively, linearize the resulting equations in $\delta a_n, \delta a_n^*$, δb_n, δb_n^*, and introduce transverse modulations similar to (14) to obtain the desired equation for the eigenvalues,

$$g^4 - \alpha_2 g^2 + \alpha_0 = 0 , \tag{17}$$

where

$$\alpha_2 = 4\bar{\gamma}^2 \left(b^2 - 2a^2 \right) - f_1^2 - f_2^2 ,$$
$$\alpha_0^2 = (f_1 f_2)^2 - (2\bar{\gamma} b f_2)^2 - 8 (\bar{\gamma} a)^2 f_1 f_2 + (2\bar{\gamma} a)^4 ,$$

and

$$f_1 = 2\bar{\gamma} b + 2\sigma c_{\mathrm{a}} (1 - \cos Q) , \qquad f_2 = 4\bar{\gamma} b + 4\sigma c_{\mathrm{a}} - 2c_{\mathrm{b}} \cos Q - \beta .$$

The MI gain follows from (17) as

$$G(Q) = \frac{\sqrt{2}}{2} \left| \Re \sqrt{\alpha_2 \pm \sqrt{\alpha_2^2 - 4\alpha_0}} \right| . \tag{18}$$

In contrast to the cubic case, where the sign of the nonlinearity is the only parameter that controls MI, the quadratic scenario appears much richer, i.e. the amplitude of the SH, b, and the wavevector mismatch, β, come into play as additional control parameters. The maximum MI gain as a function of these two parameters is displayed in Fig. 3 for both types of solutions. Unstaggered solutions are stable only for negative SH amplitudes, i.e. soliton solutions with a nonzero background and flat-top solitons can be expected to exist in that domain. Conventional bright solitons, however, will mainly emerge for positive SH amplitudes, although there is also a very small domain of their potential existence for $b < 0$.

On the other hand, staggered solutions are modulationally stable regardless of the sign of the SH amplitude, provided that the *negative* mismatch exceeds a certain threshold. Figure 3 contains the information required to identify domains where particular kinds of discrete soliton solutions can be expected to form.

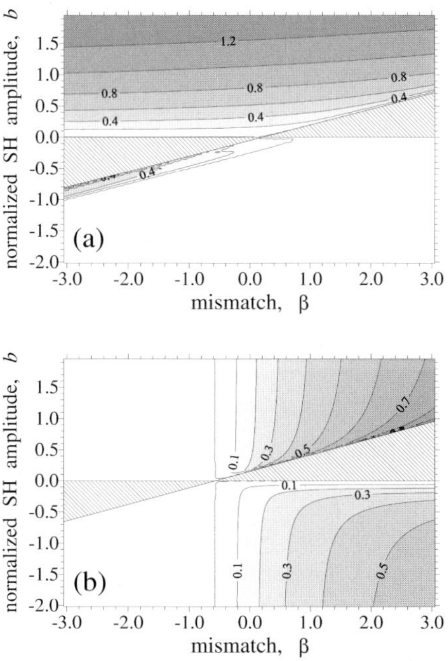

Fig. 3. MI gain as a function of mismatch β and SH amplitude b: (**a**) unstaggered solution, (**b**) staggered solution. No plane wave solutions exist in the *shaded areas*

4 Solitons in Cubic Nonlinear Media

As mentioned above, we are primarily concerned with strongly localized or narrow discrete solitons, i.e. stationary solutions of (3) which are at rest and involve only a very few excited waveguides. We are going to show how such soliton solutions can be found analytically by taking advantage of a method reported previously in the literature [2,3,5]. We shall categorize the solutions, and it will turn out that because of the discreteness, solitons with unusual topologies may form. We shall also investigate these stationary localized solutions for stability. The zoology of solutions is so rich that we can provide the reader with only an overview; for further details the reader is referred to the literature [19–21].

4.1 Types of Soliton Solutions

A stationary solution to (3) is of the form $\boldsymbol{a} = \{a_n\}$, where

$$a_n(z) = u_n \exp\left(\mathrm{i}k_{\mathrm{sol}}z\right) .$$

Here the u_n are solutions of the infinite algebraic system

$$-k_{\mathrm{sol}}u_n + c(u_{n+1} + u_{n-1}) + \gamma u_n^3 = 0 . \tag{19}$$

By solving this infinite system approximately for strong localization by reducing it to a finite number of equations, we can identify localized solutions of different topologies. We start with conventional bright solitons and proceed then to novel solutions which are a consequence of discreteness.

4.1.1 Even (Intersite) Bright Solitons

Assuming strong localization, we proceed from the ansatz

$$\boldsymbol{u}_{\mathrm{e}} = \{u_n\}_{\mathrm{e}} = U\left(..., 0, \alpha_3, \alpha_2, 1, s, s\alpha_2, s\alpha_3, 0, ...\right) , \tag{20}$$

where $s = \pm 1$ is a symmetry parameter, and the subscript $n = 0$ has been dropped for symmetry reasons. Inserting (20) into (19), we find the dispersion relation,

$$k_{\mathrm{sol}} \doteq k^{\mathrm{e}} = \gamma U^2 + sc + \frac{c^2}{\gamma U^2} \approx \gamma U^2 + sc , \tag{21}$$

and the corresponding small secondary amplitudes of the soliton solution,

$$\alpha_2 = \frac{c}{\gamma U^2} - s\left(\frac{c}{\gamma U^2}\right)^2 , \quad \alpha_3 = \left(\frac{c}{\gamma U^2}\right)^2 .$$

It is clear that strong localization requires $c \ll \left|\gamma U^2\right|$, i.e. the self-phase modulation has to exceed the linear coupling, and we may restrict ourselves to

$$\alpha_2 \approx \alpha = \frac{c}{\gamma U^2}\,, \quad \alpha_3 = 0\,.$$

As could be expected from studies of MI and the continuous limit, we find an unstaggered soliton when $\gamma > 0$ and $s = 1$,

$$\boldsymbol{u}_{\mathrm{e}}^{\mathrm{u}} = \{u_n\}_{\mathrm{e}}^{\mathrm{u}} = U\left(...,0,\alpha,1,1,\alpha,0,...\right) \tag{22}$$

and a staggered soliton when $\gamma < 0$ and $s = -1$

$$\boldsymbol{u}_{\mathrm{e}}^{\mathrm{s}} = \{u_n\}_{\mathrm{e}}^{\mathrm{s}} = U\left(...,0,-\left|\alpha\right|,1,-1,\left|\alpha\right|,0,...\right). \tag{23}$$

Here U is an arbitrary constant that determines the power and thus the degree of localization.

Unique, novel soliton solutions emerge as a result of the discreteness and have been termed *twisted solitons* [19]. They form when $\gamma s < 0$, viz., twisted unstaggered solitons $(s = -1, \gamma > 0)$,

$$\boldsymbol{u}_{\mathrm{e}}^{\mathrm{tu}} = \{u_n\}_{\mathrm{e}}^{\mathrm{tu}} = U\left(...,0,\alpha,1,-1,-\alpha,0,...\right)\,, \tag{24}$$

and twisted staggered solitons $(s = 1, \gamma < 0)$,

$$\boldsymbol{u}_{\mathrm{e}}^{\mathrm{ts}} = \{u_n\}_{\mathrm{e}}^{\mathrm{ts}} = U\left(...,0,-\left|\alpha\right|,1,1,-\left|\alpha\right|,0,...\right)\,. \tag{25}$$

Before investigating these solutions for stability, we derive the family of odd solutions.

4.1.2 Odd (On-Site) Bright Solitons

Relying on the arguments above, we may use the ansatz

$$\boldsymbol{u}_{\mathrm{o}} = \{u_n\}_{\mathrm{o}} = U\left(...0,\beta_2,\beta_1,\beta_0,s\beta_1,s\beta_2 0,...\right)\,.$$

The dispersion relation now reads

$$k_{\mathrm{sol}} \doteq k^{\mathrm{o}} = \gamma U^2 + 2\frac{c^2}{\gamma U^2} \approx \gamma U^2\,. \tag{26}$$

Again, unstaggered and staggered solitons exist, where each type splits into symmetric and antisymmetric solutions. More specifically, we can write for the unstaggered $(\gamma > 0)$ symmetric $(s = 1)$ soliton, with $\beta_0 = 1, \beta_1 = \alpha, \beta_2 = \alpha^2 \approx 0$,

$$\boldsymbol{u}_{\mathrm{o}}^{\mathrm{us}} = \{u_n\}_{\mathrm{o}}^{\mathrm{us}} = U\left(...,0,\alpha,1,\alpha,0,...\right)\,. \tag{27}$$

For the antisymmetric soliton ($s = -1$), with $\beta_0 = 0$, $\beta_1 = 1, \beta_2 = \alpha$, we can write

$$\boldsymbol{u}_{\mathrm{o}}^{\mathrm{ua}} = \{u_n\}_{\mathrm{o}}^{\mathrm{ua}} = U\,(...,0,\alpha,1,0,-1,-\alpha,0,...)\;. \tag{28}$$

Similarly, we obtain, for staggered solitons ($\gamma < 0$),

$$\boldsymbol{u}_{\mathrm{o}}^{\mathrm{ss}} = \{u_n\}_{\mathrm{o}}^{\mathrm{ss}} = U\,(...,0,-\left|\alpha\right|,1,-\left|\alpha\right|,0,...) \tag{29}$$

and

$$\boldsymbol{u}_{\mathrm{o}}^{\mathrm{sa}} = \{u_n\}_{\mathrm{o}}^{\mathrm{sa}} = U\,(...,0,-\left|\alpha\right|,1,0,-1,\left|\alpha\right|,0,...)\;. \tag{30}$$

It is interesting to note that twisted solitons have no odd analogue. This has important consequences for their stability.

4.1.3 Topological and Flat-Top Solitons

In addition to twisted bright solitons, there are other types of strongly localized solutions, unique to discrete systems [20]. First, we are concerned with topological solitons that exhibit two different asymptotic amplitudes, viz., zero and an arbitrary constant amplitude. To ensure MI of the background, we have to choose $\gamma\sigma < 0$. Here we restrict ourselves to unstaggered solutions and thus have to assume a defocusing nonlinearity ($\gamma < 0$). Because we are concerned with strong localization, we seek solutions which are characterized by a steep transition between the two asymptotic states. Hence we insert the ansatz

$$\boldsymbol{u}_{\mathrm{TO}} = \{u_n\}_{\mathrm{TO}} = U\,(...,0,0,0,u_1,u_2,u_3,u_4,1,1,1,...) \tag{31}$$

into (19) and end up with the dispersion relation

$$k_{\mathrm{sol}} \doteq k^{\mathrm{TO}} = 2c + \gamma U^2 \tag{32}$$

and the amplitudes

$$\begin{aligned} u_1 &= \alpha^2 \approx 0\,, \quad u_2 = -\alpha - \frac{3}{2}\alpha^2 \approx -\alpha\,, \\ u_3 &= 1 - \frac{\alpha}{2} - \frac{3}{8}\alpha^2 \approx 1 - \frac{1}{2}\alpha\,, \quad u_4 = 1 - \frac{\alpha^2}{4} \approx 1\,. \end{aligned}$$

Thus it turns out that, as a matter of fact, the localization is extremely strong and covers essentially two waveguides. Topological solitons with wider transition domains of this form do not exist, i.e. they require a nonmonotonic transition region [20]. Another feature of this soliton is that the part of the transition domain that approaches the nonzero background is unstaggered, whereas the domain near the zero background is staggered.

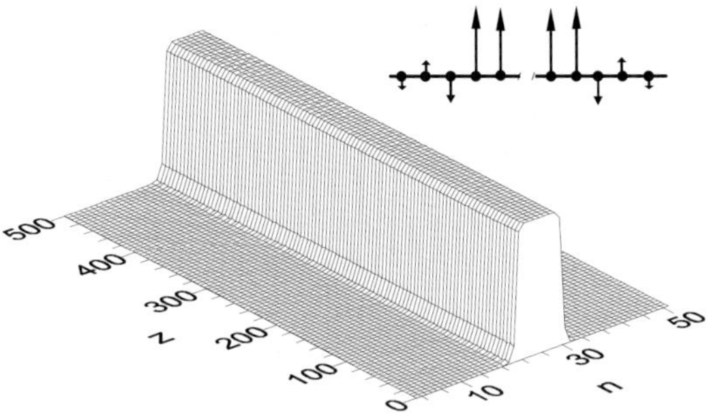

Fig. 4. Robust evolution of a flat-top soliton ($\alpha = 0.1$, width = 9 guides). The *inset* shows the topology of the soliton

Now the question arises as to whether two such topological solitons can be combined to form a *flat-top* soliton. This is indeed feasible when the width of this bright excitation is arbitrary. To the best of our knowledge, this is the first observation of an unstaggered bright soliton in a medium with a defocusing nonlinearity. This type of soliton is very robust, as shown in Fig. 4, but decays eventually, after a very long propagation distances which is likely to be irrelevant under typical experimental conditions. The robustness increases with soliton width.

4.2 Stability Analysis

The stability of the various discrete soliton solutions is of particular interest for two reasons. First, the solutions derived are approximate solutions of the nonintegrable equation (3). Thus the question arises as to whether they transform into exact stationary, stable solutions of (3). Secondly, implications for possible switching and steering operations can be derived by identifying the boundaries in the parameter space between stable and unstable solutions. In the literature, various approaches have been used to investigate the stability behavior of strongly localized solutions. Obviously, a complete scan of the parameter space by means of a direct numerical integration of (3) is inappropriate. Another method of evaluation of the stability relies on the Peierls–Nabarro (PN) potential. According to this concept, odd and even solitons of the same topology can be considered as two realizations of a common soliton, centered either on-site or between sites. Now, for odd and even solitons of equal intensities, the PN potential determines which of the two realizations corresponds to the minimum Hamiltonian and hence is stable. To correctly interpret the results, a negative effective mass has to be introduced for staggered solutions [7]. Being only a qualitative method, the PN approach provides no information about the instability gain and fails to analyze the stability of twisted modes because they lack an odd counterpart. Hence we prefer to identify the onset of instability and the initial instability gain by a

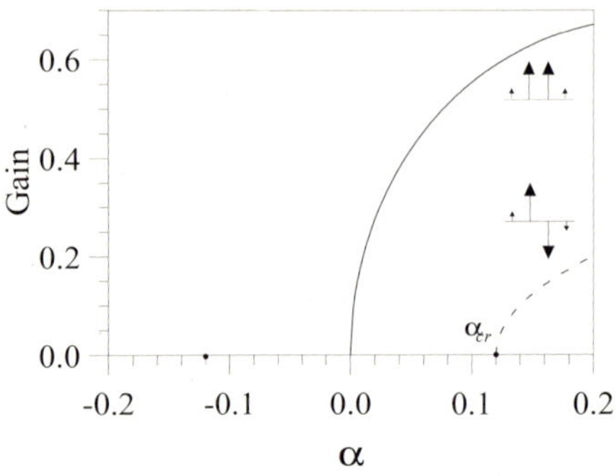

Fig. 5. Instability gain of an unstaggered conventional (*full line*) and twisted (*dashed line*) soliton. The *insets* show the topology

conventional linear stability analysis and to pursue the further evolution by numerically solving (3) with the appropriate initial condition. Here we sketch only the essential lines of the procedure, and refer for details to the original papers [19,21,20]. As in Sect. 3.2, we impose small complex perturbations on all nonvanishing amplitudes of the stationary solution in the form

$$a_n(z) = [u_n + \delta u_n(z)] \exp[ik_{\rm sol} z] \ .$$

4.2.1 Bright Solitons

The eigenvalue equation for the linear modes can be solved analytically, provided that the perturbations are separated into symmetric and antisymmetric contributions [19,21]. As could be expected from Peierls–Nabarro arguments, the even unstaggered solitons (22) and staggered solitons (23) are unstable, move slightly across the array, and transform eventually into the corresponding stationary odd solutions (27) or (29). This effect has been experimentally confirmed and exploited for discrete beam steering [12].

In contrast to these findings, twisted unstaggered solitons (24) and twisted staggered solitons (25) are stable, provided that the localization is sufficiently strong, i.e. the modulus of the secondary amplitude $|\alpha|$ must be less than some critical value $|\alpha| < \alpha_{\rm cr} = 0.12$ [19,21]. The corresponding instability gain (see Fig. 5) is given by

$$G \approx \sqrt{(-s\alpha - \alpha_{\rm cr})/2} \ , \qquad s\alpha + \alpha_{\rm cr} < 0 \ .$$

The stability of twisted solitons can be explained by the fact that they have no topological counterpart among odd solitons. Hence they cannot transform into odd solitons, and stability arguments based on the PN potential do not apply here. Beyond the critical value $\alpha_{\rm cr}$, the onset of instability manifests itself in a spreading of the mode and sets in at a weaker localization

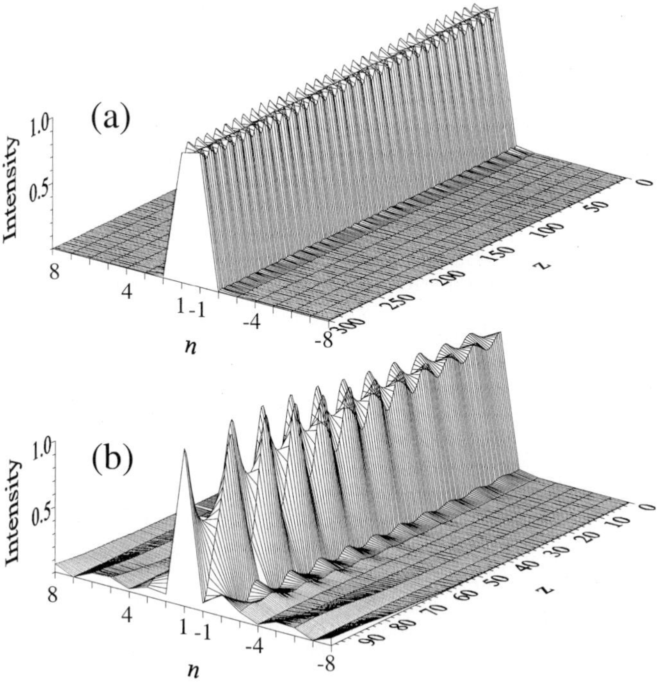

Fig. 6. Evolution of unstaggered, twisted solitons with different degrees of localization: (**a**) $\alpha = 0.11 < \alpha_{\mathrm{cr}}$, (**b**) $\alpha = 0.16 > \alpha_{\mathrm{cr}}$

because of an increasing secondary amplitude $|\alpha|$. The evolution of initially perturbed stable and unstable unstaggered twisted solitons is displayed in Fig. 6.

4.2.2 Topological and Flat-Top Solitons

We proceed to the stability analysis of the topological solitons (31). In a strict linear stability analysis, a perturbation in *any* waveguide of the infinite array has to be taken into account. To permit an analytical approach, we split the array into three parts: the zero background, the transition region, and the nonzero background (see Fig. 8). We assume that the zero background, since it is in fact a linear system, is stable, and that the nonzero background is modulationally stable because $\gamma\sigma < 0$. Thus the onset of instability is only possible in the transition region. We term this type of instability *front instability.* Following this reasoning, we impose a complex perturbation $\delta_j(z)$ ($j = 1, ..., 4$) on each *nontrivially* excited waveguide u_j only and insert the perturbed profile into (3). We have confirmed the validity of this truncated analysis by direct numerical integration of (3). The linearized eigenvalue problem for the real-valued perturbation vector $\boldsymbol{\delta}$ =

$(\delta_{1R}, \delta_{1I}, \delta_{2R}, \delta_{2I}, \delta_{3R}, \delta_{3I}, \delta_{4R}, \delta_{4I})^T$, where $\delta_j = \delta_{jR} + i\delta_{jI}$, can be written as

$$\frac{d\boldsymbol{\delta}}{dz} = \hat{\mathbf{M}}\boldsymbol{\delta}\,. \tag{33}$$

Looking for exponentially growing solutions of (33), we take $\delta_j \propto \exp(gz)$ and determine the complete set of eigenvalues $g_l(\alpha)$ ($l = 1, ..., 8$). Hence, the positive real part of $g_l(\alpha)$ corresponds to a nonzero instability gain $G(\alpha) \sim \max \Re[g_l(\alpha)]$. By inspecting the characteristic polynomial derived from the eigenvalue problem (33), we find a critical value $\alpha_{cr} \approx 0.16$, above which the instability gain

$$G(\alpha) \sim \sqrt{\frac{2(\alpha - \alpha_{cr})}{5}}\,.$$

becomes nonzero. These analytical findings are in good agreement with the numerical simulations shown in Fig. 7.

These results provide further evidence that topological solitons require strong localization. As in the case of twisted solitons, the solution (31) becomes unstable with decreasing localization and thus cannot have a counterpart in the continuous limit.

As already noted, flat-top solitons with various widths propagate quite robustly (see Fig. 4), provided that $\alpha < \alpha_{cr}$. The decay scenario is qualitatively different from that shown in Fig. 7b but can be easily understood. The soliton center transfers energy to the zero background, which grows. Because this transition region is staggered, the growing background is staggered, and becomes modulationally unstable because $\gamma\sigma > 0$. We have found numerically that the stable propagation distance of a flat-top soliton increases with the number of amplitudes that correspond to the top, and approaches infinity in the limit of a topological soliton. Thus it is reasonable to distinguish between two types of instability for topological and flat-top solitons. Short-term instability appears if the localization degree is less than the critical value, i.e. $\alpha > \alpha_{cr}$. This type of instability is predicted by the linear analysis. Long-term instability, occurring at propagation distances about ten times

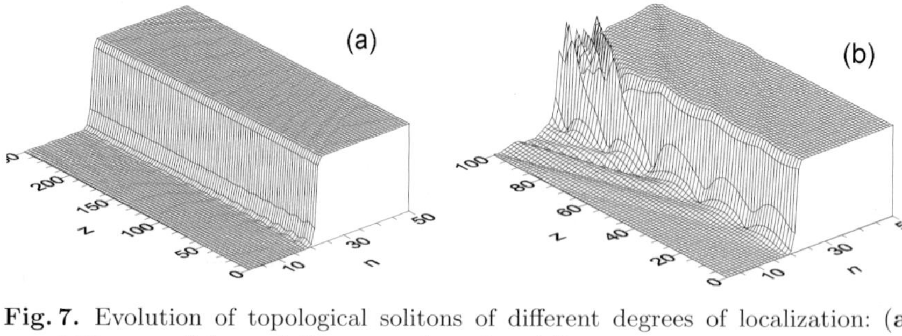

Fig. 7. Evolution of topological solitons of different degrees of localization: (**a**) $\alpha = 0.15 < \alpha_{cr}$, (**b**) $\alpha = 0.17 > \alpha_{cr}$ – front instability

larger, only develops for flat-top solitons, owing to power diffusion to the zero background.

We conclude this section with a few remarks about discrete solitons in Kerr media with self- and cross-phase modulation (vectorial nonlinear interaction). In that case additional types of discrete solitons, so-called *shifted* and *asymmetric* solitons, with a particular stability behavior [21], and *bright–dark* and *dark–antidark* pairs [22], emerge. Operating the system near the stability boundary of these solutions offers new options for all-optical switching.

5 Solitons in Quadratic Nonlinear Media

The formation of spatial and temporal solitons in continuous quadratic media has recently attracted steadily growing interest (for a recent review, see [26]). These solitons differ in many respects from their counterparts in media with a cubic nonlinearity. Thus it is worthwhile to look for discrete solitons in waveguide arrays with a quadratic nonlinearity, i.e. to seek localized solutions of the system (4a), (4b) and to study their behavior. We remind the reader that we have already shown in Sect. 3 (see Fig. 3) that even nonlinear plane waves exhibit a fairly complex destabilization behavior in this environment. As before, we focus our attention exclusively on strongly localized (narrow) solitons, which have been extensively discussed in [23,24], and refer to the literature [25] as far as weak and moderate localization are concerned.

5.1 Types of Soliton Solutions

A mutually locked stationary solution to (4a, 4b) is of the form $\boldsymbol{a} = \{a_n\}, \boldsymbol{b} = \{b_n\}$, where

$$
\begin{aligned}
a_n(z) &= u_n U \exp\left(\mathrm{i}k_{\mathrm{sol}} z\right) , \\
b_n(z) &= v_n V \exp\left(2\mathrm{i}k_{\mathrm{sol}} z\right) .
\end{aligned}
$$

Here U and V can have either sign, and V is a free parameter; u_n and v_n are solutions of the infinite algebraic system

$$
\begin{aligned}
-k_{\mathrm{sol}} u_n + c_{\mathrm{a}}(u_{n+1} + u_{n-1}) + 2\bar{\gamma} V u_n v_n &= 0 , \\
(\beta - 2k_{\mathrm{sol}})\, v_n + c_{\mathrm{b}}(v_{n+1} + v_{n-1}) + \bar{\gamma}\frac{U^2}{V} u_n^2 &= 0 .
\end{aligned}
$$

As before, we look for strongly localized solutions and begin with bright solitons.

5.1.1 Even Bright Solitons

Following essentially the same procedure as in the previous section, we introduce again the symmetry parameter $s = \pm 1$ and look first for even, or intersite, solitons of different topologies, such as

$$\begin{aligned} \boldsymbol{u}_{\mathrm{e}} &= \{u_n\}_{\mathrm{e}} = (..., 0, \alpha, 1, s, s\alpha, 0, ...) \ , \\ \boldsymbol{v}_{\mathrm{e}} &= \{v_n\}_{\mathrm{e}} = (..., 0, \rho, 1, 1, \rho, 0, ...) \ . \end{aligned}$$

The dispersion relation can be written in a similar way to (11), viz.

$$\begin{aligned} k_{\mathrm{sol}} &= k^{\mathrm{e}} = 2\left(\bar{\gamma} V_{\mathrm{e}} + s c_{\mathrm{a}}\right), \\ U_{\mathrm{e}}^2 &= 4V_{\mathrm{e}}^2 + \frac{V_{\mathrm{e}}}{\bar{\gamma}}\left(2 s c_{\mathrm{a}} - c_{\mathrm{b}} - \beta\right) > 0 \, , \end{aligned} \tag{34}$$

and the secondary amplitudes are given by

$$\alpha = \frac{c_{\mathrm{a}}}{2\bar{\gamma} V_{\mathrm{e}}} \ , \qquad \rho = \frac{c_{\mathrm{b}}}{4\bar{\gamma} V_{\mathrm{e}} - \beta} \ . \tag{35}$$

As in the cubic case, the FH component can exhibit four different topologies, viz., unstaggered ($\alpha > 0, s = 1$), staggered ($\alpha < 0, s = -1$), unstaggered twisted ($\alpha > 0, s = -1$), and staggered twisted ($\alpha < 0, s = 1$). The corresponding SH component conserves its shape regardless of the topology of the FH component.

5.1.2 Odd Bright Solitons

Odd solitons are of the form

$$\begin{aligned} \boldsymbol{u}_{\mathrm{o}} &= \{u_n\}_{\mathrm{o}} = (..., 0, \alpha, 1, \alpha, 0, ...) \ , \\ \boldsymbol{v}_{\mathrm{o}} &= \{v_n\}_{\mathrm{o}} = (..., 0, \rho, 1, \rho, 0, ...) \ , \end{aligned}$$

where the dispersion relation now reads

$$\begin{aligned} k_{\mathrm{sol}} &= k^{\mathrm{o}} = 2\bar{\gamma} V_{\mathrm{o}} \ , \\ U_{\mathrm{o}}^2 &= 4V_{\mathrm{o}}^2 - \beta \frac{V_{\mathrm{o}}}{\bar{\gamma}} > 0 \, , \end{aligned} \tag{36}$$

and the secondary amplitudes are the same as in (35) except that V_{e} is replaced by V_{o}. Depending on the sign of V_{o}, the solitons are staggered ($V_{\mathrm{o}} < 0$) or unstaggered ($V_{\mathrm{o}} > 0$).

5.1.3 Topological and Flat-Top Solitons

Similarly to the cubic case, we find also topological and flat-top solitons (for a detailed study and a discussion of possible applications, see [24]). The quadratic unstaggered topological soliton has the shape

$$\begin{aligned} \boldsymbol{u}_{\mathrm{TO}} &= \{u_n\}_{\mathrm{TO}} = (..., 0, 0, 0, u_1, u_2, u_3, u_4, 1, 1, 1...) \, , \\ \boldsymbol{v}_{\mathrm{TO}} &= \{v_n\}_{\mathrm{TO}} = (..., 0, 0, 0, v_1, v_2, v_3, v_4, 1, 1, 1...) \, . \end{aligned}$$

These solitons obey the above dispersion relation (36) but with an SH amplitude equal to V_{TO}. From Fig. 3 it is evident that this amplitude has to be less than zero to obtain a modulationally stable background. The amplitudes $u_n, v_n, n = 1, ..., 4$, are given by

$$\begin{aligned}
u_1 &= 1 - \frac{\alpha^2}{4} - \frac{3\alpha\rho}{4}\,, \qquad u_2 = 1 + \frac{\alpha+\rho}{2} - \frac{3\alpha^2}{8} + \frac{1-\bar{\beta}}{2-\bar{\beta}}\alpha\rho - \frac{3\rho^2}{8}\,, \\
u_3 &= \alpha - \frac{3\alpha^2}{2} + \frac{3\alpha\rho}{2}\,, \qquad u_4 = \alpha^2\,, \\
v_1 &= 1 - \frac{\alpha^2}{2} - \frac{\alpha\rho}{2}\,, \qquad v_2 = 1 + \alpha - \frac{\alpha^2}{2} + \frac{\alpha\rho}{2}\,, \\
v_3 &= \rho + \alpha^2 - \frac{\bar{\beta}}{2-\bar{\beta}}\alpha\rho\,, \qquad v_4 = \rho^2\,,
\end{aligned} \tag{37}$$

where α and ρ follow from (35), where V_{e} is replaced by V_{TO} and $\bar{\beta} = \beta/(2\bar{\gamma}V_{\mathrm{TO}})$ is the effective mismatch. As in the cubic case, two such topological solitons can form a bound state, which manifests itself as a flat-top soliton. A schematic representation of such a soliton is shown in Fig. 8.

We mention here that quadratic nonlinearities also admit stable *dark*, *double-hump dark*, and *antidark* soliton solutions [23].

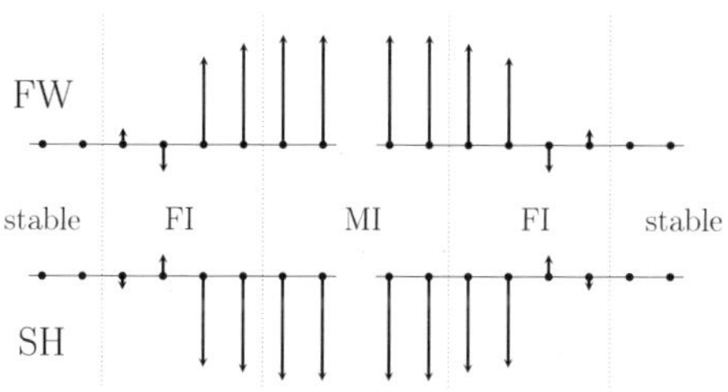

Fig. 8. Topology of a topological and a flat-top quadratic soliton. *Upward arrows*, zero phase; *downward arrows*, π phase

5.2 Stability Analysis

5.2.1 Bright Solitons

For bright quadratic solitons, a linear stability analysis is not straightforward. Hence we have used essentially numerical means to identify instabilities. We found that odd solitons are stable against fairly strong perturbations, whereas even, nontwisted solitons are always unstable, regardless of the sign of the product $\bar{\gamma}V_{\mathrm{e}}$. This finding is supported by the fact that the Hamiltonian of even solitons is always larger than that of odd solitons. In contrast to the cubic case, where the unstable even soliton transforms to a stable odd solution, we

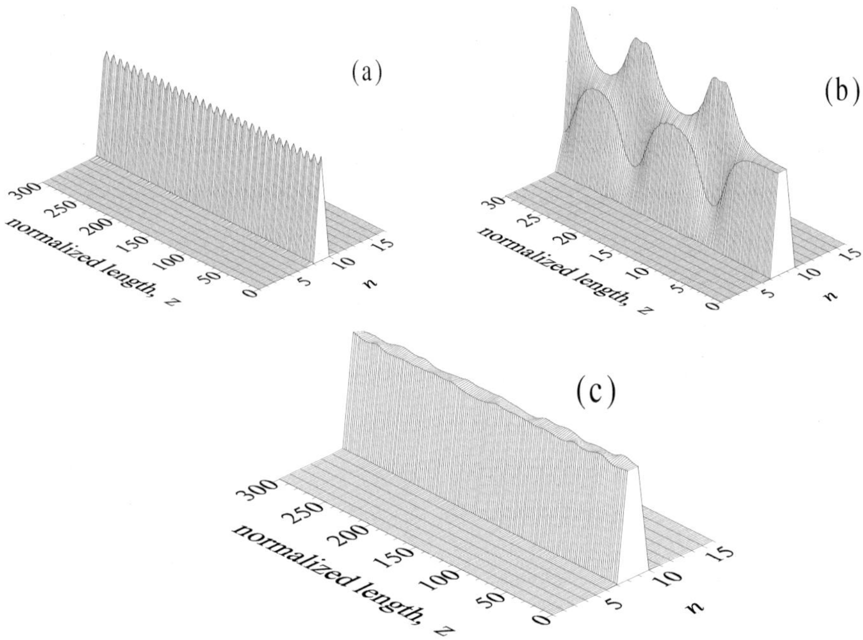

Fig. 9. Evolution of quadratic solitons: (**a**) stable odd soliton, (**b**) unstable even soliton and (**c**) stable twisted soliton. The FH component is shown. In all cases the initial condition deviated from the soliton solution. The parameters are $c_{\mathrm{a}} = c_{\mathrm{b}} = 0.15$, $\beta = 0$, $\bar{\gamma} = 1, b = 1$

observed large persistent oscillations here. These oscillations are also known in continuous quadratic media when the initial conditions are not consistent with the soliton solution [30] and exist because of an internal mode of the soliton. This instability does not occur for twisted solitons above a certain amount of localization; they appear to be stable, as in the cubic case. The results are displayed in Fig. 9.

5.2.2 Topological and Flat-Top Solitons

As we have emphasized in Sect. 3, modulational instability is more complex in a quadratic array. From Fig. 3 we can conclude that for unstaggered plane wave solutions to be stable, we require at least a negative amplitude of the second-harmonic. But even in this case MI can emerge. Thus we may anticipate that both types of solitons can destabilize via MI of the background or of the flat-top, as well as via front instability as in the cubic case. A typical flat-top soliton consisting of two topological solitons is schematically sketched in Fig. 8.

As before we assume that the zero background is essentially linear and therefore stable. Then it remains to investigate the flat area and the fronts

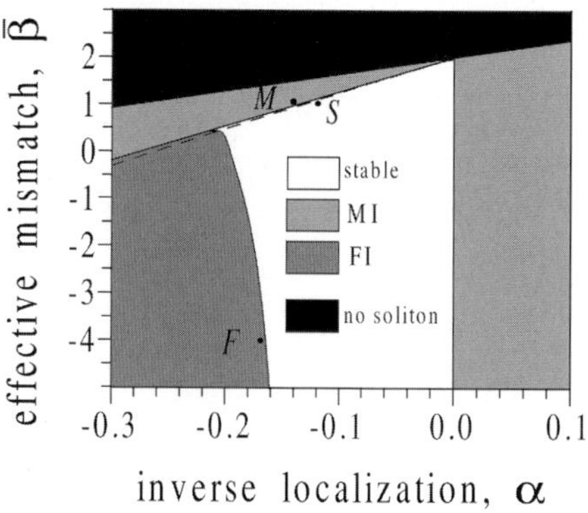

Fig. 10. Domains of stability, MI and front instability (FI) for flat-top quadratic solitons as a function of inverse localization α and effective mismatch $\bar{\beta}$; $c_b/c_a = 0.2$

for stability. We performed a stability analysis as usual. Details can be found elsewhere [24]. The main results are summarized in Fig. 10.

Indeed, both of the expected instability scenarios occur; usually, the MI gain exceeds that of the front instability. These findings are confirmed by numerical simulations. In Fig. 11 the evolution of the FH soliton component

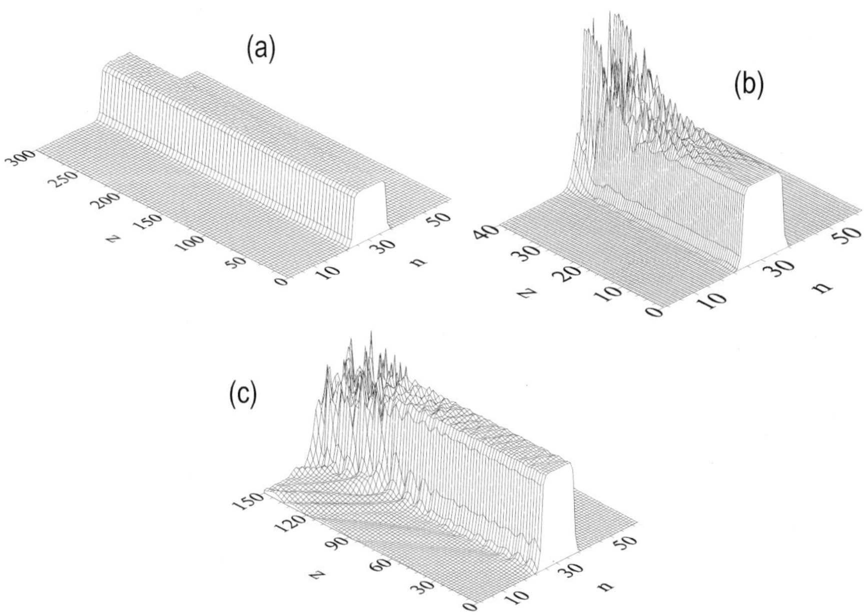

Fig. 11. Evolution of the FH component of flat-top quadratic solitons for $c_b/c_a = 0.2$, $b = -1$. (**a**) Stable evolution (point S in Fig. 10; $c_a = 0.24\bar{\gamma}$, $\bar{\beta} = -2\bar{\gamma}$). (**b**) Decay due to MI (point M; $c_a = 0.28\bar{\gamma}$, $\bar{\beta} = -2\bar{\gamma}$). (**c**) Decay due to FI (point F; $c_a = 0.33\bar{\gamma}$, $\bar{\beta} = 8\bar{\gamma}$)

is shown for a stable soliton, a soliton which decays because of MI, and another soliton which destabilizes because of front instability.

Flat-top quadratic solitons may have some potential for discrete steering . They can be excited by a homogeneous beam, and the output may be changed drastically by small changes of the FH component (for a practical example, see [24]).

6 Conclusions

Waveguide arrays can be considered as an implementation of a typical discrete system on an accessible spatial scale. Even in arrays of linear waveguides, the excitations undergo genuine discreteness effects such as controllable discrete diffraction. The nature of discrete diffraction has important consequences for the modulational instability of plane wave solutions in nonlinear arrays. More complex modulational-instability scenarios lead to novel spatial-soliton solutions. In particular, solitons without a continuous analogue emerge for strong localization. Twisted, topological, and flat-top solitons are among these types of soliton. The diversity of soliton solutions is appreciably richer if the nonlinear interaction exhibits a vectorial nature, i.e. for the vectorial Kerr and quadratic cases. Different destabilization scenarios can be exploited to achieve controllable discrete steering and switching operations. The recent experimental verification, in the cubic case, of many effects predicted previously should encourage the nonlinear-guided-wave community to undertake similar efforts for quadratic waveguide arrays.

Acknowledgments

The authors gratefully acknowledge a long-term grant from the Deutsche Forschungsgemeinschaft, generously supporting research activities in the field of discrete nonlinear systems in the framework of the Sonderforschungsbereich 196. We appreciate valuable collaborations with coauthors of our original papers, namely T. Pertsch, T. Peschel, U. Peschel, and E. Schmidt.

References

1. E. Fermi, J. Pasta, S. Ulam, *Studies of Nonlinear Problems* (Los Alamos National Laboratory, Los Alamos, NM, 1955) Report No. LA1940.
2. A.C. Scott, L. Macneil, Binding energy versus nonlinearity for a "small" stationary soliton, Phys. Lett. A **98**, 87–90 (1983).
3. J.C. Eilbeck, P.S. Lomdahl, A.C. Scott, The discrete self-trapping equation, Physica D **14**, 318–338 (1985).
4. A.J. Sievers, S. Takeno, Intrinsic localized modes in anharmonic crystals, Phys. Rev. Lett. **61**, 970–973 (1988).

5. J.B. Page, Asymptotic solutions for localized vibrational modes in strongly anharmonic periodic systems, Phys. Rev. B **41**, 7835–7838 (1990).
6. Y.S. Kivshar, D.K. Campell, Peierls–Nabarro potential barrier for higly localized nonlinear modes, Phys. Rev. E **48**, 3077–3081 (1993)
7. D. Cai, A.R. Bishop, N. Gronblech-Jensen, Localized states in discrete nonlinear Schrödinger equation, Phys. Rev. Lett. **70**, 591–595 (1993).
8. D.N. Christodoulides, R.I. Joseph Discrete self-focusing in nonlinear arrays of coupled waveguides, Opt. Lett. **13**, 794–796 (1988).
9. A. Aceves, C. De Angelis, T. Peschel, R. Muschall, F. Lederer, S. Trillo, S. Wabnitz, Discrete self-trapping, soliton interactions, and beam steering in nonlinear waveguide arrays, Phys. Rev. E **53**, 1172–1189 (1996).
10. P. Millar, J.S. Aitchison, J.U. Kang, G.I. Stegeman, A. Villeneuve, G.T. Kennedy, W. Sibbett, Nonlinear waveguide arrays in AlGaAs, J. Opt. Soc. Am. B **14**, 3224–3231 (1997).
11. H.S. Eisenberg, Y. Silberberg, R. Morandotti, A.R. Boyd, J.S. Aitchison, Discrete spatial solitons in waveguide arrays, Phys. Rev. Lett. **81**, 3383–3386 (1998).
12. R. Morandotti, U. Peschel, J.S. Aitchison, H.S. Eisenberg, Y. Silberberg, Dynamics of discrete solitons in optical waveguide arrays, Phys. Rev. Lett. **81**, 2726–2729 (1999).
13. T. Pertsch, P. Dannberg, W. Elflein, A. Bräuer, F. Lederer, Optical Bloch oscillations in temperature tuned waveguide arrays, Phys. Rev. Lett. **83**, 4752–4755 (1999).
14. R. Morandotti, U. Peschel, J.S. Aitchison, H.S. Eisenberg, Y. Silberberg, Experimental observation of linear and nonlinear optical Bloch oscillations, Phys. Rev. Lett. **83**, 4756–4759 (1999).
15. W. Krolikowski, Y.S. Kivshar, Soliton-based optical switching in waveguide arrays, J. Opt. Soc. Am. B **13**, 876–887 (1996).
16. O. Bang, P. Miller, Exploiting discretness for switching in waveguide arrays, Opt. Lett. **21**, 1105–1107 (1996).
17. T. Peschel, R. Muschall, F. Lederer, Power-controlled beam steering in nonequidistant arrays of nonlinear waveguides, Opt. Commun. **136**, 16–21 (1997).
18. F. Lederer, J.S. Aitchison, Discrete solitons in nonlinear waveguide arrays, in *Optical Solitons, Theoretical Challenges and Industrial Perspectives*, ed. by V.E. Zakharov, S. Wabnitz (Springer, Berlin, Heidelberg, 1999) pp. 349–365.
19. S. Darmanyan, A. Kobyakov, F. Lederer, Stability of strongly localized excitations in discrete media with cubic nonlinearity, JETP **86**, 682–685 (1998).
20. S. Darmanyan, A. Kobyakov, F. Lederer, L. Vazquez, Discrete fronts and quasirectangular solitons, Phys. Rev. B **59**, 5994–5997 (1998).
21. S. Darmanyan, A. Kobyakov, E. Schmidt, F. Lederer, Strongly localized vectorial modes in nonlinear waveguide arrays, Phys. Rev. E **57**, 3520–3529 (1998).
22. A. Kobyakov, S. Darmanyan, F. Lederer, E. Schmidt, Dark spatial solitons in discrete cubic media with self- and cross-phase modulation, Opt. Quantum Electron. **30,** 795–808 (1998).
23. S. Darmanyan, A. Kobyakov, F. Lederer, Strongly localized modes in discrete systems with quadratic nonlinearity, Phys. Rev. E **57**, 2344–2349 (1998).
24. A. Kobyakov, S. Darmanyan, T. Pertsch, F. Lederer, Stable discrete domain walls and quasi-rectangular solitons in quadratically nonlinear waveguide arrays, J. Opt. Soc. Am B **16**, 1737–1742 (1999).

25. T. Peschel, U. Peschel, F. Lederer, Discrete bright solitary waves in quadratically nonlinear media, Phys. Rev. E **57**, 1127–1133 (1998).
26. C. Etrich, F. Lederer, B. Malomed, T. Peschel, U. Peschel, *Optical Solitons in Media with a Quadratic Nonlinearity*, Progress in Optics (Elsevier, Amsterdam, 2000).
27. Y.S. Kivshar, M. Peyrard, Modulational instabilities in discrete lattices, Phys. Rev. A **46**, 3198–3205 (1992).
28. S. Darmanyan, I. Relke, F. Lederer, Instability of continuous waves and rotating solitons in waveguide arrays, Phys. Rev. E **56**, 7662–7668 (1997).
29. P.D. Miller, O. Bang, Macroscopic dynamics in quadratic nonlinear lattices, Phys. Rev. E **57**, 6038–6049 (1998).
30. C. Etrich, U. Peschel, F. Lederer, B. Malomed, Y. Kivshar, Origin of persistent oscillations of solitary waves in nonlinear quadratic media, Phys. Rev. E **54**, 4321–4324 (1996).

Optical Vortex Solitons

Grover A. Swartzlander, Jr.

Summary. After describing some elements of the theory of linear optical vortices I discuss experimental and theoretical aspects of optical vortex solitons. A simple computer code is included for generating the field of an arbitrary distribution of vortices

1 Introduction

Optical vortices have many analogues in nature. Systems such as hurricanes, tornadoes, eddies in normal fluids and superfluids, and optical vortices all exhibit a characteristic eye (or core). In nonlinear systems, the diameter of the core is often inversely related to a system parameter such as the energy [1]. A small eye signifies a robust core and vigorous dynamics. Figure 1 shows a side-by-side comparison of the eye of a typhoon [2] having an exceptionally small eye (about 5 miles wide) and the core of an optical vortex (roughly 450 μm in diameter). The size of the core is easily measured, and the value

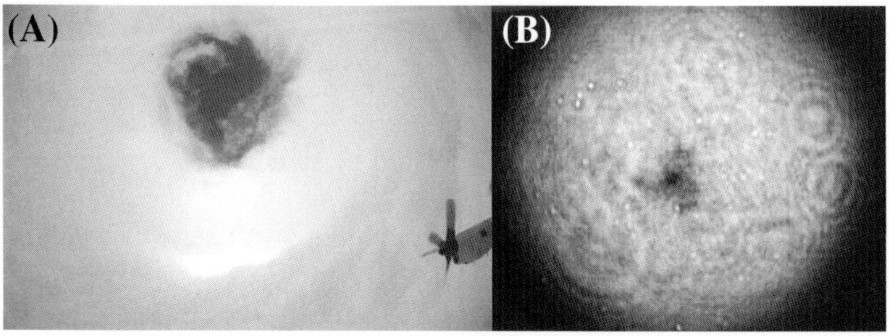

Fig. 1. (**a**) Eyewall of supertyphoon Forrest, September 1983, 350 miles northeast of Manila. Photographer Scott A. Dommin states, "In this shot, the airplane is in a tight left turn and we're looking straight up. I can also recall flying out of this storm, through the eyewall and watching the wings flapping up and down like a bird (they are not supposed to do that!)." Published with permission of Scott A. Dommin. (**b**) Laser beam containing an optical vortex soliton in a self-defocusing material

may be used to gauge the dynamic strength of the system. Indeed, the small eye of an approaching hurricane or typhoon is a harbinger of severe weather. In contrast, gusting photon fluxes merit little concern in optics; nevertheless, differences between the propagation dynamics of small- and large-core vortices can be significant [3]. These differences may be used to develop new optical devices or to explore fundmental issues in physics.

It is remarkable that systems governed by different laws of physics should produce similar patterns. Maxwell's equations and the constitutive relations dictate the propagation of light through matter, while fluid dynamics is governed by the Navier–Stokes equation. Perhaps even more remarkable is the fact that seemingly different physical systems, namely superfluids (Bose–Einstein condensates) [4–6] and nonlinear refractive materials, are both governed, to first order, by the same equation, for example the nonlinear Schrödinger (NLS) equation. When noise is added to these systems the wavefunction loses coherence, resulting in a phase-transition to a normal liquid in the condensate and laser speckle in optics. The parallels between optical and hydrodynamic systems are intriguing. Linear optics corresponds to a collisionless gas where the pressure is zero. When the effects of nonlinear refraction are added, the system resembles an inviscid compressible fluid with an intensity-dependent pressure. Experimental nonlinear optics provides advantages for studying the physics of systems governed by the NLS equation since initial conditions, environmental factors, and measurement techniques are easy to control. For example, varying the power of a beam transmitted through a nonlinear optical material allows one to observe differences between linear (low-power) and nonlinear (high power) vortex-related phenomena. To understand such differences, it is fruitful to first understand the formation and propagation of vortices in linear optics.

2 Vortex Wavefunctions

Students often encounter linear optical vortices in classical textbooks on electromagnetism [7], where vortices appear in the guise of cylindrical eigenmodes of the wave equation [8]. And anyone fascinated by the illusions and textures cast by laser speckle has unwittingly witnessed the beauty of a bounty of vortices in a beam [9]. For our purposes, we start by describing a single linear optical vortex embedded in a laser beam of finite extent, propagating in the z direction. Cylindrical coordinates (r, θ, z) are useful to describe the field, where $r = (x^2 + y^2)^{1/2}$ is the radial coordinate in a transverse plane ($z = \text{const}$) and $\theta = \arctan(y/x)$ is the azimuthal coordinate. At a given instant of time, an optical vortex whose core is centered at the origin ($r = 0$) can be written

$$E(r, \theta, z) = A_m(r, z) \exp\left[\mathrm{i}\Phi_m(r, z)\right] \exp(\mathrm{i}m\theta) \exp(-\mathrm{i}kz) , \qquad (1)$$

where $k = 2\pi/\lambda$ is the wavenumber, m is a signed integer called the topological charge, and A_m and Φ_m are radial amplitude and phase functions that

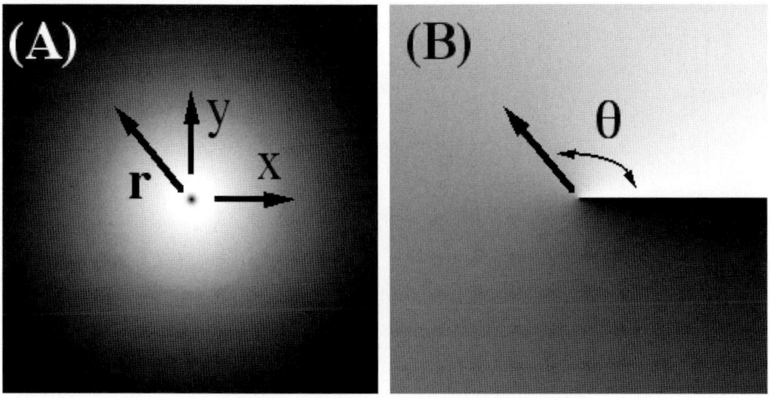

Fig. 2. Transverse amplitude (**a**) and phase (**b**) profiles of a beam containing an optical vortex. The phase increases from zero (*white*) to 2π (*black*) in this example, and the vortex core has been placed at the center of a Gaussian beam

describe the diffraction of the vortex core as the beam propagates in the z direction. The field in (1) is azimuthally harmonic as long as m is a nonzero integer; consequently, there is no diffraction (or propagation dynamics) in the θ direction. Figure 2 shows the transverse amplitude and phase profile of a singly charged vortex on a Gaussian background beam.

For convenience we introduce an explicit background amplitude G_{bg} and phase Φ that help to account for the evolution of the background beam separately from that of the vortex core. Additionally, we define the slowly varying envelope of the entire electric field, u:

$$E(r,\theta,z) = G_{\mathrm{bg}}(r,z)\exp\left[\mathrm{i}\Phi(r,z)\right]\left[A_m(r,z)\exp(\mathrm{i}m\theta)\right]\exp(-\mathrm{i}kz) \qquad \text{(2a)}$$
$$= u(r,\theta,z)\exp(-\mathrm{i}kz)\ . \qquad \text{(2b)}$$

In Fig. 2, the factor G_{bg} is a Gaussian profile having a flat phase front $\Phi = 0$; this is often the assumed initial field for theoretical problems.

3 Vortex Wavevectors

An interesting topological property of the vortex phase can be found by examining a wavefront. Neglecting diffraction, a surface of constant phase is given by $m\theta - kz = \mathrm{const}$, which maps out m nested helicoids, each having a pitch λ. The phase front appears as a right-handed or left-handed screw when the topological charge is positive or negative, respectively. The wavefront for the case $m = 1$ is shown in Fig. 3a. As the beam propagates in the z direction, the surface of constant phase curls along a helix in the direction of one's fingers when one's thumb is pointing along the z axis. Rays of light (which are directed perpendicular to the wavefront) therefore appear to spiral around the core in the $-m\hat{\theta}$ direction (i.e. clockwise when $m > 0$). The

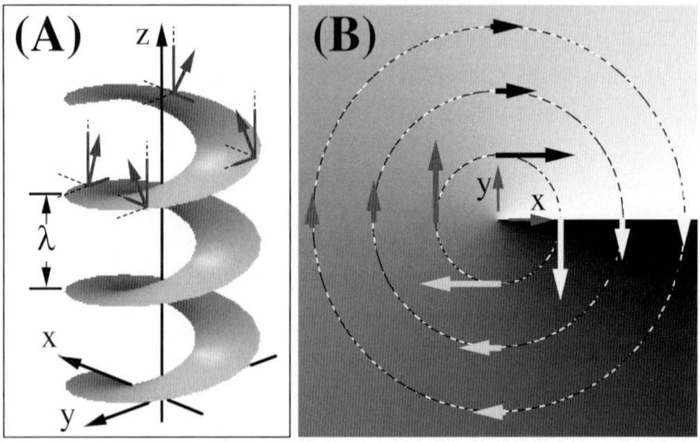

Fig. 3. (**a**). Surface of constant phase for a vortex that has a topological charge of $m = 1$. The *arrows* pointing perpendicular to the wavefront indicate the direction of the propagating rays. (**b**) Transverse phase profile (*white* $= 0$, *black* $= 2\pi$) showing circulation of transverse wavevectors $\boldsymbol{k}_\perp$ around the vortex core

transverse components of these rays (i.e. the projection onto the xy plane) point tangentially to circles centered on the vortex core, as shown in Fig. 3b.

Thus far we have neglected diffracted wavevectors in the radial direction. To specify the entire wavevector in a given transverse plane, we write the ansatz

$$u = f(r, \theta, z) \exp[\mathrm{i}s(r, \theta, z)] \ . \tag{3}$$

From (2b), we identify $s = m\theta + \Phi(r, z)$ and $f = G_{\mathrm{bg}}(r, z)A(r, z)$. Excluding the singular point at the center of the vortex ($r = 0$), where $\boldsymbol{k}_\perp$ is undefined, we find

$$\boldsymbol{k}_\perp = -\boldsymbol{\nabla}_\perp s = -\hat{r}\,\partial s/\partial r - \hat{\theta} r^{-1}\,\partial s/\partial \theta = -\hat{r}\,\partial \Phi(r, z)/\partial r - \hat{\theta} m/r \ . \tag{4}$$

Therefore, a vortex of charge m produces circulating wavevectors about the center of the vortex, and the radially dependent phase Φ shifts light toward or away from the vortex core.

The circulation around a path l enclosing the vortex is given by

$$\int_l \boldsymbol{k}_\perp \cdot \mathrm{d}\boldsymbol{l} = -2\pi m \ . \tag{5}$$

Those who are familiar with harmonic functions in quantum mechanics will not be surprised that the circulation is quantized. The topological charge, m, is identical to a quantum number for orbital momentum. In fact, it has been calculated [10] that the orbital momentum measured about the vortex core is $m\hbar$ per photon. At this point one may wonder whether the circulating photons violate Fermat's principle (e.g. that rays of light travel in a straight line). The reader is reminded that an object traveling along a straight line possesses a constant angular momentum with respect to a point that does not lie on the trajectory. Hence, optical circulation may be established by the simple pairwise intersection of three beams (or their transverse wavevectors) at the vertices of a triangle. It is therefore not surprising that scattered coherent light contains many localized vortices [9,11].

Another interesting aspect of vortices is that the phase is undefined at the center of this core. This point may be viewed as a topological singularity. This does not pose an immediate physical problem, because the field amplitude vanishes at the center of the vortex for all values of z owing to destructive interference. This cancellation occurs because the phases at opposite sides of the core are 180° out of phase. Although an accurate description of the core requires a consideration of polarization effects to ensure that Maxwell's equations are satisfied (e.g. that the divergence of the field should vanish), we shall ignore polarization effects and assume we are within the paraxial limit, for the present discussion. That is, we assume the transverse components of the wavenumber are much smaller than the longitudinal components: $k_\perp \ll k_z$, where $k^2 = k_\perp^2 + k_z^2$. Note that in the nonparaxial limit, the waves within a certain radius may become evanescent. This occurs because $k_z = k[1 - (k_\perp/k)^2]^{1/2}$ becomes imaginary for $k_\perp = |m|/r > k$. Therefore, the cutoff radius for propagation is on the order of the wavelength: $R_\mathrm{c} = |m|/k = \lambda|m|/2\pi$. In this sense, the central region of an optical vortex may contain not just a point of zero intensity, but a black dot having a quantized size.

4 r-Vortex Solutions

Explicit vortex solutions may be found by inserting (1) into the linear wave equation,

$$\nabla^2 E + k^2 E = 0\,, \tag{6}$$

where $\nabla^2 E = (1/r)\,[\partial/\partial r(r\,\partial E/\partial r)] + (1/r^2)(\partial^2 E/\partial\theta^2) + \partial^2 E/\partial z^2$ is the Laplacian in cylindrical coordinates. The reader may verify that the self-similar solutions,

$$E_m = E_0(r/L)^{|m|}\exp(\mathrm{i}m\theta)\exp(-\mathrm{i}kz), \tag{7}$$

satisfy (6), where L is an arbitrary constant having no physical meaning, and E_0 is a constant. These so-called "r-vortex" solutions diverge at infinity, but are useful for representing a beam near the vortex core. Note that the special case $m = 0$ is the plane wave solution, having no vorticity. For cylindrical beams of finite extent, Bessel and Neumann functions are commonly used as eigenmodes. Near the vortex core, the mth-order Bessel function varies as $r^{|m|}$. In contrast, Neumann functions diverge as $r^{-|m|}$ near the origin for $|m| > 0$ and, thus, do not generally appear in the discussion of free-space propagation phenomena, since these solutions are not physical.

Although it is convenient to describe a vortex in the transverse plane, in three-dimensional space the trajectory is generally a curved line (or string). Curvature may be caused by torsion (as seen in tornadoes) or by initial conditions (as in a smoke ring, where the vortex string is closed). A vortex string is more appropriately denoted by a vorticity vector $\boldsymbol{m}$, whose direction varies with propagation distance and, by convention obeys the right-hand rule. If the optical vortex vector $\boldsymbol{m}$ points parallel or anti-parallel to the direction of propagation, the topological charge is said to be positive or negative, respectively. If $\boldsymbol{m}$ lies in the transverse plane, a dark line appears across the beam, with the phase varying by π between the sides of the line (topologically, this is an edge dislocation in the phase profile). In this chapter we assume the vortex vector is paraxial (i.e. nearly parallel or antiparallel to the optical axis), and hence we treat the vorticity as a scalar quantity.

A measure of the size of an r-vortex in a beam of finite size may be determined at the half-maximum of the intensity profile. For example, the maximum intensity of a Gaussian beam containing a vortex, whose field is given by [12]

$$E_m = E_0(r/L)^{|m|}\exp[-(r/w_0)^2]\exp(im\theta)\exp(-ikz)\,, \tag{8}$$

is $I_{\max} = E_0^2(m/2\mathrm{e})^m(w_0/L)^{2m}$ (assuming $m > 0$). The peak occurs at $r = (m/2)^{1/2}w_0$, where w_0 is the characteristic size of the Gaussian beam. The half-width at half-maximum values of the core for the cases $m = 1$, 2, and 3 are factors of 0.29, 0.43, and 0.50 times smaller, respectively, than w_0. Thus for all values of $|m|$ the core size of an r-vortex is comparable to the characteristic size of the Gaussian beam. In particular, the half-width at half-maximum of a singly charged r-vortex will be written as $w_{\mathrm{rv}} = 0.29w_0$.

5 Hankel Transform into Vortex Modes

It is well known that an arbitrary field profile $f(x, y)$ may be decomposed into a series of rectilinear plane waves via a Fourier transform. Equivalently, the field can decomposed into vortex beams using a Hankel transform [13]. These two integral transformations are related through coordinate transformations. As an example, let us consider a vortex profile

$$f(x, y) = g_m(r)\exp(-im\theta)\,. \tag{9}$$

The Fourier integral is given by

$$F(k_x, k_y) = \int_{-\infty}^{\infty}\int_{-\infty}^{\infty} f(x, y)\exp[-\mathrm{i}(k_x x + k_y y)\,\mathrm{d}x\,\mathrm{d}y\,. \tag{10}$$

Substituting variables according to $x = r\cos\theta$ and $y = r\sin\theta$ and using the relation $J_m(a) = (2\pi)^{-1}\int_0^{2\pi}\mathrm{d}\theta\exp[-im(\theta+\pi/2)]\exp[-\mathrm{i}a\cos\theta]$ allows us to write (10) in circular coordinates:

$$F(k_r, \phi) = \exp[-im(\phi+\pi/2)]H_m\{g(r)\}\,, \tag{11}$$

where $k_r = \left(k_x^2 + k_y^2\right)^{1/2}$, $\phi = \arctan(k_y/k_x)$, and H_m is the mth order Hankel transform, given by

$$H_m \{g(r)\} = \int_0^\infty g_m(r) J_m(k_r r) r \, \mathrm{d}r \; . \tag{12}$$

6 Point Vortex

Let us consider a point vortex, which is the limiting case of an optical vortex soliton. The Fourier transform represents the field in the far-field region, or, equivalently, in the focal plane of a lens of focal length f. An ideal point vortex on a Gaussian beam may be represented by the envelope function

$$u(r, \theta) = \exp(-r^2/w_0^2) \exp(im\theta) \; , \tag{13}$$

and its Fourier transform may be expressed in terms of modified Bessel functions [12]. In the case $m = 1$, the transformed wavefunction can be approximated by the simple function

$$u_{\mathrm{ft}}(\rho, \phi) = \left[(\xi/1.06)^{2.60} + (1.53/\xi)^{0.86}\right]^{-1} \exp(i\phi) \; , \tag{14}$$

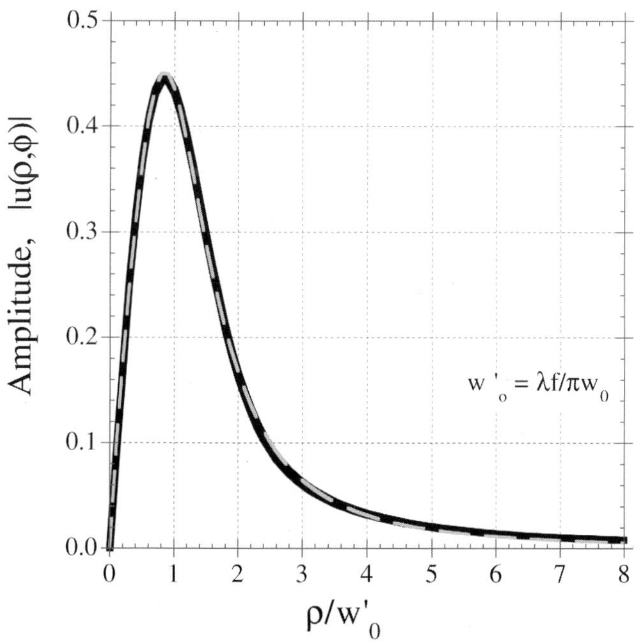

Fig. 4. Amplitude of the Fourier transform of a point vortex of topological charge $m = 1$ on a Gaussian beam. The *dashed line* corresponds to the approximate solution given in (14)

where $\xi = \rho/w_0'$ and ϕ are circular coordinates in the transformed space, and $w_0' = \lambda f/\pi w_0$ is the diffraction-limited spot size of a regular Gaussian beam ($m = 0$) in the transformed plane. Note that the constants in (14) have been numerically determined from a least-squares fit program. Figure 4 shows both the exact solution and the approximate solution. We notice the slowly decaying tail, which decreases as $\rho^{-2.6}$ as $\rho \to \infty$. In contrast, a self-similar r-vortex on a Gaussian beam decays exponentially. The tail on the far-field point vortex demonstrates the strong scattering ability of small-core vortices in linear media. As we shall see below, nonlinear refraction inhibits this scattering in a self-defocusing medium.

7 Multiple Vortices

While the presence of a vortex seems natural in a cylindrically symmetric system, it is somewhat surprising that vortices can also occur in an unbounded beam lacking symmetry. In fact, early experiments in optics and acoustics found that field vortices are generated when waves scatter from rough surfaces [9,11,14,15]. In their landmark 1974 paper, Nye and Berry [11] showed that wave pulses reflected from such surfaces contain defects similar to the ones found in imperfect crystals: edge, screw, and mixed edge–screw dislocations. Nye and Berry emphasized that their findings applied to any limited coherent wavetrains which traveled in different directions and interfered. Indeed, optical caustics [16,17] and laser speckle [9] contain many vortices distributed across the beam.

The field of a single Gaussian beam containing an arbitrary number of vortices M having arbitrary topological charges m_j, may be expressed as a simple product:

$$E(r,\theta,z=0) = E_0 \exp\left(-r^2/w_0^2\right) \prod_{j=1}^{M} A_j(r_j, z=0) \exp(\mathrm{i} m_j \theta_j) , \tag{15}$$

where the jth vortex has a core function A_j and coordinates (X_j, Y_j). The coordinates (r_j, θ_j) indicate the position with respect to the center ($r_j = 0$) of the jth vortex (see Fig. 5). The center of the Gaussian beam coincides with the origin, $r = 0$. The Cartesian and circular coordinates of the jth vortex, (X_j, Y_j) and (R_j, Θ_j), are related by the relations $X_j = R_j \cos\Theta_j$ and $Y_j = R_j \sin\Theta_j$. Furthermore, the relative positions measured from the center of the jth vortex are given by $x_j = x - X_j = r_j \cos\theta_j$ and $y_j = y - Y_j = r_j \sin\theta_j$. Lastly, we note the relation $r\exp(\pm\mathrm{i}\theta) = r_j \exp(\pm\mathrm{i}\theta_j) + R_j \exp(\pm\mathrm{i}\Theta_j)$. Computer-generated holography allows one to easily produce beams containing multiple vortices [12,18–21]. The intensity and phase profile of (15) is easily generated on a computer with the Fortran code in Table 1.

The intensity and phase profiles of two vortices separated by a distance d and having the same charge $m_j = -1$ are shown in Fig. 6. We note that

Table 1. Fortran code for determining the complex field of a system of M vortices

```
program vort
c Vortex core one pixel in size with tanh(r) profile on uniform background.
parameter (n=64)
complex jj, efield(n,n)
jj=(0.0,1.0)
print *, ' *** Vortex Field Program ***'
print *, 'N= ', n, ' NxN Pixel Area'
print 5
5 format(1x,'ENTER Number of Vortices ', $)
read(5,*) M
c Output Date File
open(unit=2,file='vortexfield.d',status='unknown',form='formatted')
rewind 2
c Initialize planar background field (change to Gaussian profile if you like)
do 10 jx=1,n
do 10 jy=1,n
10 efield(jx,jy)=(1.0,0.0)
c Calculate Field Sequentially
do 1000 i=1,M
print 999
read(5,*) xi,yi,msubi
do 100 jx=1,n
do 100 jy=1,n
x=jx-xi
y=jy-yi
r=sqrt(x**2+y**2)
if (r.eq. 0.0) y=1.0e-10
theta=atan2(y,x)
100 efield(jx,jy)=efield(jx,jy)*tanh(r)*exp(jj*msubi*theta)
1000 continue
do 300 j=1,n
do 300 i=1,n
300 write(2,*) efield(i,j)
999 format(1x,'ENTER grid point coordinates (1 to N) '
+ ' of vortex and its charge ', $)
stop
end
```

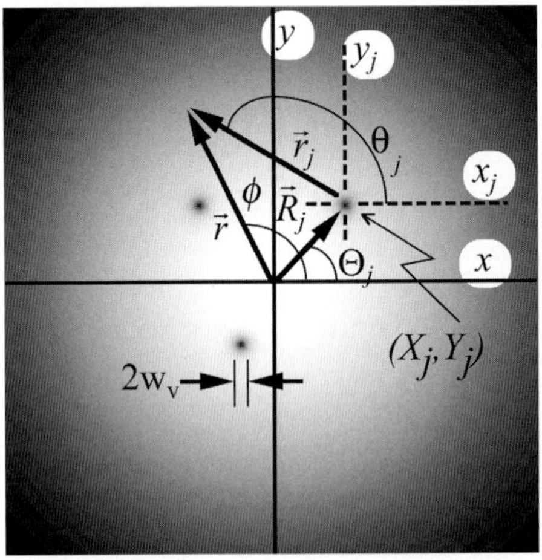

Fig. 5. Coordinates of the jth vortex in a beam

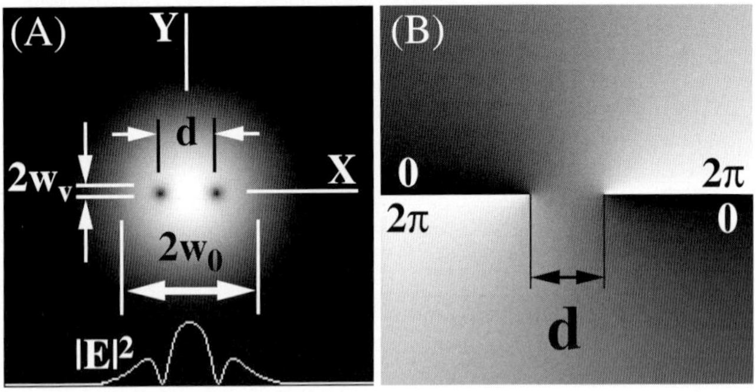

Fig. 6. Intensity (**a**) and phase (**b**) profiles of a Gaussian beam containing two symmetrically placed vortices of charge $m = -1$. The phase varies from 0 (*black*) to 2π (*white*) in the clockwise direction, and hence the circulation of each vortex is counter-clockwise

the z axis is directed out of the page and the transverse wavevectors are (initially) directed in the $\hat{\theta}_j$ directions: $\boldsymbol{k}_{\perp,\text{net}} = r_1^{-1}\hat{\theta}_1 + r_2^{-1}\hat{\theta}_2$, except at the centers of the vortices. The ray at the center of vortex 2 has the value $\boldsymbol{k}_{\perp,\text{net}} = k_{\text{perp},1} = (1/d_{\text{v}})\hat{\theta}_1$. Neglecting all other effects, this ray (and hence vortex 2) circles vortex 1 after propagating a distance z_{p} according to the relation $2\pi d_{\text{v}} = (k_\perp/k)z_{\text{p}}$ [22]. Thus, the rate of rotation of vortex 2 about vortex 1 is given by $2\pi/z_{\text{p}}$. We must also account for the similar rotation of vortex 1 about vortex 2. Both vortices rotate in the counterclockwise sense about a common point (or "center of mass") located halfway between the vortices. (The rotation would be clockwise if m_j were equal +1.) The rotation rate around this point is given by

$$\Omega_{\text{v}} = \Delta\Theta/\Delta z = -m\lambda/\pi d_{\text{v}}^2 \equiv -m/z_{\text{v}} \ , \tag{16}$$

where z_v is the characteristic diffraction distance of a feature of size d_v. This "angular velocity" is measured in units of radians per meter in the laboratory frame, but can easily be converted to radians per second in the reference frame of the photons by multiplying by the speed of light. According to this analysis, which neglects diffraction [3], the angular position of the vortex pair is expected to rotate by m radians after propagating a distance z_v.

We note that particles with a central force proportional to r^{-3} also exhibit an angular velocity that depends on the squared distance between the bodies. Optical vortices, however, are topological features and do not enjoy the degrees of freedom of particles. What is more, there is no actual force between optical vortices, and instead we speak of an "effective interaction" between vortices.

8 Nonlinear Refraction

Light propagating through a self-defocusing material experiences a lower refractive index in bright regions of the beam. Consequently, light is refracted toward dark regions, such as a vortex core or the perimeter of a beam. In an unbounded system, there is nothing to prevent light from "blooming" outward; however, the complete collapse of a vortex cannot occur, owing to the phase singularity. We have found that vortices that have a tanh-like profile may propagate through a self-defocusing material without diffracting outward or collapsing inward [23,24]. This particular solution is called an optical vortex soliton (OVS). As with other solitons in nonlinear optics, the width of an OVS decreases with increasing beam intensity.

Various materials may exhibit self-defocusing, including atomic vapors (when the frequency of the light is slightly below an atomic resonance) and semiconductors (slightly below a band edge). Laser heating may produce large decreases in the refractive index of slightly absorbing liquids [25–27]. The refractive index in such materials is often written

$$n = n_0 + \Delta n(I) , \tag{17}$$

where n_0 is the linear refractive index and Δn is an intensity-dependent term. For a so-called Kerr material, $\Delta n = n_2 I$, where $I = |E|^2$ is the beam intensity. Most nonlinear refractive materials exhibit a Kerr-like nonlinearity (to first order). The nonlinear coefficient n_2 is negative in a self-defocusing material and positive in a self-focusing material. In a slightly absorbing liquid, hot spots in a beam heat the liquid inhomogeneously, causing the density and refractive index to decrease. Thermal media can be modeled using a Kerr nonlinearity, although Δn actually changes with time (it is fluence-dependent) and is diffusive (Δn is coupled to the heat equation). Most materials exhibit some degree of diffusion and time dependence. Others, such as photorefractive crystals, may require additional controls, such as a DC electric field or background illumination [28]. At high intensities the Kerr approximation generally fails in all materials, and the value of Δn often reaches a saturation value. Thermal liquids have the highest known saturation values, $|\Delta n| \approx 0.1$, which occurs when the beam raises the temperature near the boiling point.

The governing equations for the propagation of light through a material are the wave equation and the constitutive relations describing the interaction of light with matter. For example, a bound electron perturbed by a weak optical field experiences a linear restoring force (Hooke's law), and the atomic polarization is then linearly proportional to the optical electric field. Strong optical fields distort the restoring force, which introduces a nonlinearity into the system. From a macroscopic point of view, the material polarization gives rise to a nonlinear refractive index (17). The propagation of light through a material that has an inhomogeneous refractive index may be determined, in the paraxial approximation, by the equation for the slowly varying envelope:

$$-2\mathrm{i}k\frac{\partial u}{\partial z} + \nabla_{\perp}^2 u - 2k^2\frac{\Delta n(x,y)}{n_0}u = 0\,. \tag{18}$$

Equation (18) is generally called the (2+1)-dimensional NLS equation, owing to its resemblance to the time-varying Schrödinger equation with a nonlinear potential term Δn. From a mathematical point of view, (18) has two degrees of freedom, x and y, whereas the z direction is a restricted dimension describing the beam evolution. From a physical point of view, (18) describes the propagation in a three-dimensional system. Hence (18) is also loosely referred to as the 2-D or the 3-D NLS. There are no known closed-form analytic solutions of the (2+1)-D NLS equation (unless stated otherwise, a Kerr nonlinearity will be assumed throughout this chapter), and, thus, approximate methods or numerical techniques are generally used to determine the propagation dynamics of a given initial beam profile $u(z=0)$. The split-step Fourier transform method is frequently the preferred algorithm to compute the evolving field. Approximate analytic solutions of (18) have been investigated in the context of superfluids , where the NLS describes a system of weakly interacting bosons [4–6].

9 Optical Vortex Solitons

The distinctive characteristic of an OVS is the size of its circular dark core. This hole decreases in size with increasing intensity, as demonstrated in Figs. 7 and 8. A closed-form analytic solution for the OVS does not exist, but experiments and numerical solutions [23,29] support the following approximate solution (see Fig. 9):

$$u(r,\theta,z=0) = u_0\tanh(r/w_{\mathrm{OVS}})\exp(-\mathrm{i}\theta)\,, \tag{19}$$

where

$$w_{\mathrm{OVS}} = 1.270w_{\mathrm{nl}} = 1.270k^{-1}\left(n_0/\Delta n_{\mathrm{bg}}\right)^{1/2}\,. \tag{20}$$

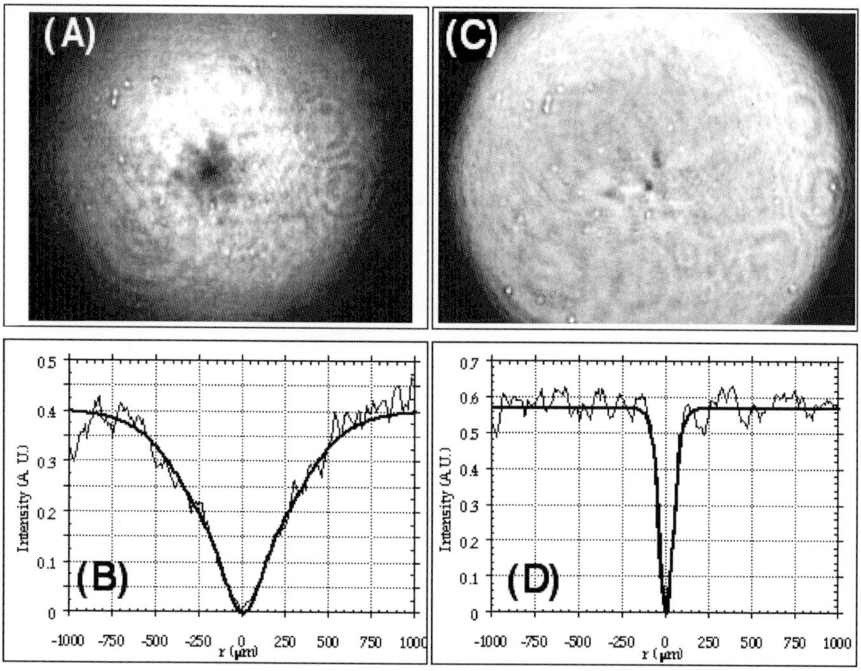

Fig. 7. Experimentally obtained intensity profiles of optical vortices at the output face of a thermal nonlinear medium for (**a**) linear and (**b**) nonlinear propagation

Here $\Delta n_{\mathrm{bg}} = n_2 I_{\mathrm{bg}}$ is the characteristic refractive-index change of the background beam (of intensity I_{bg}) supporting the vortex. The exact soliton solution, whose intensity profile ($|u|^2$) is shown in Fig. 9 is nearly indistinguishable from the tanh solution, as evidenced by the small amount of radiation in the propagated beam.

Although there is no critical power required to create an OVS, there is a practical threshold needed to discern the difference between an r-vortex and an OVS. This power may be determined by applying the condition $w_{\mathrm{rv}}/w_{\mathrm{OVS}} > 1$. Assuming the background beam is a Gaussian profile and that the topological charge $|m| = 1$, we obtain

$$P_{\mathrm{OVS}} > \pi^2 n_0/(n_2 k^2) \,. \tag{21}$$

The power of the beam P is related to the value of the background intensity via the relation $P = IA$, where A is an effective area of the beam. For a Gaussian beam, $I_{\mathrm{bg}} \approx 2P/(\pi w_0^2)$ (assuming $w_{\mathrm{OVS}} \ll w_0$). If $n_2 = 10^{-6}\,\mathrm{cm}^2/\mathrm{W}$, then only a few milliwatts is needed for one to start seeing nonlinear effects. According to (20), the core size is expected to fall as $P^{-1/2}$ in a defocusing Kerr material. Obviously, other forms for Δn produce different responses. For example, in a thermal medium, the steady-state core width contracts exponentially with power, owing to thermal diffucion (see Fig. 8).

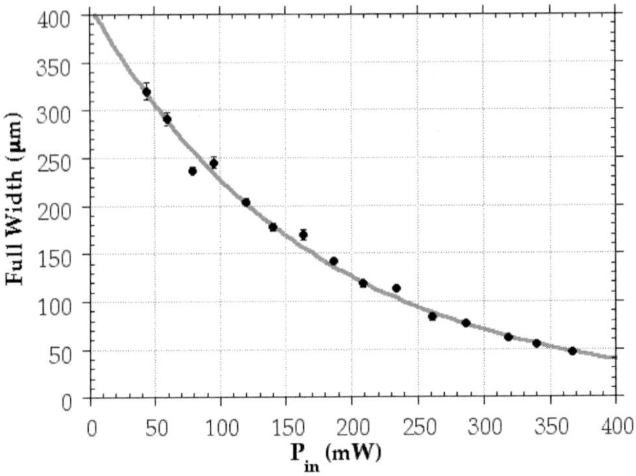

Fig. 8. Experiments in thermal defocusing media found an exponential decrease in the core size as the beam power was increased

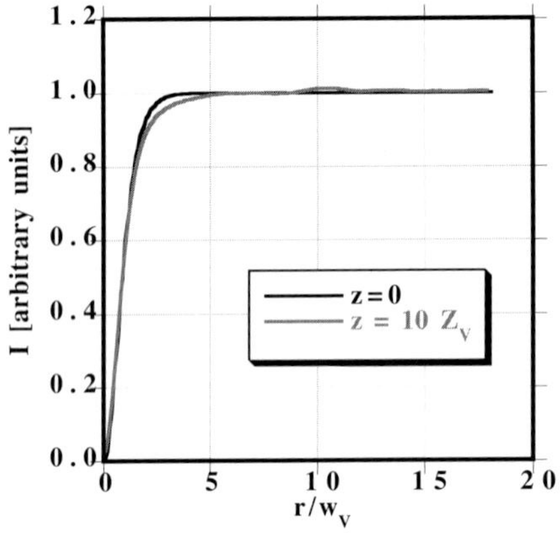

Fig. 9. Numerical verification of the $\tanh^2$ OVS intensity profile. Comparing the profiles at $z = 0$ (*black*) and $z = 10Z_\mathrm{v}$ (*gray*), we find little radiation as the vortex settles into its exact soliton profile. Here $w_\mathrm{v} = w_\mathrm{OVS}$ (see (20)) and $Z_\mathrm{v} = kw_\mathrm{v}^2/2$

10 Pinched Optical Vortex

We point out that a vortex with an arbitrary core size or shape exhibits significant radiation if the field differs from (19) and (20). An example of this is shown in Fig. 10. Nonlinear refraction induces a lower refractive index in the bright regions, and hence light is refracted into the dark core. This is analogous to matter falling into a black hole. In the first phase of the collapse, the core size begins to shrink, approaching the value of w_OVS. The core may then continue to collapse to a subsoliton size for some distance, while at the same time it develops an intensity spike. This constitutes a "pinched"

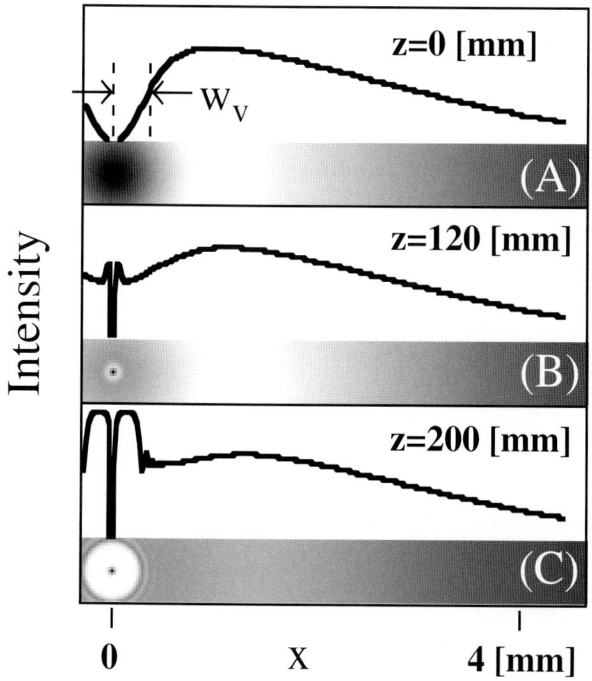

Fig. 10. Implosion and shock developing during the nonlinear propagation of an optical vortex whose core size exceeds the soliton size: $w_{\mathrm{v}} \gg w_{\mathrm{OVS}}$. The intensity profiles and images show (**a**) an initial vortex on a Gaussian background, (**b**) the collapse of the vortex core, and (**c**) the radiated optical shock wave

optical vortex soliton. Except for the spike, the intensity near the core may remain low enough for nonlinear refraction to continue drawing light toward the core, as seen in Fig. 10b. This is analogous to an implosion, where air is first sucked inward and then reflected outward as a shock front. Indeed, we see an optical shock from this event in Fig. 10c. The high intensity of light in the shock region may exceed the peak intensity of the initial beam; hence, the vortex core may be anomalously small during this phase of propagation. Eventually, the shock disperses with further propagation (like a supernova?), and since the initial beam has a finite size, the entire beam, including the vortex core, expands. This phenomenon is easily observed in thermal defocusing media [30].

11 Conservation of Momentum

Since a vortex is essentially a phase object, and since refraction affects the phase of a beam, it is natural to enquire whether vortices can be created or destroyed in a nonlinear medium. This question is of fundamental importance because a vortex is a wavefunction that possesses orbital angular momentum. The principle of the conservation of momentum for vortices is usually expressed as the conservation of topological charge. That is, if M is the net topological charge in a beam at the input face of a nonlinear medium, one expects the output beam to also have a net charge of M. Consequently, vortices may be created or annihilated in oppositely charged pairs [23,29,31–33], but single vortices cannot be generated. (We assume momentum is not transferred to the nonlinear material.) As an example, consider a single vortex with a field given by (1). Propagation over a small distance δz changes the phase of the field, i.e. (1) is multiplied by the factor $\exp\{i(2\pi/\lambda)[n_0 + \Delta n(I)]\delta z\}$. The intensity profile (the squared modulus of (1)) is azimuthally symmetric about the vortex core, and thus the propagation factor does not affect the harmonic phase of the vortex. Thus the vortex remains intact under nonlinear propagation. We note that the conservation of momentum and topological charge has not been thoroughly tested. Violations of the latter would not be surprising, since topological charge is an abstraction defined for perfectly harmonic waves – it does not include the entire momentum of the beam. As an example, consider a "nearly" harmonic wavefunction, $\exp[i(1-\epsilon)\theta]$, where $\epsilon \ll 1$. In a strict sense, the topological charge of this beam is zero. In a nonlinear material the wavefunction is attracted to the stable vortex soliton state [23], and thus the topological charge at the output may be different from that at the input face of the material. Although the topological charge has changed from zero to unity in this thought experiment, the total momentum of the beam has not changed.

12 Closing Thoughts

Optical vortices in linear and nonlinear media provide new frontiers for exploring fundamental issues in physics and novel applications in optics. Interactions between vortices [3,22,34] display interesting propagation dynamics that may be used as dynamically controlled optical interconnections. Optical limiting devices that darken under high-intensity illumination may be possible using vortex phase masks. Optical equivalents of field-effect transistors based on the waveguiding ability of optical vortex solitons have been proposed [30]. Issues involving the conservation of angular momentum, wave–particle duality and quantum phase, and phase transitions between coherent and incoherent states are presently being explored. Hopefully this chapter has provided a foundation that helps address these and other ideas involving optical vortices.

I would like to thank Chiu Tai Law (University of Wisconsin, Milwaukee) for many insightful discussions on optical vortices, and several of my former students, David Rozas, Zachary Sacks, and Anton Deykoon, for their contributions and for helping to break frontiers in the area of optical vortex physics.

References

1. H.J. Lugt *Vortex Flow in Nature and Technology* (Wiley, New York, 1983).
2. S.A. Dommin, http://members.aol.com/hotelq/. (Date of last access: June 30, 2001).
3. D. Rozas, C.T. Law, G.A. Swartzlander, Jr., Propagation dynamics of optical vortices, J. Opt. Soc. Am. B **14**, 3054–3065 (1997).
4. V.L. Ginzburg, L.P. Pitaevskii, On the theory of superfluidity, Zh. Eks. Teor. Fiz. **34**, 1240–1245 (1958) [Sov. Phys. JETP **7**, 858–861 (1958)].
5. L.P. Pitaevskii, Vortex lines in an imperfect Bose gas, Zh. Eks. Teor. Fiz. **40**, 646–651 [(1961) Sov. Phys. JETP **13**, 451–454 (1961)]
6. A.L. Fetter, Vortices in an imperfect Bose gas. I. The condensate, Phys. Rev. **138**, A429–A437 (1965).
7. S. Ramo, J.R. Whinnery, T. Van Duzer, *Fields and Waves in Communication Electronics* (Wiley, New York 1965).
8. G. Goubau, F. Schwering, On the guided propagation of electromagnetic wave beams, IRE Trans. Antennas Propag. **9,** 248–256 (1961).
9. N.B. Baranova, B.Ya. Zel'dovich, A.V. Mamaev, N.F. Pilipetskii, V.V. Shkunov, Dislocations of the wavefront of a speckle-inhomogeneous field (theory and experiment), Pis'ma Zh. Eks. Teor. Fiz. **33**, 206–210 (1981) [JETP Lett. **33**, 195–199 (1981)].
10. L. Allen, M.W. Beijersbergen, R.J.C. Spreeuw, J.P. Woerdman, Orbital angular momentum and the transformation of Laguerre-Gaussian modes, Phys. Rev. A **45,** 8185–8189 (1992).
11. J.F. Nye, M.V. Berry, Dislocations in wave trains, Proc. R. Soc. Lond. A **336,** 165–190 (1974).
12. Z.S. Sacks, D. Rozas, G.A. Swartzlander, Jr., Holographic formation of optical vortex filaments, J. Opt. Soc. Am. B **15,** 2226–2234 (1998).
13. J.D. Gaskill, *Linear Systems, Fourier Transforms and Optics* (Wiley, New York, 1978), p. 320.
14. M.V. Berry (1981) Singularities in waves and rays in *Physics of Defects, Les Houches Sessions XXXV*, ed. by R. Balian, M. Kleman, J.-P. Poirier, (North-Holland, Amsterdam, 1981), pp. 453–543.
15. N.B. Baranova, A.V. Mamaev, N.F. Pilipetsky, V.V. Shkunov, B.Ya. Zel'dovich Wave-front dislocations: topological limitations for adaptive systems with phase conjugation, J. Opt. Soc. Am. **73**, 525–528 (1983).
16. J.F. Nye, Optical caustics in the near field from liquid drops, Proc. R. Soc. Lond. A **361**, 21–41 (1978)
17. J.F. Nye, The catastrophe optics of liquid drop lenses, Proc. R. Soc. Lond. A **403**, 1–26 (1986)
18. V.Y. Bazhenov, M.V. Vasnetsov, M.S. Soskin, Laser beams with screw dislocations in their wavefronts, Pis'ma Zh. Eks. Teor. Fiz. **52**, 1037–1039, (1990) [JETP Lett. **52**, 429–431 (1990)].
19. V.Y. Bazhenov, M.S. Soskin, M.V. Vasnetsov, Screw dislocations in light wavefronts, J. Mod. Opt. **39**, 985–990 (1992).
20. N.R. Heckenberg, R. McDuff, C.P. Smith, A.G. White, Generation of optical phase singularities by computer-generated holograms, Opt. Lett. **17**, 221–223 (1992)

21. Z.S. Sacks, *Construction of Vortex Phase Holograms*, Master's thesis, Worcester Polytechnic Institute, Worcester, MA, USA (1995).
22. D. Rozas, Z.S. Sacks, G.A. Swartzlander, Jr., Experimental observation of fluidlike motion of optical vortices, Phys. Rev. Lett. **79**, 3399–3402 (1997).
23. G.A. Swartzlander, Jr., C.T. Law, Optical vortex solitons observed in Kerr nonlinear media, Phys. Rev. Lett. **69**, 2503–2506 (1992)
24. A.W. Snyder, L. Poladian, D.J. Mitchell, Stable black self-guided beams of circular symmetry in a bulk Kerr medium, Opt. Lett. **17**, 789–791 (1992).
25. R.C.C. Leite, R.S. Moore, J.R. Whinnery, Low absorption measurements by means of the thermal lens effect using a He–Ne Laser, Appl. Phys. Lett. **5**, 141–143 (1964)
26. J.P. Gordon, R.C.C. Leite, R.S. Moore, S.P.S. Porto, J.R. Whinnery, Long-transient effects in lasers with inserted liquid samples, J. Appl. Phys. **36**, 3–8 (1965)
27. G.A. Swartzlander Jr., B.L. Justice, A.L. Huston, A.J. Campillo, C.T. Law, Characteristics of a low f-number broadband visible thermal optical limiter, Int. J. Nonlinear Opt. Phys., **2**, 577–611 (1993).
28. G. Duree, M. Morin, G. Salamo, M. Segev, B. Crosignani, P. DiPorto, E. Sharp, A. Yariv, Dark photorefractive spatial solitons and photorefractive vortex solitons, Phys. Rev. Lett. **74**, 1978–1981 (1995).
29. C.T. Law, G.A. Swartzlander, Jr., Polarized optical vortex solitons: instabilities and dynamics in Kerr nonlinear media, Chaos Solitons Fractals **4**, 1759–1766 (1994).
30. G.A. Swartzlander, Jr., D.L. Drugan, N. Hallak, M.O. Freeman, C.T. Law, Optical transistor effect using an optical vortex soliton, Laser Phys. **5**, 704–709 (1995).
31. C.T. Law, G.A. Swartzlander, Jr., Optical vortex solitons and the stability of dark solitons stripes, Opt. Lett. **18**, 586–588 (1993).
32. A.V. Mamaev, M. Saffman, A.A. Zozulya, Propagation of dark stripe beams in nonlinear media: snake instability and creation of optical vortices, Phys. Rev. Lett. **76**, 2262–2265 (1996).
33. A.V. Mamaev, M. Saffman, D.Z. Anderson, A.A. Zozulya, Propagation of light beams in anisotropic nonlinear media: from symmetry breaking to spatial turbulence, Phys. Rev. A **54**, 870–879 (1996).
34. D. Rozas, G.A. Swartzlander, Jr., Observed rotational enhancement of nonlinear optical vortices, Opt. Lett. **25**, 126–128 (2000).

Solitons of the Complex Ginzburg–Landau Equation

Nail Akhmediev and Adrian Ankiewicz

Summary. This chapter deals with the cubic-quintic complex Ginzburg-Landau equation as a model for describing localised solutions in various optical systems with gain and loss. Qualitative differences of solitons in dissipative systems from those in Hamiltonian ones are revealed. Various examples of solitons and their bound states are considered.

1 Introduction

A soliton is a localized solution of a partial differential equation (PDE) describing the evolution of a nonlinear system with an infinite number of degrees of freedom. Solitons are usually considered to be attributes of systems which possess the property of integrability. In this instance, solitons remain unchanged during interactions, apart from a phase shift. They can be viewed as "modes" of the system, and, along with radiation modes, they can be used to solve initial-value problems using a nonlinear superposition of modes [1].

Reductions to integrable systems are extreme simplifications of the complex systems existing in nature. They can be considered as a subclass of the more general Hamiltonian systems (see Fig. 1). Such a simplification allows us to analyze the systems quantitatively and to completely understand the behavior of the solitons. Solitons in Hamiltonian (but nonintegrable) systems can also be regarded as nonlinear modes, in the sense that they allow us to describe the behavior of systems with an infinite number of degrees of freedom in terms of a few variables, thus allowing us effectively to reduce the number of degrees of freedom. Solitons in these systems collide inelastically and interact with radiation waves, thus showing that they are qualitatively different from those in integrable systems. However, as in the integrable case, the solitons are still a one- (or few-) parameter family of solutions.

The next level of generalization is to consider solitons in systems far from equilibrium. The main feature of these systems that they include energy exchange with external sources. These are no longer Hamiltonian, and the solitons in these systems are also qualitatively different from those in Hamiltonian systems. However, solitons, when they exist, can again be considered as "modes" of these systems. A rough classification of systems admitting soliton solutions is presented in Fig. 1.

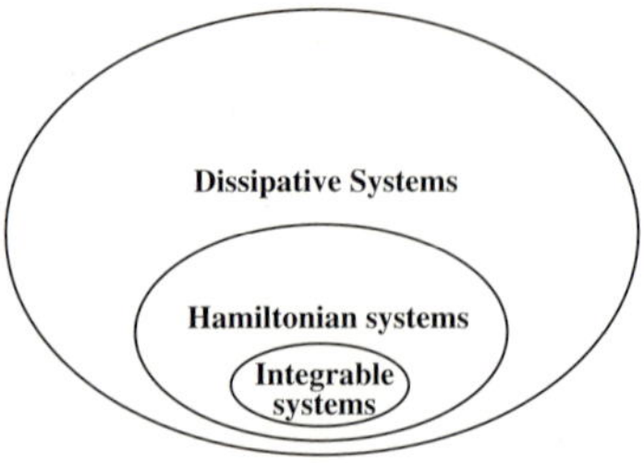

Fig. 1. A rough classification of nonlinear systems with an infinite number of degrees of freedom admitting soliton solutions. Hamiltonian systems can be considered as a subclass of dissipative systems, while integrable systems can be viewed as a subclass of Hamiltonian systems. In the particular case of (1), the system is Hamiltonian when $\delta = \epsilon = \beta = \mu = 0$. The system is integrable, when, in addition, $\nu = 0$

Many nonequilibrium phenomena, such as convection instabilities [2], binary fluid convection [3], and phase transitions [4], can be described by the complex Ginzburg–Landau equation (CGLE). In optics, this equation (or a generalization of it) describes the essential features of processes in lasers [5–9], optical parametric oscillators [10] , spatial-soliton lasers [11,12], Fabry–Perot cavities filled with nonlinear material and driven by an external field [13–15], free-electron laser oscillators [16], and all-optical transmission lines [17]. Planar soliton systems with gain [18,19] or with light-guiding-light phenomena [20] are also particular examples which can be described by the CGLE. This equation is essentially the nonlinear Schrödinger equation (NLSE) with gain and loss , where both gain and loss are frequency- and intensity-dependent.

Laser systems are usually described by a combined set of Maxwell–Bloch equations. In laser systems, a soliton propagates in the presence of the periodic loss and amplification that occur during each round trip. However, in certain limits, these equations can be reduced to the complex Ginzburg–Landau equation [6–9,21,22]. Solitons in passively mode-locked lasers appear either in time, as ultrashort pulses [23], or in space, as transverse modes [6–9,11,24]. Here there is a linear gain element (an erbium-doped fiber amplifier, for example), as well as nonlinear and dispersive elements, and losses. In addition, a soliton laser incorporates some element with an intensity-dependent transmission function, such as a Kerr cell [11] or a semiconductor saturable absorber [25]. To describe this element mathematically, a nonlinear loss must be introduced into the averaged equation [26].

There is a significant difference between solitons in Hamiltonian and dissipative systems. In Hamiltonian systems, soliton solutions appear as a result of a balance between diffraction (or dispersion) and nonlinearity. Diffraction spreads the beam, while nonlinearity focuses it and makes it narrower. The balance between the two results in a stationary solution, which usually forms a one-parameter family. In systems with gain and loss, in order to have stationary solutions, gain and loss must also be balanced. This additional balance results in solutions which are fixed. The shape, amplitude, and width are all fixed and depend on parameters of the equation. This situation is presented qualitatively in Fig. 2. The rigidity of the soliton may provide efficient suppression of noise and stop any drift in the soliton parameters.

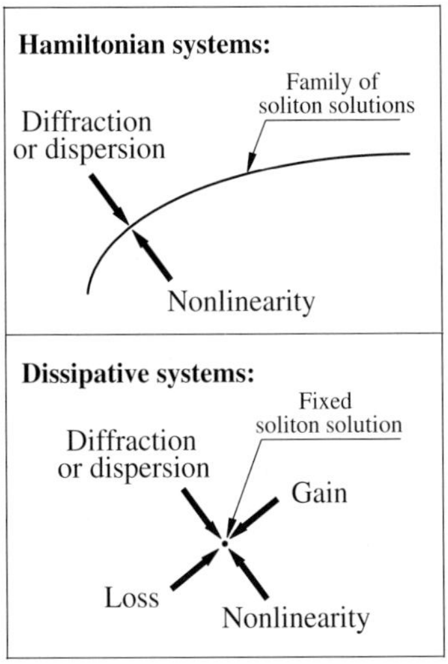

Fig. 2. Qualitative difference between the soliton solutions in Hamiltonian and dissipative systems. In Hamiltonian systems, soliton solutions are the result of a single balance, and comprise one- or few-parameter families, whereas, in dissipative systems, the soliton solutions are the result of a double balance and, in general, are isolated. There can be exceptions to this rule [27–29], but, usually, the solutions are fixed (i.e. isolated from each other). On the other hand, it is quite possible for several isolated soliton solutions to exist for the same equation parameters. This is valid for (1 + 1)-dimensional as well as for (2 + 1)-dimensional cases. In the latter case, the terms "localized structures" [12], "bullets" [13,14], and "patterns" [21] are also used along with the term "solitons" [15]

Another simple qualitative picture is presented in Fig. 3. In order to be stationary, solitons in dissipative systems need to have regions where they extract energy from an external source, as well as regions where energy is dissipated to the environment. A stationary soliton is the result of a dynamic process of continuous energy exchange with the environment and of redistribution of energy between various parts of the soliton. Hence the soliton, by itself, is an object which is far from equilibrium. In this sense, it is more like a living thing than an object of the inanimate world. It is like a species in biology which is fixed (or isolated) in its properties.

In one of the forms used in nonlinear optics, the quintic CGLE can be written as

$$\mathrm{i}\psi_\xi + \frac{D}{2}\psi_{\tau\tau} + |\psi|^2\psi = \mathrm{i}\delta\psi + \mathrm{i}\epsilon|\psi|^2\psi + \mathrm{i}\beta\psi_{\tau\tau} + \mathrm{i}\mu|\psi|^4\psi - \nu|\psi|^4\psi \;, \qquad (1)$$

where τ is the normalized transverse spatial coordinate (or retarded time, in temporal problems), ξ is the normalized propagation distance or the normalized number of round trips, and ψ is the complex envelope of the electric

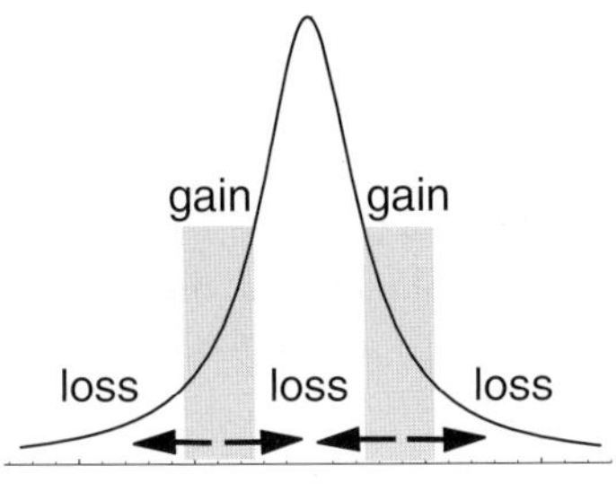

Fig. 3. Qualitative description of a soliton in a dissipative system. The soliton has areas of consumption as well as dissipation of energy which can be both dependent on frequency (spatial or temporal) and intensity. The *arrows* show the energy flow across the soliton. The soliton is a result of complicated dynamical processes of energy exchange with the environment and between its own parts

field. The coefficients δ, β, ϵ, μ, and ν are real constants (we do not require them to be small). The physical meaning of these quantities is the following: δ is the linear gain (or loss) at the (spatial or temporal) central frequency, β describes spatial or temporal spectral filtering ($\beta > 0$) and is sometimes called the "diffusion coefficient" [12], ϵ represents nonlinear gain/absorption processes (a particular example is two-photon absorption), and μ is a higher-order correction to the nonlinear amplification/absorption. The term containing ν is a higher-order correction term to the nonlinear refractive index. The parameter D determines the lowest order diffraction or group-velocity dispersion . In problems related to pulse propagation, positive D corresponds to anomalous dispersion and negative D to normal dispersion.

Equation (1) has been written in such a way that if the right-hand side of it is set to zero, we obtain the standard NLSE. For spatial solitons in wide-aperture laser cavities, the form of the coefficients in this equation can be different [11]. The equation can also have additional terms related to a finite aperture [11] and to other forms of local [12] or nonlocal [30,31] nonlinearity. The theory of phase-sensitive amplification (or "parametrically amplified" optical systems) [32] also uses a different form of the CGLE, which is called a "parametric Ginzburg–Landau equation" . In this review, we shall retain the form presented above as the basic one, as it gives the main properties of solitons in dissipative systems. We shall also concentrate on the $(1+1)$-dimensional case, as it is the fundamental case which allows us to understand some of the features of solitons in $(2+1)$-dimensional cases.

Equation (1) is nonintegrable, and only some particular exact solutions can be obtained. In general, initial-value problems with arbitrary initial conditions can only be solved numerically. The cubic CGLE, obtained by setting $\mu = \nu = 0$ in (1), has been studied extensively [33–41]. Exact solutions to this equation can be obtained using a special ansatz [34], the Hirota bilinear method [40], or reduction to systems of linear PDEs [42]. However, it was realized many years ago that the soliton-like solutions of this equation are unstable.

The case of the quintic CGLE has been considered in a number of publications, using numerical simulations, perturbative analysis, and analytic solutions. Originally, this equation was used mainly as a model for binary fluid convection [43–45]. The existence of soliton-like solutions of the quintic CGLE in the case $\epsilon > 0$ has been demonstrated numerically [44,45]. A qualitative analysis of the transformation of the regions of existence of the soliton-like solutions when the coefficients on the right-hand-side change from zero to infinity has been given in [46]. An analytic approach, based on the reduction of (1) to a three-variable dynamical system, which allows us to obtain exact solutions for the quintic equation, has been developed in [28,47]. The most comprehensive mathematical treatment of the exact solutions of the quintic CGLE, using Painlevé analysis and symbolic computations, is given in [48]. The general approach used in that work is the reduction of the differential equation to a purely algebraic problem. The solutions include solitons, sinks, fronts, and sources . The great diversity of possible types of solutions requires a careful analysis of each class of solutions separately. In this relatively short review, we concentrate solely on soliton-like solutions.

2 Balance Equations and Perturbation Theory

The CGLE has no known conserved quantities. Instead, we consider the energy (power for spatial solitons) associated with a solution ψ, $Q = \int_{-\infty}^{\infty} |\psi|^2 \, \mathrm{d}\tau$, and its rate of change with respect to ξ [49]:

$$\frac{\mathrm{d}}{\mathrm{d}\xi} Q = F[\psi] \,, \tag{2}$$

where the functional $F[\psi]$ is given by

$$F[\psi] = 2 \int_{-\infty}^{\infty} \left(\delta |\psi|^2 + \epsilon |\psi|^4 + \mu |\psi|^6 - \beta |\psi_\tau|^2 \right) \, \mathrm{d}\tau \,. \tag{3}$$

Similarly [49], the momentum is $M = \mathrm{Im} \left(\int_{-\infty}^{\infty} \psi_\tau^* \psi \, \mathrm{d}\tau \right)$, and its rate of change is defined by

$$\frac{\mathrm{d}}{\mathrm{d}\xi} M = J[\psi] \,, \tag{4}$$

where the real functional $J[\psi]$ is given by

$$J[\psi] = 2 \ \mathrm{Im} \int_{-\infty}^{\infty} \left[(\delta + \epsilon |\psi|^2 + \mu |\psi|^4) \psi + \beta \psi_{\tau\tau} \right] \ \psi_\tau^* \ \mathrm{d}\tau \,. \tag{5}$$

By definition, this functional is the force acting on a soliton along the τ-axis. There are only two rate equations, viz. (2) and (4), which can be derived for the CGLE. They can be used for solving various problems related to CGLE solitons. Two examples will be given in this review.

If the coefficients δ, β, ϵ, μ and ν on the right-hand side are all small, then soliton-like solutions of (1) can be studied by applying perturbation theory to the soliton solutions of the NLSE. Such an analysis of the solitons of the cubic–quintic CGLE in the NLSE limit has often been used [34,43,50–53]. Following this approach, let us consider the right-hand side of (1), with $D = +1$, as a small perturbation and write the solution as a soliton of the NLSE, viz.

$$\psi(\tau, \xi) = \frac{\eta}{\cosh[\eta(\tau + \Omega\xi)]} \exp[-\mathrm{i}\Omega\tau + \mathrm{i}(\eta^2 - \Omega^2)\xi/2] \,. \tag{6}$$

In the presence of the perturbation, the parameters of the soliton, viz. the amplitude η and frequency (or velocity) Ω, change adiabatically. The equations for them can be obtained from the balance equations for the energy and momentum. Using (6) and (2), we have the equation for the evolution of $\eta(\xi)$, which is proportional to the soliton energy:

$$\frac{\mathrm{d}\eta}{\mathrm{d}\xi} = 2\eta\left[\delta - \beta\Omega^2 + \frac{1}{3}(2\epsilon - \beta)\eta^2 + \frac{8}{15}\mu\eta^4\right] . \tag{7}$$

Similarly, using (6) and the equation for the momentum (4), we can derive the equation for $\Omega(z)$:

$$\frac{\mathrm{d}\Omega}{\mathrm{d}\xi} = -\frac{4}{3}\,\beta\Omega\eta^2 . \tag{8}$$

The dynamical system of equations (7) and (8) has two real dependent variables and the solutions can be represented on a plane. An example is given in Fig. 4. The solution has a line of singular points at $\eta = 0$, and, depending on the equation parameters, may have one or two singular points on the semi-axis $\Omega = 0$, $\eta > 0$. The values of η^2 for the singular points are defined by finding the roots of the biquadratic polynomial in the square brackets in (7). When the roots are negative (and hence η is imaginary), there are no singular points and hence no soliton solution. If both roots of the quadratic polynomial (in η^2) are positive (so that both values of η are real), then there are two fixed points and two corresponding soliton solutions. Both roots are positive when either $\beta < 2\epsilon$, $\mu < 0$, and $\delta < 0$, or $\beta > 2\epsilon$, $\mu > 0$, and $\delta > 0$. The stability of at least one these fixed points requires $\beta > 0$. Moreover, the stability of the background requires $\delta < 0$. In the latter case we necessarily have $\beta < 2\epsilon$ and $\mu < 0$, and the upper fixed point is a sink (as shown in Fig. 4) which defines the parameters of a stable approximate soliton solution of the quintic CGLE. The background $\psi = 0$ is also stable, so that the whole solution (soliton plus background) is stable. Finally, when only one of the roots is positive, there is a singular point in the upper half-plane and there is a corresponding soliton solution. However, either the background or the soliton itself is unstable, so that the total solution is unstable. The term containing ν in the CGLE does not influence the location of the sink. It only introduces an additional phase term, $\exp(8\mathrm{i}\nu\eta^4\xi/15)$, into the solution of (6).

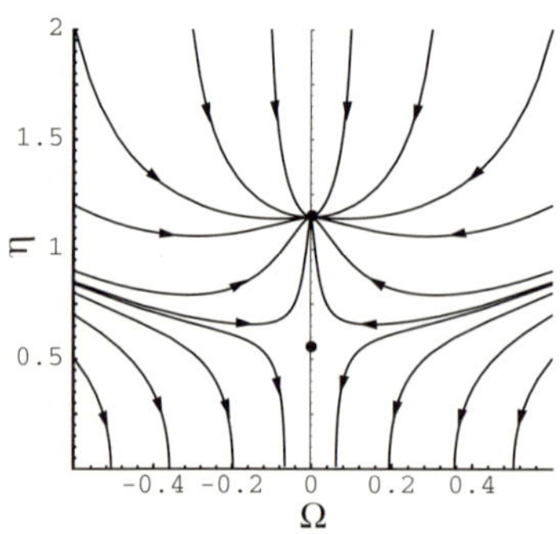

Fig. 4. The phase portrait of the dynamical system (7) and (8) for $\delta = -0.03$, $\beta = 0.1$, $\epsilon = 0.2$, and $\mu = -0.11$. The upper fixed point is a sink which defines the parameters of a stable, approximate, soliton-like solution of the quintic CGLE. Any soliton-like initial condition in close proximity to a fixed point will converge to a stable stationary solution. The points on the line $\eta = 0$ are stable when $\delta < 0$ and $\beta > 0$. This condition is needed for the background state $\psi = 0$ to be stable

In the case of the cubic CGLE, $\mu = 0$ and $\nu = 0$. The stationary point is then

$$\eta = \sqrt{3\delta/(\beta - 2\epsilon)}\,, \qquad \Omega = 0\,. \tag{9}$$

This point is stable provided that $\delta > 0$, $\beta > 0$, and $\epsilon < \beta/2$. Clearly, in this case the soliton and the background cannot be stable simultaneously. Hence, this approach shows that to make both the soliton and the background stable, we need to have quintic terms in the CGLE (see also [54]).

This simple perturbative approach shows that, in general, the CGLE has stationary soliton-like solutions, and that for the same set of equation parameters there may be two of these solutions simultaneously (one stable and one unstable). Moreover, this approach shows that the soliton parameters are *fixed* as depicted in Fig. 2. This occurs because the dissipative terms in (1) break the scale invariance associated with a conservative system.

Despite its simplicity and advantages in giving the stability and other properties of solitons, the perturbative analysis has some serious limitations. Firstly, it can be applied only if the coefficients on the right-hand side of (1) are small, and this is not always the case in practice. Correspondingly, it describes the convergence correctly only for initial conditions which are close to the stationary solution. Secondly, the standard perturbative analysis cannot be applied to the case $D < 1$, when the NLSE itself does not have bright-soliton solutions. However, (1) has stable soliton solutions for this case as well.

3 Exact Solutions

Various types of soliton-like solutions, described by a special ansatz, have been presented and classified in [55]. Owing to the restrictions imposed by the ansatz, these solutions do not cover the whole range of parameters, but they can serve as a basis for further generalizations. Firstly, we consider the stationary solutions of (1) that have zero transverse velocity. This occurs when $\beta \neq 0$. The solution can be written in the form

$$\psi(\tau, \xi) = A(\tau)\exp(-\mathrm{i}\omega\xi)\,, \tag{10}$$

where ω is a real constant. The complex function $A(\tau)$ can always be written in the following explicit form:

$$A(\tau) = a(\tau)\exp[\mathrm{i}\phi(\tau)]\,, \tag{11}$$

where a and ϕ are real functions of τ. Moreover, it is assumed that

$$\phi(\tau) = \phi_0 + d\ln[a(\tau)]\,, \tag{12}$$

where d is the chirp parameter and ϕ_0 is an arbitrary phase; we can take $\phi_0 = 0$ for simplicity. For the cubic case, this ansatz covers all soliton-like solutions. In the quintic case, however, (12) is a restriction imposed on $\phi(\tau)$, because the chirp could have a more general functional dependence on τ. This ansatz allows us to find some classes of solutions in analytical form.

After some involved transformations, the equation for $a(\tau)$ is found to be the following:

$$(a')^2 + \frac{2\nu}{\chi} a^6 + \frac{2(2\beta - \epsilon)}{3d\lambda^2} a^4 - \frac{\delta}{d - \beta + \beta d^2} a^2 = 0 \,, \tag{13}$$

where, here and henceforth, $\chi = 8\beta d - d^2 + 3$ and $\lambda = \sqrt{1 + 4\beta^2}$. The parameter d is expressed, in forms of β and ϵ only, as

$$d = d_{\pm} = \frac{3(1 + 2\epsilon\beta) \pm \sqrt{9(1 + 2\epsilon\beta)^2 + 8(\epsilon - 2\beta)^2}}{2(\epsilon - 2\beta)} \,. \tag{14}$$

The expression for d is the same for both the cubic and the quintic CGLE. The expression for ω is the following:

$$\omega = -\frac{\delta(1 - d^2 + 4\beta d)}{2(d - \beta + \beta d^2)} \,. \tag{15}$$

If both coefficients μ and ν are nonzero, then

$$2\nu/\chi = \mu/(3\beta - 2d - \beta d^2) \tag{16}$$

gives the relation between the four parameters ϵ, β, μ, and ν when a solution exists in the form of (11) and (12). Equation (13) is an elliptic equation, and its solutions are relatively simple.

3.1 Solutions of the Cubic CGLE

Firstly, we consider the cubic CGLE, that is, (1) with $\nu = \mu = 0$. Then (13) reduces to

$$(a')^2 + \frac{2(2\beta - \epsilon)}{3d\lambda^2} a^4 - \frac{\delta}{d - \beta + \beta d^2} a^2 = 0 \,, \tag{17}$$

which has the solution [34,39]

$$a(\tau) = BC \operatorname{sech}(B\tau) \,, \quad C = \sqrt{\frac{3d\lambda^2}{2(2\beta - \epsilon)}} \,, \quad B = \sqrt{\frac{\delta}{d - \beta + \beta d^2}} \,, \tag{18}$$

and d is given by (14), where the minus sign is chosen in front of the square root. The other value of d leads to an unphysical solution, as the expression under the square root for C then becomes negative. An important feature of the solution (18) is that its amplitude and width depend specifically on the

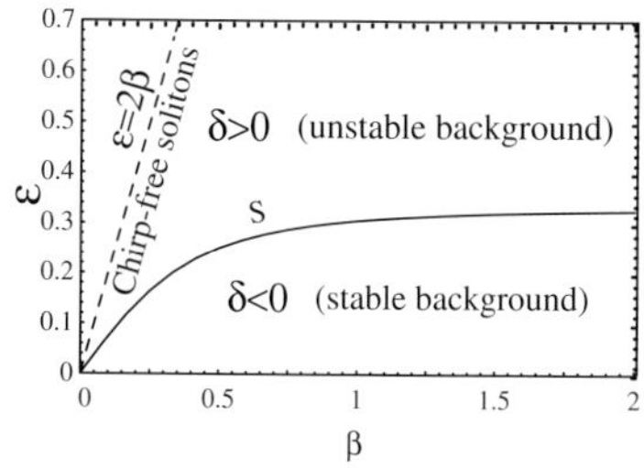

Fig. 5. The curve S given by (19) on the (β, ϵ) plane where the solutions with fixed amplitude, (18) and (25), become singular, and where the classes of special solutions with arbitrary amplitude, (20) and (29), exist. This plot applies to both the cubic and the quintic cases

parameters of the equation. In other words, (18) is a solution with a fixed amplitude.

The solution (18) depends on three parameters: δ, β, and ϵ. However, the parameter δ appears only in the expression for B. It can be seen that a variation of δ leads merely to a rescaling of the soliton amplitude and width. So, in what follows, we analyze the solution (18) on the (β, ϵ) plane. To find the range of existence of the solution (18), we note that, on the (β, ϵ) plane, the denominator in the expression for B is positive below the curve S (see Fig. 5) given by

$$\epsilon_S = \beta\,(3\lambda - 1)/(4 + 18\beta^2)\,, \tag{19}$$

and negative above it. Hence, for the solution (18) to exist, the value of δ must be positive below S and negative above it. As this solution exists almost everywhere on the (β, ϵ) plane, we call it the general solution. S itself is the line where this solution becomes singular, i.e. its amplitude BC tends to infinity, while the width $1/B$ vanishes.

From numerical simulations [55], it follows that S separates the regions of stable and unstable solitons in the (β, ϵ) plane. Thus, the solution (18) exists and is stable below S for $\delta > 0$. This is presented schematically in Fig. 5. We note, from perturbation theory, that the solution is stable provided $\delta > 0$ and $\epsilon < \beta/2$ [52]. Recall that perturbation theory can be applied only for $|\delta|, |\beta|, |\epsilon| \ll 1$. S has two limits, $\epsilon \approx \beta/2$ for $\beta \ll 1$ and $\epsilon \to 1/3$ for $\beta \gg 1$, so, at small β, it coincides with the stability threshold given by perturbation theory (9).

For positive δ, the background state ($\psi = 0$) is unstable. If the initial condition is close to the exact solution (18) and $\delta \ll 1$, this instability develops slowly and the soliton can propagate for distances up to $\xi_0 \sim \delta^{-1}$. Beyond that distance, radiation waves, growing linearly from the noise, become appreciable and can distort the soliton itself. The general conclusion is that either the soliton itself or the background state is unstable at any point in the (β, ϵ) plane. This means that the total solution is always unstable. We need to emphasize the importance of the curve S. For the solutions with fixed amplitude, it gives the range of existence, singularity, and stability. Moreover, another significant class of solutions exists on this line.

3.1.1 The Solution with Arbitrary Amplitude

The solution (18) does not exist on the curve given by (19). However, if we also impose the condition $\delta = 0$, then a new solution, valid only on the curve given by (19), can be found:

$$a(\tau) = GF\,\mathrm{sech}(G\tau)\,. \tag{20}$$

Here G is an arbitrary positive parameter, and d, ω and F are given by

$$d = \frac{\lambda - 1}{2\beta}\,, \qquad \omega = -\,d\,\frac{\lambda^2}{2\beta}\,G^2\,, \qquad F = \sqrt{\frac{d\lambda}{2\epsilon_S}} = \frac{1}{\beta}\sqrt{\frac{(2+9\beta^2)\lambda\,(\lambda-1)}{2(3\lambda-1)}}\,,$$

where $\lambda = \sqrt{1+4\beta^2}$, as before.

Arbitrary-amplitude solitons (20) exist because, when $\delta = 0$, the cubic CGLE becomes invariant relative to the scaling transformation $\psi \to G\psi$, $\tau \to G\tau$, $\xi \to G^2\xi$. Hence, if we know a particular solution of this equation, then the whole family can be generated using this transformation. The class of solutions (20) can be considered as a limiting case of the solution with fixed amplitude (18) when $\delta \to 0$ and, simultaneously, ϵ and β are moving towards S. At the line of singularity, S, the amplitude–width product C remains finite for the general solution (18) and has a finite limit on the line. The values C and F coincide on S.

The class of arbitrary-amplitude solutions is stable relative to small perturbations at any point of S. The most important feature of these solutions is that the background state, $\psi = 0$, is also stable, because $\delta = 0$. This means that, by removing the linear gain from the system and imposing a particular relation between the coefficients ϵ and β, we can achieve stable propagation of these solitons on a stable background. It is interesting to note that this class of solutions is the only family of stable bright solitons in the cubic model.

3.1.2 Chirp-Free Soliton

Apart from at the singularity on S, the solution (18) does not apply on the line $\epsilon = 2\beta$, as C then becomes indeterminate ($d \to 0$ when $\epsilon \to 2\beta$). However, the soliton amplitude remains finite near this line on the (β, ϵ) plane for finite, fixed δ. It follows from (14) that, for $\epsilon = 2\beta$, the chirp parameter d is 0, and, from (15), that $\omega = \delta/2\beta$. The solution to equation (13) then becomes

$$a(\tau) = \sqrt{-\frac{\delta}{\beta}}\,\mathrm{sech}\left(\sqrt{-\frac{\delta}{\beta}}\,\tau\right)\,. \tag{21}$$

The coefficients δ and β must have opposite signs for this solution to exist. As $d = 0$, the solution (21) has no chirp, in contrast to other soliton solutions of the CGLE. This occurs because of the special choice of the coefficients. In this case the complex constant $(1 - 2\mathrm{i}\beta)$ can be factored out of (1) when it is reduced to an ordinary differential equation in terms of $a(\tau)$. The soliton itself is unstable, as solutions of this type are located on the (β, ϵ) plane above S (see Fig. 5).

3.2 Dark-Soliton Solution of the Cubic Equation

For the cubic CGLE, dark solitons can also be written in analytic form. Thus, we consider solutions of (1) with $\mu = \nu = 0$. The analytic solution has been found in [27] and refined in [28,56]. It is a traveling solution of (1) with transverse velocity V:

$$\psi(\tau, \xi) = \left[\kappa u \tanh(\kappa\zeta) + (d + 2\mathrm{i})\, \frac{w}{d}\right] \exp[\mathrm{i}\phi(\zeta)] \exp(\mathrm{i}K\zeta - \mathrm{i}\Omega\xi) \,, \quad (22)$$

where u, w, d, κ, K, and Ω are real constants, and $\phi(\zeta) = d \ln[\cosh(\kappa\zeta)]$ is a real function of $\zeta = \tau - V\xi$. The phase modulation parameter, d, is given by

$$d = \frac{3\,(2\epsilon\beta + 1) + \sqrt{9\,(2\epsilon\beta + 1)^2 + 8\,(2\beta - \epsilon)^2}}{2\,(2\beta - \epsilon)} \,.$$

The parameter u^2 depends on ϵ and β only, because d depends on ϵ and β:

$$u^2 = \frac{3d\lambda^2}{2(2\beta - \epsilon)} = \frac{\lambda^2(d^2 - 2)}{2(1 + 2\epsilon\beta)} \,. \quad (23)$$

Equation (22) represents a one-parameter family of solutions, contrary to the usual rule. This occurs owing to the presence of a hidden symmetry in the solution [28].

It is most convenient to take κ as the independent parameter. Then the other parameters of the solution (22) will depend on κ. Equations for these parameters and a plot of all these parameters versus κ are given in Fig. 6.

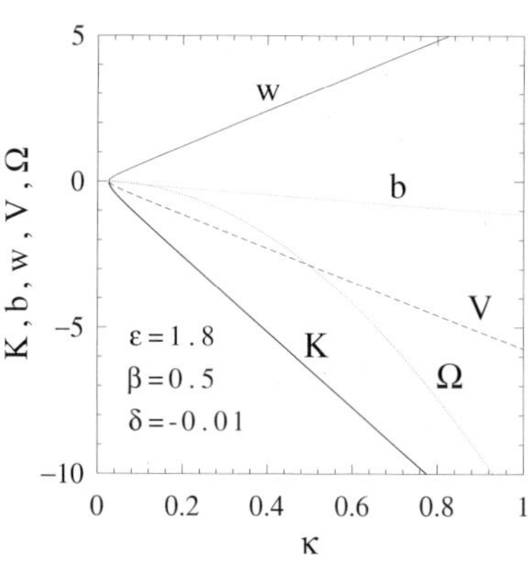

Fig. 6. Parameters in the solution (22) representing the family of gray dark solitons, versus κ. In general,

$$\Omega = \frac{\delta}{2\beta} + \frac{2\beta - \epsilon}{2\epsilon} \left(K^2 - \kappa^2 d^2 + \frac{\delta}{\beta}\right),$$

$$K = (\epsilon u/\beta)\ (w/d) \,, \quad V = K(\epsilon - 2\beta)/\epsilon \,,$$

$$\left(\frac{w}{d}\right)^2 = \beta\, \frac{\kappa^2(3d + 4\beta) - 2\delta}{3\epsilon(d + 4\beta)} \,.$$

For the parameters given in the plot, we have $w = V = K = 0$ when $\kappa = 0.02579$. This point corresponds to the black soliton

Note that the rest of the parameters, apart from Ω, are zero at finite κ. This is one of the reasons for taking κ as the arbitrary, independent parameter. A special case, the black soliton, which has been found in [40], is especially simple. Setting $K = w = V = 0$ in the above equations, we then have fixed values of κ^2 and Ω, viz. $\kappa^2 = 2\delta/(3d + 4\beta)$ and $\Omega = \kappa^2(1 - 3\beta d)$.

Stationary solitons are not the only solitons that exist for the cubic CGLE. An exact solution for a soliton which periodically changes its shape during propagation has been obtained in [41]. Numerical simulations [67] show that pulsating solutions also exist for other forms of the CGLE.

3.3 Solitons of the Quintic CGLE

3.3.1 Relation Between Coefficients

Soliton solutions of the quintic CGLE exist for a wide range of values of the coefficients β, ϵ, μ, and ν. The ansatz (12) is the condition that restricts this range. At least one of the coefficients μ or ν must be nonzero, and the solutions can be expressed in terms of β, ϵ, and ν. Using (16), the solutions can, alternatively, be expressed in terms of β, ϵ, and μ.

3.3.2 Solutions with Fixed Amplitude

By using the substitution $f = a^2$, we can rewrite (13) in the form

$$(f')^2 + \frac{8\nu}{\chi} f^4 + \frac{8(2\beta - \epsilon)}{3d\lambda^2} f^3 - \frac{4\delta}{d - \beta + \beta d^2} f^2 = 0 \,. \tag{24}$$

This is again an elliptic-type differential equation. Bounded soliton-like solutions exist only if $4\delta/(d - \beta + \beta d^2) > 0$. The positive solution of (24) is

$$f(\tau) = \frac{2 f_1 f_2}{(f_1 + f_2) - (f_1 - f_2)\cosh(2\alpha\sqrt{f_1|f_2|}\,\tau)} \,, \tag{25}$$

where

$$\alpha = \sqrt{\left|\frac{2\nu}{\chi}\right|} = \sqrt{\left|\frac{\mu}{3\beta - 2d - \beta d^2}\right|} \,, \tag{26}$$

and f_1 and f_2 are the roots of the quadratic equation:

$$\frac{2\nu}{\chi} f^2 + \frac{2(2\beta - \epsilon)}{3d\lambda^2} f - \frac{\delta}{d - \beta + \beta d^2} = 0 \,, \tag{27}$$

viz.

$$f_{1,2} = \frac{\chi}{6d\nu\lambda^2}\left([-(2\beta - \epsilon) \pm \sqrt{(2\beta - \epsilon)^2 + \frac{18\,\delta\, d^2\nu\lambda^4}{\chi(d - \beta + \beta d^2)}}\right) \,. \tag{28}$$

Clearly, one of the roots (we choose f_1) must be positive. The second one can have either sign. If it is also positive, we choose $f_1 < f_2$. More detailed analysis is given in [49].

3.3.3 The Solution with Arbitrary Amplitude

A singularity appears when $d - \beta + \beta d^2 = 0$. This occurs on the same curve S in the (β, ϵ) plane as in the cubic case (19). The singularity exists when the roots f_1 and f_2 have opposite signs. If β and ϵ satisfy (19) and $\delta = 0$, a class of soliton solutions with arbitrary amplitude exists. This class is described by:

$$f(\tau) = 3d\lambda^2 P \Big/ \Big[(2\beta - \epsilon) + D\cosh(2\sqrt{P}\,\tau)\Big], \tag{29}$$

where $D = \sqrt{(2\beta - \epsilon)^2 + 18\, d^2 \nu \lambda^4 P/\chi}$, and P is an arbitrary positive parameter. The values of d and ω are given by $d = (\lambda - 1)/(2\beta)$, $\omega = -\, d\lambda^2 P/(2\beta)$. This class of solutions is stable at any point on the special curve S and for any value of P in (29). The background state is also marginally stable, as $\delta = 0$. These solitons are the only analytic solutions which are stable for their whole range of parameters.

3.3.4 Flat-Top Solitons

The soliton (25) becomes wider and flatter as the two positive roots approach each other. When $f_1 = f_2$, the soliton splits into two fronts with zero velocity. Each of them can be written in the form

$$f(\tau) = f_1/[1 + \exp(\pm\alpha f_1 \tau)]\ , \quad f_1 = \chi(\epsilon - 2\beta)/\big(6d\nu\lambda^2\big)\ , \tag{30}$$

and the sign in (30) determines the orientation of the front. The two roots f_1 and f_2 become identical when $(2\beta - \epsilon)^2 = -18\,\delta\nu d^2\lambda^4/[(d - \beta + \beta d^2)\chi]$. This condition involves all the parameters of the equation. Depending on δ and ν, this condition can be satisfied at any point in the (β, ϵ) plane. The transition from the general solution (25) to the flat-top solution (30) as $f_1 \to f_2$ is shown in Fig. 7.

If $f_1 = f_2$ exactly, the width of the soliton goes to infinity and the soliton is described by two fronts. In the region of nonzero intensity, the soliton phase $\phi(\tau)$ tends to a constant value exponentially. So, if we combine the two fronts with opposite orientations in (30) to form a wide rectangular soliton of finite width, then the influence of each front on the other is exponentially small. Stable stationary flat-top solitons have been observed experimentally in binary fluid convection [57].

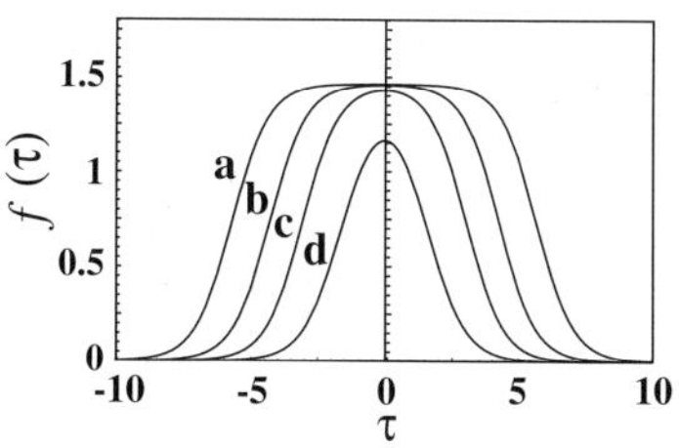

Fig. 7. The shapes of the solutions (25) when the two roots f_1 and f_2 are close to each other. The top of the soliton becomes flatter as the roots approach each other. The parameters chosen are $\beta = 0.5$, $\nu = -0.5$, $\delta = -0.1$, and $d = d_-$, which gives $\mu = -0.227$, while ϵ is (a) 0.4005969, (b) 0.4006, (c) 0.4007, (d) 0.41

3.3.5 Rational Solution

If $\delta = 0$ and (β, ϵ) is not located on the curve (19), then $\omega = 0$, and the solution has the form of a Lorentz function,

$$f(\tau) = \frac{f_1}{1 + k_0 f_1^2 \tau^2}, \quad f_1 = \frac{\chi(\epsilon - 2\beta)}{3d\nu\lambda^2}, \tag{31}$$

where $k_0 = 2\nu/\chi$. The values of f_1 and k_0 must be positive, which restricts the allowed values of the coefficients of the equation.

The rational soliton is unstable over the whole range of parameters in which it exists. The rational solution represents a special, weakly localized limit of the solution with fixed amplitude (25). Rational solitons exist, and play an important role, in other integrable and nonintegrable systems, including the NLSE and its generalizations.

3.3.6 Chirp-Free Soliton

Another degenerate case occurs when $\epsilon = 2\beta$ and $\mu = 2\beta\nu$. The solution to (24) then reduces to

$$f(\tau) = \frac{-2\delta}{\beta}\left[1 - \sqrt{1 - \frac{8\delta\nu}{3\beta}}\cosh\left(2\sqrt{-\frac{\delta}{\beta}}\tau\right)\right]^{-1}. \tag{32}$$

Clearly, this solution exists when δ/β is negative and ν positive. Here $d = 0$, so this solution to the CGLE has no phase chirp. This solution arises because the coefficients of the equation are chosen in such a way that a complex constant can be factored out from (1) when it is reduced to an ordinary differential equation in terms of $f = a^2$. The latter then becomes purely real. Numerical simulations show that chirp-free solitons are unstable for all values of the parameters.

If $\beta = 0$, then it is possible to have solitons with nonzero velocity (i.e. traveling solitons). These solutions can be obtained using a simple transformation, because, for $\beta = 0$, (1) has an additional symmetry – it is invariant relative to a Galilean transformation [49]. As a result, traveling soliton-like solutions can be obtained from zero-velocity ones using the Galilean transformation. Hence, we can use the above fixed-amplitude solution, put $\beta = 0$, and use the Galilean transformation to obtain the whole family of traveling solitons. If $\beta \neq 0$, then we have to use other methods to find solitons with nonzero velocity.

3.4 Reduction to a Set of Ordinary Differential Equations

Analytic solutions of the quintic CGLE can be found only for certain combinations of parameters. In other regions, a variational approach [58], other approximate methods [59], or numerics must be used to study the solutions. One way to find stationary solutions of (1) numerically is to reduce it to a set of ordinary differential equations (ODEs). Thus, we seek solutions in the form

$$\psi = a(t)\,\exp[\mathrm{i}\phi(t) - \mathrm{i}\omega\xi]\;, \tag{33}$$

where a and ϕ are real functions of $t = \tau - v\xi$, v is the soliton velocity and ω is the nonlinear shift of the propagation constant. Substituting (33) into (1), we obtain an equation for two coupled functions, a and ϕ. Separating real and imaginary parts, we obtain, after some algebra, the following set of ODEs:

$$\begin{aligned}
\frac{\lambda^2}{2}M' &= (D\delta - 2\beta\omega)a^2 + (D\epsilon - 2\beta)a^4 + (D\mu - 2\beta\nu)a^6 - 2\beta vM + Dvay\;,\\
y' &= \frac{M^2}{a^3} - \frac{2(D\omega + 2\beta\delta)}{\lambda^2}a - \frac{2(D + 2\beta\epsilon)}{\lambda^2}a^3 - \frac{2\theta}{\lambda^2}a^5 - \frac{4\beta v}{\lambda^2}y - \frac{2Dv}{\lambda^2}\frac{M}{a}\;,\\
a' &= y\;,
\end{aligned} \tag{34}$$

where $\theta = D\nu + 2\beta\mu$, $M = a^2\phi'$, and each prime denotes a derivative with respect to t. The solutions of this set contain all stationary and uniformly translating solutions. The parameters v and ω are the eigenvalues of (34). In the (M, a) plane, the solutions corresponding to solitons are closed loops starting and ending at the origin.

If we are only interested in zero-velocity ($v = 0$) solutions, (34) can be simplified:

$$\begin{aligned}
\frac{\lambda^2}{2}M' &= (D\delta - 2\beta\omega)a^2 + (D\epsilon - 2\beta)a^4 + (D\mu - 2\beta\nu)a^6\;,\\
y' &= \frac{M^2}{a^3} - \frac{2(D\omega + 2\beta\delta)}{\lambda^2}a - \frac{2(D + 2\beta\epsilon)}{\lambda^2}a^3 - \frac{2\theta}{\lambda^2}a^5\;,\\
a' &= y\;.
\end{aligned} \tag{35}$$

This set of first-order ODEs can be solved numerically. The asymptotic behavior of (35) at small a is given by

$$a = a_0\,\exp(g\tau)\;,\quad M = [(D\delta - 2\beta\omega)/(g\lambda^2)]a_0^2\,\exp(2g\tau)\;, \tag{36}$$

where the soliton tail exponent g can be found from the biquadratic equation

$$g^4 + \left[2(D\omega + 2\beta\delta)/\lambda^2\right]g^2 - (D\delta - 2\beta\omega)^2/\lambda^4 = 0\;. \tag{37}$$

Thus $g^2 = \pm\sqrt{\omega^2 + \delta^2}/\lambda - (D\omega + 2\beta\delta)/\lambda^2$. Using this approximation for the tails, it is possible to find the rest of the soliton solution with a shooting method. Examples are given below.

4 Stability Issues

An important issue is the stability of the exact solutions. Stability can be studied analytically near the NLSE limit [60,61], where perturbation theory is valid. Otherwise, stability has to be studied numerically [19,62]. Such an analysis includes the solution of the linearized problem, i.e. the calculation of the perturbation eigenmodes and their growth rates. A perturbed solution has to be written in the form

$$\psi(\tau,\xi) = [A_0(\tau) + \gamma g(\tau,\xi)] \exp(-\mathrm{i}\omega\xi) , \tag{38}$$

where $A_0(\tau)$ is the stationary solution (11) under study, γ is a small parameter, and $g(\xi,\tau)$ is a perturbation function. Inserting (38) into (1), and linearizing in the small parameter γ, we obtain

$$\begin{aligned} &\mathrm{i}g_\xi + \omega g + (1/2 - \mathrm{i}\beta)\, g_{\tau\tau} + 2|A_0|^2(1-\mathrm{i}\epsilon)g + A_0^2(1-\mathrm{i}\epsilon)g^* \\ &\quad + (\nu - \mathrm{i}\mu)(3|A_0|^4 g + 2|A_0|^2 A_0^2 g^*) = 0 . \end{aligned} \tag{39}$$

This equation for the perturbation function g may have many possible types of solutions. Moreover, because the linear operator in (39) is not Hermitian, its eigenvalues are, in general, complex.

Figure 8 shows, on a logarithmic scale, the growth rate of perturbation as a function of ϵ, for $(\beta,\nu) = (0.5, -0.5)$, with $\delta = -0.1$ and -0.001. Each curve is labeled with its value of δ. For each δ, there are two curves, corresponding to the two possible solutions associated with the values of d. For $|\delta| = 0.001$, we have the solid curve ($d = d_-$) and the dotted one ($d = d_+$), while for $|\delta| = 0.1$, the dot–dashed and dashed curves correspond to $d = d_-$ and $d = d_+$, respectively. Clearly, in all cases the solutions obtained for $d = d_+$ have much higher growth rates of the instability than those associated with d_-.

For negative δ (and $d = d_-$), at any given ϵ, the growth rate of the perturbation decreases as $|\delta|$ decreases. Another interesting feature of the growth rate curves in Fig. 8 is that, at $d = d_-$, the solution becomes more stable as we move toward its smallest allowed value of ϵ. This happens when

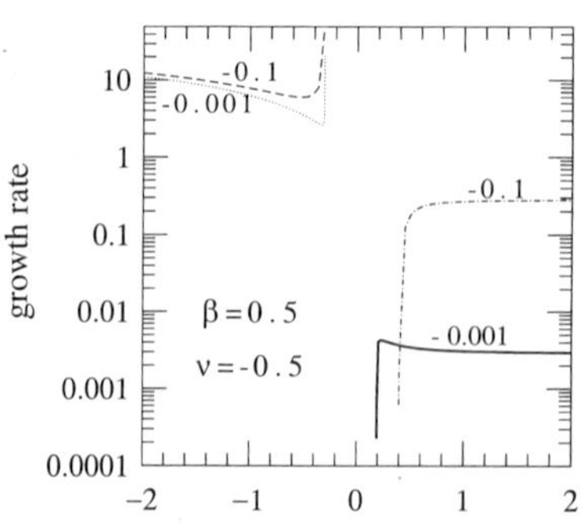

Fig. 8. The growth rate of the predominant perturbation eigenmode associated with the soliton solution, as a function of ϵ for $\beta = 0.5$, $\nu = -0.5$, and $\delta < 0$. The *solid curve* is for $|\delta| = 0.001$ and $d = d_-$, the *dot–dashed curve* is for $|\delta| = 0.1$ and $d = d_-$, the *dotted curve* is for $|\delta| = 0.001$ and $d = d_+$, and the *dashed curve* is for $|\delta| = 0.1$ and $d = d_+$. Each curve is labeled with its value of δ

f_1 approaches f_2. It can be shown that solutions obtained for negative δ and $d = d_-$ are stable when f_1 is close to f_2 (dot–dashed and dotted curves in Fig. 8). The transformation of the solitons into a pair of zero-velocity fronts occurs on these lines. Similar results for the stability of these solutions are valid for other values of β and ν.

Although exact solutions (25) to the quintic CGLE can be found when a particular relation between the parameters is satisfied, almost all of them are unstable. An exception appears in the vicinity of the boundary that separates solitons from pairs of fronts. The growth rate of perturbations of these soliton solutions falls to zero as this limit is approached. These stable solutions have flat tops, indicative of the transition from solitons to fronts.

4.1 Stable Solitons

The fact that exact solutions are not stable does not mean that stable solutions do not exist. They do exist but may not be expressible analytically. Figure 9 shows schematically the space of parameters of the CGLE and the regions where various solutions can be located. The first presentation of strictly stable soliton-like solutions was reported in [44]. The authors found some points in the parameter space where stable solitons exist. Rough estimates of the locations of the boundaries between fronts and soliton-like solutions of the CGLE have been given in [46]. However, there is no sharp boundary between the two classes of solutions. It has been found [28] that, for some values of the parameters, a variety of fronts and soliton-like solutions exist.

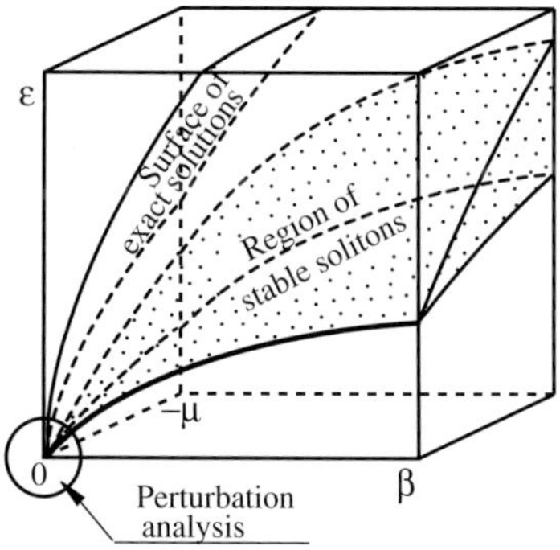

Fig. 9. Qualitative diagram of the parameter space of the quintic CGLE showing the locations of exact solutions, stable solitons, and solutions which can be found using perturbations of the NLSE. Note that the CGLE (1) has five parameters, but in the figure we can show only three of them. A five-dimensional space of parameters would give the complete picture

Thus, it is important to know the values of the coefficients where the quintic CGLE has stable soliton-like solutions. This must be the region of global stable-soliton propagation, i.e. the region in the parameter space where a broad class of initial conditions converge to a stationary soliton, which therefore represents a stable soliton-like solution of the quintic CGLE.

Comprehensive numerical results have been obtained in [62]. These give the values of the coefficients $(\delta, \beta, \epsilon, \mu, \nu)$ (i.e. the subspace of the parameter space) of the quintic CGLE where stable solitons exist. Stable solitons exist in a certain region, and it is interesting to compare this region with the region of lower dimensionality where the analytical solutions given by (25) exist.

The parameter space where stable solitons exist has certain limitations. The parameter β must be nonnegative, in order to stabilize the soliton in the frequency domain. The linear gain coefficient δ must be zero or negative to provide stability of the background. In this case, for $\mu = 0$, stable solitons can exist only for values of ϵ above the curve S. The parameter μ must be negative in order to stabilize the soliton against collapse. The parameter ν can have either sign.

Stable solitons can be found numerically from the propagation equation (1), taking a Gaussian shape of arbitrary amplitude and width as the initial condition. The shape of the initial condition appears to be of little importance. If the solution converges to a stationary solution, it can be viewed as a stable solution, and the chosen set of parameters can be deemed to belong to the class of those which permit the existence of solitons. Figure 10 shows three examples of the soliton solutions found with this method. By repeating these calculations systematically for other sets of parameters, it is possible to construct the regions in the parameter space where stable propagation of bounded solutions is possible.

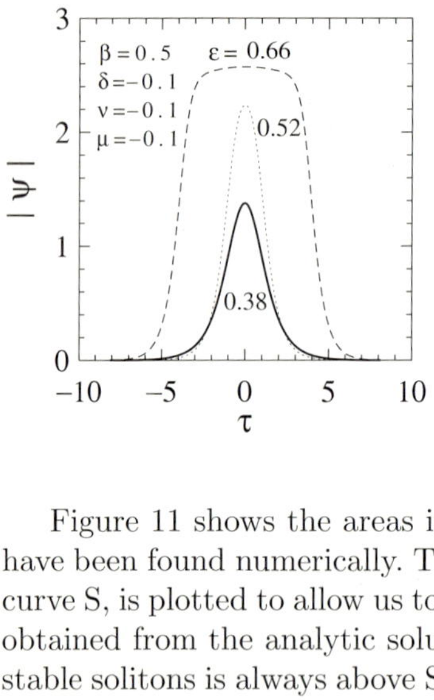

Fig. 10. Amplitude profile $|\psi|$ of soliton solutions found numerically for $\beta = 0.5$, $\nu = \mu = \delta = -0.1$, and ϵ equal to 0.38 (*solid curve*), 0.52 (*dotted*) and 0.66 (*dashed*). Such a solution is a "sink" in the sense depicted in Fig. 4, and any localized initial condition with an amplitude above a certain limit (such as that in Fig. 4) will converge to it. Wide initial conditions may create several solitons

Figure 11 shows the areas in the (β, ϵ) plane have been soliton solutions have been found numerically. The lower curve (dashed), which represents the curve S, is plotted to allow us to make some comparisons with the conclusions obtained from the analytic solutions. First of all, we note that the region of stable solitons is always above S, and that the lower boundary of the stability region (solid line) is roughly parallel to S. The distance between this lower boundary and S depends on δ, μ, and ν. For small μ, ν, and δ, this distance is small. For given values of ν and δ, the hatched regions become wider as $|\mu|$ increases, and the lower boundary becomes higher. For fixed ν and μ, the lower boundary approaches S as δ goes to zero. We would expect that, at zero δ, S would denote the onset of instability.

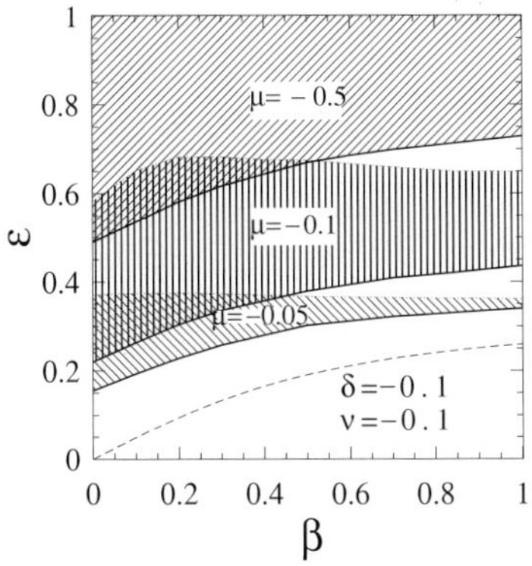

Fig. 11. Regions in the (β, ϵ) plane where stable soliton-like solutions are found. *Differently hatched areas* are for different values of μ, as indicated in each area. All these areas are located above S (*dashed curve*). Here $\nu = -0.1$ and $\delta = -0.1$

Figure 11 gives a rough idea of how the regions of existence of stable solitons in the plane (β, ϵ) change when μ is changed. Stationary solitons must balance loss and gain. For systems whose parameters are located below the lower boundary of the hatched regions, solitons attenuate as they propagate. The energy flux added to the initial soliton owing to a positive ϵ is less than the energy decrease due to linear ($\beta < 0$, $\delta > 0$) and nonlinear ($\mu < 0$) losses. The physical processes on the upper boundary are different. In general, the upper boundary in the (β, ϵ) diagram coincides with the curve where fronts have zero velocity. Above that curve, two fronts of a wide soliton diverge, while below it, two fronts converge, forming a stable soliton at the end of this process. Thus, stable solitons can exist only below that line.

Figure 12 shows the area of stable solitons in the plane (μ, ϵ) for fixed values of ν, δ, and β. As $|\mu|$ increases, the interval of allowed values of ϵ becomes wider, and its central value larger. This last observation is also expected, as it indicates that larger fifth-order nonlinear losses must be compensated for by increasing the third-order nonlinear gain. The width of the strip becomes infinitesimally small at $\mu \approx -0.04$.

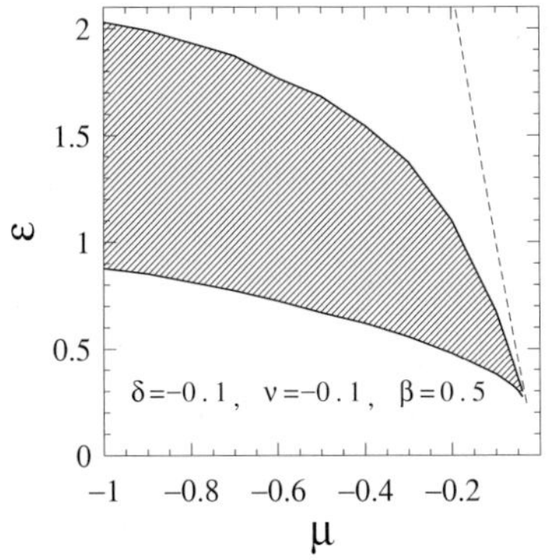

Fig. 12. Regions in the plane (μ, ϵ) where stable solitons exist. Here $\beta = 0.5$ and $\delta = -0.1$. The *dashed line* is the curve where the exact analytical solutions given by (25) are located for the given values of (δ, ν, β). This line is not in the area of stable solitons. This shows that the analytical solutions are beyond that region and therefore are unstable

5 Multiplicity of Solutions

5.1 Composite Solitons

Stable plain solitons exist only when δ and μ are negative, and β and ϵ are positive. The parameter ν can be either positive or negative. These conditions mean that one term (that with coefficient ϵ) provides gain for the soliton, while three other terms (those with coefficients δ, β, and μ) produce losses. At certain values of the parameters, a soliton-like initial condition converges to a new type of solution, namely, a composite soliton (CS) [63] (see also [64]). We may view this as a combination of a source and two fronts.

The intensity profile of a CS is shown in Fig. 13. The curve for the plain soliton solution (for the same set of parameters) is also given for comparison. The CS consists of two fronts with a small "hill" between them. This hill is the domain boundary between the two fronts, as they have nonzero wavevectors. This hill should be counted as a source, because it follows from the phase profile that energy flows from the center to the wings of the CS. Note that the flat regions between the source and the fronts are relatively small. The typical width of the source is the same as the typical width of the front.

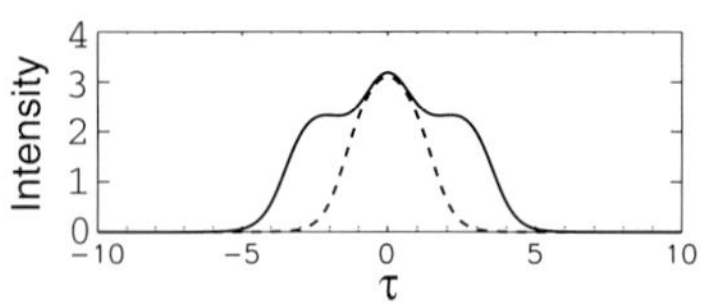

Fig. 13. Intensity profiles of a composite soliton (*solid curve*) and plain soliton (*dashed curve*) with $\delta = -0.1$, $\beta = 0.5$, $\epsilon = 1.75$, $\mu = -0.6$, and $\nu = -0.1$

To compare the CS with a plain soliton, we use the (a, M) plane. From Fig. 14 we see that that there are many similarities between the two solutions. In particular, the top part of the CS (i.e. the source) has the same shape as the top part of the soliton. The wings of the plain soliton and of the CS are also very similar. The difference between the nonlinear propagation constants ω of the two structures is around 10%. The similarity between the asymptotes follows from the linearized version of (1).

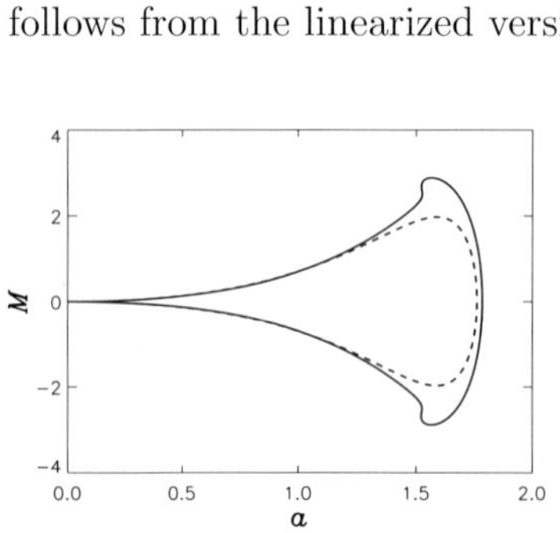

Fig. 14. Plain soliton (*dashed curve*) and composite soliton (*solid curve*) in the (a, M) plane, obtained by the method outlined in Sect. 3.4. The parameters of the equation are the same as in Fig. 13

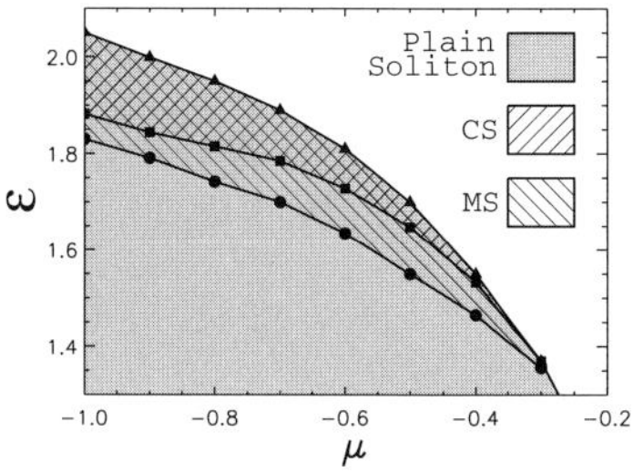

Fig. 15. Range of existence of stationary stable solitons on the (μ, ϵ) plane. The *solid curve marked by triangles* gives the threshold for zero front velocity, and *filled areas* give the ranges of existence of the plain soliton, the composite soliton, and the moving soliton. Parameters are $\delta = -0.1$, $\beta = 0.5$, and $\nu = -0.1$.

Figure 15 shows the ranges of parameters where stable plain solitons and CSs exist. The range of existence of the CSs is limited by the threshold for a positive front velocity. The upper boundary on this plot corresponds to this threshold and, hence, to the transition into a pair of fronts. The lower boundary corresponds, in general, to the transition from the composite soliton to the plain soliton – this can be quite complicated.

5.2 Moving Solitons

Moving solitons (MSs) can be observed as a result of the instability of the CS at the lower boundary of the region of stable CSs in the (μ, ϵ) plane. If an antisymmetric perturbation has a high growth rate, then, instead of a transformation into a plain soliton, we obtain a spontaneous transformation into an asymmetric MS (Fig. 16).

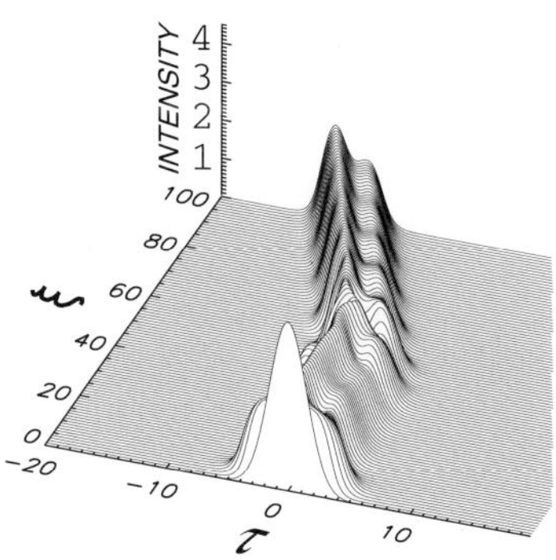

Fig. 16. Spontaneous transformation of a composite soliton into a moving soliton for $\delta = -0.1$, $\beta = 0.5$, $\nu = -0.1$, $\epsilon = 1.8$, and $\mu = -0.8$. A moving soliton can also be formed by choosing a moving initial condition. In this case, both left- and right-moving solitons can be created. They are mirror images, owing to the τ-reversal symmetry of (1)

The intensity profile of an MS is given in Fig. 17. The intensity profile is indeed very close to the profiles of the plain soliton and the composite soliton. In other words, an MS can be considered as a bound state (i.e. nonlinear superposition) of a plain soliton and a front, or as a CS with one of the fronts missing. The spectrum of the MS is also asymmetric.

It follows from the representation of the MS as a bound state of a soliton and a front that its energy can be roughly written as $E_{\mathrm{MS}} = E_{\mathrm{P}} + E_{\mathrm{F}}$, while

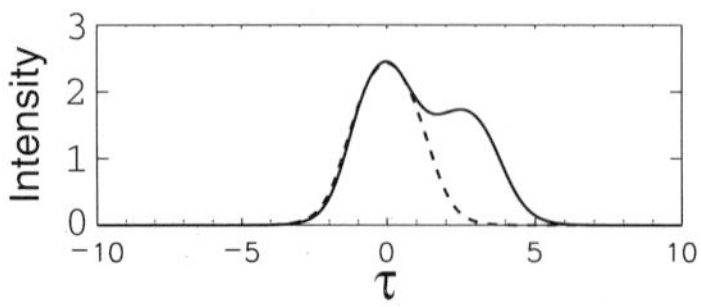

Fig. 17. Intensity profile of a moving soliton (*solid curve*) and the plain soliton (*dashed curve*). The parameters are the same as those in Fig. 16

the energy of the composite soliton is roughly $E_{\mathrm{CS}} = E_{\mathrm{P}} + 2E_{\mathrm{F}}$, where E_{P} and E_{F} are the energies of the plain soliton and the front, respectively. In other words, the difference between the energy of the MS and the energy of the plain soliton (for the same set of parameters) is half of the difference between the energy of the CS and of the plain soliton. Numerical calculations show qualitative agreement with this prediction, as Fig. 18 shows.

The MS always moves with the soliton leading (Fig. 16). It exists in the region of parameters where the front velocity is negative. The front pushes the soliton from one side. Another important feature is that the velocity of an MS is always smaller than the velocity of the front for the same set of parameters. For the MS, the front tends to move with its own velocity but the soliton tends to be stationary, owing to the spectral filtering . The resulting velocity of the MS is determined by competition between these two processes.

The range of existence of the MS is even larger than that of the CS, as can be seen from Figs. 15 and 18. The upper boundary in terms of ϵ is almost the same for plain solitons, the CS, and the MS, while the range of existence of the MS is approximately twice as wide as that of the CS. However, neither the CS nor the MS exists when $\mu > -0.3$. Apparently, this is the threshold where double matching (between amplitudes and wavevectors) of the soliton and front becomes possible.

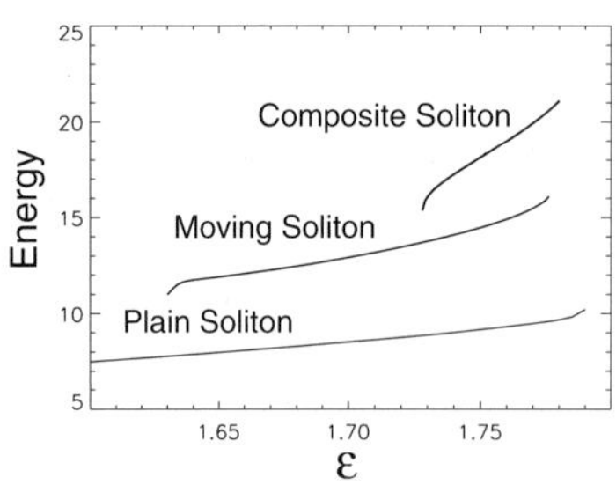

Fig. 18. The energies of the plain soliton, the composite soliton and the moving soliton versus ϵ. Here $\delta = -0.1$, $\beta = 0.5$, $\mu = -0.6$, and $\nu = -0.1$

5.3 Coexistence of Solitons

Solitons, fronts, and sources can be considered as elementary building blocks which can be combined to form more complicated structures. Stable structures arise as a result of the strong interaction between solitons and fronts. However, such structures require the matching conditions for the amplitude and the wavevector to be fulfilled. This is the reason why these structures exist only in a relatively narrow range of parameters. Two different stable, stationary solutions of the quintic CGLE, can coexist for the same set of parameters. This result is quite natural if we view fronts, solitons, and sources as elementary building units which can be combined to form more complicated structures.

The CS solutions exist in the range of parameters where the front has a small negative velocity. Hence, the CS exists owing to repulsion between the source and the front. When the front velocity is small, fronts tend to move toward each other, but they encounter repulsion from the source, which keeps them at a fixed distance. For parameters where the front velocity is large enough, they overcome this repulsion and a plain soliton is formed. For the range of parameters where the front has a zero wavevector, the plain soliton and the CS become essentially identical.

In other regions of the parameter space, the number of varieties of soliton solutions may be more then three [65], with branches located higher in energy in Fig. 18. At this stage, it is difficult to tell how many types of soliton solutions can exist simultaneously for a given set of parameters. This is true even for the simplest cases of the quintic CGLE. In more complicated systems, the number of solutions may be infinite. This situation again resembles the world of biology, where the number of species is apparently huge.

Composite and moving solitons, although they seem esoteric, may play a pivotal role in the dynamics of some particular systems. We have seen that the exact solution of the quintic CGLE becomes stable in the vicinity of the transition from solitons to fronts. At the same time, this is the very range of existence of the CSs and MSs. So, as the system parameters are moved toward the stability region, the appearance of CSs and MSs becomes possible. For such interesting forms of solution, we have neither an exact nor an approximate solution.

5.4 Collisions

There are four stable soliton solutions in some regions of the parameter space, viz. a plain soliton, a composite soliton, and left- and right-moving solitons. This gives three possibilities for pair interaction, i.e. interaction of a moving soliton with a plain soliton, with a composite soliton, and with another moving soliton. The result of the collision also depends on the relative phase of the interacting solitons. If a moving soliton collides with a stationary soliton, and the propagation constants are unequal, then the relative phase at the collision point depends on both the initial phase difference and the initial separation, so it is difficult to control the relative phase in numerical simulations. If two moving solitons collide, the relative phase between them is determined only by the initial phase difference, so it can be controlled easily.

Figures 19 and 20 show some possible types of interaction. Collisions may result in several outcomes. Figure 19 is plotted for $\epsilon = 1.75$. For smaller ϵ,

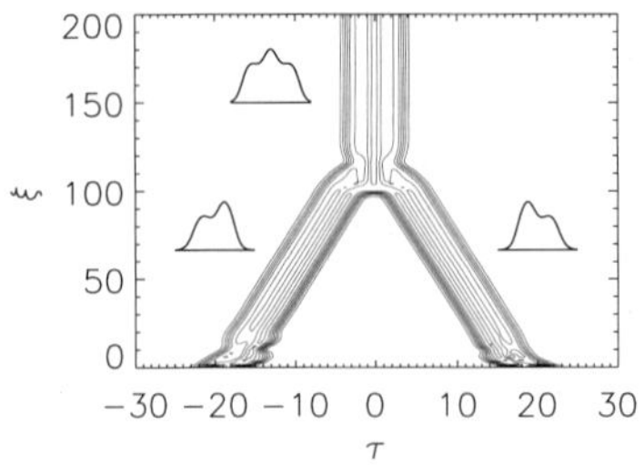

Fig. 19. Interaction between in-phase MSs for $\epsilon = 1.75$, $\delta = -0.1$, $\beta = 0.5$, $\nu = -0.1$, and $\mu = -0.6$. The profiles near the contour plots show the orientation of the MS. In general, the formation of either a CS (for in-phase MSs), a plain soliton, or another MS (for out-of-phase solitons) can be observed

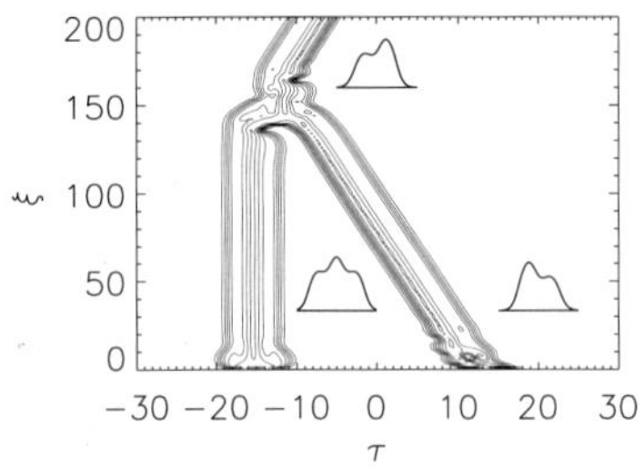

Fig. 20. Interaction between MS and CS for $\epsilon = 1.73$, $\delta = -0.1$, $\beta = 0.5$, $\nu = -0.1$, and $\mu = -0.6$. If the MS changes its direction of motion after collision, then it also changes its orientation, because an MS always moves with the pulse leading and the front trailing

say $\epsilon \approx 1.73$, the CS still exists, but is less stable. For this set of parameters, an MS is formed after the collision, and the direction of propagation of this MS depends on the phase difference between the solitons. Potentially, there are other possible types of interaction, including complete annihilation of interacting pulses, and the tunneling of one pulse through the other.

6 Soliton Bound States

For the nonlinear Schrödinger equation, two solitons have zero binding energy. Hence, any nonlinear superposition of two solitons is neutrally stable, and can be made unstable with a very small perturbation. On the other hand, for the NLSE, there is no stationary solution in the form of two solitons with equal amplitudes and velocities and with a fixed separation. For Hamiltonian generalizations of the NLSE, the interaction between the pulses becomes inelastic, so that two-soliton solutions of the perturbed NLSE (when they exist) are unstable owing to energy exchange between the pulses [49].

The situation changes dramatically for nonconservative systems. Each soliton then has its own internal balance of energy which maintains its constant amplitude. Fixing the amplitudes effectively reduces the number of degrees of freedom in the system of two solitons and can make it stable. As a result, plain pulses in the CGLE model can form bound states [45,66–68], and these are also stationary solutions of the CGLE. The pulses in such bound states overlap only weakly.

Bright-soliton solutions of the CGLE form a discrete set, so that, if the values of the parameters of the equation are specified, then the amplitude and width of the soliton are fixed. This fact implies that, during the interaction of two solitons, basically only two parameters may change: their separation ρ, and the phase difference ϕ between them. Thus the phase space here is truly 2-D, and we may analyze the bound states, their stability, and their global dynamics in this 2-D space, which we call the "interaction plane". The possibility of this reduction in the number of degrees of freedom is a unique feature of systems with gain and loss. It does not apply for nonintegrable Hamiltonian systems, where the amplitudes of the solitons can also change, and therefore more sources of instability of the bound states appear [49].

For stationary solutions, the energy and momentum in (2) and (4) do not change, and the corresponding solutions must satisfy the set of two equations

$$F[\psi] = 0 \,, \; J[\psi] = 0 \,. \tag{40}$$

The first identity indicates the necessary balance that must exist between losses and gain for any stationary solution, while the second guarantees a balance between the transverse forces acting on solitons. Trivially, $J[\psi] = 0$ for any symmetric ($\psi(\tau) = \pm\psi(-\tau)$) solution, but $J[\psi]$ may also be zero for other solutions (which can have nonzero velocity). For any given set of coefficients of the equation, we call the corresponding "plain" soliton solution $\psi_0(\tau)$. A bound state of two plain solitons is well approximated by

$$\psi(\tau) = \psi_0(\tau - \rho/2) + \psi_0(\tau + \rho/2)\exp(i\phi) \,, \tag{41}$$

where the values of the distance between the solitons, ρ, and of their relative phase, ϕ, are those which satisfy the equations (40), and we assume that ρ is not too small.

The zeros of these functionals are presented in the interaction plane in Fig. 21 for the parameters given in the figure, in the interval $0.4 < \rho < 4$. The separation ρ must be of the same order as, but larger than, the width of a single soliton (indicated by a dashed circle in the figure). Smaller ρ

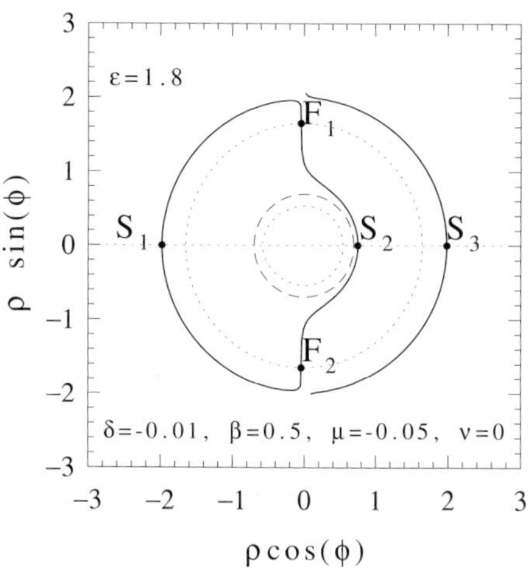

Fig. 21. Zeros of F (*solid lines*) and J (*dotted lines*) in the interaction plane. The parameters of the CGLE are shown in the figure. The points of intersection of the *solid curves* with the *dotted curves* correspond to bound states of two solitons. These are shown as *bold points*. The radius of the *dashed circle* indicates the full width at half maximum of a single soliton. For separations less than that distance, the interaction is too strong to be accurately described by the model

corresponds to merging of the solitons, and at larger ρ the interaction between the solitons is too weak. The solid lines in Fig. 21 show the locus of the points where $F[\psi] = 0$, while the dotted lines show the points where $J[\psi] = 0$. It can be seen from this figure that the functional $F[\psi]$ has two zeros in the interval $-\pi/2 < \phi < \pi/2$ but only one in the interval $\pi/2 < \phi < 3\pi/2$.

$J[\psi]$ is zero on the horizontal axis of the interaction plane, so every intersection of a solid curve with the horizontal axis corresponds to a two-soliton bound state. There are three examples of this type of bound state in Fig. 21, viz. $S_i, i = 1, 2, 3$. For these, the component solitons are in phase or out of phase. $J[\psi]$ also has zeros along two almost circular arcs. The intersections of the outer circle with the solid curve (points F_1 and F_2) correspond to new bound states where the phase difference between the solitons is close to $\pi/2$. The phase profile of the solution is necessarily asymmetric, owing to this phase difference.

These predictions have been confirmed numerically [68] by solving the propagation equation. The general dynamics of the interaction of two solitons can also be described using the interaction plane. An initial condition (41), in the form of two stable solitons with arbitrary separation (ρ) and phase difference (ϕ), will result in a trajectory on this plane. Bound states are the singular points of this plane, with the type of singular point defining the stability of the state. Figure 22 gives an example of these numerical simulations. This figure indicates that, for the given parameters, there are five singular points. Within the accuracy of the method, these coincide with the solutions that can be found using the balance equation. Three of these singular points (S_1, S_2 and S_3) are saddles, where the phase difference between the solitons is zero or π. These are unstable bound states of two solitons.

In addition, there are two symmetrically located stable foci (F_1 and F_2), and these correspond to stable bound states of two solitons in quadrature,

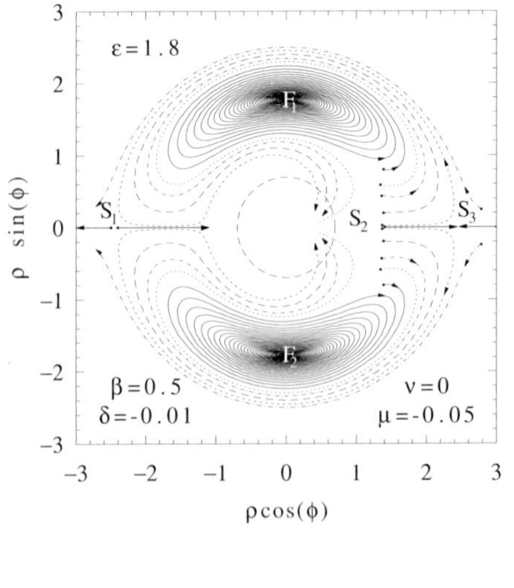

Fig. 22. Trajectories showing the evolution of two-soliton solutions in the interaction plane for the same parameters as those used in Fig. 21. The five singular points correspond to the five bound states depicted in Fig. 21. Only two of them (F_1 and F_2) are stable. The central part of the figure, where ρ is less than a single soliton width, does not describe a valid bound state. Trajectories converging to the *center* describe the merging of two solitons

i.e. their phase difference is $\pm\pi/2$. These are the bound states with asymmetric phase profiles. The spectra are also asymmetric owing to the phase asymmetry, as expected for this type of solution [70]. A consequence of this asymmetry is that a two-soliton solution moves with a constant velocity. We should note that asymmetric bound states are not always stable. Changing the parameters in the CGLE may convert stable foci into centers (or elliptic points) and, further, into unstable foci. They can even disappear [68], in full agreement with the predictions of the balance equations.

As a consequence of the existence of two-soliton solutions, solutions with three or more solitons also exist. An example of a multisoliton solution is shown in Fig. 23. As a result of the above-mentioned asymmetry, multisoliton solutions are also asymmetric and move with the same constant velocity along the τ axis. Periodic solutions of the CGLE [69] can clearly be constructed this way; such a "train" will move with a constant velocity. On the other hand, more complicated periodic solutions exist, even for the cubic equation [41].

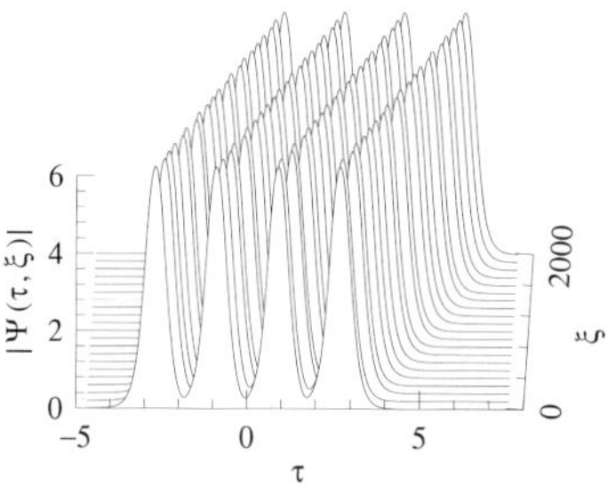

Fig. 23. Stable propagation of a four-soliton bound state [68]. The equation parameters are the same as those in Fig. 22. Similar solutions but with a phase difference of π between the solitons have also been found for a different form of the CGLE [67]

7 Conclusions

The cubic–quintic CGLE is an important model for describing various optical systems with gain and loss. These include spatial- and temporal-soliton lasers, parametric amplifiers, and optical transmission lines. The model can also be applied in other fields of physics. Therefore, knowledge of the general properties of its solutions is of vital importance. We hope that, in this short review chapter, we have covered at least a relevant selection of the problems related to this subject.

References

1. V.E. Zakharov, A.B. Shabat, Exact theory of two dimensional self focusing and one dimensional self modulation of nonlinear nonlinear media, Sov. Phys. JETP, **34**, 62–69 (1971).
2. C. Normand, Y. Pomeau, Convective instability: a physicist's approach, Rev. Mod. Phys. **49**, 581–623 (1977).

3. P. Kolodner, Collisions between pulses of travelling-wave convection, Phys. Rev. A, **44**, 6466–6479 (1991).
4. R. Graham, *Fluctuations, Instabilities and Phase Transitions*, ed. by T. Riste (Springer, Berlin, Heidelberg 1975), pp. 215–279.
5. H. Haken, *Synergetics* (Springer, Berlin, Heidelberg 1983).
6. C.O. Weiss, Spatio-temporal structures. Part II. Vortices and defects in lasers, Phys. Rep., **219**, 311–338 (1992).
7. K. Staliunas, Laser Ginzburg–Landau equation and laser hydrodynamics, Phys. Rev. A, **48**, 1573–1581 (1993).
8. P.K. Jakobsen, J.V. Moloney, A.C. Newell, R. Indik, Space–time dynamics of wide-gain-section lasers, Phys. Rev. A, **45**, 8129–8147 (1992).
9. G.K. Harkness, W.J. Firth, J.B. Geddes, J.V. Moloney, E.M. Wright, Boundary effects in large-aspect-ratio lasers, Phys. Rev. A, **50**, No 5, 4310–4322 (1994).
10. P.-S. Jian, W.E. Torruellas, M. Haelterman, S. Trillo, U. Peschel, F. Lederer, Solitons of singly resonant optical parametric oscillators, Opt. Lett., **24**, 400–402 (1999).
11. A.M. Dunlop, E.M. Wright, W.J. Firth, Spatial soliton laser, Opt. Commun., **147**, 393–401 (1998).
12. V.B. Taranenko, K. Staliunas, C.O. Weiss, Spatial soliton laser: localized structures in a laser with a saturable absorber in a self-imaging resonator, Phys. Rev. A, **56**, 1582–1591 (1997).
13. W.J. Firth, A.J. Scroggie, Optical bullet holes: robust controllable localized states of a nonlinear cavity, Phys. Rev. Lett., **76**, 1623–1626 (1996).
14. N.A. Kaliteevstii, N.N. Rozanov, S.V. Fedorov, Formation of laser bullets, Opt. Spectrosc., **85**, 533–534 (1998).
15. D. Michaelis, U. Peschel, F. Lederer, Oscillating dark cavity solitons, Opt. Lett., **23**, 1814–1816 (1998).
16. C.S. Ng, A. Bhattacharjee, Ginzburg–Landau model and single-mode operation of a free-electron laser oscillator, Phys. Rev. Lett., **82**, 2665–2668 (1999).
17. A. Hasegawa, Y. Kodama *Solitons in Optical Communications* (Oxford University Press, New York, 1995).
18. G. Khitrova, H.M. Gibbs, Y. Kawamura, I. Iwamura, T. Ikegami, J.E. Sipe, L. Ming, Spatial solitons in a self-focusing gain medium, Phys. Rev. Lett., **70**, 920–923 (1993).
19. C. Paré, L. Gagnon, P.-A. Bélanger, Spatial solitary wave in a weakly saturated amplifying/absorbing medium, Opt. Commun., **74**, 228–232 (1989).
20. A.W. Snyder, D.J. Mitchell, L. Poladian, F. Ladouceur, Self-induced optical fibers: spatial solitary waves, Opt. Lett., **16**, 21–23 (1991).
21. G.-L. Oppo, G. D'Alessandro, W.J. Firth, Spatiotemporal instabilities of lasers in models reduced via center manifold techniques, Phys. Rev. A **44**, 4712–4720 (1991).
22. P. Coullet, L. Gil, F. Rocca, Optical vortices, Opt. communications, **73**, 403–408 (1989).
23. H.A. Haus, J.G. Fujimoto, E.P. Ippen, Structures for additive pulse mode locking, J. Opt. Soc. Am. B, **8**, 2068–2076 (1991).
24. N.N. Rozanov, *Optical Bistability and Hysteresis in Distributed Nonlinear Systems* (Physical and Mathematical Literature Publishing Company, Moscow, 1997).

25. F.X. Kärtner, U. Keller, Stabilization of soliton-like pulses with a slow saturable absorber, Opt. Lett. **20**, 16–18 (1995).
26. H. Haus, Theory of mode locking with a fast saturable absorber, J. Appl. Phys., **46**, 3049 (1975).
27. N. Bekki, K. Nozaki, Formations of spatial patterns and holes in the generalized Ginzburg–Landau equation, Phys. Lett., **110A**, 133 (1985).
28. W. Van Saarloos, P.C. Hohenberg, Fronts, pulses, sources and sinks in generalized complex Ginzburg–Landau equations, Physica D **56**, 303–367 (1992).
29. N.N. Akhmediev, V.V. Afanasjev, Novel arbitrary-amplitude soliton solutions of the cubic–quintic complex Ginzburg–Landau equation, Phys. Rev. Lett., **75**, 2320–2323 (1995).
30. V.S. Grigoryan, T.C. Muradyan, Evolution of light pulses into autosolitons in nonlinear amplifying media. J. Opt. Soc. Am. B **8**, 1757–1765 (1991).
31. N.N. Akhmediev, M.J. Lederer, B. Luther-Davies, Exact localized solution for nonconservative systems with delayed nonlinear response, Phys. Rev. E, **57**, 3664–3667 (1998).
32. S. Longhi, G. Steinmeyer, W.S. Wong, Variational approach to pulse propagation in parametrically amplified optical systems, J. Opt. Soc. Am. B, **14**, 2167–2173 (1997).
33. N.R. Pereira, Soliton in the damped nonlinear Schrödinger equation, Phys. Fluids, **20**, 1735–1743 (1977).
34. N.R. Pereira, L. Stenflo, Nonlinear Schrödinger equation including growth and damping, Phys. Fluids, **20**, 1733–1734 (1977).
35. D.R. Nicholson, M.V. Goldman, Damped nonlinear Schrödinger equation, Phys. Fluids, **19**, 1621–1625 (1976).
36. N.R. Pereira, F.Y.F. Chu, Damped double solitons in the nonlinear Schrödinger equation, Phys. Fluids, **22**, 874–881 (1979).
37. J. Weiland, Y.H. Ichikawa, H. Wilhelmsson, A perturbation expansion for the NLS with application to the influence of nonlinear Landau damping, Phys. Scr., **17**, 517–522 (1978).
38. P.-A. Bélanger, L. Gagnon, C. Paré, Solitary pulses in an amplified nonlinear dispersive medium, Opt. Lett., **14**, 943–945 (1989).
39. L.M. Hocking, K. Stewartson, On the nonlinear response of a marginally unstable plane parallel flow to a two-dimensional disturbance, Proc. R. Soc. London A, **326**, 289–313 (1972).
40. K. Nozaki, N.J. Bekki, Exact solutions of the generalized Ginzburg–Landau equation, J. Phys. Soc. Jpn., **53**, 1581–1582 (1984).
41. A.V. Porubov, M.G. Velarde, Exact periodic solutions of the complex Ginzburg–Landau equation, J. Math. Phys., **40**, 884–896 (1999).
42. R. Conte, M. Musette, Linearity inside nonlinearity: exact solutions to the complex Ginzburg–Landau equation, Physica D **69**, 1–17 (1993).
43. B.A. Malomed, Evolution of nonsoliton and "quasi-classical" wavetrains in nonlinear Schrödinger and KdV equations with dissipative perturbations, Physica D, **29**, 155–172 (1987).
44. O. Thual, S. Fauve, Localized structures generated by subcritical instabilities, J. Phys. (France), **49**, 1829–1833 (1988).
45. H.R. Brand, R.J. Deissler, Interaction of localized solutions for subcritical bifurcations, Phys. Rev. Lett. **63**, 2801–2804 (1989).
46. V. Hakim, P. Jakobsen, Y. Pomeau, Fronts vs. solitary waves in nonequilibrium systems, Europhys. Lett., **11**, 19–24 (1990).

47. W. Van Saarloos, P.C. Hohenberg, Pulses and fronts in the complex Ginzburg–Landau equation near a subcritical bifurcation, Phys. Rev. Lett., **64**, 749–752 (1990).
48. P. Marcq, H. Chaté, R. Conte, Exact solutions of the one-dimensional quintic complex Ginzburg–Landau equation, Physica D, **73**, 305–317 (1994).
49. N. Akhmediev, A. Ankiewicz, *Solitons, nonlinear pulses and beams* (Chapman & Hall, London, 1997).
50. S. Fauve, O. Thual, Solitary waves generated by subcritical instabilities in dissipative systems, Phys. Rev. Lett., **64**, 282–285 (1990).
51. Y. Kodama, M. Romagnoli, S. Wabnitz, Soliton stability and interactions in fibre lasers, Electron. Lett., **28**, 1981–1982 (1992).
52. V.V. Afanasjev, Soliton singularity in the system with nonlinear gain, Opt. Lett., **20**, 704–706 (1995).
53. A. Ankiewicz, Simplified description of soliton perturbation and interaction using averaged complex potentials, J. Nonlinear Opt. Phys. Mater., **4**, 857–870 (1995).
54. J.D. Moores, On the Ginzburg–Landau laser mode-locking model with fifth-order saturable absorber term, Optics Commun., **96**, 65–70 (1993).
55. N.N. Akhmediev, V.V. Afanasjev, J.M. Soto-Crespo, Singularities, special soliton solutions of the cubic–quintic complex Ginzburg–Landau equation, Phys. Rev. E, **53**, 1190–1198 (1996).
56. N. Akhmediev, A. Ankiewicz, J.M. Soto-Crespo, Stable soliton pairs in optical transmission lines and fiber lasers, J. Opt. Soc. Am. B, **15**, 515–523 (1998).
57. P. Kolodner, D. Bensimon, C.M. Surko, Travelling-wave convection in an annulus, Phys. Rev. Lett., **60**, 1723–1726 (1988).
58. D.J. Kaup, B.A. Malomed, The variational principle for nonlinear waves in dissipative systems, Physica D, **87**, 155–159 (1995).
59. J.M. Soto-Crespo, L. Pesquera, Analytical approximation of the soliton solutions of the quintic complex Ginzburg–Landau equation, Phys. Rev. E, **56**, 7288–7293 (1997).
60. T. Kapitula, B. Sandstede, Instability mechanism for bright solitary-wave solutions to the cubic–quintic Ginzburg–Landau equation, J. Opt. Soc. Am. B **15**, 2757–2762 (1998); Stability of bright solitary-wave solutions to perturbed nonlinear Schrödinger equations, Physica D **124**, 58–103 (1998).
61. T. Kapitula, B. Sandstede, Stability criterion for bright solitary waves of the perturbed cubic–quintic Schrödinger equation, Physica D **116**, 95–120 (1998).
62. J.M. Soto-Crespo, N.N. Akhmediev, V.V. Afanasjev, Stability of the pulselike solutions of the quintic complex Ginzburg–Landau equation, J. Opt. Soc. Am. B, **13**, 1439 (1996).
63. V.V. Afanasjev, N.N. Akhmediev, J.M. Soto-Crespo, Three forms of localized solutions of the quintic complex Ginzburg–Landau equation, Phys. Rev. E, **53**, 1931–1939 (1996).
64. N.N. Rozanov, S.V. Fedorov, G.V. Khodova, A.A. Zinchik, The regime of the leading center for transversely unidimensional laser autosolitons, Opt. Spectrosc., **83**, 370–371 (1997).
65. N.N. Akhmediev, J.M. Soto-Crespo, Plethora of soliton-like pulses in passively mode-locked fiber lasers, in *Nonlinear Guided Waves and their Applications*, Technical digest series, Cambridge, 1996, paper SaD10, pp. 197–199.
66. V.V. Afanasjev, N.N. Akhmediev, Soliton interaction and bound states in amplified–damped fiber systems, Opt. Lett., **20**, 1970–1972 (1995).

67. N.N. Rozanov, S.V. Fedorov, G.V. Khodova, Pulsating and bound transversely unidimensional laser autosolitons, Optics and Spectroscopy, **81**, 896–898 (1996).
68. N.N. Akhmediev, A. Ankiewicz, J.M. Soto-Crespo, Multisoliton solutions of the complex Ginzburg–Landau equation, Phys. Rev. Lett., **79**, 4047–4051 (1997).
69. W. Schöpf, L. Kramer, Small-amplitude periodic and chaotic solutions of the CGLE, Phys. Rev. Lett., **66**, 2316–2319 (1991).
70. N.N. Akhmediev, A. Ankiewicz, Generation of a train of solitons with arbitrary phase difference between neighboring solitons, Opt. Lett., **19**, 545–547 (1994).

Existence, Stability, and Properties of Cavity Solitons

William J. Firth and Graeme K. Harkness

Summary. The history, theory, and physical interpretation of cavity solitons are reviewed. These are bright, stable, nondiffracting spots of light in a driven optical cavity. The cavity must contain a nonlinear medium, which need not, however, support ordinary (propagating) spatial solitons. We use the Kerr cavity as an example to describe methods to find them and analyze their stability, demonstrating a sizeable domain of stability of two-dimensional cavity solitons in a Kerr cavity. Cavity solitons in semiconductor microresonators may have properties interesting for applications to optical information processing, and we outline some results from models of such systems.

1 Introduction

Spatial optical solitons are beams of light in which nonlinearity counterbalances diffraction, leading to a robust structure which propagates without change of form. Among the many interesting spontaneous spatial structures which have been predicted and/or observed in nonlinear optical systems [1] are localized bright spots in driven optical cavities [2–28] (see also the chapter by Weiss et al. in this book). These bright spots share some properties with spatial solitons, and we shall refer to them as *cavity solitons* (CSs). Such structures could be natural "bits" for parallel processing of optical information, especially if they exist in semiconductor microresonators. Recent work ([16,17,21] and the chapter by Weiss et al.) is encouraging in that respect.

In this chapter we briefly review the history and underlying nonlinear optics of cavity solitons. We then describe in more detail some models which show these soliton-like structures. It is somewhat paradoxical that stable cavity solitons can exist in media with properties different from, even opposite to, those required for Kerr solitons. We therefore consider several candidate physical interpretations of these structures: "soliton-in-a-box", "self-trapped switching wave", "pattern element", and "local nonlinear resonance". We derive and examine the perturbation eigenmodes of CSs in sample systems, which give information on the soliton's stability and response to external influences such as noise, neighboring solitons, and gradients of the holding field. Cavity solitons can be created by localized pulses of light [8], and can thus be used to store images or information. The ability to control and manipulate these solitons offers potential advantages over competing systems, and we mention some device ideas which might capitalize on these advantages. Experimental observations on cavity solitons are still sparse, but as described in the chapter by Weiss et al., substantial progress is being made.

This article concentrates mainly on our own work, primarily to illustrate points of general relevance, and we have tried to present a fair and reasonably comprehensive list of relevant references.

2 History

The story of cavity solitons can perhaps be divided into three Ages: the Stone Age, the Space Age, and the Information Age. The Stone Age lasted roughly until 1990. During it a few pioneers [2–4] considered relevant spatial phenomena in cavities. A special issue of *Journal of the Optical Society of America B* [29] ended that stage, and roughly marks the transition into the Space Age. In the Space Age several favorable circumstances combined to stimulate a huge expansion of interest in spatio-temporal nonlinear optics. Workstation development made dynamical simulation of two-dimensional field patterns widely available. In addition, simple mean-field models of such systems provided a convenient framework which readily matched studies of pattern formation in fluids and other fields. A later special issue, in 1994 [1], shows how dramatically the field changed in just a few years. This special issue includes some papers predicting mean-field cavity solitons [12,13], a trickle which has become a flood in the last few years. As yet, the Information Age has barely dawned: in it the emphasis will switch from the mere existence of cavity solitons to their engineering and applications.

The history of optical cavity solitons probably began with the seminal paper of Moloney and coworkers [2],[1] who used split-step fast Fourier transform methods to simulate transverse effects in optical bistability (OB) [30,31]. The model system was a ring cavity, driven by a Gaussian beam, and containing a self-focusing Kerr-like medium. The field was propagated (in z) around the cavity, and added coherently to the driving beam at the input beam splitter. The transverse simulation was one-dimensional (1D), i.e. the intracavity field was described by $E_n(x, z)$, with n counting the cavity round trips. When the input field was ramped up to exceed the OB switch-up threshold, the beam center switched, and a *switching wave* moved out, switching up most of the beam (Fig. 1). Then something unexpected happened: a new instability. The interface between the "on" and "off" domains spawned what would now be termed a modulational instability (MI) of the "on" region, which broke up into a set of distinct peaks [2]. These were interpreted as a group of spatial solitons circulating in the medium, perturbed by the output coupling losses and sustained by the input field. Here, therefore, the model was clearly "soliton-in-a-box". Such an interpretation is tenable only for a medium which could sustain solitons.

[1] See also Fig. 5.16 on p. 225 of [2a], and the associated text.

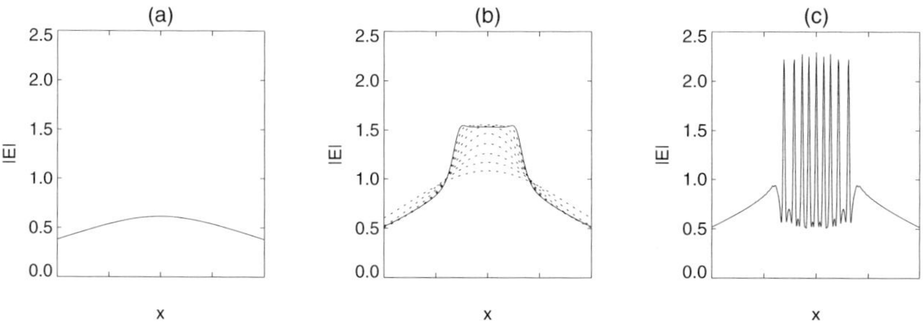

Fig. 1. Gaussian beam switching in an OB cavity. The central section of a smooth broad beam (**a**) switches; a switching wave then forms and moves out (*dotted curves* in (**b**); the high-intensity central region then breaks up into spikes (**c**). (After Moloney et al. [2])

In this early cavity soliton work [2] the beam either contains no solitons or is full of solitons: just two states, and therefore only a one-bit memory in applications terms. McDonald and Firth [4] showed that it was possible to make the individual solitons independently switchable by using a pump beam with a spatially varying amplitude. They modeled a 20-bit memory of this kind, and also showed that it was possible to switch solitons "off" as well as "on" (with an out-of-phase address pulse) [6].

The other key Stone Age pioneer was Rosanov, who from study of OB switching waves developed the idea of "diffractive autosolitons" [3] in nonlinear optics. Switching waves between coexistent stable states are known in many fields, such as reaction–diffusion systems. Purely diffusive switching waves have monotonic profiles, and the more stable state simply wipes out the less stable, as was shown in an OB model with pure diffusion [32]. When diffraction is present, however, the switching wave typically has ripples. These can trap other switching waves, and thus two can trap each other. Or, in 2D, one switching wave might bend around and close on itself, forming a stable island of one phase surrounded by the other – a diffractive autosoliton (DAS) [3]. Here, then, is a second physical interpretation of cavity solitons: self-trapped switching waves. Note that here there is no requirement for bulk solitons: indeed, Rosanov showed that DAS can occur even in a saturable absorber. Nor is OB required: one can have switching waves even in the absence of OB, e.g. between homogeneous and patterned states.

Rosanov has made many significant contributions to OB and related fields, too many to describe here: the interested reader is referred to his review article [33]. He has also investigated DAS in lasers [34], in which context we remark that we shall not deal here with cavity solitons in lasers, nor with the self-imaging oscillators in which *single* soliton-like structures have been experimentally observed ([35,36] and the chapter by Weiss et al. in this book).

3 Mean-Field Models and Cavity Solitons

We now enter the Space Age, characterized by so-called mean-field cavity models, in which alternation of propagation around the cavity with coherent addition of the input field is replaced by a single partial differential equation with a driving term. In the context of spatial pattern formation, this approach is ascribed to the seminal paper of Lugiato and Lefever [37]. Here we give a heuristic derivation of the Lugiato–Lefever (LL) equation , starting from the nonlinear Schrödinger equation (NLS), which is well described in the introductory chapter by Chiao in this book.

The NLS assumes an *infinite* nonlinear medium. Real nonlinear optical media have finite dimensions and, except in glass optical fibers , solitons can rarely propagate more than a few centimeters before running out of material. This, perhaps, makes it natural to put mirrors around the medium, confining the soliton into a finite slab of material. With perfect reflection and zero absorption, one could indeed confine a soliton in a box. Real mirrors and materials are lossy, but we can make good the loss by 'feeding' the caged soliton with an input field. We are thus led to consider a perturbed NLS:

$$\mathrm{i}\frac{\partial E}{\partial t} + \frac{1}{2}\frac{\partial^2 E}{\partial x^2} + |E|^2 E = \mathrm{i}\varepsilon(-E - \mathrm{i}\theta E + E_{\mathrm{in}}) . \tag{1}$$

The three terms on the right-hand side are perturbations of the NLS, all small if ε is. The first is just a linear loss ($\varepsilon > 0$), and the last is the driving field E_{in} needed to sustain E against that loss. Less obvious is the middle term, in θ, but we must remember that coherent light confined between mirrors lies within an optical cavity, and so the response to the driving field will strongly depend on whether or not it is in resonance with the cavity. Hence, therefore, the presence of θ, the *cavity mistuning.* If we ignore the *left* side of (1), then $E = E_{\mathrm{in}}/(1 + \mathrm{i}\theta)$, showing that the cavity has a resonance Lorentzian in θ, which is appropriate for high finesse, where a single longitudinal mode may be considered.

There is one further change from the usual spatial-soliton NLS: propagation (in z) is replaced by evolution (in t). This is natural: the soliton is now in a box, and not going anywhere.

In the limit $\varepsilon \to 0$, (1) recovers the NLS, with a soliton solution with a sech profile in x, time-independent except for a phase rotation. We might expect, therefore, that for finite ε it would have sech-like *cavity soliton* solutions for suitable E_{in}, and indeed it has. It is usual to consider E_{in} to be a plane wave, independent of x (and of y in 2D), in which case the soliton sits on a homogeneous nonzero background field E_{s}. A finite but relatively broad driving beam supports cavity solitons qualitatively similar to those predicted for the simpler plane-wave input case.

We now set $\varepsilon = 1$, which is equivalent to a rescaling. This yields the Lugiato–Lefever equation, which was originally introduced [37] as a model for pattern formation. Note that in the LL equation the time t is scaled to the cavity loss time. The NLS limit is recovered as $\theta \to \infty$. The LL equation is also appropriate as a mean-field model for OB with transverse effects. It is "mean-field" because it is usually derived by assuming a high finesse, so that the cavity field is approximately constant along the cavity axis. The high finesse allows the Airy function response of the cavity to be approximated by a single longitudinal mode, giving the Lorentzian resonance mentioned above. For a plane-wave pump, the plane-wave cavity field obeys $E_\mathrm{s} = E_\mathrm{in}/[1+\mathrm{i}(\theta-|E_s|^2)]$, which is three-valued for $\theta \geq \sqrt{3}$, showing that the model exhibits OB [37]. In fact, Lugiato and Lefever showed that E_s is stable if $|E_\mathrm{s}| < 1$, but unstable, usually because of spontaneous pattern formation, above that threshold [37]. We can now admit that Fig. 1 was generated by simulating the LL equation (for $\theta = 2.1$, and $E_\mathrm{in}(0) = 1.5$ for Fig. 1a and $E_\mathrm{in}(0) = 2.5$ for Figs. 1b,c), rather than the original model of Moloney et al. [2]. The strong similarity of the respective results shows how such mean-field models can capture the essential features of a full cavity model while being both cheaper to simulate and easier to analyze.

Such analysis is still by no means simple. Unlike the case for the NLS, exact analytical solitons are not known for the LL equation. The usual approach has been to perturbatively derive amplitude equations or to resort to numerical integration of the model, neither of which is ideal. The approaches based on perturbation theory, by their very nature, cannot provide quantitative results, and integration gives a very restricted view of the model's bifurcation behavior, by only finding solutions which are dynamically stable. Another technique, which we describe here, is to use numerical methods to find the system's stationary solutions, their stability, and their response to perturbations. We shall apply the method first to the LL equation and later to a model for a semiconductor microcavity.

We look for stationary solutions ($\partial/\partial t = 0$) of (1), after setting $\varepsilon = 1$ as discussed:

$$0 = -(1+\mathrm{i}\theta)E + \mathrm{i}|E|^2 E + \mathrm{i}\nabla^2 E + E_\mathrm{in} , \tag{2}$$

where the exact form for ∇^2 depends on whether we consider 1D, cylindrically symmetric, or fully 2D geometries. We discretize the space variable(s) on N grid points, apply periodic boundary conditions, and use a fast Fourier transform (FFT) algorithm to evaluate the spatial derivatives. This gives a highly accurate, $O(N)$ set of coupled algebraic equations which can be solved using an iterative Newton method. Given a suitably close initial guess, this method rapidly converges to a stationary solution of the original LL equation. These solutions can then be tracked in parameter space, tracing out branches.

The use of a Newton method is advantageous because, as a by-product of this process, it also finds the linearization, in the form of a Jacobian matrix, around the solution found. The resultant eigenvalues, β, give the solutions' stability, and the eigenvectors $\{\boldsymbol{u}\}$ the associated modes.

Figure 2 shows three solution branches tracked in this way. The lower line shows the plane-wave solution discussed already to be stable below $|E_\mathrm{s}|^2 = 1$

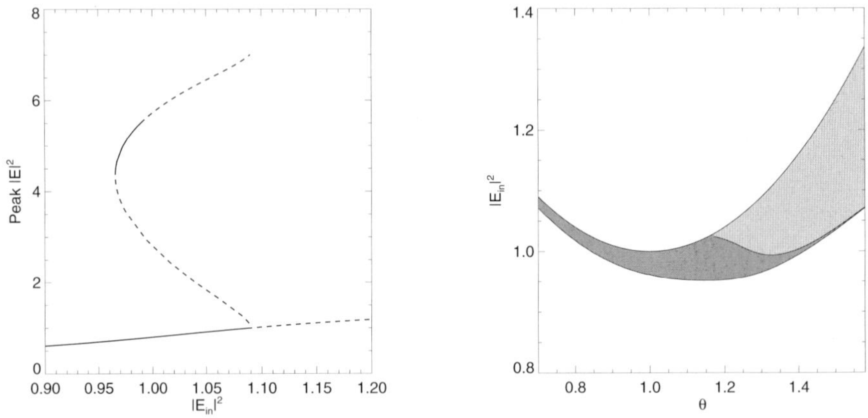

Fig. 2. *Left*: branches showing the homogeneous solution and the cylindrically symmetric cavity soliton solution for a 2D Kerr cavity ($\theta = 1.3$). *Right*: the existence domain of the 2D Kerr cavity soliton (*shaded region*), stable only in the *darker region*

and unstable above. At this point, a branch of localized, cylindrically symmetric ($\nabla^2 = \partial^2/\partial r^2 + (1/r)(\partial/\partial r)$) solutions bifurcates subcritically before bending around to form a positive-slope branch of finite amplitude. This is the cavity equivalent of the unstable soliton-like solution to the 2D NLS. This bifurcation structure is typical of cavity solitons in many systems. For values of $|E_{\rm in}| > E_{\rm in}^{(\rm sn)}$, the two branches coexist with the homogeneous background, to which the solitons asymptote at large radius. Note that only for $\theta > \sqrt{3}$ is the homogeneous solution multivalued and so, over a broad range, any interpretation of CSs as self-trapped switching waves must relate to the interface between the homogeneous solution and a pattern, rather than simply between homogeneous solutions.

It is now instructive to compare these cavity solitons with the system's pattern solutions. The presence of quadratic terms in the amplitude equations means that the bifurcation leading to hexagons is subcritical, of transcritical type. In other words, the branch has a finite slope at the bifurcation point. This is reminiscent of the case in Fig. 2 and leads to an interpretation of a CS as a single spot of a pattern solution, the "pattern element" mentioned in the introduction to this chapter. By plotting both the hexagonal and the soliton solutions on a phase plot (the imaginary part of the field versus its real part), this equivalence can be made even more convincing [8].

The requirement to have a subcritically bifurcating pattern solution is particularly relevant in 1D, where the only possible pattern is rolls. It has been shown [37] that the bifurcation to rolls is subcritical only for $\theta > 41/30$, and so, we should not expect to find 1D cavity solitons below this value, and do not. Other issues relating to 1D solitons have been discussed by Barashenkov and Zemlyanaya in related models [26].

We now turn to the stability of these cavity soliton solutions. As might be expected, the lower branch of the loop is always unstable but, unlike the 2D NLS case, the upper branch may be stable [10]. The shaded area in Fig. 2 shows where CSs exist, and the darker region shows where they are stable. For small θ, they are always stable, but for large θ, the NLS limit, the domain of stability shrinks. We find that the onset of instability is due to the presence of a Hopf bifurcation, and not, as in bulk Kerr media, collapse. Figure 3 shows how the eigenvalues with the largest real parts change with $|E_{\rm in}|^2$. A complex conjugate pair of eigenvalues cross the imaginary axis for a value of $|E_{\rm in}|^2$ just less than one. Direct simulation confirms the stability analysis. A perturbed cavity soliton exhibits damped oscillations in the stable domain, which become undamped as the stability boundary is crossed. Just inside the instability region, CSs can show periodic oscillations. Indeed, Fig. 3 shows phase portaits of the upper- and lower-branch solitons and the extent of the oscillation due to the Hopf unstable mode. Note that larger oscillations may cause a collision with the lower-branch soliton, which generally results in the soliton decaying. For large θ, the cavity soliton either decays or becomes very narrow, suggestive of collapse [9], much as 2D Kerr solitons do in the bulk material.

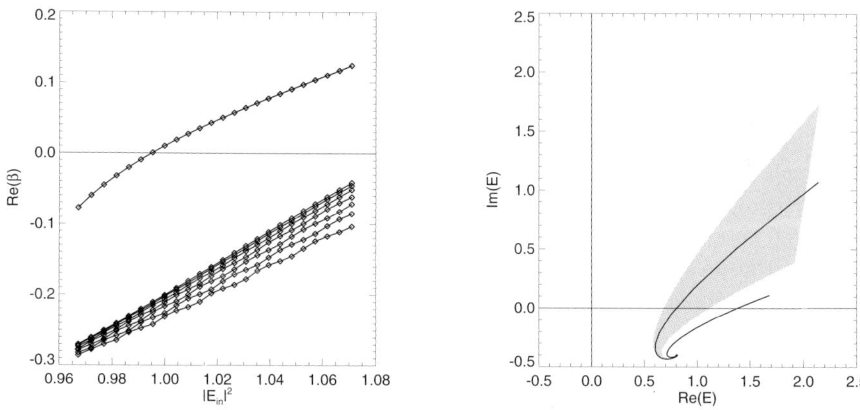

Fig. 3. *Left*: how the eigenvalues with largest real parts change with $|E_{\rm in}|^2$. *Right*: phase portraits (imaginary part versus real part of the field) for the upper- and lower-branch cavity solitons. The *shaded area* shows the extent of the oscillations due to the Hopf instability. The parameters are $\theta = 1.3, |E_{\rm in}|^2 = 1$, giving a value of $E_{\rm s} = 4/5 - 2/5{\rm i}$

Thus we find stable 2D cavity solitons in the Lugiato–Lefever mean-field model, a behavior qualitatively different from its bulk-medium equivalent. This encourages a closer look at the LL equation, to see which terms we can vary and which we should maintain, in exploring further the world of cavity solitons. Comparing the LL equation with the NLS, we note that the time derivative and all three terms on the right-hand side are features associated with the cavity, and thus ought to be maintained. This leaves the diffraction term and the nonlinearity as terms we can play with. We have already modified the diffraction term, when we considered the 2D Kerr cavity. We can go further and add *group-velocity dispersion* to make '3D' cavity solitons [27]. Or we can replace diffraction with dispersion, and consider, for example, fibre cavities, where Steinmeyer et al. [15] have found evidence of soliton-like structures in synchronously pumped fibre loops, and Wabnitz [5] has examined data storage issues. This case is equivalent to diffraction in the 1D geometry of the LL equation, and the existence and stability of CSs in 1D is perhaps to be expected. It turns out that CSs exist only for $\theta > 41/30$, below which the stripe pattern exists only for $|E_\mathrm{s}|^2 > 1$, and thus does not coexist with the homogeneous solution.

We have assumed a simple cavity with planar mirrors: a nonlinear medium placed within a curved-mirror cavity can be described using a generalized mean-field model involving terms in, for example, $(x^2 + y^2)E$, dependent on the ABCD matrix of the cavity [38]. We shall not consider such cases [35,36] (see also the chapter by Weiss et al. in this book) here.

The nonlinearity also offers considerable scope for variation. Perhaps the obvious generalization from the Kerr nonlinearity is to a two-level atom-like response, which becomes Kerr-like far from the atomic resonance [3,7,8]. For exact atomic resonance the medium is just a saturable absorber, with no nonlinear refractive index contribution. It nonetheless supports stable, robust cavity solitons [8]. A further generalization is to consider a nonlinearity mediated by a material excitation. This opens up the further possibility that the medium can have its own dynamics and spatial (usually diffusive) coupling. Semiconductors are particularly interesting among such media, and it has been shown [17,18] that cavity solitons extend to semiconductor models, even in the presence of such "soliton-antagonistic" effects as diffusion and a measure of self-defocusing. Even more surprisingly, Michaelis et al. [21] found bright solitons in a cavity model with a purely defocusing, diffusive saturable Kerr medium, such as is found in semiconductors just below the band edge. We present below a summary of some recent results [39] on the existence, stability, and dynamical properties of CSs in a semiconductor microresonator model.

A different approach is to couple several *optical fields* through, for example, a $\chi^{(2)}$ nonlinearity. This has been shown in mean-field models to support cavity solitons in both second-harmonic-generation [20] and optical-parametric-oscillator [14,22,23,25,40] configurations. We shall not discuss $\chi^{(2)}$ phenomena, because they are comprehensively analyzed in the chapter by Trillo and Haelterman in this volume. A vector Kerr medium also involves coupled fields, and exhibits *polarized* CSs for appropriate parameters [19], even in a medium with a defocusing Kerr nonlinearity.

Localized states with properties very similar to the CSs discussed here have been found in single-mirror feedback systems [41], and in liquid-crystal light-valve systems, which are somewhat equivalent [42,43]. We do not have space to discuss feedback systems in detail, but remark that it seems that they share with CSs the feature of being dissipative structures, and thus can form stable attractors in the state space of the system in question.

4 Cavity Solitons in Semiconductor Microresonators

The existence and stability of CS in semiconductor microresonators has previously been established and investigated using numerical integration or a shooting method [17,18,44,45]. Here we summarise recent work [39] which exploits the stationary solution method described above to confirm and extend previous results. We establish here the existence of cavity solitons in a semiconductor microresonator with a bulk GaAs or a multiple-quantum-well (MQW) GaAs/AlGaAs active layer. We show how to relate the speed with which a CS moves under external perturbations to the projection of the perturbations onto the neutral mode and give some examples, including phase and amplitude gradients of the driving field and interaction with other CSs. Finally we characterize the locus of unstable CSs as a separatrix between two stable, coexisting solutions: the homogeneous solution and the CS. The latter results are important in view of future applications of CSs to optical information processing.

The system we consider consists of an optical cavity containing a nonlinear medium and driven by an external coherent field; the nonlinear medium is either a multiple quantum well or a bulk sample of GaAs [39]. In a notation similar to earlier models, the coupled dynamical equations governing the electric field inside the cavity and the carrier density of the active material take the form

$$\frac{\partial E}{\partial t} = -(1+\eta+\mathrm{i}\,\theta)E + E_{\mathrm{in}} + \mathrm{i}\,\Sigma\chi_{\mathrm{nl}}E + \mathrm{i}\,\nabla_\perp^2\, E\ , \tag{3}$$

$$\frac{\partial N}{\partial t} = -\gamma\left[N + \beta N^2 - \mathrm{Im}\,(\chi_{\mathrm{nl}})|E|^2 - d\,\nabla_\perp^2\, N\right]\ , \tag{4}$$

where η is the linear absorption coefficient due to the material in the regions between the semiconductor and the mirrors; Σ is a "bistability parameter"; N is the carrier density scaled to its transparency value; γ and β are the normalized decay rates of the carrier density that describe the nonradiative and radiative carrier recombination, respectively; and d is the diffusion coefficient. The complex susceptibility χ_{nl} describes the nature of the radiation–matter interaction and differs between the MQW and the bulk cases [18,44]. Whereas in MQW structures a Lorentzian excitonic resonance and a linear dependence on the carrier density N can be assumed [39], the bulk case is more complex and numerical simulations more demanding, and the susceptibility's dependence on N is highly implicit [39,44].

5 Stationary Solutions and Stability

Both models described above consist of three coupled, nonlinear, time-dependent partial differential equations. Cylindrically symmetric cavity soliton solutions have been found numerically for both models using the method described above [39].

For the bulk case, the soliton branch shows an intricate spiraling behavior (Fig. 4), which might suggest a region of bistability between solitons of different intensity. We should also, however, consider azimuthally varying perturbations. So let us analyze perturbations of the form $\varepsilon = R(r)r^{|m|}e^{im\theta}$. The stability analysis is easily generalized to allow for nonzero m, yielding a separate eigenspectrum for each azimuthal index. We can see in Fig. 5 that for $m = 0, 1$ we have no unstable modes, but there is a neutral mode for $m = 1$. It is easy to show that this neutral mode, which here has a mixed field–photocarrier character, is related to the translational invariance or symmetry of the governing equations. Finally, for $m = 2$ we see that the upper CS branch loses its stability at $E_{\text{in}} = 38.5$, in accordance with the dynamical simulations.

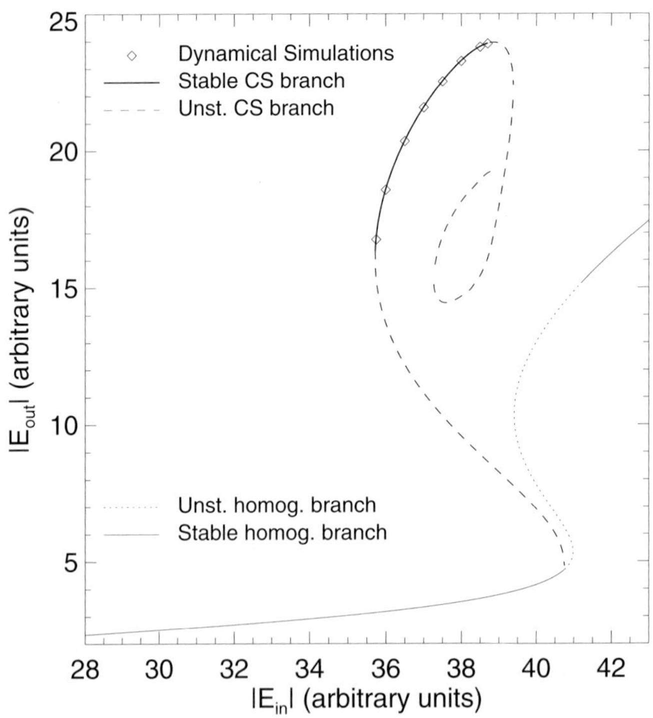

Fig. 4. Results from the bulk model in two transverse dimensions [39], showing CS and homogeneous solutions and their stability

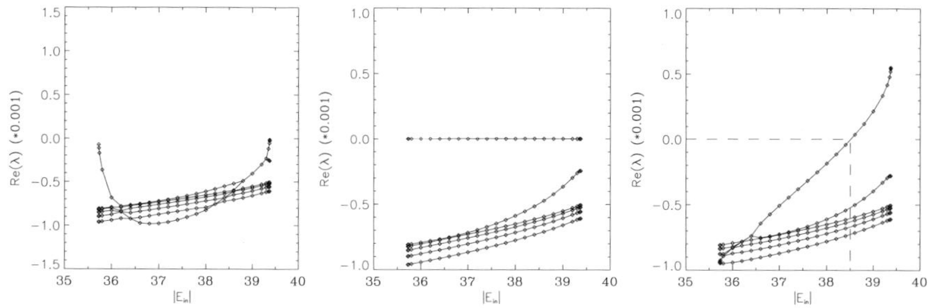

Fig. 5. 2D bulk model. Perturbation eigenvalues for a cylindrically symmetric CS as a function of the input field. The perturbations have azimuthal index $m = 0$ (*left*); $|m| = 1$ (*center*); $|m| = 2$ (*right*)

In Fig. 6 we display frames from a dynamical simulation which correspond to the destabilization of a cavity soliton for an input field slightly above $|E_{\mathrm{I}}| = 38.5$, and we clearly see that it loses its stability via an asymmetric deformation of $m = 2$ type.

Now, consider the effect of perturbations on stationary CS solutions, important in view of possible applications to optical processing. We identify three types of perturbation: those due to imposed modifications to the external driving field, those due to noise, and those due to interactions between cavity solitons. The first can be used to *manipulate* CSs, the other two are undesirable in most applications (but of scientific interest nevertheless).

For a stable stationary solution, all eigenvalues have negative real part apart from the neutral mode $\boldsymbol{u}_0$. This means that, as $t \to \infty$, the amplitude a_0 of the neutral mode dominates over all other a_i. Thus the dynamical effect of any perturbation $\boldsymbol{\mathcal{P}}$ on a stationary stable state is primarily determined by its projection onto the neutral mode, which yields the following equation [39]:

$$\frac{\mathrm{d}a_0}{\mathrm{d}t} = \frac{1}{\langle \boldsymbol{v}_0 \mid \boldsymbol{u}_0 \rangle} \langle \boldsymbol{v}_0 \mid \boldsymbol{\mathcal{P}} \rangle \ . \tag{5}$$

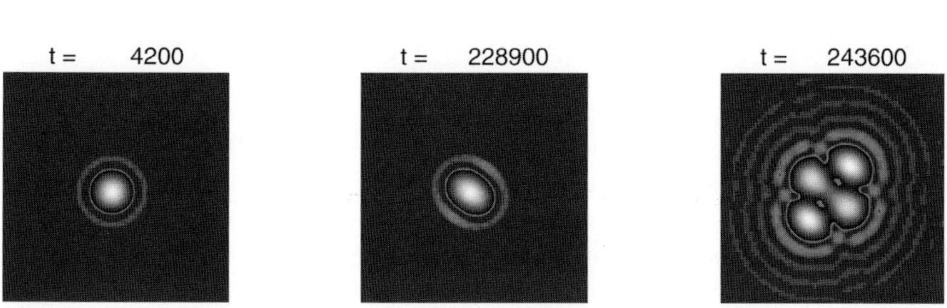

Fig. 6. 2D bulk model. Dynamical evolution of a CS for input field $|E_{\mathrm{I}}| = 38.6$; the soliton destabilizes via an $m = 2$ azimuthal instability [39]

Here $\boldsymbol{v}_0$ is the neutral mode of the corresponding adjoint problem. Because the neutral mode is just the gradient of the CS, physically da_0/dt is the translational velocity of the CS under the influence of the perturbation.

Among the various types of perturbation, three of particular relevance are a phase or an amplitude gradient of the driving field, and a perturbation of one soliton by another. For a weak phase gradient, the input field around a CS at $x = 0$ can be locally approximated by $E_{\rm in} = E_0(1 + ikx)$. Inserting the perturbation part $ikxE_0$ into (5), we can calculate the drift velocity of a cavity soliton due to the phase gradient (or, similarly, an amplitude gradient). The results obtained for both the MQW and the bulk models are in very good agreement with direct simulations [39].

Consider now the perturbation of a soliton by another soliton. The oscillations of the phase and amplitude of the CS field as it dies away into the background field can be regarded as somewhat equivalent, in their effects on a second nearby soliton, to imposition of varying gradients on the input field. The resulting induced relative motion of the two CSs will be positive or negative, depending on their separation. We can expect to find equilibrium positions where the relative velocity is zero, and that these define stable or unstable bound states of two CSs. Such behavior is indeed found in both semiconductor models [39]. Analogous CS interaction phenomena have been reported for other systems [25,26].

Finally, an important result concerning the role of the unstable CS branch, acting as a separatrix, can be obtained by exploiting the considerations developed so far. The dynamics of the unstable CS, both in the MQW and in the bulk model, are governed by a single eigenvalue, with a corresponding eigenvector with a shape similar to the unstable soliton [39]. We can infer that the unstable CS is actually *metastable*, in that it will attract "nearby" field configurations of broadly similar shape and strength. These will reshape towards the unstable CS configuration along its stable manifold, before escaping along its *one-dimensional* unstable manifold. Given this essentially single-mode behavior, we can anticipate that the locus of the unstable CS will act as a separatrix of the two stable coexisting solutions: the homogeneous solution and the CS. Dynamical simulations confirm this, for both the MQW and the bulk, in 1D and in 2D [39]. It follows that writing of stable CSs can be achieved even with relatively imprecise address pulses.

Experimental confirmation of such separatrix behavior has been obtained for "feedback mirror solitons" in Na, vapor [41] where the behavior was used to map out the unstable soliton branch. In a cavity situation, indications of separatrix behavior have been observed in "localized switching" of pattern elements in a semiconductor microresonator (see the chapter by Weiss et al. in this book). It is likely that many similar systems will show corresponding dynamics. This shows both how unstable states may be experimentally mapped out and that knowledge of such states can be very significant in the understanding and control of spatial structures in nonlinear optical systems. This makes the analysis of stationary solutions an even more valuable technique in the study of model systems.

6 Applications

Arrays of cavity solitons may have applications in parallel information processing [16]. They can be created at any location by a suitable address pulse, and are nondiffracting and dynamically stable, all of which makes them suitable "bits" for capture, storage, and processing of images or data. This has been demonstrated numerically [8] and confirmed in a prototype experiment [16] (see also the chapter by Weiss et al. in this book). A simple binary array memory is unlikely to be competitive with electronic storage. Cavity solitons, however, also offer functionalities which are beyond any microstructured material array, whether optical or electronic. In particular, we have seen that they can be optically manipulated, for example by imposing a spatial phase profile on the driving field [8,18,39]. Processing schemes which take advantage of their unique properties may avert the unequal competition with silicon which has plagued other all-optical switching and processing schemes.

The plasticity of cavity solitons can be linked to the "neutral mode" identified above, which is associated with the translational invariance of the underlying equations defining the cavity solitons. Any perturbation to the pump field which has a finite gradient at the soliton location will couple to the neutral mode and cause the soliton to move. As already mentioned, this has implications for the response of the cavity solitons to noise and to any stray gradients, and also for interaction between solitons. It can also be used postively, to *control* the motion and location of the CS through the spatial phase profile of the pump field. One can think of the solitons in a "landscape" determined by the phase of the holding (control) field, a landscape in which they move in response to phase gradients. Thus a simple memory array [8] consists of a regular landscape of "hills" and "valleys", with the solitons attracted to the peaks. Unlike one formed from machined pixels, however, this landscape is reconfigurable by changing the phase profile of the control field. This allows cavity soliton bits to be manipulated, by either global or local reconfigurations of the control field. This plasticity opens up possibilities for novel processing functions and applications [46]. One can also envisage useful applications in digital image processing, for example in feature extraction.

7 Conclusion

We have discussed a class of stable soliton-like structures predicted to exist in driven optical cavities containing any of a wide variety of nonlinear materials. This class includes semiconductor microresonators, which is promising for possible applications of these cavity solitons. Cavity solitons can be formed into two-dimensional arrays of information bits which can be written, stored, read, erased [11], and spatially manipulated in various ways. They can thus act as the basis of a new kind of all-optical parallel processor, with functionalities not available to other processing and storage devices in information technology.

The first experimental verifications of these solitons have been performed [16] (see also the chapter by Weiss et al. in this book), while they have also been found theoretically in quite a wide variety of cavity systems containing a nonlinear optical medium. We believe that in the coming years they will find an important role both in optics and in optoelectronic technology.

Acknowledgments

This work is partially supported by ESPRIT project 28235 PIANOS and EPSRC grant GR/M 19727. We thank our PIANOS partners for many helpful discussions and insights. In particular, we are extremely grateful to Tommaso Maggipinto and Massimo Brambilla for allowing us to use extracts from our joint work [39] prior to publication. We also thank Angus Lord and Andrew Scroggie for important contributions to this work.

References

1. L.A. Lugiato (ed.) Chaos Solitons Fractals **4** 1307 (1994), Special issue on "Nonlinear Optical Systems, Chaos, Noise".
2. D.W. McLaughlin, J.V. Moloney, A.C. Newell, Solitary waves as fixed points of infinite dimensional maps in an optical bistable ring cavity, Phys. Rev. Lett. **51**, 75 (1983).
2. (a) J.V. Moloney, A.C. Newell, *Nonlinear Optics* (Addison-Wesley, Redwood City, 1992).
3. N.N. Rosanov, G.V. Khodova, Opt. Spektrosk. **65**, 1375 (1988); Diffractive autosolitons in nonlinear interferometers, J. Opt. Soc. Am. B **7**, 1057 (1990).
4. G.S. McDonald, W.J. Firth, Spatial solitary-wave optical memory, J. Opt. Soc. Am. B (1990). **7**, 1328
5. S. Wabnitz, Suppression of interactions in a phase-locked soliton optical memory, Opt. Lett. **18**, 601 (1993).
6. G.S. McDonald, W.J. Firth, Switching dynamics of spatial solitary wave pixels, J. Opt. Soc. Am. B **10**, 1081 (1993).
7. M. Tlidi, P. Mandel, R. Lefever, Localized structures and localized patterns in optical bistability, Phys. Rev. Lett. **73**, 640 (1994).
8. W.J. Firth, A.J. Scroggie, Optical bullet holes: robust controllable localized states of a nonlinear cavity, Phys. Rev. Lett. **76**, 1623 (1996).
9. W.J. Firth, A. Lord, Two dimensional solitons in a Kerr cavity, J. Mod. Opt. **43**, 1071 (1996).
10. W.J. Firth, A. Lord, A.J. Scroggie, Optical bullet holes, Phys. Scr. **T67**, 12 (1996).
11. M. Brambilla, L.A. Lugiato, M. Stefani, Interaction and control of optical localized structures, Europhys. Lett. **34**, 109 (1996).

12. M. Tlidi, P. Mandel, Spatial patterns in nascent optical bistability, Chaos Solitons Fractals **4**, 1475 (1996).
13. A.J. Scroggie, W.J. Firth, G.S. McDonald, M. Tlidi, R. Lefever, L.A. Lugiato, Pattern formation in a passive Kerr cavity, Chaos Solitons Fractals **4**, 1323 (1996).
14. S. Longhi, Dark solitons in degenerate optical parametric oscillators, Opt. Lett. **21**, 860 (1996).
15. G. Steinmeyer, A. Schwache, F. Mitschke, Quantitative characterization of turbulence in an optical experiment, Phys. Rev. E **53**, 5399 (1996).
16. C.O. Weiss, et al., ESPRIT LTR Project 21112 PASS, Report (1997).
17. M. Brambilla, L.A. Lugiato, F. Prati, L. Spinelli, W.J. Firth, Spatial soliton pixels in semiconductor devices, Phys. Rev. Lett. **79**, 2042 (1997).
18. L. Spinelli, G. Tissoni, M. Brambilla, F. Prati, L.A. Lugiato, Spatial solitons in semiconductor microcavities, Phys. Rev. A **58**, 2542 (1998).
19. R. Gallego, M. San Miguel, R. Toral, Self-similar domain growth, localized structures, and labyrinthine patterns in vectorial Kerr resonators, Phys. Rev. E **61**, 2241 (2000).
20. C. Etrich, U. Peschel, F. Lederer, Solitary waves in quadratically nonlinear resonators, Phys. Rev. Lett. **79**, 2454 (1997).
21. D. Michaelis, U. Peschel, F. Lederer, Multistable localized structures and superlattices in semiconductor optical resonators, Phys. Rev. A **56** R3366 (1997).
22. K. Staliunas, V.J. Sanchez-Morillo, Localized structures in degenerate optical parametric oscillators, Opt. Commun. **139**, 306 (1997).
23. M. Tlidi, P. Mandel, M. Haelterman, Spatiotemporal patterns and localized structures in nonlinear optics, Phys. Rev. E **56**, 6524 (1997).
24. D. Michaelis, U. Peschel, F. Lederer, Oscillating cavity solitons with a saturating defocusing nonlinearity, *OSA Topical Meeting*: Nonlinear Guided Waves, Victoria, Canada, 1988, paper NWA3 (1998).
25. D.V. Skryabin, W.J. Firth, Interaction of cavity solitons in degenerate optical parametric oscillators, Opt. Lett. **24**, 1056 (1999).
26. I.V. Barashenkov, E.V. Zemlyanaya, Stable complexes of parametrically driven, damped nonlinear Schrödinger solitons, Phys. Rev. Lett. **83**, 2568 (1999).
27. M. Tlidi, M. Haelterman, Three dimensional structures in diffractive and dispersive nonlinear ring cavities, *Euroconference on Patterns in Nonlinear Optical Systems*, Alicante, Spain, 1998.
28. W.J. Firth, G.K. Harkness, Cavity solitons, Asian J. of Phys. **7**, 665 (1998).
29. N.B. Abraham, W.J. Firth, Overview of transverse effects in nonlinear-optical systems, J. Opt. Soc. Am. B **7**, 951 (1990).
30. H.M. Gibbs, *Optical Bistability – Controlling Light with Light*, (Academic Press, Orlando, 1985).
31. L.A. Lugiato, Theory of optical bistability, in *Progress in Optics* Vol. 21, ed. by E. Wolf (North-Holland, Amsterdam, 1984), p. 69
32. W.J. Firth, I. Galbraith, Diffusive transverse coupling of bistable elements – switching waves and crosstalk, IEEE J. Quantum Electron. **21**, 1399 (1985).
33. N.N. Rosanov, Trasverse patterns in wide-aperture optical systems, in *Progress in Optics* Vol. 35, ed. by E. Wolf (North-Holland, Amsterdam, 1996), p. 1
34. N.N. Rosanov, S.V. Fedorov, G.V. Khodova, J. Exp. Theor. Phys. **107**, 376 (1995); N.N. Rosanov, S.V. Fedorov, G.V. Khodova, Characterization of localized transverse structures in wide-aperture lasers, Physica D **96**, 272 (1996).

35. V.Yu Bazhenov, V.B. Taranenko, M.V. Vasnetsov, Proc. SPIE **1840**, 183 (1992); V.B. Taranenko, K. Staliunas, C.O. Weiss, Spatial soliton laser: localized structures in a laser with a saturable absorber in a self-imaging resonator, Phys. Rev. A **56**, 1582 (1997).
36. M. Saffman, D. Montgomery, D.Z. Anderson, Collapse of a transverse mode continuum in a self-imaging photorefractively pumped ring resonator, Opt. Lett. **19**, 518 (1994).
37. L.A. Lugiato, R. Lefever, Spatial dissipative structures in passive optical systems, Phys. Rev. Lett. **58** 2209 (1987).
38. A.M. Dunlop, W.J. Firth, E.M. Wright, Master equation for spatiotemporal beam propagation and Kerr lens mode-locking, Opt. Commun. **138**, 211 (1997).
39. T. Maggipinto, M. Brambilla, G.K. Harkness, W.J. Firth, Cavity Solitons in semiconductor microresonators: existence, stability and dynamical properties, Phys. Rev. E **62**, 8726–8739 (2000).
40. G.-L. Oppo, A.J. Scroggie, W.J. Firth, From domain walls to localized structures in degenerate optical parametric oscillators, J. Opt. B – Quantum and Semiclassical Optics **1**, 133 (1999).
41. B. Schäpers, M. Feldmann, T. Ackemann, W. Lange, Phys. Rev. Lett. **85**, 748 (2000).
42. R. Neubecker, G.-L. Oppo, B. Thuering, T. Tschudi, Pattern formation in a liquid crystal light valve with feedback, including polarization, saturation and internal threshold effects, Phys. Rev. A **51**, 791 (1995).
43. A. Schreiber, B. Thüring, M. Kreuzer, T. Tschudi, Experimental investigation of solitary structures in a nonlinear optical feedback system, Opt. Commun. **136**, 415 (1997).
44. G. Tissoni, L. Spinelli, M. Brambilla, T. Maggipinto, I. Perrini, L.A. Lugiato, Cavity solitons in passive bulk semiconductor microcavities: I: Microscopic model and modulational instabilities, J. Opt. Soc. Am. B **16** 2083 (1999).
45. G. Tissoni, L. Spinelli, M. Brambilla, T. Maggipinto, I. Perrini, L.A. Lugiato, Cavity solitons in passive bulk semiconductor microcavities: II: Dynamical properties and control, J. Opt. Soc. Am. B **16** 2095 (1999).
46. W.J. Firth, Processing information with arrays of spatial solitons, Proc. SPIE **4016**, 388 (2000).

Parametric Solitons in Passive Structures with Feedback

Stefano Trillo and Marc Haelterman

Summary. This chapter is aimed at reviewing our understanding of localization phenomena in passive feedback structures with quadratic nonlinearities. We analyze first feedback due to passive resonantors, discussing the existence, stability, and control of solitons in the transverse plane of optical parametric oscillators and intracavity second-harmonic generation. Then, we consider briefly the case of Bragg gratings responsible for distributed feedback, revealing the existence of new types of longitudinally localized two-color gap solitons.

1 Introduction

A well-known means to enhance the effects of any parametric mixing interaction is to recirculate the fields in an optical cavity, thus introducing a linear positive-feedback mechanism into a nonlinear process that would otherwise occur in a traveling-wave configuration. A device of this kind is a passive nonlinear resonator, which, strictly speaking, behaves as an oscillator whenever it has a definite threshold of operation. The first demonstration of an optical parametric oscillator (OPO) was given in the early days of nonlinear optics by generating the subharmonic of a pump field via an intracavity $\chi^{(2)}$ parametric process [1]. Soon thereafter, intracavity second-harmonic generation (IC-SHG) was also demonstrated [2]. Since then, intracavity parametric mixing and, in particular, OPOs have attracted steadily growing interest from the point of view of applications, as they are becoming widespread as tunable and versatile sources [3]. From a different perspective, intracavity parametric mixing offers the opportunity to investigate and understand more fundamental aspects, such as those related to the physics of nonequilibrium systems [4], chaos [5], and quantum optics (see, e.g., [6–8]). In particular, intracavity mixing is described by mathematical models which belong to the general class of nonlinear dissipative dynamical systems with driving (optical pumping) and dissipation (cavity losses), which in turn describing the physics of nonequilibrium systems. It is well known that this kind of system exhibits the phenomenon of dissipative pattern formation (see the reviews by Cross and Hohenberg [9] and Walgraef [10] for an interdisciplinary exhaustive approach, and by Lugiato et al. for this phenomenon in optics [11]). Nevertheless, it was only in 1994 that the possibility of pattern formation driven by diffraction in broad-area (or large-aspect-ratio) passive $\chi^{(2)}$ cavities was pointed out

[12], though such mechanisms had been addressed earlier in non-$\chi^{(2)}$ passive resonators [13–15]. Even more recent is the idea of searching for fields localized (i.e. trapped) in the transverse plane of $\chi^{(2)}$-filled resonators [16–18], owing to the interplay of diffraction and parametric nonlinearity as well as pumping and losses. Such entities, namely dissipative parametric solitons or parametric cavity solitons (PCSs), appear to be a particular nonperiodic case of pattern formation, which also forms a bridge between this area of research and that concerning parametric solitons in a traveling-wave configuration (see the chapter by Torruellas et al. in this book). In the present work, our aim is to review the main ideas and results obtained in the young and rapidly developing field of PCSs.

Linear feedback, however, can also be introduced in a distributed way without employing mirrors. A Bragg grating or dielectric stack made from materials which respond nonlinearly constitutes a nonlinear distributed-feedback (NLDFB) structure. NLDFB structures sustain a different type of self-trapped wave envelope, namely Bragg or gap soliton, discussed in detail, for the case of Kerr nonlinearities by de Sterke et al. in this book. Owing to the mutual compensation of nonlinearity and dispersion originating from distributed feedback, these solitons are localized along the propagation direction rather than in the transverse plane. While referring the reader to that specific chapter of this book and references therein for the basic concepts concerning gap solitons, we discuss below how quadratic nonlinearities can be exploited to trap electromagnetic energy at different (parametrically mixing) carrier frequencies in an NLDFB device in the form of parametric gap solitons (PGSs).

The chapter is organized as follows. In Sect. 2 we review the models which govern the formation of transverse coherent structures in passive quadratic cavities. Section 3 is devoted to a general discussion of instabilities in such cavities, while Sects. 4 and 5 specifically address OPOs and IC-SHG, respectively. Finally, in Sect. 6, we review the main results underlying the physics of PGSs.

2 General Models for Quadratic Passive Cavities

A simple and widely used approach can be developed for monochromatic fields, in the good-cavity, single-mode limit, starting from coupled mean-field partial differential equations (PDEs) obtained by continuation of the map which couples the cavity boundary conditions with the propagation equations. Dealing with single longitudinal modes, the nondegenerate parametric mixing interaction of three envelopes E_j, $j = 1, 2, 3$, at frequencies ω_j where $\omega_3 = \omega_1 + \omega_2$ can be described by the following standard paraxial equations (for the sake of simplicity we assume noncritical phase-matching or walk-off-compensated geometries [19] here, and hence we neglect spatial walk-off for the time being):

$$
\begin{aligned}
-\mathrm{i}\frac{\partial E_1}{\partial Z} &= d_1 \nabla^2 E_1 + \chi_1 E_3 E_2^* \exp(\mathrm{i}\Delta k Z) , \\
-\mathrm{i}\frac{\partial E_2}{\partial Z} &= d_2 \nabla^2 E_2 + \chi_2 E_3 E_1^* \exp(\mathrm{i}\Delta k Z) , \\
-\mathrm{i}\frac{\partial E_3}{\partial Z} &= d_3 \nabla^2 E_3 + \chi_3 E_1 E_2 \exp(-\mathrm{i}\Delta k Z) .
\end{aligned} \tag{1}
$$

Here $\nabla^2 = \partial^2_{XX} + \partial^2_{YY}$ is the transverse Laplacian ($\nabla^2 = \partial^2_{XX}$ in a cavity with waveguiding confinement along Y), $\delta k = k_3 - k_1 - k_2$ is the wavevector mismatch, $k_j = \omega_j n_j / c$ ($j = 1,2,3$), $d_j = 1/(2k_j)$, and χ_j stands for the nonlinear coefficient $\omega_j d_{\mathrm{eff}}/(c n_j)$, d_{eff} being the effective second-order nonlinear susceptibility. In a ring cavity with flat mirrors (or a Fabry–Perot cavity), the fields after two successive round trips, say the nth and $(n+1)$th, are coupled by the boundary conditions

$$
E_j^{(n+1)} = R_j \exp(\mathrm{i}\Phi_j)\, E_j^{(n)} + T_j E_j^{(\mathrm{p})} , \; j = 1,2,3 , \tag{2}
$$

where $E_j^{(\mathrm{p})}$ are the external pump field amplitudes, and $R_j = R_j(\omega_j)$, and $T_j = T_j(\omega_j)$, where $R_j^2 + T_j^2 = 1$, are the reflectivity and transmissivity of the input/output coupler.

2.1 Mean-Field Models

A standard continuation of the map yields (see [20,21] for an explicit derivation)

$$
\mathrm{i}\frac{\partial u_j}{\partial z} + \rho_j \nabla^2 u_j + u_3 u_{3-j}^* + (\mathrm{i}\alpha_j - \Delta_j) u_j = \mathrm{i}\, S_j , \quad j = 1,2 ,
$$

$$
\mathrm{i}\frac{\partial u_3}{\partial z} + \rho_3 \nabla^2 u_3 + u_1 u_2 + (\mathrm{i}\alpha_3 - \Delta_3) u_3 = i\, S_3 , \tag{3}
$$

where we have introduced, for convenience, the normalized distance $z = Z/Z_{\mathrm{c}}$ in units of the mean photon propagation length in the cavity $Z_{\mathrm{c}} = L/\gamma_1$, associated with losses $\gamma_1 = T_1^2/2$ at ω_1. Note that $z = \gamma_1 N_{\mathrm{rt}}$ is physically accessible only for integer values of the number of round trips $N_{\mathrm{rt}} = Z/L$, and can be equivalently regarded as a normalized time $t = \gamma_1 T/T_{\mathrm{rt}} = \gamma_1 N_{\mathrm{rt}}$, T and T_{rt} being the observation time in the lab frame and the round-trip time, respectively. The choice of using t is more common in the literature concerning parametric cavities, although the choice of z is more natural if one wishes to compare the results with parametric effects in traveling-wave geometries. Furthermore, the transverse coordinates (X, Y) have been scaled to $x = (Z_{\mathrm{d}}/Z_{\mathrm{c}})^{1/2} X/X_0$ and $y = (Z_{\mathrm{d}}/Z_{\mathrm{c}})^{1/2} Y/X_0$, where X_0 is a reference transverse width and Z_{d} is the corresponding diffraction length at frequency ω_1, i.e. $Z_{\mathrm{d}} = X_0^2/d_1$. In (3) the rescaled Laplacian thus reads $\nabla^2 = \partial^2_{xx} + \partial^2_{yy}$, and $\rho_j = d_j/d_1$ are diffraction coefficients. The rescaled cavity losses $\alpha_j = (T_j/T_1)^2$, together with the normalized cavity detunings $\Delta_j = (\omega_{cj} - \omega_j)2n_j L/(cT_1^2)$, where ω_{cj} is the cavity resonance frequency closest to ω_j, describe *linear* steady-state Lorentzian resonances $u_j = S_j/(\alpha_j + \mathrm{i}\Delta_j)$. The new dimensionless intracavity field envelopes are conveniently written, by introducing the common factor $\eta = Z_\mathrm{c}\,\mathrm{sinc}(\Delta k\, L/2)$, as $u_j = (\chi_3\chi_{3-j})^{1/2}\eta E_j$, where $j = 1, 2$, and $u_3 = (\chi_1\chi_2)^{1/2}\eta E_3\, \exp(\mathrm{i}\Delta k L/2)$. The rescaled pump amplitudes are $S_j = 2T_j(\chi_3\chi_{3-j})^{1/2}\eta E_j^{(\mathrm{p})}/T_1^2$ $(j = 1, 2)$ and $S_3 = [2T_3(\chi_1\chi_2)^{1/2}E_3^{(\mathrm{p})}/T_1^2]\exp(\mathrm{i}\,\Delta k\, L/2)$.

In practice, the cavity is usually pumped at one frequency only. By pumping the ω_3 field ($S_3 \neq 0, S_{1,2} = 0$), one obtains a nondegenerate OPO (hereafter referred to as an "NDOPO"), where u_1 and u_2 are the signal and idler fields, respectively. While the cavity is always made to resonate at the signal field (singly resonant case), whenever it is also resonant (i.e. low-loss) at one or both of the other frequencies, one has a doubly or triply resonant OPO, respectively. Moreover, (3) describes also the frequency-degenerate case, in which u_1, u_2 are distinguished by their polarizations in the so-called type II phase-matching configuration. In this case (3), with $S_3 = 0$, also governs type II IC-SHG.

Of great importance is the completely degenerate case, i.e. $\omega_1 = \omega_2 = \omega = \omega_3/2$ in the type I phase-matching configuration, for which the fields at the fundamental frequency (FF) ω become indistinguishable. In the case of IC-SHG, following the procedure which leads to (3), it is easy to find that the envelopes u_1, u_2 (at frequencies $\omega, 2\omega$, respectively) obey the system

$$-\mathrm{i}\frac{\partial u_1}{\partial z} = \rho_1\nabla^2 u_1 + u_2 u_1{}^* + (\mathrm{i} - \Delta_1)u_1 - \mathrm{i}\,S_1\,,$$

$$-\mathrm{i}\frac{\partial u_2}{\partial z} = \rho_2\nabla^2 u_2 + \frac{u_1^2}{2} + (\mathrm{i}\alpha - \Delta_2)u_2\,, \tag{4}$$

whereas the model governing a degenerate OPO (DOPO) reads

$$-\mathrm{i}\frac{\partial u_1}{\partial z} = \rho_1\nabla^2 u_1 + u_2 u_1{}^* + (\mathrm{i} - \Delta_1)u_1\,,$$

$$-\mathrm{i}\frac{\partial u_2}{\partial z} = \rho_2\nabla^2 u_2 + \frac{u_1^2}{2} + (\mathrm{i}\alpha - \Delta_2)u_2 - \mathrm{i}\,S_2\,. \tag{5}$$

Since (5) can be thought of as a limiting case of an NDOPO, we shall continue to refer to it as a triply resonant case (whenever $\alpha \sim 1$), to avoid confusion with a doubly resonant (signal and idler) NDOPO. In (4) and (5) $\rho_1 = 1$, and $\rho_2 \simeq 1/2$ owing to phase-matching constraints. Therefore, the cavity is characterized by three parameters $(\alpha, \Delta_1, \Delta_2)$ which can be controlled together with pump intensity S_1^2 or S_2^2. Note, however, that the same mean-field models hold true for a nearly confocal cavity if the coefficients in front of the transverse Laplacian are rescaled, i.e. $\rho_2 \neq 1/2$ (see [22,23] and references therein).

Since a large body of the literature on PCSs deals with DOPOs, it is worth pointing out that (5) is also presented in several different equivalent forms. For instance, the mixing nonlinear terms might appear with different signs, following the trivial rescaling $u_1/\sqrt{2} \to u_1$, $u_2 \to -\mathrm{i}u_2$, $S_2 \to -\mathrm{i}S_2$. A less trivial set of new variables $U_1 = u_1 \exp(\mathrm{i}\phi_\mu/2)$, $U_2 = u_2 \exp(-\mathrm{i}\phi_\mu) - \mu$, where $\mu = S/\sqrt{\alpha_2^2 + \Delta_2^2}$ and $\phi_\mu = \pi - \tan^{-1}(\alpha_2/\Delta_2)$ ($u_1 = 0$, $u_2 = \mu \exp(\mathrm{i}\phi_\mu)$ is the trivial homogeneous solution of (5)), yields the following equivalent model (see, e.g. [16,25]):

$$-\mathrm{i}\frac{\partial U_1}{\partial z} = \rho_1 \nabla^2 U_1 + (U_2 + \mu)\, U_1{}^* + (\mathrm{i} - \Delta_1) U_1 \, ,$$

$$-\mathrm{i}\frac{\partial U_2}{\partial z} = \rho_2 \nabla^2 U_2 + \frac{U_1^2}{2} + (\mathrm{i}\alpha - \Delta_2) U_2 \, , \tag{6}$$

where μ measures the pump strength.

It is important to realize that one major difference exists between cavity models and the case of traveling-wave (cavityless) mixing in transparent media discussed by Torruellas et al. in this book. In the latter case the underlying models possess a Hamiltonian structure $-\mathrm{i}(\partial u_j/\partial z) = \delta H/\delta u_j^*$, where $\delta/\delta u_j^*$ is the variational derivative with respect to the modal amplitude $u_j = u_j(z, \boldsymbol{r})$, and $\boldsymbol{r} = (x, y)$. The Hamiltonian (electromagnetic energy) H is conserved together with mass (power) as a consequence of symmetries of the model (see, e.g., [24]). In (4)–(6) the cavity introduces a perturbation (detuning, pumping, and losses) to the combined effect of diffraction and mixing, which is described in traveling-wave geometries by the Hamiltonian $H = \int_{-\infty}^{+\infty} (1/2)(u_2^* u_1^2 + \text{c.c.}) - \rho_1 |\nabla u_1|^2 - \rho_2 |\nabla u_2|^2 \,\mathrm{d}\boldsymbol{r}$; this can be introduced thanks to the scaling $u_1^2 \to u_1^2/2$ implicitly adopted in (4)–(6). Although detuning ($\Delta_{1,2}$) and pumping (S_1 or S_2) terms can be easily included in the Hamiltonian (see, e.g., [25]), this is not the case for the unavoidable presence of cavity losses. These destroy the invariance of the Hamiltonian and the power, thus spoiling the Hamiltonian structure, as generally occurs in dissipative systems. The analysis of the cavity models is further complicated by the fact that they do not possess a gradient form (for a detailed discussion, see p. 117 of [10]) and hence their evolution is not governed by the existence of a real Lyapunov functional.

2.2 Order-Parameter Equations

The great majority of theoretical studies of solitons and pattern formation in quadratic passive cavities are based on the mean-field models discussed above. We mention, however, that the improved performance of personal computers has made accessible direct numerical studies of infinite-dimensional maps such as those described by (1) and (2) [26,27].

Yet another approach (oriented more towards analytical techniques) consists in reducing, when possible and under some constraints, the mean-field starting models to a unique equation called an order-parameter equation (OPE). The derivation of OPEs is of great fundamental and practical importance since, on the one hand, they allow a simplified description of the spatiotemporal dynamics of the intracavity frequency conversion processes and, on the other hand, they provide connections with other pattern-forming systems in other branches of nonlinear optics (e.g. laser physics) and in nature in general (e.g. in hydrodynamics, chemistry, and biology [9,10]). In the following we briefly review the main outcomes of the perturbation analysis leading to an OPE, while referring the reader to the original literature for the derivation.

Staliunas has pioneered this approach for the description of transverse pattern formation in quadratic passive cavities [28]. He proceeded through adiabatic elimination of the field variables for singly and doubly resonant NDOPOs. In a singly resonant NDOPO, $\alpha_2, \alpha_3 \gg \alpha_1 = 1$, and in (3) the field derivatives $\partial_z u_2, \partial_z u_3$ can be dropped, together with diffraction for $u_{2,3}$ (which plays a significant role only over several passages, i.e. for recirculating fields), so as to obtain an OPE for the complex signal amplitude [28]):

$$\frac{\partial u_1}{\partial z} = g u_1 + \mathrm{i}(\Delta + a_1 \nabla^2) u_1 - |u_1|^2 u_1 \,, \tag{7}$$

where g (parametric gain slightly above threshold), Δ (effective detuning), and a_1 are real coefficients (note that the quantities in (7) are rescaled with respect to those in (3)). The OPE (7) is thus nothing but the complex Ginzburg–Landau equation (CGLE). This equation is well known, in particular in hydrodynamics and laser physics, where it has been shown to describe transverse effects in class A lasers in the limit of wide spectral gain bandwidth (see, e.g., [29]). The structures emitted by a singly resonant OPO are thus, in principle, identical to those of this class of lasers. In particular, vortices can be expected in resonators with a large Fresnel number [30].

In a doubly resonant OPO one has $\alpha_3 \gg \alpha_2, \alpha_1 = 1$, and adiabatic elimination of the field u_3 in (3) leads to [28]

$$\frac{\partial u_1}{\partial z} = g u_1^* + (\mathrm{i}\Delta - 1 + \mathrm{i}a_1 \nabla^2) u_1 - (a_2 + \mathrm{i}\Delta')|u_1|^2 u_1 \,, \tag{8}$$

where all coefficients are real, and the variables have been suitably rescaled. This equation has also been derived independently by another method [31]. The OPE (8) has the form of a CGLE but includes a parametric-gain term ($g u_1^*$) that introduces phase sensitivity into the system. This equation has a rather universal nature, as it can be derived in different settings, such as the evolution of temporal solitons in fiber chains with parametric amplifiers [32], and hydrodynamic solitons [33] in the so-called Faraday resonance problem [34]; it can even be derived in the context of OPOs, where it describes situations as different as an antiresonant OPO where PCSs arise from balancing of spectral filtering and pump depletion [35], and a triply resonant DOPO in the limit of large pump detuning [16,18]. It has stable soliton solution, which will be discussed in detail in Sects. 4.1 and 4.2, as well as periodic pulse train solutions [36]. The phase sensitivity in (8) drastically affects the transverse dynamics through the appearance of standing-wave patterns (instead of the traveling-wave patterns of class A lasers and singly resonant OPOs) in the form of rolls or hexagons. It has also been shown that, close to the oscillation threshold, (8) can be reduced to a real Swift–Hohenberg equation (SHE) [28,37],

$$\frac{\partial u_1}{\partial z} = (S-1)u_1 - u_1^3 - \frac{1}{2}(\nabla^2 - \Delta)^2 u_1 \, , \tag{9}$$

that also supports rolls and hexagons. The approach can be generalized to the nondegenerate case, which, under the constraint of small detunings $\Delta_{1,2}$, leads to the following complex SHE [37,38]:

$$\frac{\partial u_1}{\partial z} = u_1 + \mathrm{i}(\Delta - a_1\nabla^2)u_1 - (a_2\nabla^2 + \Delta)^2 u_1 - |u_1|^2 u_1 \, . \tag{10}$$

The real SHE model (9) has later been derived for the triply resonant DOPO described in the mean-field limit by (5) [39]. A more rigorous mathematical approach based on a multi-scale expansion was used in this case. Subsequently, on the basis of a stability analysis of the roll solutions of the real SHE (9), localized structures, or PCSs, were identified in the form of isolated stripes and filaments (i.e. bright PCSs) similar to those discussed in detail in Sect. 4 [17]. Generally speaking, the importance of the SHE was recognized previously in the context of transverse localization, as a model for nascent bistability [40] and lasers [41].

In a triply resonant DOPO, another real OPE which plays an important role in the PCS dynamics can be derived close to threshold. In this case the rescaled signal field amplitude u_1 at the FF obeys the Fisher–Kolgomorov equation (FKE) [42] (or real Ginzburg–Landau equation, sometimes simply referred to as the Landau equation [10]), derived in [16,43].

$$\frac{\partial u_1}{\partial z} = \nabla^2 u_1 - u_1^3 + 2\gamma_1 u_1 \, , \tag{11}$$

where $\gamma_1 > 0$ accounts for the deviation from threshold and is responsible for signal growth, which is saturated by the nonlinear effect under suitable constraints on the detunings [16,43].

Other OPE models are also convenient for the description of other types of PCSs. In particular, the following universal cubic–quintic CGLE (see also the chapter by Akhmediev and Ankiewicz in this book) describes soliton dynamics in a singly resonant NDOPO in a presence of a Kerr nonlinearity [27]:

$$\frac{\partial u_1}{\partial z} = (g - \alpha)u_1 - \mathrm{i}\nabla^2 u_1 + \gamma_3|u_1|^2 u_1 + \gamma_5|u_1|^4 u_1 \, , \tag{12}$$

where γ_3 is complex and all other coefficients are real.

The order-parameter approach has also been extended to obtain OPEs of the Ginzburg–Landau [44,45] or SHE [46] type, which accounts for the additional effect of walk-off in OPOs.

It is worth pointing out that, in general, the formation of the same physical localized structure might follow from different reductions to an OPE, for example, fronts or domain walls can be described analytically by either the parametric CGLE (8) [18], the SHE (10), or the FKE (11) [47–49]. On the other hand the same OPE with different values of the coefficients can describe different structures depending on the parameter region where the OPO operates, for example (8) describes domain walls (see Sect. 4.1) as well as bright PCSs below the oscillation threshold (see Sect. 4.2).

All these studies illustrate the practical interest of OPE models for the description and physical interpretation of transverse effects, in particular of soliton formation, in quadratic passive cavities. However, to end this section, we would like to draw the reader's attention to the fact that OPE models must be used with great care. These models are in general based on strong assumptions that are, most of the time, difficult to meet in practice. As a result, the OPE models are strictly valid only in very restricted parameter ranges corresponding to very specific physical situations. This has two consequences. On the one hand, OPE models should be derived very carefully, without neglecting parameters that are liable to play a relevant role in the cavity dynamics. For this purpose, it is imperative to acquire a good physical insight into the physical problem before formulation of the basic assumptions that will lead to the OPE model. On the other hand, even if derived in this spirit, OPEs such as those presented above should be considered only as convenient tools providing a qualitative description of the phenomenology involved in quadratic passive cavities. Despite this restriction, thanks to their generic nature, OPEs provide interesting links with other domains of nonlinear science and, in this way, they have the great potential to provide deeper physical insights into the complex phenomenology of intracavity frequency conversion processes.

3 Competing Instabilities in Intracavity Parametric Mixing

The preliminary study of steady-state ($\partial_z = 0$) homogeneous or plane wave (PW, $\nabla^2 \to 0$) solutions of cavity models, with their bifurcations and stability, is crucial because they represent the background on which any localized wave decays at infinity (in practice, for large transverse radii). In general, the linear stability analysis (LSA) of steady-state PWs in intracavity parametric mixing shows the occurrence of three different types of instabilities, two of which are known from early studies (see, e.g., [4,5,15,50]):

- *Bistability*, associated with subcritical branching bifurcation (reminiscent of a second-order phase-transition) of nontrivial PW solutions. Bistability is a well-known and deeply studied feature of any nonlinear structure with feedback. Except for the fact that the negative-slope branches are always unstable, IC-SHG (4) and DOPOs (5) exhibit qualitatively different features as a consequence of their PW bifurcation pictures. IC-SHG is thresholdless, i.e. a 2ω PW is generated at any pumping level. Bistability occurs in (4) whenever $\Delta_1\Delta_2 > \alpha$ and $(4/3)(\alpha - \Delta_1\Delta_2)^2/[(1+\Delta_1^2)(\alpha^2 + \Delta_2^2)] > 1$, and results in a typical S-shaped input–output response at both ω and 2ω. Figure 1a shows the intracavity intensities $P_j = |u_j|^2$, $j = 1, 2$, as a function of pump intensity S_1^2. Conversely, the DOPO model (5) has a trivial PW solution $(u_1, u_2) = (0, S_2/(\alpha + \mathrm{i}\Delta_2)$ which bifurcates at the threshold $S_2^2 = S_{\mathrm{th}}^2 = (1 + \Delta_1^2)(\alpha + \Delta_2^2)$, exchanging its stability with a nontrivial branch $(u_1 \neq 0)$. Under the conditions for such a bifurcation to be subcritical, i.e. $\Delta_1\Delta_2 > \alpha$ and $S_2^2 > S_{\mathrm{sub}}^2 = (\Delta_2 + \alpha\Delta_1)^2$, the DOPO exhibits signal (FF) bistability, whereas the second-harmonic (SH) beam grows linearly with S_2^2 up to threshold, above which it remains locked to its maximum value (see Fig. 1b).

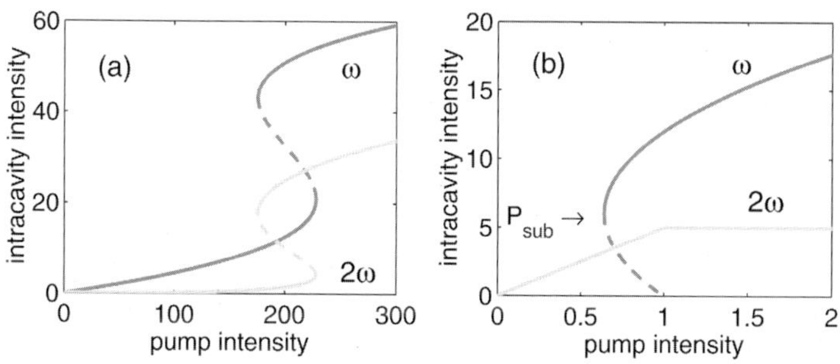

Fig. 1. Plane-wave bistable response of $\chi^{(2)}$ passive cavities. The intracavity intensities $P_1 = |u_1|^2$ at ω and $P_2 = |u_2|^2$ at 2ω are plotted versus pump intensity: (**a**) S_1^2 in IC-SHG; (**b**) S_2^2/S_{th}^2 in a DOPO. The *solid* and *dashed curves* represent steady-state PW branches which are stable and unstable, repectively, against PW perturbations. The parameters are as follows: (**a**) $\alpha = 1$, $\Delta_1 = \Delta_2 = 5$; (**b**) $\alpha = 1$, $\Delta_1 = -4$, $\Delta_2 = 1$

- *Self-pulsing instability*, associated with a Hopf bifurcation of a positive-slope branch of steady-state solutions [5,50,51]. Periodic self-pulsation can be observed in the case of a supercritical Hopf bifurcation which leads to a stable limit cycle. Above the bifurcation point, the limit cycle exhibits further bifurcations, which lead to period-N dynamics and eventually to disordered evolution (fully developed chaos [5]). Importantly, the systems have high dimensionality (four real variables for a DOPO or type I IC-SHG device), and the cascade of successive bifurcations does not follow a Feigenbaum-like route to chaos involving $N = 2^m$, where m is an integer. Self-pulsing usually occurs at a relatively low intracavity intensity above the bifurcation point $P_1 = |u_1|^2 > P_{\rm H}$, and only in highly detuned cavities. Close to resonance, the Hopf bifurcation intensity $P_{\rm H}$ usually becomes very large.

- *Dissipative modulational instability* (MI), also known as Turing [14] or Newell–Benjamin–Feir instability [52], which entails exponential growth of modulated waves with a 1D transverse spatial frequency (wavenumber) Ω ($\Omega = (\Omega_x, \Omega_y)$ in 2D) in a given bandwidth. This instability is the basic mechanism behind the formation of periodic structures (spatial patterns) in parametric cavities [12], which occurs thanks to the driven–damped (attractive) nature of the evolution equations for the relevant unstable modes. However, unlike the other two mechanisms, it survives (generally with recursive behavior) in the cavityless case also [53], giving rise to the observed filamentation phenomena [54]. In the intracavity configuration, after the pioneering paper by Oppo et al. [12], pattern formation has been the object of detailed theoretical studies in Fabry–Perot OPOs [11,28,36,39,55–59]. These studies have been extended to nearly confocal OPOs [22,60] for which recent experimental results have been obtained [61], and to IC-SHG in type I [62–65] and type II phase-matching configurations [66,67]. These studies have shed light on the occurrence of traveling-wave states, hexagons and other types of patterns, robust mixed Hopf–Turing modes, and turbulence, as well as the role played by universal OPE reductions.

  In a DOPO, MI is particularly important because it leads to threshold lowering for $\Delta_1 < 0$, i.e. the formation of a stationary roll pattern below the PW threshold for oscillation $S_{\rm th}$ [12]. The striped pattern has a periodicity corresponding to the most unstable wavenumber $\Omega = \sqrt{-\Delta_1/\rho_1}$ and is formed by a perturbation which grows exponentially with gain $g = -1 + \sqrt{P_2 - (\Delta_1 + \rho_1 \Omega)^2}$ on top of the trivial solution (where the intracavity power $P_2 = |u_2|^2$).

These three mechanisms compete. Interestingly enough, the LSA shows that MI can prevail over self-pulsing and bistability. In other words PW hysteresis and self-pulsing are unlike scenarios, at least in wide regions of parameter space. This is indeed the case as shown in Fig. 2 for a DOPO, where the MI gain at a finite spatial frequency Ω exceeds the gain at $\Omega = 0$ associated either with self-pulsing (Fig. 2a) or with negative-slope-branch instability of the bistable cycle (Fig. 2b). In Fig. 2a the high-gain branch is well separated from the low-frequency Hopf branch, which appears above the Hopf bifurcation power $P_{\rm H} = 1.25$. In the case of Fig. 2b bistability originates from the subcritical nature of the DOPO bifurcation shown in Fig. 1b, leading in turn to a negative-slope branch for an intracavity signal power $0 < P_1 < P_{\rm sub}$, which appears to be unstable against PW ($\Omega = 0$) perturbations. In the presence of transverse effects (diffraction), however, the

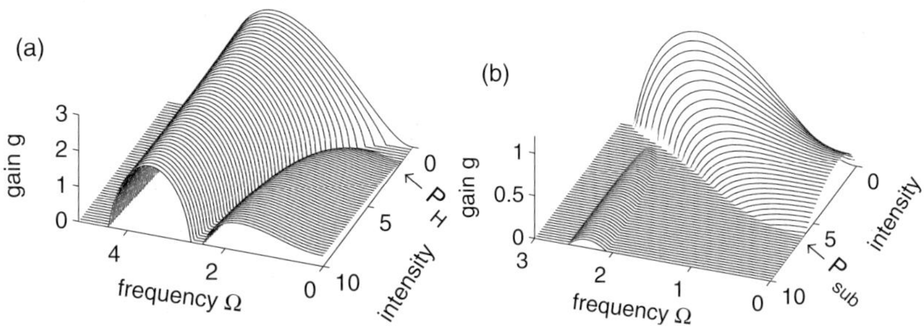

Fig. 2. MI gain g versus modulation frequency Ω and intracavity intensity $P_1 = |u_1|^2$ of a nontrivial PW branch in a DOPO. (**a**) For $\Delta_1 = -4$, $\Delta_2 = 1$, $\alpha = 1$, an MI branch with high frequency prevails over the low-frequency branch responsible for self-pulsing above the Hopf bifurcation power ($P_1 > P_{\mathrm{H}} = 1.25$). (**b**) For $\Delta_1 = \Delta_2 = -2$, $\alpha = 1$, bistability (gain at $\Omega = 0$ for $0 < P_1 < P_{\mathrm{sub}}$) is overwhelmed by MI at higher frequency

fastest-growing frequency is finite and, unlike in the self-pulsing case, belongs to the same branch, which accounts for the instability of the negative-slope branch at $\Omega = 0$.

As far as the existence and formation of PCSs is concerned, self-pulsing and chaos are simply detrimental, whereas bistability and MI play a more fundamental role. Indeed, we shall see that the permitted types of PCS are deeply related to the features of the PW bifurcation diagram, whereas the importance of MI is twofold. If MI coexists with PCS branches, it affects the stability of a PCS by breaking up its background. On the other hand, owing to the dissipative nature of the cavity problem, stable spatially periodic structures behave as attractors of the resonator dynamics past the early stage of exponential amplification of periodic perturbations, as shown in Fig. 3 (left panel) for a 1D case (in a 2D environment, the case shown is equivalent to a striped pattern). Roughly speaking, any single hump of the stationary pattern formed via MI can be viewed as a localized envelope. However, the spatial period of the pattern is fixed by the fastest-growing perturbation, and thinking about a PCS as the limit of the pattern for an infinite period is not legitimate. Nevertheless, the correspondence between spatial patterns and localized waves can be well founded when continuity exists between the branch of PCSs and, for example, the pixels of a hexagonal pattern [68].

To understand better how diffraction can alter substantially the evolution scenario shown in Fig. 3, we also report the ideal evolution which would be observed in the absence of diffraction (see right panel). As shown, the intracavity intensity exhibits disordered oscillations, since the DOPO is well above the self-pulsing threshold ($P_1 = 10 \gg P_{\mathrm{H}}$). These dynamics are, however, strongly stabilized by coherent self-organization into an MI-induced pattern-like output (left panel).

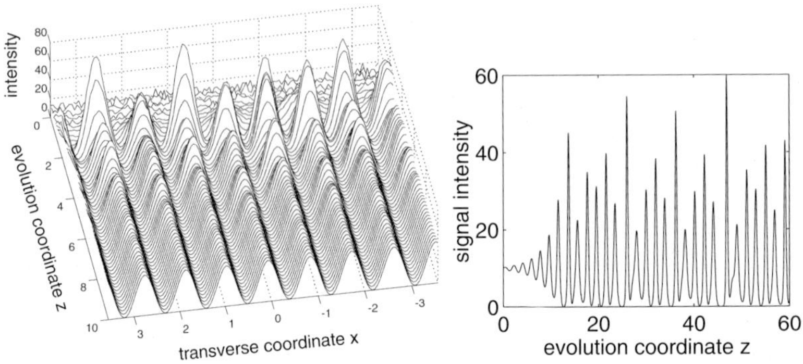

Fig. 3. Evolution of signal field intensity $P_1 = |u_1(x, z)|^2$ in a DOPO. *Left*: transverse dynamics showing self-organized spatial pattern formation in the presence of diffraction. *Right*: erratic self-pulsing well above the Hopf bifurcation ($P_H = 1.25$) in the absence of diffraction (PW dynamics). The parameters are as in Fig. 2a, and the DOPO is initially set up with a noisy steady-state PW, corresponding to a nontrivial supercritical branch with $P_1 = 10$

4 Optical Parametric Oscillators

As far as transverse solitary structures are concerned, OPOs may exhibit different features depending on the chosen configuration (basically, the number of resonant fields and properties of the degeneracy) and, for any particular configuration considered, on the parameter values. In the following two subsections, we shall focus on the triply resonant DOPO, which has been the object of the great majority of studies. The remaining subsections are devoted to discussion of nondegeneracy and walk-off effects.

4.1 Localized Phase Defects or Domain Walls in DOPOs

The PW bifurcation diagram of a DOPO (shown only for the intensity in Fig. 1b) shows that, *above threshold*, in the whole parameter space $(\Delta_1, \Delta_2, \alpha)$, the device can in fact reach two different nontrivial steady states, with equal field intensity but opposite phases (in other words, a given value of signal intensity in Fig. 1b corresponds to two permitted solutions with opposite amplitudes). These phases are likely to be randomly selected for a signal that is built up from noise. Diffraction can be envisaged to act as the physical mechanism which permits the formation of a signal domain wall (DW) solitary structure or kink linking the two stable homogeneous solutions, a generic phenomenon in dissipative systems (see, e.g., [10,48,69–71]). These DWs, or fronts (note, however, that a front is often distinguished from a DW as representing a localized wave which links a trivial or linear state with a nontrivial one [72]), represent phase defects which separate the random patches of different phase in the transverse plane. Since they have a zero in the signal field, they appear as dark structures in the intensity pattern, and because of this feature they are also referred to as dark solitons [18] or Ising walls [73]. Symmetry considerations suggest that one should search for such DWs in 1D, as done in [16,18] (in a 2D environment, this type of solution obviously represents DW stripes). From the viewpoint of dynamical systems, DWs correspond in phase space to heteroclinic trajectories linking the two fixed points that correspond to the homogeneous solutions with equal amplitudes and opposite phases [70]. In the DOPO, such DWs can be found from (5) by setting the evolution derivatives $\partial/\partial z$ to zero, and solving numerically the second-order coupled problem by means of efficient numerical techniques such as relaxation relaxation method or shooting methods [18]. Nevertheless, a reduction of the problem to an OPE provides an effective means to ascertain analytically the existence of these DWs. An obvious reduction exploits the cascading limit, for which repeated and mismatched up- and down-conversions induce an effective Kerr phase shift (see also the chapter by Torruellas et al. in this book). In the resonator, the cavity detuning replaces the role played by the mismatch in the traveling-wave configuration. The limit of high pump detuning $|\Delta_2| \gg 1$ and $\alpha \ll |\Delta_1 \Delta_2|$ (5) leads to a CGLE with parametric gain (i.e. an OPE of the type of (8)), which, for definiteness, we repeat below [18,81]:

$$\mathrm{i}\frac{\partial u_1}{\partial z} + \nabla^2 u_1 + \chi |u_1|^2 u_1 + (\mathrm{i} - \Delta_1) u_1 = \mathrm{i} g u_1^* \,, \tag{13}$$

where the effective nonlinear coefficient is $\chi = (2\Delta_2)^{-1}$ and the parametric gain $g = S_2/\Delta_2$. When the nonlinearity is defocusing, i.e. for $\Delta_2 < 0$, (13) has the following DW solution:

$$u_1(x) = u_0 \tanh(x/x_0) \exp(\mathrm{i}\phi) \,, \tag{14}$$

where $u_0 = \sqrt{P_{\mathrm{GL}}^{\pm}}$, $P_{\mathrm{GL}}^{\pm} = |u_1^{\pm}|^2 = 2(\Delta_1 \Delta_2 \pm \sqrt{S^2 - \Delta_2^2})$ approximates the intracavity power of the OPO nontrivial PW solution, $x_0 = 2\sqrt{-\Delta_2}/u_0$ is the DW width, and $\phi = (1/2)\cos^{-1}(\Delta_2/S)$ is a constant phase. On a different basis, a DW solution with a real tanh profile can be obtained close to the threshold S_{th}, where the dynamics are governed by the FKE (11) [16,43]. This equation supports the DW solution for positive finite signal detunings $\Delta_1 = O(1) > 0$ in the monostable case $\alpha - \Delta_1\Delta_2 = O(1) > 0$ (i.e. the DW bifurcates supercritically above S_{th}), without further explicit restriction on the value of the SH detuning Δ_2.

The analytical solution (14) can be continued numerically to explore the existence of a DW against change of parameters. In general, signal DWs have oscillating tails, and are accompanied by a pump beam with a twin-hole structure or a bright spot surrounded by radially decaying oscillations in the intensity [18].

Remarkably, DWs can be observed to form spontaneously from noise, as shown in Fig. 4, where a DW stripe in 2D is observed to emerge from a noisy

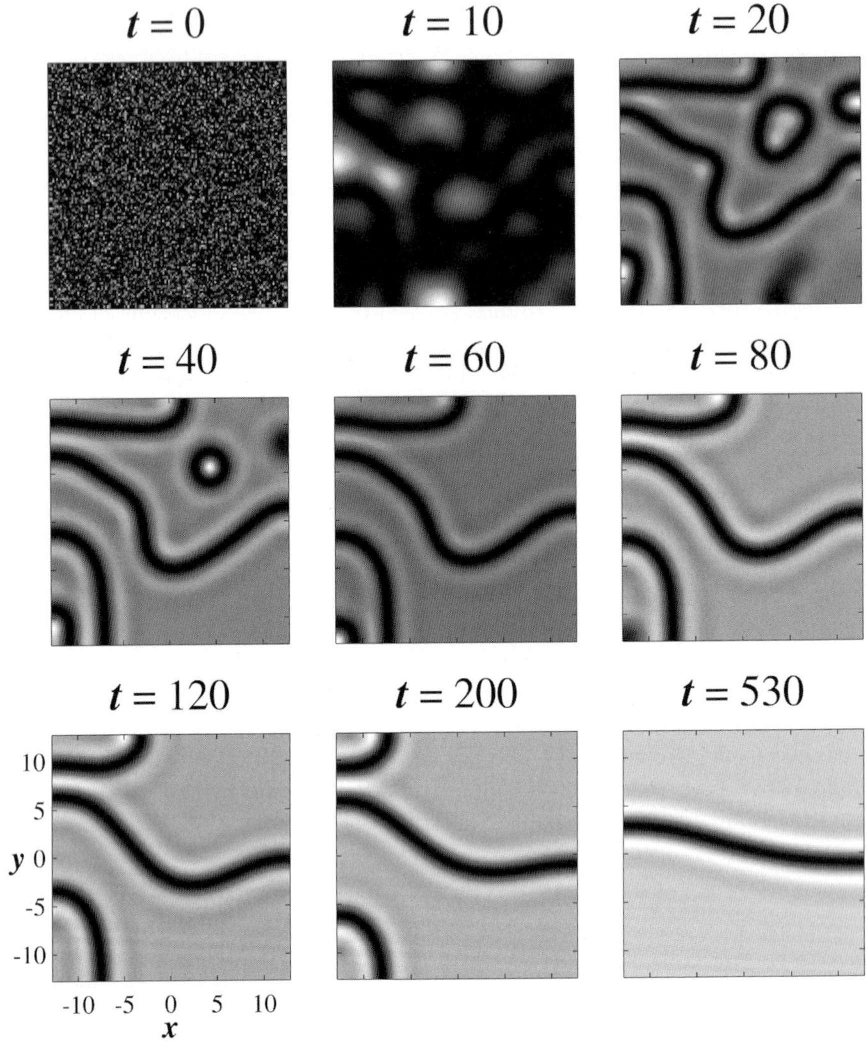

Fig. 4. Spontaneous formation of a DW soliton stripe in the transverse distribution of the signal field intensity in a DOPO on resonance (numerical simulation based on (5)). Snapshots of the signal intensity distribution in the transverse plane taken at different normalized times (distances) t are shown. *Darker regions* correspond to lower intensities

signal pattern. The experimental observation of DWs in the transient evolution of a closely related system (see [71] and the chapter by Weiss et al. in this book) encourages one to believe that it may be possible to observe them in a DOPO. To this end, it is natural to wonder about their stability. In particular, one may argue that the evolution of randomly curved walls into stripes (Fig. 4) could be artificially forced by the imposed boundary conditions. This problem has recently been addressed in [43], where preliminary results have shown that small 1D perturbations decay exponentially close to the threshold where the DWs are neutrally stable. Moreover, in the framework of the same perturbative approach, the effect of different physically perturbing terms corresponding to different physical processes in the underlying model can be established. For instance, straight DWs are shown to drift under the action of walk-off, giving rise to traveling DWs. Importantly, for a curved DW the curvature obeys a diffusive heat equation, which supports the observed progressive evolution into line defects. A different approach to the generation of stable DWs in 2D, by means of an input Gauss–Laguerre pump mode with unity topological charge, has been proposed in [74].

However, not only stripe defects can be observed in a DOPO. Ring-shaped DWs can separate an internal and a surrounding field that possess different phases. They can be thought of as being formed by closing a DW stripe on itself. Usually, this type of switching wave shrinks owing to its strong curvature until it disappears (see panels for $t = 20, 40, 60$ in Fig. 4). However, as shown in [47], stable formation of ring DWs occurs in the region of negative signal detuning Δ_1 that corresponds to MI-induced threshold lowering. In this case DWs exhibit at least two dips (or more for a DW with concentric rings) in their transverse radial profile. The stabilization of these structures has been attributed to the change of the effective diffusion (in an OPE reduction) from the normal regime to antidiffusion [47]. More recently, however, ring DWs have also been shown to exist for positive signal detunings in a suitable range of pumping [75], and in the non-mean-field evolution as well [26]. The origin of stable DWs would appear be due to the same mechanism that is behind the formation of stable twin hole dark solitons in the traveling-wave configuration [76], namely matching of overlapping oscillatory tails over the transverse section [75]. As a result, the existence of DWs is strongly affected by the diffraction coefficient ρ in (5), which can be altered in nonplane resonators [23]. It is also likely that these oscillatory tails will affect deeply the properties of the interaction and clustering behavior, as recently observed in other pattern-forming optical systems [77]. It is clear, however, that a thorough understanding of the physical mechanisms behind these types of fascinating objects and a methodical classification of them will require further efforts. Since they play an important role in other optical contexts [48,49], a useful exchange of ideas can take place.

To conclude this section, we mention that a qualitatively different type of DWs, namely Bloch-type walls, is allowed to exist, for which the field amplitude is constant and only the phase rotates between the two homogeneous domains. Such type of DWs arise generically in nonequilibrium systems which exhibit symmetry breaking in the phase of a complex OPE description (i.e. chiral interfaces in GL-type OPEs [10,73]). In OPOs the occurrence of such walls requires a type II configuration, owing to symmetry breaking induced by birefringence or dichroism, as predicted recently by Izús et al. [78].

4.2 Below-Threshold Bright Signal Solitons in DOPOs

Since for large radii in the transverse plane PCSs must tend asymptotically to steady-state PWs, the first requirement for a stable bright signal soliton to exist is that a vanishing signal PW is a steady-state solution which is stable against PW perturbations, at least. To fulfill this requirement, a DOPO must necessarily operate below threshold, where the trivial steady-state solution is stable. Whenever the PW bifurcation is subcritical (i.e. the OPO is bistable), for driving in the range $S^2_{\mathrm{sub}} < S^2 < S^2_{\mathrm{th}}$, the trivial PW solution coexists with two branches of nontrivial PW solutions which correspond to parametric oscillation (see Fig. 1b). Under such conditions, localized solutions or cavity solitons are generic [79,80].

Roughly speaking, in the OPO, the localized solution can be thought of as linking the oscillating solution (at the soliton peak) with the trivial solution (at infinity, or, in practice, at a finite distance much larger than the characteristic soliton width but still smaller than the resonator width). Such a type of PCS has been investigated in [16,25,17,81,82]. Formally, the PCS profiles can be found from (1), setting the evolution derivative to zero ($\partial_z = 0$), and setting $\nabla = \partial_x^2$ in the 1D case, or $\nabla = \partial_r^2 + r^{-1}\partial_r$, with $r^2 = x^2 + y^2$ in the 2D case. Then bifurcation diagrams can be drawn by plotting the peak signal intensity (or, equivalently, the soliton signal power) versus the normalized driving $X = S^2/S^2_{\mathrm{th}}$, for fixed values of the other parameters (detunings and absorptions). An example of the outcome of such a calculation is shown in Fig. 5, which compares soliton and PW bifurcation diagrams. As shown, the soliton branches bifurcate subcritically (compare with the Lugiato–Lefever equation in the chapter by Firth and Harkness in this book), and hence, in both 1D and 2D, solitons are bistable. In the 1D case, solitons are amenable to analytical description, where (13) holds and yields the PCS solution

$$u_1 = u_1^{\pm} = \sqrt{2P^{\pm}_{\mathrm{GL}}}\,\mathrm{sech}(x/x_0)\,\exp(\mathrm{i}\phi)\,, \tag{15}$$

where $\phi = (1/2)\cos^{-1}(\Delta_2/S)$, the soliton width is $x_0 = \sqrt{\Delta_2/(2P^{\pm}_{\mathrm{GL}})}$, and $P^{\pm}_{\mathrm{GL}} = |u_1^{\pm}|^2 = \Delta_1\Delta_2 \pm \sqrt{S^2 - \Delta_2^2}$ approximates the OPO nontrivial PW solution. The PCS solutions (15) not only give a good approximation (see Fig. 5) to the solutions obtained numerically from (5), but also permit one to establish the existence region as that pertaining to positive detunings $\Delta_{1,2} > 0$, which yields a focusing effective nonlinearity in (13). The bifurcation structure of Fig. 5 is related to single-hump PCSs. Actually, double humps [17] and higher-order branches corresponding to multihump PCSs exist, leading to soliton multistability [25].

Once the PCS family has been characterized, it is natural to wonder about the excitation, stability, and interactions of such structures, which are key issues for their observability. First, as expected, the negative-slope branch is

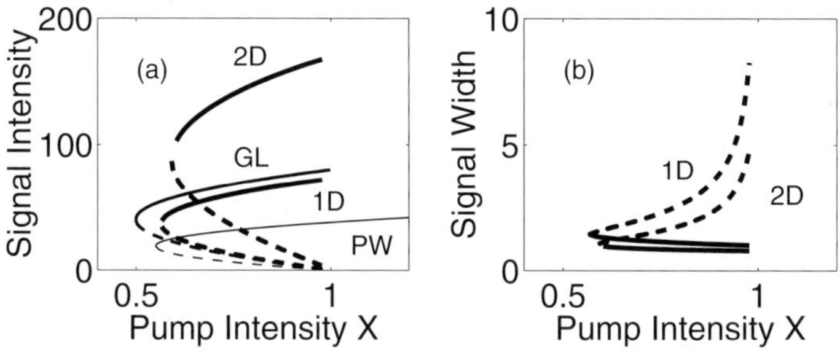

Fig. 5. Bifurcation diagram for PWs and PCSs in a DOPO. (**a**) Signal intensity versus PW pump intensity $X = S^2/S_{\rm th}^2$: the *solid* and *dashed lines* refer to stable and unstable branches, respectively. The signal intensity of the homogeneous or PW nontrivial branch (*curve* denoted by PW) is compared with the peak intensity obtained from the 1D CGLE ((13), *curve* denoted by GL), and from numerical integration of the full mean-field model (5) in 1D and 2D. (**b**) Signal width (FWHM of intensity) of 1D and 2D branches, as obtained from the full model

found to be unstable [25,81]. The positive-slope branch of single-hump PCSs, however, is not entirely stable. The stability of the OPE reduction (13) has been addressed in [83,84]. In the full model ((5) and (6)) instabilities arise, owing to a Hopf mechanism in which pairs of purely imaginary eigenvalues of the linearized problem collide and enter the right-hand half of the complex plane [25]. Hopf instabilities are generic in dissipative systems: for instance, dissipative perturbations of the nonlinear Schrödinger equation lead to unstable complex eigenvalues moving out of the continuous spectrum (i.e. edge bifurcation [85]). In the OPO, the colliding eigenvalues have been identified [25] as those originating, in the Hamiltonian limit of negligible losses and driving, from the neutral mode (a zero eigenvalue, associated with conservation of total power) and the internal mode (an imaginary eigenvalue corresponding to bound oscillations around the soliton; see also the chapter by Kivshar and Sukhorukov in this book). The different evolution scenarios of unstable PCSs range from the excitation of a stable oscillating bound solution close to the instability threshold, to soliton destruction [25].

The spontaneous buildup of stable bright PCSs from noise is more difficult than DW formation because the OPO is below threshold. However, the formation of stable bright PCSs can be easily observed by superimposing a sufficiently bright central spot on the PW pump; the role of this spot is, roughly speaking, to bring the OPO above threshold locally. If this central spot is then switched off to leave PW pumping, it has been shown that the transverse beam profile relaxes to a PCS [81].

Finally, PCS interactions have been studied [82] by employing a perturbative technique pioneered in [86]. While out-of-phase PCSs repel, in-phase PCSs might coalesce, form an oscillatory or stable bound state, or even generate patterns.

4.3 Solitons in NDOPOs

When degeneracy is removed, OPOs exhibit some degree of diversity with respect to DOPOs. For instance, Longhi has shown that, in NDOPOs, traveling-wave states are preferred to the standing-wave patterns (rolls) distinctive of a DOPO [56]. The latter are in fact due to interferences between two simultaneously generated traveling waves.

Importantly, however, PCSs similar to those discussed in Sect. 4.2 for the DOPO can be sustained also in NDOPOs. In this respect, triply resonant devices allow a low operating threshold, and hence are very attractive. A CW-operating semimonolithic OPO has permitted the observation of bistability (including dynamical bistability or delayed bifurcation, induced by a temporal sweep of the pump power in the stationary monostable region of the parameters) [87]. The subcritical PW bifurcation associated with bistability is the origin of the type of PCS discussed in Sect. 4.2. Specific attention has been directed to this kind of PCS in the nondegenerate, triply resonant case in [38,88–90], to which we refer the reader for details. The usual approach of deriving an OPE with a Kerr-equivalent nonlinearity can also be extended to the three-wave problem at the cost of dealing with a pair of coupled equations for the signal and idler envelopes [90].

Among the phenomena which cannot have any similar phenomenon in a DOPO, of particular interest is the role played by PCSs in the frequency selection rules for the idler and signal [89]. Also, we remark that bright PCSs do not necessarily require the device to be operated below threshold. In a singly resonant OPO, at high fluences, the parametric and intrinsic (material) Kerr nonlinearities cooperate to support a self-trapped signal wave [27]. In this case a perturbative expansion of (3) leads to (12), that is, a cubic–quintic CGLE. In this model the $\chi^{(2)}$ mixing term is responsible, above the signal generation threshold, for the parametric gain g which compensates for the resonator loss α, for gain saturation (related to $\mathrm{Re}(\gamma_3)$) and for a quintic term (γ_5). The quintic term is indeed important in explaining the stability of chirped PCSs and their zero background, which are observed in non-mean-field numerical simulations [27].

4.4 Role of Walk-Off

The study of the effects of walk-off on pattern formation dynamics in OPOs has revealed intriguing features of this class of devices. Transverse walk-off between the pump and the generated waves is present in OPOs owing to the intracavity crystal birefringence exploited to achieve phase matching of the frequency down-conversion process. The effects of walk-off have been studied theoretically in the case of a triply-resonant DOPO that is governed, in the absence of walk-off, by the mean-field model (5). Walk-off appears in this model in the form of a simple convection (or drift) term of the type $\mathrm{i}\delta\,\partial_y u_1$ in the first equation of (5). Here δ accounts for the magnitude of the walk-off. In the presence of an instability such as the pattern-forming instability described in [12], convection can lead to a situation where local perturbations are advected away more rapidly than their rate of spreading. In this case the system enters the so-called convectively unstable regime, in which no patterns are formed even above the instability threshold. However, as was shown in different contexts [91,92], when noise is continuously applied to the system, a macroscopic pattern is sustained (continuosly regenerated) that is reminiscent of the absolute instability regime.

Experimentally, the role of convection in a (bistable) optical system has been studied in the temporal regime in a synchronously pumped fiber cavity, where it originates from imperfect synchronization [93]. Other aspects related to drift in optical systems were addressed recently in [94].

In OPOs, noise-sustained structures would find their ultimate origin in quantum fluctuations of the signal field. And, since the generated patterns constitute a macroscopic manifestation, obtained through an amplification process, of the originating microscopic noise, the OPO can be seen as a useful tool for the study of the frequency conversion process at the quantum level [95]. In a DOPO, the transition from the convective to the absolute regime can be characterized by the LSA [95], the subcritical (bistable) case requiring nonlinear analysis instead [45]. In [44], the triply resonant NDOPO has been studied analytically and numerically by means of a reduction of the mean-field model to a generalized CGLE (see Sect. 2.2). In that work the analysis was performed in the absence of MI (positive signal detuning), but the same conclusions were drawn as regards the existence of noise-sustained patterns induced by walk-off. The particular case of the triply resonant OPO with type II phase matching has been considered in [96]. In this configuration, walk-off was shown to induce, for each polarization component, a competition between two phase-stripe patterns (traveling waves) of different wavelengths. In the convective regime of the instability, the dynamical amplification of noise in the two competing modes leads to noise-sustained patterns analogous to those reported in [95]. In this case, however, standing-wave intensity patterns due to interference between traveling waves are observed [96], unlike what was found without walk-off [56].

Walk-off in the triply resonant DOPO was also shown to be responsible for the spontaneous formation of traveling periodic arrays made up of the dark-soliton stripes presented in Sect. 4.1 [97]. This result is rather counter-intuitive, since in the domain of existence of the dark-stripe solitons the homogeneous nontrivial solution of (5) is modulationally stable, and one would thus not expect periodic pattern formation. The generation of dark stripes occurs when the OPO is driven by a finite-width pump beam. The stripes are generated at the upstream edge of the pump beam where its intensity reaches the oscillation threshold. The stripes are oriented orthogonally to the walk-off direction, while the pattern moves in the walk-off direction. This phenomenon has been described theoretically by means of a standard method developed for the study of front propagation into unstable states [42]. It is based on a linear analysis of exponentially growing fronts with nonzero wavevector. The wavevector corresponding to the most unstable mode gives the periodicity of the pattern selected. A remarkably good agreement has been obtained between the theoretical predictions and numerics [97]. Recently, the transition between nonuniformly translating (pattern-forming) and uniformly translating (leaving behind a homogeneous state [44]) fronts has been characterized analytically in terms of the dependence of the critical signal detuning on the walk-off strength, by means of an OPE reduction to a convective real SHE [46].

This pattern formation mechanism is intimately related to the phase sensitivity of the intracavity frequency conversion process. More precisely, it occurs owing to the π phase indeterminacy of the steady-state PW solutions, which is behind the existence of the quadratic dissipative DW described in Sect. 4.1. For this reason, it does not appear in devices that do not exhibit phase sensitivity, such as the singly resonant OPO.

5 Intracavity Second-Harmonic Generation

Thanks to recent progress in the design and fabrication of quadratic nonlinear optical resonators, highly efficient SHG has been made possible with CW pumping in the milliwatt range [98,99]. The most efficient configuration is a triply resonant cavity that allows confinement of both the pump and the signal field, which naturally leads to strong wave interaction and thus to potentially high frequency conversion efficiency. For this purpose, monolithic total-internal-reflection resonators have been designed that made conversion efficiencies as high as 50% with pump powers as low as 5 mW possible [98]. Besides its obvious technological interest, doubly resonant IC-SHG is also considered as a paradigmatic problem of fundamental research in both classical and quantum optics. SHG in a high-finesse doubly resonant cavity is indeed described by the generic mean-field model (4), which, as discussed in Sect. 3, exhibits a large spectrum of complex nonlinear behaviors, including self-pulsing, chaos, and MI. As regards quantum effects, peculiar properties of the stationary state (CW regime), such as squeezing, have been predicted [6] and observed [7].

5.1 Pattern Formation and Localized Structures

Pattern formation in doubly resonant IC-SHG devices was first studied theoretically in [62]. In that work, the authors considered temporal patterns resulting from the interplay between group-velocity (chromatic) dispersion and quadratic nonlinearity of the cavity material. The treatment of dispersion instead of diffraction was motivated by the fact that diffraction is controlled in practical CW IC-SHG configurations in order to optimize the cavity confinement effect, so as to obtain efficient conversion at low pump powers. Spherical mirrors, designed so as to allow single-transverse-mode (TEM_{00}) operation, are used [98,99], which naturally suppresses any significant transverse effect. Conversely, chromatic dispersion cannot be controlled so easily, especially in the monolithic configuration [98], and one may expect that temporal pattern formation will be the dominant complex behavior in practical IC-SHG systems. On the basis of an LSA of the mean-field model (4), it was shown that dispersion is indeed responsible for the occurrence of MI, which competes with and prevails (over a wide parameter range) over the other instability mechanisms, such as self-pulsing [62], as mentioned in Sect. 3. This result can obviously be applied to the spatial domain by virtue of the formal equivalence between dispersion and diffraction in the mean-field model (4).

An exhaustive investigation of 2D spatial pattern formation (neglecting dispersion) in doubly resonant IC-SHG devices has been reported in [63]. This study reveals the formation, through MI, of stable hexagonal patterns characterized by a high contrast, which, together with their coexistence with the homogeneous solutions (bistability), suggests the existence of PCSs. These are identified in the mean-field model (4) as a family of radially symmetric bright solutions with oscillating tails, which are characterized by a bistable bifurcation branch, as shown in Fig. 6 [68]. Note that even though IC-SHG devices have no PW threshold (unlike an OPO), the formation of PCSs occurs only above a driving intensity threshold (critical point) corresponding to the instability threshold.

In a theoretical study that included both dispersion and diffraction in the mean-field model, IC-SHG was shown to support three-dimensional periodic structures (optical lattices) analogous to those found in [100] for Kerr-type cavities. Lamellae, 3D hexagons, and body-centered cubic structures were shown to be stable in some parameter ranges. The coexistence of these periodic structures with the homogeneous solutions indicates the existence of structures localized in 3D, namely 3D PCSs. This was confirmed numerically in [101], which reports the existence of 3D PCSs in the form of drops (3D bright solitons) and cylinders. It should be noted, however, that this theoretical work is based on an approximate OPE model (the real SHE, see Sect. 2.2) in which neither spatial walk-off nor group-velocity differences between the interacting waves have been taken into account. The first-derivative terms corresponding to these effects were dropped in the derivation of the OPE by assuming that they were quantities of order one in terms of the smallness parameter of the reduction procedure. In view of the determining role played by walk-off effects in quadratic cavities (see Sect. 4.4), we can anticipate that this assumption is too strong to be applicable in practice to the type II phase-matching configuration considered in [101]. To give practical sense to the theoretical predictions of [101], i.e. to the existence of 3D parametric optical lattices and solitons, one must consider a very specific phase-matching condition that still remains to be determined. Dissipative PCSs constitute

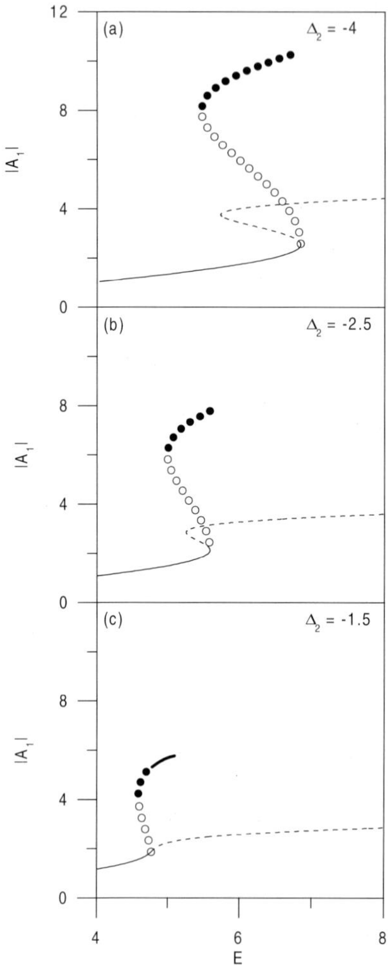

Fig. 6. PCS bifurcation diagram obtained numerically for doubly resonant IC-SHG (from [68], courtesy of the authors) for three different negative values of SH detuning Δ_2, and fixed $\alpha = \rho_2 = 0.5$, $\Delta_1 = -4$ in (4). The *empty* and *filled circles* show the FF soliton peak amplitude $|A_1| = |u_1|/\sqrt{2}$ of the unstable and stable branches, respectively, versus the driving amplitude $E = S_1/\sqrt{2}$ in (4). These are compared with stable (*solid line*) and unstable (*dashed line*) PW branches. In (**a**) and (**b**), IC-SHG exhibits PW bistability, and PCSs emanate from and end up at the first critical point (knee of the PW response). Panel (**c**) refers to the case where the PW response is monostable. Here PCSs continue to bifurcate subcritically at the point where PWs become modulationally unstable, and the stable branch shows a continuous transition to a hexagonal pattern whose maximum amplitude is shown as a *bold line*

bound (or symbiotic) states of three interacting waves in which a perfect balance between diffraction/dispersion and nonlinearity is realized. One can expect that walk-off between the waves dramatically affects this balance and either prevents the formation of solitons or leads to new types of solitons in a way akin to what is encountered in free-space propagation [102].

5.2 Vectorial (Type II) Effects

Peschel et al. have studied the dynamics of IC-SHG in the type II phase-matching configuration [103,104]. Considering first the homogeneous solutions of the corresponding mean-field model ((3) in Sect. 2, where $S_3 = 0$ and the equations in u_1 and u_2 have identical coefficients), they identified a symmetry-breaking bifurcation [103]. The symmetry-breaking affects the polarization state of the FF pump field, which is supposed to be linearly polarized at 45 degrees with respect to the optical axis of the crystal in the input pump beam. The corresponding bifurcation branches correspond to two equally probable FF elliptical-polarization states that are symmetric with respect to the input polarization direction. The mechanism responsible for this symmetry-breaking bifurcation is identical to that encountered in Kerr-type cavities [21]. It was indeed shown in [103] that, at steady state, the two FF polarization components are coupled through cross-phase modulation, as for the circular-polarization components in a Kerr medium. Several studies have shown that cross-phase modulation is one of the possible mechanisms responsible for symmetry breaking in cavities [21,105,106]. In the case of IC-SHG the asymmetric states are destabilized through a Hopf bifurcation, as for the symmetric branch (scalar model) [4]. This instability leads to self-pulsations characterized by antiphase dynamics between both polarization components. The authors of [103] have considered the possibility of exploiting the symmetry-breaking bifurcation for all-optical switching in the flip-flop operation mode, as suggested in [107].

Considering transverse spatial variation of the fields by including diffraction in the model, Peschel et al. have shown that the two asymmetric states of the bifurcation branches can coexist in the transverse plane [104]. In this case, the domains of opposite polarization states are separated by a wall that constitutes a DW solitary structure of the system, in a way analogous to the dark stripe soliton of the DOPO described in Sect. 4.1. In that latter case, the wall separates homogeneous domains of opposite phases, while here it separates homogeneous domains of opposite polarization states. In this respect, the localized structures of IC-SHG are similar to the so-called polarization DWs sustained in free propagation in isotropic or birefringent Kerr media [108]. On the basis of this analogy, one can expect the existence of polarization DWs in Kerr-type cavities that exhibit polarization symmetry-breaking due to cross-phase modulation [21].

Contrary to what happens in the Kerr case, both LSA and numerical simulations reveal that DWs in IC-SHG are not stable. They destabilize through

MI, which, in the fully nonlinear regime, leads to the formation of finger patterns [104]. Each of the resulting fingers, taken in isolation, can be seen as a localized structure of the system, i.e. a vector, dissipative, parametric bright (nontopological) soliton stripe. When truncated, these bright stripes grow, which leads to intriguing collision dynamics when several truncated solitons are present in the transverse field distribution of the cavity [104].

The studies of [103,104] reveal an extremely rich spectrum of nonlinear behaviors that illustrates very well the interest of nonlinear quadratic cavities for fundamental studies of pattern formation in nonequilibrium systems. As regards practical considerations, it should be noticed, however, that the results in [103,104] are based on the assumption of equal detunings for both FF polarization components. This assumption was necessary in order to make the system symmetric with respect to the input polarization state, so as to obtain the possibility of a symmetry-breaking. In a real system, however, type II phase matching implies that the two FF components see different refractive indices, which in turn implies that the detunings of both components are in general different. As a consequence, in order for the predictions of [103,104] to be compared with experiment, three detunings must be independently controlled with high accuracy, which certainly represents a major technological obstacle.

Longhi performed, independently, a study of vectorial dynamics in type II doubly resonant IC-SHG [66]. He also considered, first the ideally symmetric situation corresponding to identical FF detunings. In this case, an LSA of the steady-state PW solutions revealed the existence of two competing instabilities. The first instability preserves the FF linear-polarization state; it is the same as that found in the type I configuration [62], i.e. either a Hopf or a Turing instability, depending on the parameters of the cavity. The second instability is of the Turing type and does not preserve the polarization state; it is reminiscent of the symmetry-breaking instability described in [103,104]. For this reason, it has been called "polarization instability".

To study analytically this polarization instability and the resulting vector Turing patterns, the mean-field model (3), considered in the limit of nearly resonant FF and SH fields and close to the instability threshold, has been reduced on the basis of a weakly nonlinear analysis developed from a standard multiple-scale expansion procedure. The model was, in this way, reduced to a real SHE including a driving term (see Sect. 2.2). This equation had already been derived in a study of patterns and localized structures in nonlinear optical cavities with third-order nonlinearities [40]. A Turing bifurcation leading to hexagonal patterns or rolls was described in this model. In the case where this bifurcation is subcritical, i.e. when there is coexistence of the Turing patterns and the homogeneous solutions, localized structures were found to exist that correspond, in the present context to vector PCSs. We note that the SH OPE model proposed in [66] was obtained by relaxing the symmetry constraint on the system. In particular, a difference between the FF detunings was introduced as a quantity of second order in terms of the smallness parameter of the multiple-scale analysis. In this respect, the study reported in [66] is closer to physical reality than those in [103,104].

6 Parametric Gap Solitons in Distributed-Feedback Structures

In this section, we shall consider a different form of trapping, originating from the Bragg effect instead of diffraction, i.e. in a grating where the passive feedback is distributed along the structure. In such an NLDFB structure, stopbands or photonic bandgaps exist around multiples of the Bragg wavelength. As discussed in detail in the chapter by de Sterke et al., the Kerr nonlinearity enables propagation inside the stopband in the form of self-transparent envelopes trapped along the propagation direction, so-called Bragg or gap solitons. Note that gap soliton envelopes exist also for a medium with periodic properties directed orthogonally to the main propagation direction, that is, in transverse gratings [109].

The question as to whether localization and gap solitons arise also in quadratic media has recently been addressed [110–121]. The first hint of the existence of PGSs in SHG was obtained once again from the so-called cascading limit of SHG, in which repeated up- and down-conversion processes in the large-wavevector-mismatch limit result in effective Kerr-like self-phase and/or cross-phase modulation [110,111]. In this case the leading role is played by the beam at the FF, while the SH remains small and is needed only to induce phase changes in the beam components at the FF. More generally, PGSs of genuine parametric nature can be formed via SHG in a grating which exploits double Bragg resonance (i.e. a grating resonant at both optical carrier fields). This arises naturally from the first two harmonic components ($m = 1, 2$) in the Fourier expansion $\sum_m \Gamma_m \exp(\pm i m \beta_g z)$ of the Bragg coupling coefficient between counterpropagating waves due to an anharmonic perturbation (grating) with period $\Lambda = 2\pi/\beta_g$. Under such conditions, the forward- (subscript $+$) and backward- (subscript $-$) propagating envelopes $E_{1\pm}$ at the Bragg carrier frequency ω_{B1} (where $2k(\omega_{B1}) = \beta_g$), and the harmonic envelopes $E_{2\pm}$ at the frequency $2\omega_{B1}$ obey the following scalar coupled-mode model with four equations [113]:

$$\mathrm{i}V_1^{-1}\partial_T E_{1\pm} \pm \mathrm{i}\partial_Z E_{1\pm} + \Gamma_1 E_{1\mp} + \chi_1 E_{2\pm} E_{1\pm}^* \mathrm{e}^{\pm \mathrm{i}\,\Delta k\, Z} = 0\,,$$

$$\mathrm{i}V_2^{-1}\partial_T E_{2\pm} \pm \mathrm{i}\partial_Z E_{2\pm} + \Gamma_2 E_{2\mp} \mathrm{e}^{\mp \mathrm{i}2\,\Delta k\, Z} + \chi_2 E_{1\pm}^2 \mathrm{e}^{\mp \mathrm{i}\,\Delta k\, Z} = 0\,, \qquad (16)$$

where $V_n = (\mathrm{d}k/\mathrm{d}\omega)^{-1}|_{n\omega_{B1}}$, $n = 1, 2$, are group velocities, and $\Delta k \equiv k(2\omega_{B1}) - 2k(\omega_{B1}) = k(2\omega_{B1}) - 2\beta_g$ is the nonlinear mismatch (also equal to half the Bragg detuning $2k(2\omega_{B1}) - 4\beta_g$ at the SH frequency). Note that exact Bragg resonance for the SH occurs at a frequency ω_{B2} such that $k(\omega_{B2}) = 2\beta_g$, where $\omega_{B2} \neq 2\omega_{B1}$ unless one has perfect phase matching ($\Delta k = 0$). The location of any driving FF $\omega_{B1} + \delta\omega$ ($\delta\omega \ll \omega_{B1}$) and its SH is conveniently characterized by the normalized detunings $\delta_1 = \delta\omega/(\Gamma_1 V_1)$ and $\delta_2 = (2\delta\omega/V_2 + \Delta k)/\Gamma_1$ with respect to the Bragg frequencies ω_{B1} and ω_{B2}, respectively. Then, (16) can be recast, in a form more suitable for theoretical and numerical analysis, in terms of frequency-shifted dimensionless variables $u_1^{\pm}(z,t) \equiv \sqrt{2\chi_1\chi_2/\Gamma_1^2} E_{1\pm}\exp(\mathrm{i}\delta\omega T)$, $u_2^{\pm}(z,t) \equiv (\chi_1/\Gamma_1)E_{2\pm}\exp(\mathrm{i}2\delta\omega T \pm \mathrm{i}\,\Delta k\, Z)$, leading to

$$
\begin{aligned}
&(\pm\mathrm{i}\partial_z + \mathrm{i}v_1^{-1}\partial_t)u_1^{\pm} + \delta_1 u_1^{\pm} + \kappa_1 u_1^{\mp} + u_2^{\pm}(u_1^{\pm})^* = 0\,, \\
&(\pm\mathrm{i}\partial_z + \mathrm{i}v_2^{-1}\partial_t)u_2^{\pm} + \delta_2 u_2^{\pm} + \kappa_2 u_2^{\mp} + \frac{(u_1^{\pm})^2}{2} = 0\,,
\end{aligned}
\tag{17}
$$

where $z = \Gamma_1 Z$, $t = \Gamma_1 V_1 T$, $v_2 = V_2/V_1$, $\kappa_2 = \Gamma_2/\Gamma_1$, and $v_1 = \kappa_1 = 1$ have been introduced simply to symmetrize the two pairs of equations.

In the low-power limit, the equations (17) decouple. Following the linear analysis outlined in the chapter by de Sterke et al. in this book, we obtain a grating dispersion relation with two forbidden frequency gaps of width κ_m centered around the Bragg frequencies ω_{Bm}, $m = 1, 2$, as shown schematically in Fig. 7. The dispersive features of the doubly resonant grating are determined by the curvature of the linear dispersion relation, which is in general different for the two harmonics and for the lower branch (LB) and upper branch (UB) of the two stopbands. The counterbalance between grating dispersion and parametric nonlinearity is the origin of PGSs. Although they share with PCSs the property that they are sustained by a feedback mechanism and a nonlinearity, PGSs, unlike dissipative PCSs, are conservative

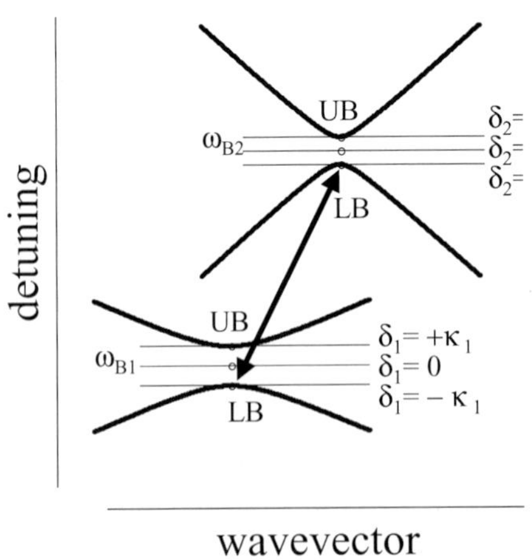

Fig. 7. Linear dispersion relation for a doubly resonant grating, showing the presence of two gaps. The *double arrow* indicates LB-LB coupling, which supports bright PGSs close to the gap edges (assuming $\kappa_{1,2} > 0$; see [113] for details)

entities. Indeed, (17) constitutes a Hamiltonian system [113], where driving and damping are absent. On the other hand, there is a marked difference between PGSs and Kerr gap solitons. Unlike the Kerr coupled-mode model ((14) of the chapter by de Sterke et al.), the conservation laws of (17) [113] are not sufficient to construct families of PGSs explicitly, i.e. by integrating analytically the system of four ordinary differential equations (ODEs) which follow from (17) when one searches for PGS envelopes $u_{1,2}^{\pm}(z,t) = u_{1,2}^{\pm}(z-vt)$, which travel locked at a common normalized velocity v. Nevertheless, PGSs can be found by following some different strategies:

- The most direct approach amounts to solving numerically the ODEs with appropriate boundary conditions. Algorithms such as shooting or relaxation methods are quite efficient in this case, too. In a search for bright (i.e. vanishing at $\pm\infty$) PGSs, the region of the parameter space where linear exponentially decaying solutions (corresponding to soliton tails) exist for both frequency components has been explored [115]. The main conclusion is that PGS solutions fill only a portion of this region. By means of a numerical approach, the existence of multidimensional PGSs (i.e. bullets), which account for the additional effect of diffraction has also been investigated [114].
- When both harmonic frequencies lie near the LB or UB of the stopbands, then by using the envelope function approach [113] or the effective-mass approach [114], one can reduce (17) to a lower-dimensional model which governs SHG in dispersive *homogeneous* media, much as the Kerr coupled-mode model reduces to the nonlinear Schrödinger equation. In this case, bright PGSs exist when both harmonics are close to the LBs of the respective stopbands, as indicated schematically in Fig. 7. In this case, if one knows the families of SHG parametric solitons in homogeneous media, four-wave PGS envelopes can be constructed [113,114,116].
- Exact solutions can be found, by using symmetry arguments to reduce the effective dimensionality of the ODE model and imposing extra resonance conditions [117]. This approach had the merit of showing analytically that bright PGSs span the whole gap at the FF and also exist, against intuition, when the SH beam is outside the gap (i.e. where the linear solutions for the SH beams are no longer damped, the decay along the soliton tails being of nonlinear origin).
- PGS solutions can easily be found explicitly when the SH beam can be adiabatically eliminated. We warn the reader, however, that this is not necessarily equivalent to the usual cascading limit of large mismatches. Indeed a reduced solvable model can be obtained, for arbitrary mismatches, in the limit of large SH Bragg coupling [115]. The opposite limit of a single Bragg resonance at the FF appears particularly promising in view of its simplicity of implementation [118–120]. In this case, one recovers, for large mismatches, a particular case of the Kerr model with only self-phase modulation, owing to cascaded conversion processes. Exact solutions ob-

tained in this limit can be extended by means of perturbative expansions where the mismatch plays the role of the smallness parameter (see [120] and a recent extension in [121]).

Although we refer the interested reader to the original literature for further details, we recall here the salient features of PGSs. Like Kerr gap solitons, they are localized self-transparent envelopes which travel with any velocity between zero and the group velocity of light in the host medium. However, unlike the Kerr case, they are formed by phase-locked and spatially (or temporally) locked envelopes at different carrier frequencies. An important issue for their experimental observability is their stability, which must be ascertained numerically, owing to the occurrence of oscillatory instabilities (see [122] and the chapter by Kivshar and Sukhorukov in this book for the general concept). Although the available results do not cover the entire existence region of PGSs, there is evidence for stable propagation in a large domain of the parameter space [119,123]. From this point of view, PGSs are more flexible than their Kerr counterpart. For instance, the cascading limit of a singly resonant grating yields, in the same physical system, stable PGSs for both the lower and the upper half of the stopband, thanks to the dependence of the effective sign of the nonlinearity on the mismatch [119], a possibility which does not exist in Kerr media.

Importantly, PGSs can be excited by illuminating the Bragg grating at the FF, as shown in Fig. 8. FF pulses in the picosecond range are injected on both sides of the NLDFB and generate enough SH intensity to form PGSs, which

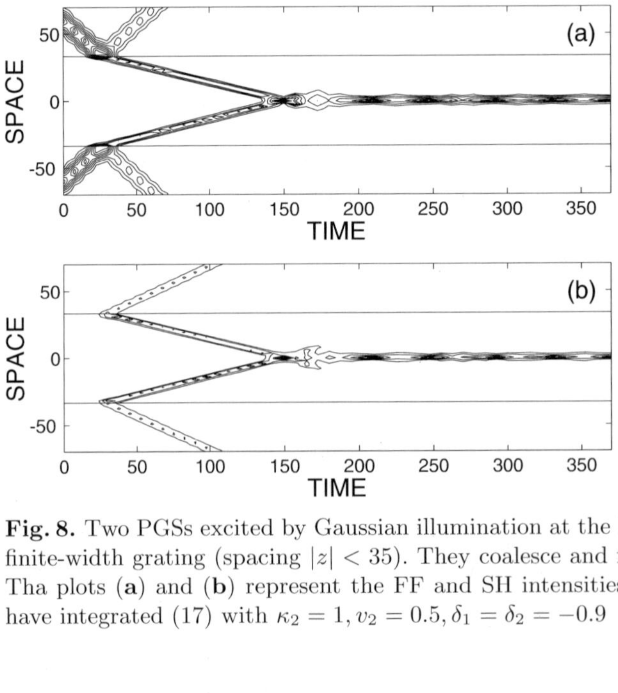

Fig. 8. Two PGSs excited by Gaussian illumination at the FF from both sides of a finite-width grating (spacing $|z| < 35$). They coalesce and form a stationary PGS. Tha plots (**a**) and (**b**) represent the FF and SH intensities, respectively. Here we have integrated (17) with $\kappa_2 = 1, v_2 = 0.5, \delta_1 = \delta_2 = -0.9$

travel slowly along the structure. If the excitation is nearly symmetric and in phase, the two colliding PGSs coalesce, giving rise to a stationary PGS, as evident from Fig. 8. We wish to mention also some other recent results related to PGSs. First, the coupled-mode theory can be reformulated by using a Bloch function approach, which has the advantage that it permits the description of deep Bragg gratings and a periodically varying nonlinearity [124,125]. In the shallow-grating approximation, the effect of optical rectification (i.e. a DC field) has also been included in the coupled-mode approach to PGSs [126]. Second, quadratic NLDFBs seem to be a suitable environment for studying solitons that exist at discrete values of their internal parameters (e.g. the detuning in this case) and are embedded in the continuous spectrum (so called embedded solitons, see [127,128]).

Other results, related to the CW nonlinear response of a quadratic NLDFB, have been obtained. MI of CW eigenmodes has been investigated [129], while the link between PGSs and the bistable response of an NLDFB has been studied [128]. In two different cascading limits, corresponding to strong Bragg resonance at either the FF or the SH, the spatial dynamics and hence the nonlinear response are affected qualitatively by the presence of zero-velocity PGSs with nonvanishing boundary conditions (corresponding to separatrices in the phase-space description of a CW NLDFB). Although these PGSs are coherent structures, sufficiently large departures from the cascading limit lead to the onset of spatial chaos, where they play a prominent role. The measurable signature of chaotic behavior is erratic multistable switching, which has no counterpart in a scalar Kerr NLDFB structure [128].

7 Summary and Further Developments

In summary, we hope to have convinced the reader that the combination of passive feedback and quadratic nonlinearities represents an area in which scientists are called on to develop new, stimulating physics and applications. Forms of feedback as different as those related to parametrically pumped cavities and Bragg gratings have been conceived.

Cavities with a large Fresnel number show an extremely rich assortment of transverse effects, including pattern formation and self-trapping in the form of fronts and strictly localized PCSs. Such stable structures represent steady-state configurations which exist relatively close to the longitudinal resonances of the cavity, and arise from a double counterbalance of diffraction and nonlinear phase-shifts (associated with parametric exchange of energy) on one hand, and cavity losses and parametric pumping on the other hand. Here, the role of the cavity is to permit paraxial nonlinear propagation along the resonator axis, with periodic losses (output coupling) and injection (parametric pumping), similar to the case of a damped line with lumped amplifiers.

A different form of localization takes place via Bragg coupling between counterpropagating waves. In this case, self-trapped waves called PGSs are confined along the structure (i.e. the Bragg grating), which provides feedback. In general these are nonstationary envelopes that move along the grating with a velocity smaller than the group velocity of light.

Undoubtedly, the most important issue related to this type of physics is the experimental demonstration. While an experimental proof of both PCSs and PGSs is still lacking, there are good reasons to believe that they may be observed in the near future, which will undoubtedly push up the interest in this topical area. Preliminary observations of pattern formation in a nearly confocal OPO [61] and phase defects in closely related systems [71] demonstrate the feasibility of experiments, at the same time calling for a more systematic understanding of transverse dynamics in parametrically driven resonators. Although it is likely that synchronously pumped cavities will be employed to meet the intensity requirements, the impressive developments in diode-pumped and laser-intracavity CW OPOs made possible by periodically poled materials (see, e.g., [130]) might also have an impact in this area. On the other hand, an experimental demonstration of PGSs can be envisaged that makes use of the availability of Bragg gratings in quadratic materials [131]. Photorefractive gratings in $LiNbO_3$ [132] and fibers with poling [133] appear to be particularly promising for this purpose.

Other directions, not covered in this chapter, might be developed further as a consequence of both theoretical and experimental progress. For instance, as far as PCSs are concerned, the issue of the temporal dynamics of the cavity leads to new ideas on the formation of spatiotemporal patterns and localized structures [58,101] with increasingly complex topology, which may also stimulate the investigation of temporal parametric solitons in analogy (or in contrast) with spatial PCSs in cavities where dispersion is relevant and can be controlled to some extent [134]. Localized waves with non-PW pumping may also be important from the point of view of applications, as reliable spatial eigenmodes of a parametric source of light.

As far as PGSs are concerned, other forms of distributed feedback are commonly used in parametric generation, such as quasi-phase matching (QPM) for backward generation. This and other possible configurations call for investigation of solitonic effects. A full understanding of the variety of possible trapping configurations due to distributed feedback and of their feasibility will require further work, and hopefully will stimulate new ideas. Furthermore, intracavity formation of PGSs may be explored, eventually leading to combination with PCSs, i.e. trapping in three dimensions.

From the strict point of view of applications, both PCSs and PGSs have potential to become vehicles for optical information storage in some way, and the ultimate challenge of turning this idea into reality remains.

Acknowledgments

We are indebted to several colleagues for many stimulating discussions on these topics. Among them we wish to thank particularly our collaborators Nathan Kutz, William Torruellas, and Gaetano Assanto. A special mention goes to Claudio Conti, a talented and hard-working PhD student, whose work has made possible every single achievement in the area of gap solitons. We also thank Dmitry Skryabin and Marco Santagiustina for critical reading of the manuscript and useful comments.

References

1. J.A. Giordmaine, R.C. Miller, Phys. Rev. Lett. **14**, 973–975 (1965).
2. J.E. Geusic, H.J. Levinstein, S. Singh, R.G. Smith, L.G. Van Uitert, IEEE J. Quantum Electron. **QE-4**, 352 (1968).
3. W.R. Bosenberg, R.C. Eckardt, (eds.), *Optical Parametric Devices*, J. Opt. Soc. Am. B **12**, (1995), special issue.
4. P.D. Drummond, K.J. McNeil, D.F. Walls, Opt. Acta **27**, 321–335 (1980).
5. C.M. Savage, D.F. Walls, Opt. Acta **30**, 557–561 (1983).
6. M.J. Collett, D.F. Walls, Phys. Rev. A **32** 2887–2892 (1985).
7. S.F. Pereira, M. Xiao, H.J. Kimble, J.L. Hall, Phys. Rev. A **38**, 4931–4934 (1988).
8. C. Fabre, Phys. Rep. **219**, 215 (1992).
9. M.C. Cross, P.C. Hohenberg, Rev. Mod. Phys. **65**, 851–1112 (1993).
10. D. Walgraef *Spatio-temporal pattern formation*, (Springer, New York, 1997).
11. L.A. Lugiato, M. Brambilla, A. Gatti, Optical pattern formation, in *Advances in Atomic Molecular and Optical Physics*, Vol. 40, ed. by B. Bederson, H. Walther (Academic Press, New York, 1998).
12. G. Oppo, M. Brambilla, L.A. Lugiato, Phys. Rev. A **49**, 2028–2032 (1994).
13. D.W. McLaughlin, J.V. Moloney, A.C. Newell, Phys. Rev. Lett. **51**, 75–78 (1983); Phys. Rev. Lett. **54**, 681–684 (1985).
14. L.A. Lugiato, R. Lefever, Phys. Rev. Lett. **58**, 2009–2012 (1987).
15. P. Mandel, *Theoretical problems in cavity nonlinear optics* (Cambridge University Press, Cambridge, 1997).
16. S. Longhi, Phys. Scr. **56**, 611–618 (1997).
17. K. Staliunas, V.J. Sánchez-Morcillo, Opt. Commun. **139**, 306–312 (1997).
18. S. Trillo, M. Haelterman, A. Sheppard, Opt. Lett. **22**, 970–972 (1997)
19. D.J. Armstrong, W.J. Alford, T.D. Raymond, A.V. Smith, M.S. Bowers, J. Opt. Soc. Am. B **9**, 460–474 (1997).
20. T. Debuisschert, A. Sizmann, E. Giacobino, C. Fabre, J. Opt. Soc. Am. B **9**, 1668–1680 (1993).
21. M. Haelterman, S. Trillo, S. Wabnitz, J. Opt. Soc. Am. B **11**, 446–456 (1994).
22. M. Marte, H. Ritsch, K.I. Petsas, A. Gatti, L.A. Lugiato, C. Fabre, D. Leduc, Opt. Express **3**, 71–80 (1998).
23. V.J. Sanchez-Morcillo, K. Staliunas, Phys. Rev. A **61**, 7076–7080 (2000).
24. A.V. Buryak, Y.S. Kivshar, S. Trillo, Phys. Rev. Lett. **77**, 5210–5213 (1996).
25. D.V. Skryabin, Phys. Rev. E **60**, R3508–R3511 (1999).

26. M. Le Berre, D. Leduc, E. Ressayre, A. Tallet, J. Opt. B **1**, 153–160 (1999).
27. P.S. Jian, W. Torruellas, M. Haelterman, S. Trillo, U. Peschel, F. Lederer, Opt. Lett. **24**, 400–402 (1999).
28. K. Staliunas, J. Mod. Opt. **42**, 1261–1269 (1995).
29. P. Coullet, C. Gil, F. Rocca, Opt. Commun. **73**, 403–408 (1989).
30. K. Staliunas, Phys. Rev. A **48**, 1573–1581 (1993).
31. S. Longhi, J. Mod. Opt. **43**, 1089–1094 (1996); J. Mod. Opt. **43**, 1569–1575 (1996).
32. A. Mecozzi, W.L. Kath, P. Kumar, C.G. Goedde, Opt. Lett. **19**, 2050–2052 (1994); C.G. Goedde, W.L. Kath, P. Kumar, J. Opt. Soc. Am. B **14**, 1371–1379 (1997).
33. J. Miles, J. Fluid Mech. **148**, 451–460 (1984); X. Wang, R. Wei, Phys. Rev. Lett. **78**, 2744–2747 (1997).
34. I. Barcholdin, A. Il'ichev, Eur. J. Mech. B **78**, 2744–2747 (1999).
35. S. Longhi, Opt. Lett. **21**, 860–862 (1996).
36. S. Longhi, A. Geraci, Appl. Phys. Lett. **67**, 3060–3062 (1995).
37. S. Longhi, A. Geraci, Phys. Rev. A **54**, 4581–4584 (1996).
38. V.J. Sanchez-Morcillo, E. Roldan, G.J. de Valcarcel, K. Staliunas, Phys. Rev. A **56**, 3237–3244 (1997).
39. G.J. de Valcarcel, K. Staliunas, E. Roldan, V.J. Sanchez-Morcillo, Phys. Rev. A **54**, 1609–1624 (1996).
40. M. Tlidi, P. Mandel, R. Lefever, Phys. Rev. Lett. **73**, 640–643 (1994).
41. J. Lega, J.V. Moloney, A.C. Newell, Phys. Rev. Lett. **73**, 2978–2981 (1994).
42. G.T. Dee, W. Van Sarloos, Phys. Rev. Lett. **60**, 2641–2644 (1988).
43. J.N. Kutz, T. Erneux, S. Trillo, M. Haelterman, J. Opt. Soc. B **16**, 1936–1941 (1999).
44. H. Ward, M.N. Ouarzazi, M. Taki, P. Glorieux, Eur. Phys. J. D **3**, 275–288. (1998).
45. M. Taki, N. Ouarzazi, H. Ward, P. Glorieux, J. Opt. Soc. Am. B **61** 997–1003 (2000).
46. M. Taki, M. San Miguel, M. Santagiustina, Phys. Rev. E **61** 2133–2136 (2000).
47. K. Staliunas, V.J. Sánchez-Morcillo, Phys. Rev. A **57**, 1454–1457 (1998).
48. M. Tidli, P. Mandel, R. Lefever, Phys. Rev. Lett. **81** 979–982 (1998).
49. K. Staliunas, V.J. Sánchez-Morcillo, Phys. Lett. A **241**, 28–31 (1998).
50. L.A. Lugiato, C. Oldano, C. Fabre, E. Giacobino, R.J. Horowicz, Il Nuovo Cimento **10D**, 959–978 (1988).
51. P. Mandel, N.P. Pettiaux, W. Kaige, P. Galatola, L. Lugiato, Phys. Rev. A **43** 424–432 (1991).
52. H.R. Brand, V. Steinberg, Phys. Rev. A **29**, 2303–2304 (1984); H.R. Brand, R.J. Deissler, Phys. Rev. A **45**, 3732–3736 (1992).
53. S. Trillo, P. Ferro, Opt. Lett. **20**, 438–440 (1995).
54. R.A. Fuerst, D.M. Baboiu, B. Lawrence, W.E. Torruellas, G.I. Stegeman, S. Trillo, S. Wabnitz, Phys. Rev. Lett. **78**, 2756–2759 (1997).
55. G. Oppo, M. Brambilla, D. Camesasca, A. Gatti, L.A. Lugiato, J. Mod. Opt. **41**, 1151–1162 (1994).
56. S. Longhi, Phys. Rev. A **53**, 4488–4499 (1996).
57. S. Longhi, Phys. Rev. E **53**, 5520–5522 (1996).
58. K. Staliunas, Phys. Rev. Lett. **81**, 81–84 (1998).
59. K. Staliunas, V.J. Sánchez-Morcillo, Opt. Commun. **177**, 389–395 (2000).

60. D. Schowb, P.F. Cohadon, C. Fabre, M.A.M. Marte, H. Ritsch, A. Gatti, L.A. Lugiato, Appl. Phys. B **66**, 685–699 (1998).
61. M. Vaupel, A. Maitre, C. Fabre, Phys. Rev. Lett. **83**, 5278–5281 (1999).
62. S. Trillo, M. Haelterman, Opt. Lett. **21**, 1114–1116 (1996).
63. C. Etrich, U. Peschel, F. Lederer, Phys. Rev. E. **56**, 4803–4808 (1997).
64. M. Tlidi, M. Haelterman, P. Mandel, Phys. Rev. E **56**, 6524–6530 (1997).
65. M. Tlidi, H. Haelterman, Phys. Lett. A **239**, 59–64 (1998).
66. S. Longhi, Opt. Lett. **23**, 346–348. (1998).
67. S. Longhi, Phys. Rev. A **59**, 4021–4040 (1999).
68. C. Etrich, U. Peschel, F. Lederer, Phys. Rev. Lett. **79**, 2454–2457 (1997).
69. P. Coullet, C. Elphick, D. Repaux, Phys. Rev. Lett. **58** 431–434 (1987).
70. P. Coullet, L. Gil, D. Repaux, Phys. Rev. Lett. **62** 2957–2960 (1989).
71. V.B. Taranenko, K. Staliunas, C.O. Weiss, Phys. Rev. Lett. **81**, 2236–2239 (1998).
72. W. Van Sarloos, P.C. Hohenberg, Phys. Rev. Lett. **64**, 749–752 (1990).
73. P. Coullet, J. Lega, B. Houchmanzadeh, J. Lajzerowicz, Phys. Rev. Lett. **65** 1352–1355 (1990).
74. G.L. Oppo, A.J. Scroggie, W.J. Firth, Domain walls in optical parametric oscillators: dynamics and stabilisation, in *Proceedings of the European Conference on Laser and Electro-Optics*, Glasgow, UK, 1998, paper QFA, p. 245.
75. G.L. Oppo, A.J. Scroggie, W.J. Firth, J. Opt. B **1**, 133–138 (1999).
76. A.V. Buryak, Y.S. Kivshar, Phys. Rev. A **51**, R41–R44 (1995).
77. B. Scäpers, M. Feldmann, T. Ackemann, W. Lange, Phys. Rev. Lett. **85**, 748–751 (2000).
78. G. Izús, M. San Miguel, M. Santagiustina, Opt. Lett. **25**, 1454–1456 (2000).
79. H.R. Brand, R.J. Deissler, Phys. Rev. Lett. **62**, 2957–2960 (1989).
80. S. Fauve, O. Thual, Phys. Rev. Lett. **64**, 282–285 (1990).
81. S. Trillo, M. Haelterman, Opt. Lett. **23**, 1514–1516 (1998).
82. D.V. Skryabin, W. Firth, Opt. Lett. **24**, 1056–1058 (1999).
83. J.N. Kutz, W.L. Kath, SIAM J. App. Math. **56**, 611–626 (1996).
84. S. Longhi, Phys. Rev. E **55**, 1060–1070 (1997).
85. T. Kapitula, B. Sandstede, J. Opt. Soc. Am. B **15**, 2757–2762 (1998).
86. K.A. Gorshkov, L.A. Ostrovskii, Physica D **3**, 428–438 (1981).
87. C. Richy, K.I. Petsas, E. Giacobino, C. Fabre, J. Opt. Soc. Am. B **12**, 456–461 (1995).
88. S. Longhi, Opt. Commun. **149**, 335–340 (1998).
89. D.V. Skryabin, A.R. Champneys, W. Firth, Phys. Rev. Lett. **84**, 463–466 (2000).
90. G.J. De Valcarcel, E. Roldan, K. Staliunas, Opt. Commun. **181**, 207–213 (2000).
91. R.J. Deissler, J. Stat. Phys. **40**, 371 (1985); K.L. Babock, G., Ahlers, S. Cannel, Phys. Rev. Lett. **67** 3388–3391 (1991).
92. M. Santagiustina, P. Colet, M. San Miguel, D. Walgraef, Phys. Rev. Lett. **79**, 3633–3636 (1998).
93. S. Coen, M. Tlidi, Ph., Emplit, M. Haelterman, Phys. Rev. Lett. **83** 2328–2331 (1999); S. Coen, M. Haelterman, Phys. Rev. Lett. **79** 4139–4142 (1997).
94. J. Bragard, P.L. Ramazza, F.T. Arecchi, S. Boccaletti, L. Kramer, Phys. Rev. E **61** R6045–R6048 (2000).
95. M. Santagiustina, P. Colet, M. San Miguel, D. Walgraef, Phys. Rev. E **58**, 3843–3853 (1998).

96. G. Izús, M. Santagiustina, M. San Miguel, P. Colet, J. Opt. Soc. Am. B **16**, 1592–1596 (1999).
97. M. Santagiustina, P. Colet, M. San Miguel, D. Walgraef, Opt. Lett. **23**, 1167–1169 (1998); Opt. Express **3**, 63–70 (1998).
98. K. Fiedler, S. Schiller, R. Paschotta, P. Kürz, J. Mlynek, Opt. Lett. **18**, 1786–1788 (1993).
99. Z.Y. Ou, S.F. Pereira, E.S. Polzik, H.J. Kimble, Opt. Lett. **17**, 640–642 (1992).
100. M. Tlidi, M. Haelterman, P. Mandel, Phys. Rev. E **56**, 6524–6530 (1998).
101. M. Tlidi, P. Mandel, Phys. Rev. Lett. **83**, 4995–4998 (1999).
102. L. Torner, D. Mazilu, D. Mihalache, Phys. Rev. Lett. **77**, 2455–2458 (1996).
103. U. Peschel, C. Etrich, F. Lederer, Opt. Lett. **23**, 500–502 (1998).
104. U. Peschel, D. Michaelis, C. Etrich, F. Lederer, Phys. Rev. E **58**, R2745–R2748 (1998).
105. M. Haelterman, P. Mandel, Opt. Lett. **15**, 1412–1414 (1990).
106. H. Kawaguchi, T. Irie, Electron. Lett. **28**, 1645–1646 (1992).
107. M. Haelterman, Opt. Commun. **86**, 189–191 (1991).
108. M. Haelterman, A.P. Sheppard, A. Snyder, Opt. Lett. **19**, 96–98 (1994).
109. R.F. Nabiev, P. Yeh, D. Botez, Opt. Lett. **18**, 1612–1614 (1993).
110. Y.S. Kivshar, Phys. Rev. E **51**, 1613–1616 (1995).
111. S. Trillo, Opt. Lett. **21**, 1732–1734 (1996).
112. M. Picciau, G. Leo, G. Assanto, J. Opt. Soc. Am. B **13**, 661–670 (1996).
113. C. Conti, S. Trillo, G. Assanto, Phys. Rev. Lett. **78**, 2341–2344 (1997); Opt. Lett. **22**, 445–447 (1997).
114. H. He, P.D. Drummond, Phys. Rev. Lett. **78**, 4311–4314 (1997); Phys. Rev. E **58**, 5024–5046 (1998).
115. T. Peschel, U. Peschel, F. Lederer, B.A. Malomed, Phys. Rev. E **55**, 4730 (1997).
116. C. Conti, S. Trillo, G. Assanto, Opt. Lett. **23**, 334–336 (1998); Electron. Lett. **34**, 689–690 (1998).
117. C. Conti, S. Trillo, G. Assanto, Phys. Rev. E **57**, R1251–R1254 (1998).
118. C. Conti, G. Assanto, S. Trillo, Opt. Lett. **22**, 1350–1352 (1997).
119. C. Conti, A. De Rossi, S. Trillo, Opt. Lett. **23**, 1265–1267 (1998).
120. C. Conti, G. Assanto, S. Trillo, Phys. Rev. E **59**, 2467–2470 (1999).
121. T. Iizuka, C.M. de Sterke, Phys. Rev. E **62**, 4246–4250 (2000).
122. V.I. Barashenkov, D.E. Pelinovsky, E.V. Zemlyanaya, Phys. Rev. Lett. **80**, 5117–5120 (1998); A. De Rossi, C. Conti, S. Trillo, Phys. Rev. Lett. **81**, 85–88 (1998); D. Mihalache, D. Mazilu, L. Torner, Phys. Rev. Lett. **81**, 4353–4356 (1998).
123. J. Schöllmann, R. Scheibenzuber, A.S. Kovalev, A.P. Mayer, A.A. Maradunin, Phys. Rev. E **59**, 4618–4629 (1999).
124. C. Conti, G. Assanto, S. Trillo, Opt. Express **3**, 389–404 (1998).
125. A. Arraf, C.M. de Sterke, Phys. Rev. E **58**, 7951–7958 (1998); A. Arraf, H. He, C.M. de Sterke, Opt. Fiber Tech. **5**, 223 (1999).
126. T. Iizuka, Y.S. Kivshar, Phys. Rev. E **59**, 7148–7151 (1999).
127. W.C.K. Mak, B.A. Malomed, P.L. Chu, Phys. Rev. E **58**, 6708–6722 (1998); A.R. Champneys, B.A. Malomed, Phys. Rev. E **61**, 886–890 (1998).
128. A.V. Buryak, I. Towers, S. Trillo, Phys. Lett. A **267** 319–325 (2000); S. Trillo, C. Conti, G. Assanto, A.V. Buryak, Chaos **10**, 590–600 (2000).

129. H. He, A. Arraf, C.M. de Sterke, P.D. Drummond, B.A. Malomed, Phys. Rev. E **59**, 6064–6078 (1999).
130. I.D. Lindsay, G.A. Turnbull, M.H. Dunn, M. Ebrahimzadeh, Opt. Lett. **23**, 1889–1891 (1998); M. Ebrahimzadeh, G.A. Turnbull, T.J. Edwards, D.J.M. Stothard, I.D. Lindsay, M.H. Dunn, J. Opt. Soc. Am. B **16**, 1499–1511 (1999).
131. J. Söchtig, R. Grob, I. Baumann, W. Sohler, H. Shultz, R. Widmer, Electron. Lett. **31**, 551–552 (1995).
132. C., Becker, A. Greiner, T., Oesselke, A. Pape, W. Sohler, H. Suche, Opt. Lett. **23** 1195–1197 (1998).
133. P.G. Kazansky, V. Pruneri, J. Opt. Soc. Am. B **14**, 3170–3179 (1997).
134. D.T. Reid, J.M. Dudley, M. Ebrahimzadeh, W. Sibbett, Opt. Lett. **19**, 825–827 (1994).

Spatial Solitons in Resonators

Carl O. Weiss, Gintas Slekys, Victor B. Taranenko, Kestutis Staliunas, and Robert Kuszelewicz

Summary. We discuss solitons in nonlinear resonators for the following different systems: laser resonators containing a nonlinear absorber, intracavity parametric mixing, and semiconductor microresonators. In these structures the following localized structures (or spatial solitons) were predicted: bright solitons, phase-defect solitons, and bright and dark coupled solitons. We report our experimental findings, briefly linking them to the main theoretical predictions. Our results confirm that these structures can be observed, paving the way to practical applications.

1 Introduction

Solitary structures can form in optics when a balance occurs between a linear and a nonlinear optical process. The best known instance of such structures in optics is that of solitary pulses propagating along optical. Here a balance occurs between the linear mechanism of dispersion, which leads to a broadening of the pulse in the propagation direction, and the nonlinear mechanism of self-phase modulation or an intensity-dependent refractive index (e.g. the Kerr effect), which tends to shorten the pulse. The result is a pulse traveling along the fiber without changing its shape. Such pulses can be well described by the soliton solutions of the (1 + 1)D nonlinear Schrödinger equation (NLSE) [1]. In this equation, the time coordinate could equally well be a spatial coordinate. It follows that spatial solitons should also exist. The linear broadening mechanism in this case is diffraction . In one spatial dimension, such spatial, or propagation, solitons do indeed exist [2]. The phenomenon manifests itself in the contraction of a light beam with propagation, until a filament of constant thickness is formed, which then propagates without further change (a self-trapped beam).

Taking, however, a normal laser beam, which diffracts or contracts in 2D (the beam cross section), the NLSE has no stable solutions of the form of a beam propagating with a constant diameter, at least in the paraxial optics approximation. The Kerr nonlinearity is stronger than the diffraction, so that catastrophic collapse of the beam cross section occurs [3], see also Bergé in this volume. It was pointed out than such a collapse can be avoided if the nonlinearity is saturable, i.e. it reduces with increased light intensity [4]. In this way a variety of experiments on propagation solitons in 2D has been made possible (see, e.g., [5]). Technological applications in information routing and field steering with such propagation solitons are under consideration.

A particular situation occurs if such a self-trapped beam propagates inside an optical resonator. The finite mirror reflectance acts in this case somewhat similarly to saturation of the nonlinearity because, in each round trip of the light, to the unfocused light which irradiates the resonator is added to the light that has already been self-focused, thus continually weakening the self-focusing. In this case, in resonators of finite finesse, stable filamentation is possible [6]

Evidently, the stability of such structures can be enhanced further by a saturability of the nonlinearity of the material filling the resonator. Thus, the occurrence of spatial resonator solitons has been predicted for a number of nonlinear materials [7]. The first observations of such solitary structures in optical resonators were made before the bulk of the theoretical work on passive resonator solitons appeared. In [8] such solitary structures were observed using a liquid-crystal film inside a resonator, and in [9] such a spatial soliton was observed in a resonator made up of two phase-conjugating mirrors which contained a saturable absorber.

Spatial resonator solitons can exist if the characteristic of the resonator shows two coexisting stable steady states (bistability). Other cases in which such solitons can exist have been discussed in [7]. In a bistable resonator, if it is of large Fresnel number, domains of the two states can exist, which are then connected by a switching front (or switching-waves). Such switching waves move into or out of domains of one of the states, depending on the difference between the background field value and the field value corresponding to the unstable steady-state solution lying between the two stable steady states. A switching front will move into the state of higher intensity if the background field is larger than that of the unstable steady state, and it will move into the domain of the lower-intensity state if the background field is smaller than that of the unstable steady state. Thus, in general, one kind of domain will shrink and the other expand.

In general, the asymptotic state will be one in which the total resonator cross section is entirely switched to one of the two states. If, however, the system is not far from a modulational instability , then the switching fronts do not connect the two states aperiodically but can be accompanied by damped spatial oscillations on either side of the fronts. Then, as a domain contracts, the switching front on one side of the domain will eventually feel the spatial oscillations of the field close to the front on the other side of the domain. The spatial field minima can then trap the front on the other side of the domain, in which case the (small) domain has attained stability (and is then called a solitary or localized structure). Such a domain can be freely moved around the resonator cross section. If the spatial field oscillations on one side of the domain trap not the front from the other side itself, but the spatial oscillations accompanying it, then a higher-order spatial soliton is formed [10] (see also [11]).

We have found, so far, that the solitons of low orders resemble Gauss–Laguerre modes with ring nodes (not the flower-like variety). There is no obvious reason why this should be so. Optical-resonator modes are the eigenfunctions of a boundary problem, with the boundaries given by the mirror surfaces. Conversely, solitons are the solutions of a self-consistent problem where the potential, constituted in the case of an optical resonator by the resonator mirrors, is instead created by the light field itself. Thus it is not obvious why these two problems should have similar solutions, and the reasons for the similarity of the solutions are an open question. Interestingly, it has been found that the potential created by a fundamental soliton can actually allow – in addition to the existence of the field of the fundamental soliton – the stable existence of a first-order soliton [11].

One can picture a spatial soliton as a small domain of one of two coexisting states, surrounded by a stationary switching front which has locked into a stable ring. Owing to the bistable character of the resonator, such spatial solitons are bistable. They can be switched on or off and are thus suitable for carrying information.

2 Solitons in a Laser with a Nonlinear Absorber

The theoretical work on solitons in active resonators dates back to the 1980s (see [12] for a summary).

This work suggested experiments with a repetitively pulsed dye laser, with an internal saturable absorber [13][1]. Figure 1 shows the output power of a dye laser with an internal bacteriorhodopsin (BR) absorption cell, as a function of pump power. The pulse repetition rate was 12 Hz; the acidity of the BR absorber solution was chosen to provide an absorber recovery time constant of about 300 ms, so that the system dynamics were controlled by emitting the absorber and the system could be treated like a continuously system. The bistability of the system is apparent. The upper part of Fig. 1 shows the output beam cross section. Evidently, the narrowest beam occurs within the bistable region. This beam represents a spatial soliton.

The resonator used was of a self-imaging type. This kind of resonator is, in its transverse-mode structure, equivalent to a plane resonator of zero length. For the precise self-imaging length, its transverse modes are completely degenerate. The diffraction losses, equally, correspond to those of a plane resonator of zero length. Thus this resonator, on the one hand, has the complete transverse-mode degeneracy of a plane resonator of zero length, as is necessary for arbitrary images to resonate, and, on the other hand, has sufficient length to house various intracavity elements without the detrimental diffraction losses of a plane resonator of the same length. (It may be noted

[1] A photorefractive version of this experiment was described in [14]

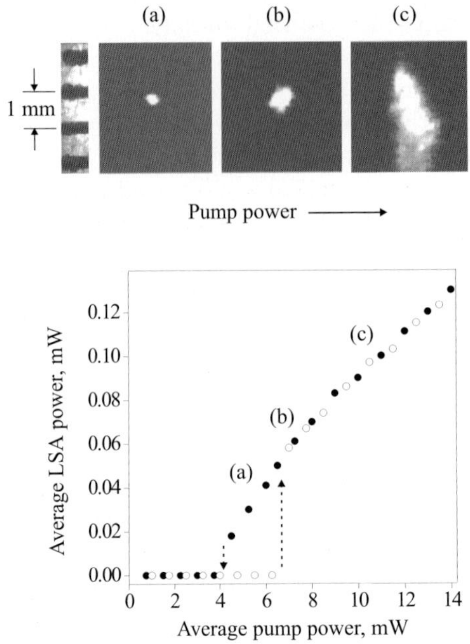

Fig. 1. Average output power of a laser with a BR absorber as a function of pump power, showing bistability. Beam profiles corresponding to the positions (**a**), (**b**), (**c**) in the graph are shown above the graph

that such a resonator also permits one to realize a negative length, as far as the transverse-mode structure is concerned).

Figure 2 shows the writing of a spatial soliton in this system in various places within the resonator cross section. The absorber cell was locally bleached for a short time (by a He–Ne laser beam). The result was a stationary spatial soliton (which remained after the external bleaching was stopped).

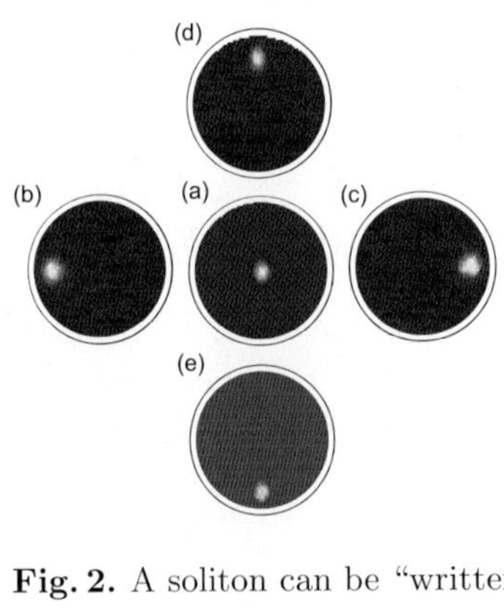

Fig. 2. A soliton can be "written" at any location in the laser cross section

Figure 2 shows, on the one hand, that the solitons are bistable (i.e. can be switched on and off) and, on the other hand, that they can exist at any location in the cross section.

The motion of solitons in field gradients was also tested. In a fluid analogy of a laser [15], a phase gradient corresponds to a flow velocity and an intensity gradient to a density (or pressure) gradient. Thus a soliton should move in such gradients. Figure 3 shows experiments with phase gradients. In Fig. 3 (left), a phase gradient across the laser cross section was created by a small tilt of one resonator mirror of the laser. The snapshots, taken at equidistant times, show the motion of the soliton induced by the phase gradient. By varying the length of the self-imaging laser resonator away from the precise self-imaging length, a phase trough was created with its minimum at the center of the resonator (Fig. 3, right). As the snapshots show, the soliton is drawn from all sides towards the center of the phase trough, where it is then trapped. This movement and trapping of solitons could be important in the use of resonator solitons in optical information processing [16].

In these initial experiments, the laser containing the nonlinear absorber emitted a large number of longitudinal modes . Thus the tuning of the resonator was of no importance, as the modes emitted adjusted to the resonator length. The resonator tuning does, however, affect the solitons, their motion, and their characteristics if emission is restricted to a single longitudinal-mode family. The simplest way to achieve such mode selection is to use an active medium with a very narrow gain spectrum. By far the narrowest gain spectra (if one interprets this term in the context of a laser physics) are provided by photorefractive gain media [17]. Therefore experiments were conducted using resonators of the self-imaging type [18] with photorefractive gain. To picture the effects of a narrow gain line, one can think of the self-imaging resonator

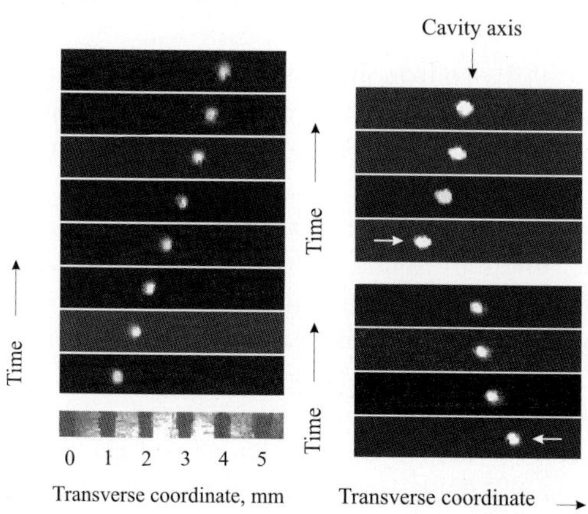

Fig. 3. A phase gradient (tilt of one resonator mirror) leads to drift of the soliton (*left*). A soliton drifts from initial all directions into a phase trough (*right*)

as a plane–plane resonator. Such a resonator has a (longitudinal) resonance if the length can accommodate an integer number N of (half) wavelengths of radiation generated – this wavelength is fairly strictly determined by the wavelength of the center of the gain line.

Then, if the length of the resonator is between $N\lambda/2$ and $(N-1)\lambda/2$, the emission can adjust to this resonator length by a tilt of the propagation direction of the radiation with respect to the resonator axis. In this way N half wavelengths can be accommodated in the resonator [19]. Such a tilted wave has a propagation component along the resonator axis and a (small) propagation component lying in the plane of the resonator mirror. Light will therefore move across the cross section of the resonator in a detuned resonator, corresponding to motion in a phase gradient. We can therefore expect stationary solitons in the case of precise resonator tuning, and moving solitons in the case of detuning.

These stationary solitons correspond to the stationary solitons described above for the broadband (dye) multimode laser, which can adjust its detuning to zero by a choice of different longitudinal modes. For the moving solitons corresponding to detuned (i.e. tilted-wave) emission, one can directly deduce the characteristics of the soliton motion from this picture, as follows.

- The direction of the tilt of the wavevector of the light generated is free. The decomposition of the wavevector into a longitudinal and a transverse component is fixed only in terms of the magnitude of the components. Thus the direction of motion of the solitons is, a priori, undetermined. The actual direction of motion is determined by spontaneous symmetry-breaking (in the same way as a single-mode laser chooses the phase of its field at laser threshold). As for the laser phase, there is no restoring force for any particular direction here. The direction of motion of a soliton can therefore change under external influences. In the presence of noise the direction will change diffusively.
- Whereas the direction of motion of a soliton and its position in the laser cross section are free, the magnitude of the soliton velocity is fixed and is determined by the detuning of the resonator (wavevector tilt is proportional to detuning).
- Different longitudinal orders can be emitted simultaneously if their wavevectors are tilted by different amounts. Thus, in a resonator of length $N\lambda/2$, which emits a stationary soliton, simultaneous emission of moving solitons is possible. The velocities of the moving solitons are quantized according to the longitudinal orders for a resonator of given length.
- If we suppose in the experiment a self-imaging resonator is used, with the nonlinear absorber in the near field (near a plane mirror) and the (photorefractive) gain medium in the far field, a strange type of competition among moving solitons is expected. The fields of two solitons overlapping inside the gain medium compete; only one soliton can then survive. The consequence is a competition of solitons in *velocity space*. They will com-

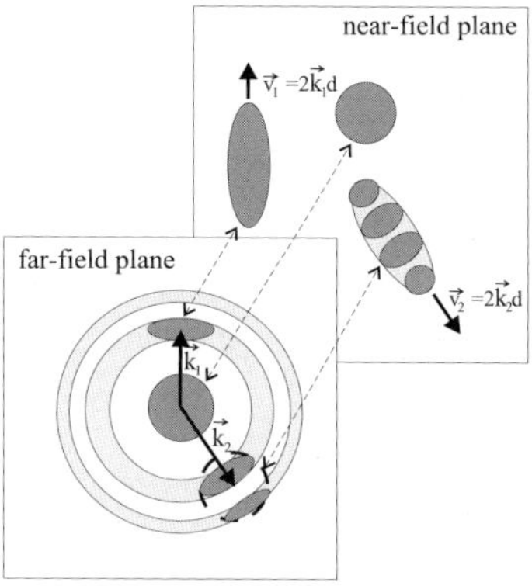

Fig. 4. Intensity distributions in the near and far field for stationary and moving solitons. A central spot in the far field corresponds to a stationary soliton with arbitrary position in the near field. A spot (*elliptical shape*) on a resonant ring corresponds to a moving soliton. Simultaneous emission on two rings corresponds to a moving soliton with (stationary) modulation of the soliton intensity ("inchworm")

pete if and only if the two solitons have the same vectorial velocity (i.e. the directions and magnitudes of the velocities are equal). Notably, this will occur even when they are far apart in the near-field plane.

- For stationary solitons, the competition condition is trivially fulfilled. Therefore only one stationary soliton can exist at a time.
- Moving solitons with different directions of motion and/or different magnitudes of velocity can coexist; thus, in general, a large number of moving solitons can coexist with one stationary soliton.

Figure 4 depicts the situation: a stationary soliton corresponds to emission along the resonator axis. Moving solitons correspond to emission at a fixed angle (determined by the detuning) to the resonator axis, in the far field. Therefore the stationary soliton corresponds to light in the central spot of the Airy function rings of the resonator, while the moving solitons correspond to emission on the rings.

A restriction on the wavelength of the light generated is set by the finite widths of the gain line of the active medium. The finite width of this gain line corresponds to the allowed spread of tilt angles of the emitted wave. Therefore the emission of the stationary soliton corresponds to a central disk of finite diameter, and that of the moving solitons to finite-area sections of a ring. For moving solitons, the wavevector of the emitted light changes faster with radial angle than with azimuthal angle. Therefore the light of a moving soliton in the far field occupies an elliptical section of an Airy ring (see Fig. 4). Its Fourier transform (a moving soliton in the near field) is of elliptical shape, with the long axis in the direction of motion (suggesting again a fluid picture).

Figure 5 shows cases of all these soliton types, as recorded at the output of a photorefractive $BaTiO_3$ oscillator, with a BR saturable absorber and a self-imaging resonator [18]. Excitation of emission segments on two rings into the same azimuthal angle results in an inchworm soliton (Fig. 5b).

In these experiments the gain elements of the lasers were placed in the conjugate plane of the near field. This leads to competition among certain

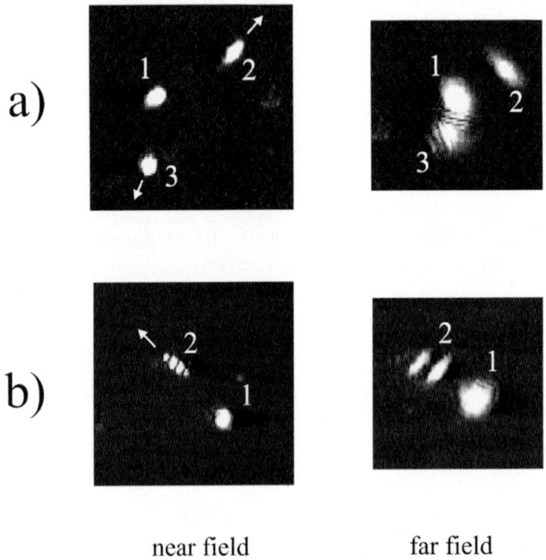

Fig. 5. Experimental realization of the solitons described in Fig. 4 (*left*, near field; *right*, far field): (**a**) one stationary soliton (1) and two moving solitons (2), (3); (**b**) one stationary soliton (1) and a moving "inchworm"-soliton (2)

solitons and, in particular, only one stationary soliton can exist. For applications where such solitons are to serve as binary elements for information storage, however, large numbers of stationary solitons are desirable.

In order to test whether this was achievable, experiments were conducted with both the gain element and the nonlinear absorber in the near-field plane. This case was extensively treated in [12] for the situation in which both elements are a plane resonator. In the experiments, the unsaturated absorption of the nonlinear absorber was so high that, even at the highest pump power available, the laser could not be made to emit. External bleaching of the absorber (by a green laser) was therefore used for complete saturation of the absorber, so that large-area laser emission occurred. In this situation, if the pump strength is then reduced, the absorber gradually unsaturates and becomes intensity dependent (nonlinear).

Figure 6 shows the formation of the solitons in the experiment. To illustrate the development of solitons out of the laser emission, Fig. 7 shows a numerical calculation of the process. Initially (top of Fig. 7), the emission typical of a tuned laser shows a number of optical vortices, which are separated by "shocks" [20] (i.e. a "vortex glass" [21]). As the pump power is reduced, the vortices develop into dark areas and the shocks convert to one-dimensional soliton structures. Further reduction of the pump power leads to shortening of the 1D solitary structures, which can then be converted to 2D bright solitons by increasing the pump power slightly. This final increase in pump power is necessary because the diffraction losses for a 2D soliton are

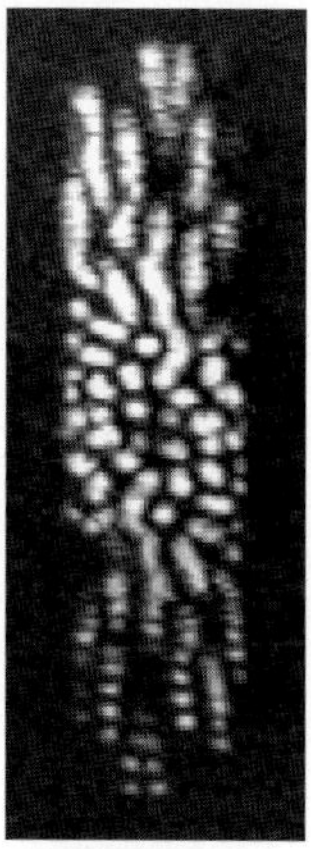

Fig. 6. Transition from 1D to 2D solitary structures

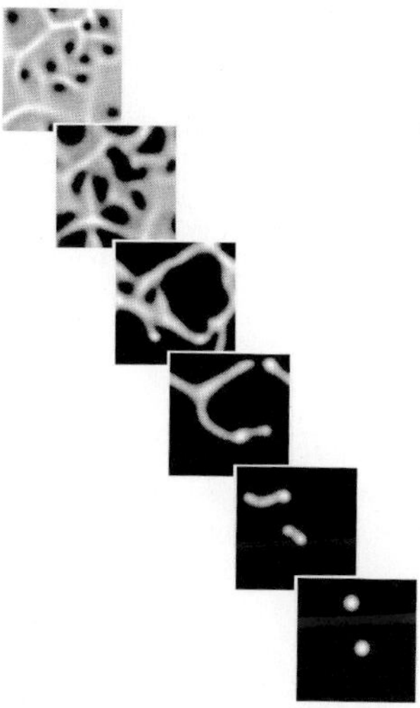

Fig. 7. Numerical calculation of the transition from laser to soliton emission

larger than for a 1D solitary line (for further details see [18]). This can be seen in Fig. 6: in the outer regions, where the pump intensity is weak, stripes prevail, while at the higher pump intensity in the center, spots dominate.

Figure 8 shows ensembles of 2D solitons in the final stage, for different pump powers. It appears that the number of solitons existing in the final state is a monotonic function of the pump power.

As the pump beam had a Gaussian intensity profile, gradients existed in the emission, which caused a slow outward motion of the solitons. Thereby some solitons would reach the edge of the emission field and be extinguished there. This loss of solitons at constant pump power was accompanied by continual splitting of solitons in the central area of the near field. It appeared that the splitting occurred so as to balance the loss of solitons at the edges. Figure 9 shows such soliton splitting.

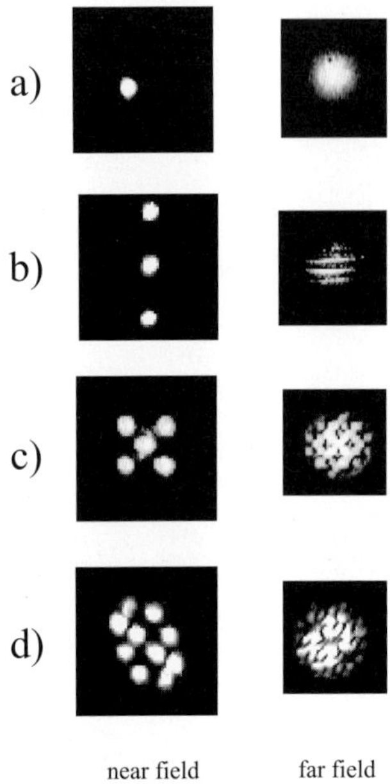

Fig. 8. Collections of solitons (as experimentally observed) corresponding to different pump powers

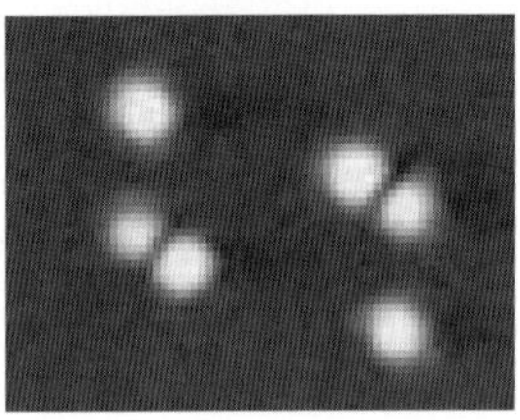

Fig. 9. Splitting of solitons in a laser with a saturable absorber (for details see text)

Time has not yet allowed us to study the interaction of solitons in this system, which must exhibit interesting phenomena. Each of the solitons here is an independent laser whose phase is arbitrary. The interaction between solitons would, on the one hand, depend on the relative phase, on the other hand, two solitons whose phase is free to change can be expected to synchronize their phases. The posibility of this synchronization being either in or out of phase, combined with the initial independence of the phases of the individual solitons, can be expected produce a complicated interaction.

3 Parametric-Mixing Solitons

Whereas the field of a laser, as used in the experiments described above, can have any phase value, in wave mixing with phase-matching the phase of the generated field is tied to the phase of the pump field. The generated field can therefore be described by a real-valued variable – as opposed to the complex-valued field of a laser.

Spatial resonator solitons require, as has been described in Sect. 1, a bistable characteristic of the resonator. The experiments described so far have utilized a subcritical bistability with a high- and a low-intensity branch. For degenerate wave mixing such as four-wave mixing (D4WM) or degenerate parametric mixing (DOPO), a phase bistability of the (real-valued) field occurs [22]. Although this is a symmetric, and supercritical bistability, one would expect that this kind of bistability would also support spatial solitons, which we would call *phase solitons.* A calculation shows [23] that spatial solitons indeed exist for a finite, small detuning.

Figure 10 gives the shapes of such solitons in intensity and phase. Inside the solitons, the phase of the field is opposite to that in the surrounding area, so that a dark, circular interference fringe forms a switching front connecting the two steady states.

The corresponding experiment was conducted using D4WM in $BaTiO_3$ [24]. Figure 11 shows the resonator used. Two pump beams, together with the generated fields, form an index grating in the material, which diffracts pump radiation into the generated fields and adjusts self-consistently to the generated field.

The two counterpropagating generated fields resonate in the same (linear) resonator, which forces them to be degenerate and, consequently, forces the generated field to be bistable and real-valued. A typical intensity distribution of the field generated experimentally is shown in Fig. 12. Small circular domains coexist whith larger domains and black domain walls, as was expected. Figure 13 shows a domain wall of complicated shape, together with an interferogram that indicates that the fields on either side of the domain wall have opposite phases. We note that the domain walls themselves are ex-

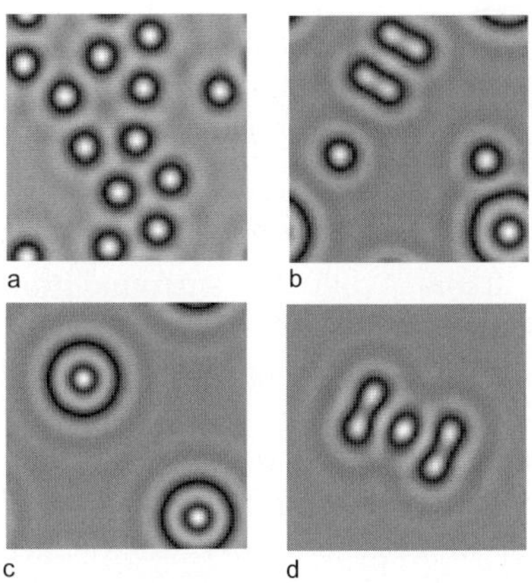

Fig. 10. Different stable localized structures calculated for a DOPO or for D4WM: (**a**) fundamental solitons; (**b**) bound states between fundamental solitons, and between fundamental and higher-order solitons; (**c**) higher-order solitons; (**d**) a complicated bound state of solitons

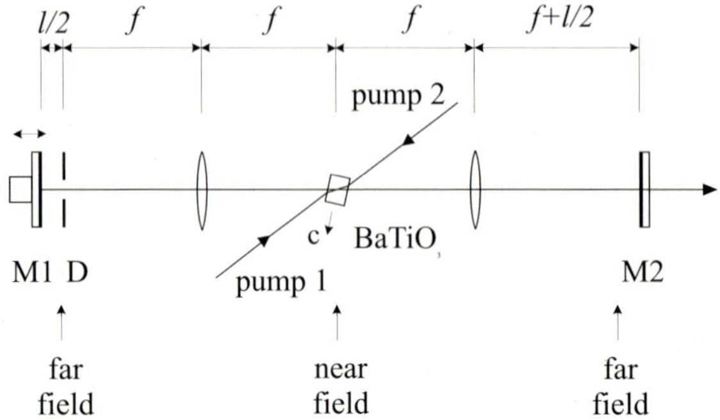

Fig. 11. Schematic illustration of the resonator used for observing phase solitons in degenerate (photorefractive) four-wave-mixing. M, mirrors; f, focal length of lenses; l, deviation from self-imaging length; D, iris for blocking high-order resonant rings

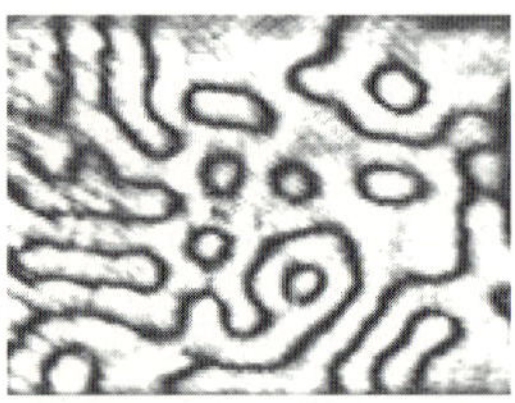

Fig. 12. A typical intensity distribution observed for a resonator of the type shown in Fig. 11. Small circular domains coexist with large, irregularly shaped domains

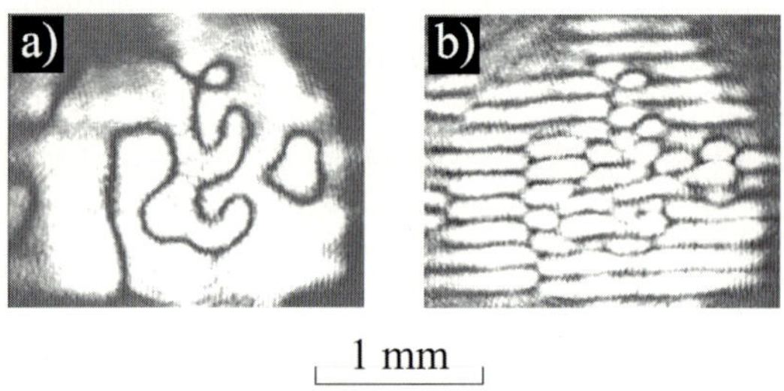

Fig. 13. (a) A domain boundary of complex shape. **(b)** Interferogram showing that the fields separated by the boundary have opposite phases (experimental)

tended 1D solitary structures [25]. They are switching waves connecting the two steady states of the resonator (fields with phases of $+\pi/2$ and $-\pi/2$). In general, such switching waves move and the domains that they surround grow or shrink, the length of the domain walls expanding or contracting. The expansion/contraction can be controlled by resonator detuning [23]. Figure 14 shows the contraction of a domain wall.

Although in the experiment small domains appeared to be stable, it is necessary to prove their stability more explicitly, since stability is hard to distinguish from slow, transient dynamics. Recordings under well-defined resonator-tuning conditions were therefore analyzed. The resonator length was, for this purpose, actively stabilized with respect to the pump light frequency in a manner similar to that described in [26].

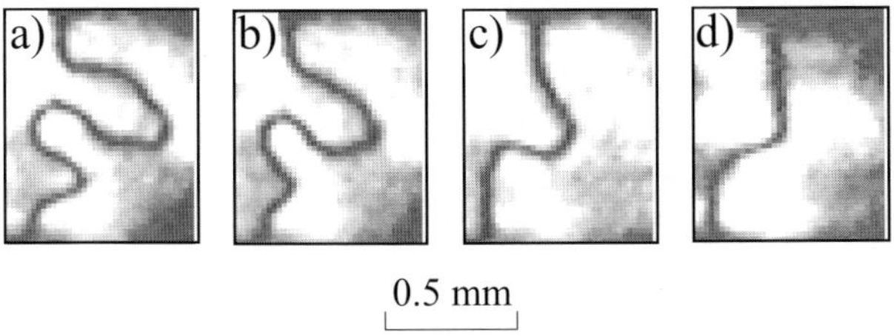

Fig. 14. Contraction of a domain boundary (experimental)

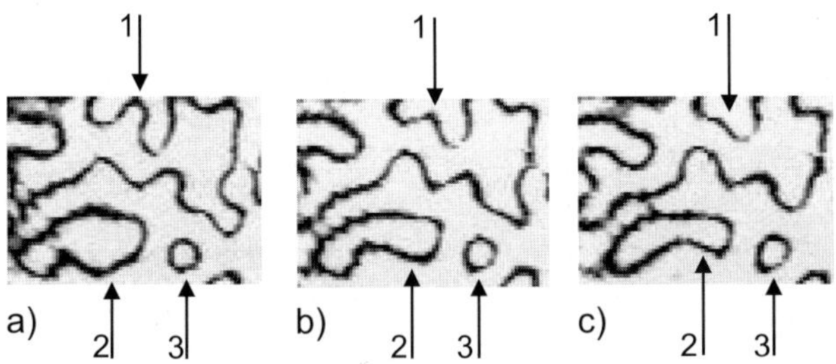

Fig. 15. Three snapshots from an evolution of domains. The large domains (1, 2) shrink, while the small domain (3) maintains a constant size (experimental)

Figure 15 shows three snapshots out of an evolution captured in 20 frames. Figure 16 shows the lengths of the boundaries of the domains denoted by 1, 2, 3 in Fig. 15 as a function of time, as measured from the 20 recorded frames. The boundary of the largest domain (1) shrinks fastest. The boundary of

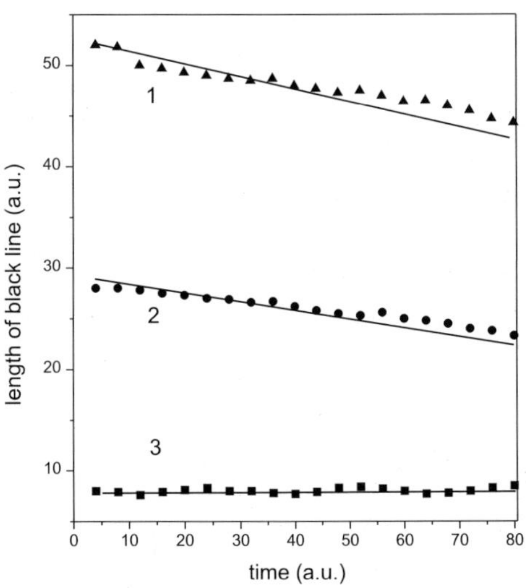

Fig. 16. Lengths of boundaries of the domains 1, 2, 3 in Fig. 15 as a function of time (experimental)

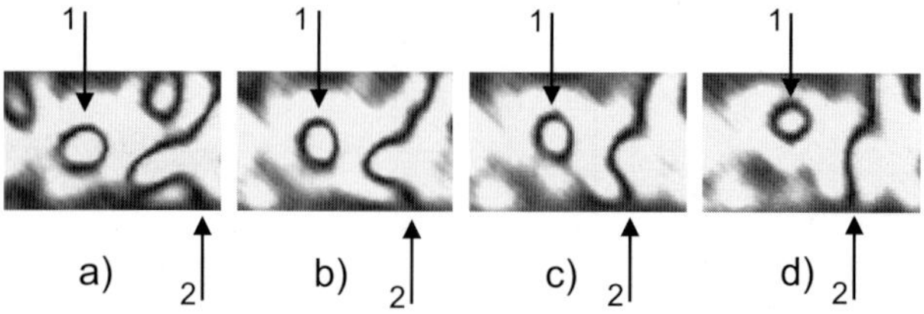

Fig. 17. The length of boundary 2 shrinks rapidly, while the length of boundary 1 is constant in time (experimental)

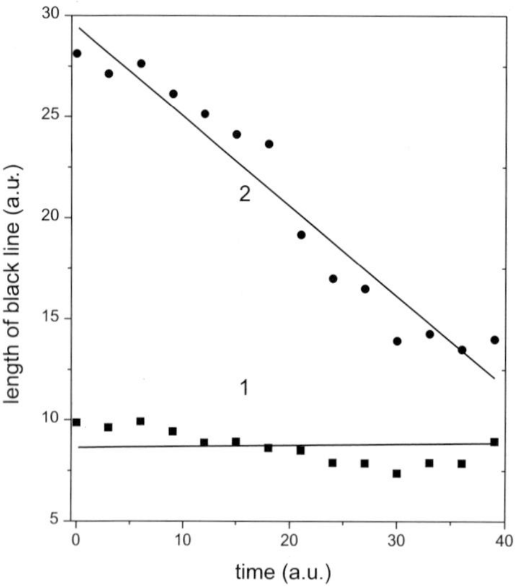

Fig. 18. Boundary lengths of Fig. 17 as a function of time (experimental)

the medium-sized domain (2) shrinks at a slower rate, while the boundary of domain 3 does not change in time. This is proof that the small circular domain 3 is stable and represents a phase soliton. The faster shrinking of the larger domain is what is expected theoretically [23].

The stability of a soliton, under conditions where the shrinkage rate of a large domain is even larger, is shown in the four snapshots in Fig. 17. Analysis of the 20 frames, out of which the snapshots shown in Fig. 17 are taken, leads to the results shown in Fig. 18, which prove again that the small domain is a phase soliton.

4 Nonlinear Semiconductor Resonators

Semiconductors are interesting nonlinear materials for technological applications. Solitons in semiconductor resonators were predicted in [27].

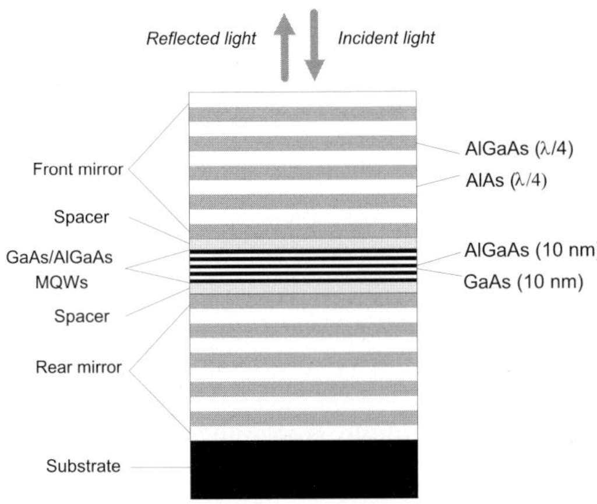

Fig. 19. Schematic illustration of a semiconductor microresonator

We have used a nonlinear semiconductor Fabry–Perot resonator for initial experiments on spatial solitons. The resonator consists of two Bragg mirrors of about 99.5% reflectivity with 18 pairs of $GaAs/Ga_{0.5}Al_{0.5}As$ quantum wells between them (Fig. 19). The thickness of this structure is a few microns while the cross section of the resonator is a few centimeters. Details of these nonlinear Fabry–Perots resonators are given in [28]. The resonance of the Bragg resonator is slightly dependent on the location on the sample, so that by a suitable choice of the area to be irradiated, the wavelength of excitation of the semiconductor material can be chosen to lie either in the interband transition, between the interband transition and the exciton line, on the exciton line, or above the exciton line. Typically, we work a few tens of nanometers above the exciton wavelength so that the nonlinearity is largely dispersive and defocusing. The finesse of the empty resonator is around 500. With the residual absorption of the semiconductor material, the resonator finesse is ~ 100. A CW $Ti{:}Al_2O_3$ laser is used for excitation.

To limit thermal effects, the observations are made during radiation pulses of a few microseconds length, which are repeated every millisecond. To create the pulses, acousto-optic modulators are used. The radiation is focused onto the semiconductor resonator surface into a spot size of 50–100 μm. This provides a reasonably large Fresnel number, since the optical resonator length is only about 3 μm. The light reflected from the sample is observed by a CCD camera or by a fast (2 ns) photodiode.

As has been theoretically predicted for such dispersively nonlinear resonators, structure forms under irradiation. Figure 20 shows that the structure is a hexagonal lattice, as expected [29].

Bistability of the resonator is easily reached (at intensities of a few hundreds of W/cm^2). Figure 21a shows the incident intensity (dashed line) and the reflected intensity (solid line) as measured by the fast photodiode.

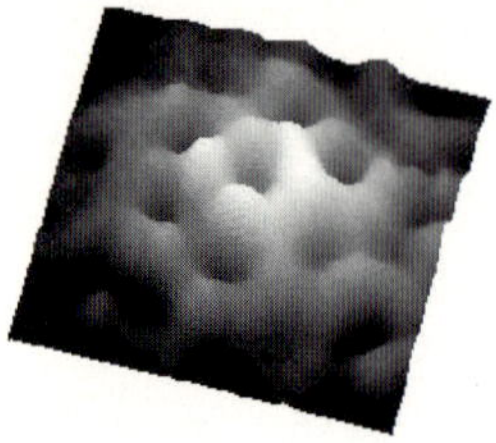

Fig. 20. Hexagonal pattern observed for the semiconductor nonlinear Fabry–Perot resonator shown in Fig. 19

The reduction of reflected light as the sample is switched on is clearly seen, as is the increase of reflected light as the sample is switched off again. From the intensities at which switching on and off occurs, the width of the bistability loop is apparent. After the resonator had been switched on at the point of observation (the image of the detector), we varied the irradiating intensity in order to observe the motion of the switching waves connecting the "on" and "off" regions. By recording curves as shown in Fig. 21a for different locations on a diameter of the laser irradiation spot on the sample, we were able to construct the time history of the resonator dynamics on this diameter. Figure 21d shows a recording thus obtained. The brightness in Fig. 21d corresponds to the reflectivity. The corresponding irradiation is shown in Fig. 21b in the form of equi-intensity lines.

As the irradiation intensity is initially increased, the switching-on threshold is reached at a certain time in the center of the laser field. A switching wave travels then outward until it becomes stationary. We call it then a switching zone. As mentioned in Sect. 1, a switching wave moves into the unswitched region if the background intensity is larger than that corresponding to the unstable steady-state solution on the unstable branch of the S-shaped resonator characteristic, and vice versa. Thus the switching wave becomes stationary at a particular intensity, corresponding to a certain distance from the maximum of the Gaussian laser beam. This is what we observe.

When the power of the laser field is reduced, one would then expect that the stationary switching wave (switching zone) would move towards the center of the laser beam. Comparison of Figs. 21b,d confirms this. The switching zone (boundary between "on" and "off" areas) moves precisely on an equi-intensity contour of the input light. Figure 21d shows for clarity the equi-reflectivity line corresponding to the switching zone, which follows the second lowest intensity contour of the incident light. If one chooses a location on the sample where the resonator resonance is further from the exciton line, the switching zone becomes accompanied, on the lower branch-side, by fringes. This is an indication that, under these conditions, the lower branch is close to a modulational instability . This is a requirement for the formation of spatial dark solitons, as described in Sect. 1.

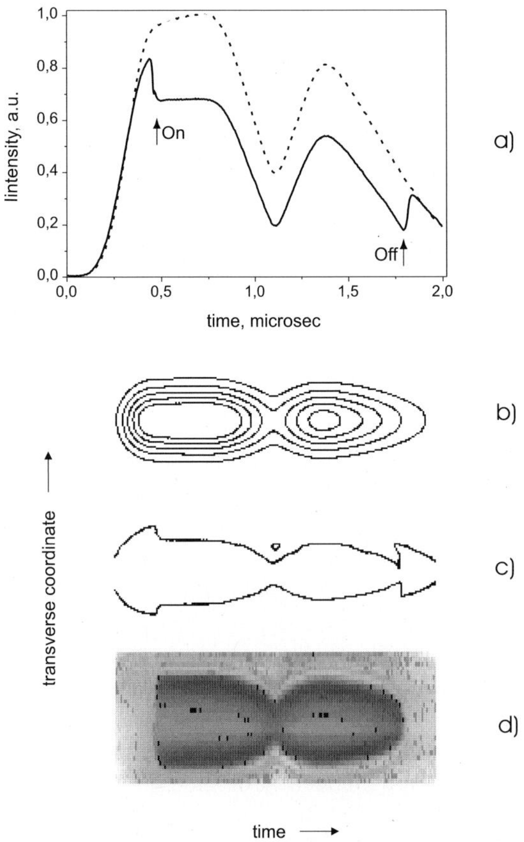

Fig. 21. Bistable switching of semiconductor microresonator and spatial behavior. (**a**) Bistable switching: input intensity, *dashed line*; reflected intensity, *solid line*. By an initial increase of irradiation, the sample is switched "on", and by a later decrease, it is switched "off". In between, the intensity is varied to demonstrate the width of the bistable region and to observe the dynamics of switching fronts. (**b**) Equi-intensity lines (input intensity) plotted for a spatial coordinate on a diameter of the illuminated region, and time. (**c**) Motion of the switching front. (**d**) Reflectivity of sample

A suitable choice of parameters appears indeed to lead to a solitary structure. Figure 22 shows a bright, narrow spot of a size of the order of the elements of the hexagonal pattern shown in Fig. 20. (Figure 22 is an average over 20 laser pulses with a rectangular profile of intensity versus time.) We can clearly show that this small structure is bistable, as predicted for a soliton: Fig. 23 shows the evidence. We used a rectangular laser pulse with a constant intensity in the middle of the bistability region. A short increase in intensity beyond the upper intensity of the bistability switches the small

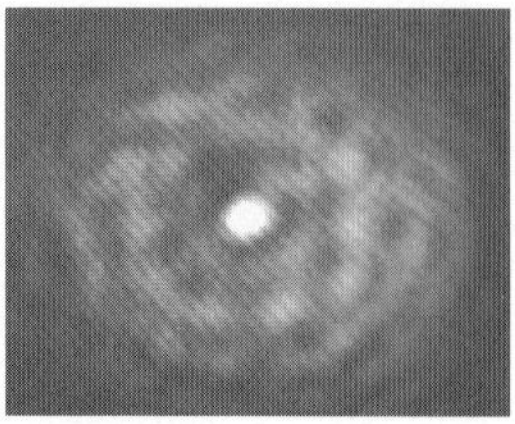

Fig. 22. A small (10 μm) bright structure giving the first indication of a soliton in a semiconductor resonator

localized structure "on". It remains "on" until the intensity is, for a short time, reduced below the lower intensity of the bistability.

This clearly demonstrates that the small structure is bistable, as expected for a spatial soliton. We have tried to show that this structure has a stable shape, as one would expect for a soliton (or a circularly locked switching zone). This is shown in Fig. 24. Figure 24a gives the equi-intensity lines of input field. During a short high-intensity period at the beginning, the central part of the beam is switched up (see the reflected intensity in Fig. 24c). Reducing the intensity then lets the switched-up region contract to the small diameter of the structure shown in Fig. 22.

We then tested the stability of this structure by a variation of the light intensity: if the small bright structure is just a circular switching zone (which is not locked and thus is not a soliton), then its diameter should follow a contour of the incident light. Figure 24b shows the contour corresponding to the switching zone. Evidently, it does not follow any of the contours of the incident light (Fig. 24a). This indicates a certain robustness of the narrow structure against changes of system parameters, for which reason Fig. 22 can be taken as the first indication of the existence of localized structures in semiconductor resonators.

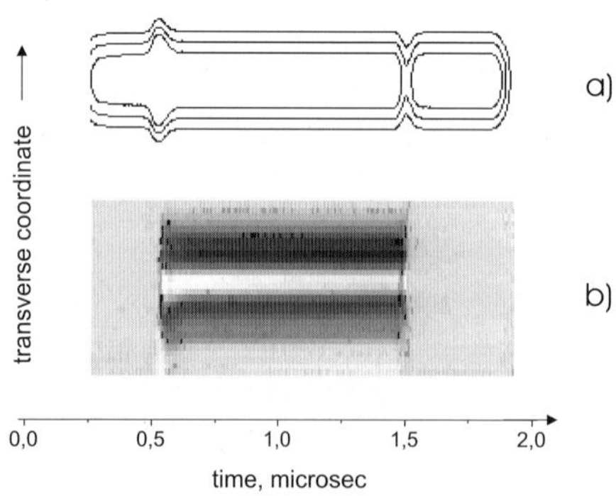

Fig. 23. Test of bistability of the small structure shown in Fig. 22: (**a**) equi-intensity lines of input light; (**b**) sample reflectivity

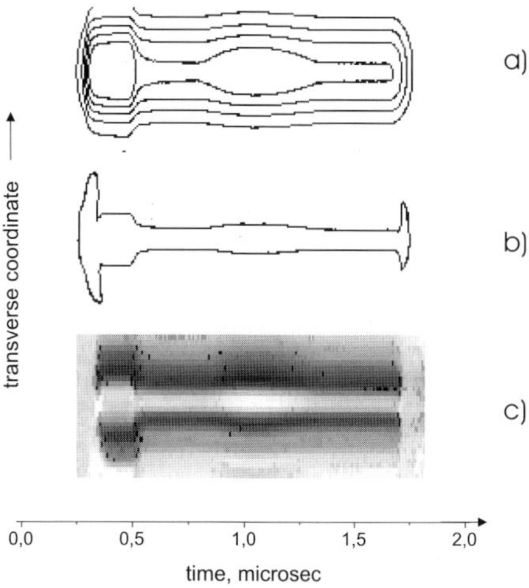

Fig. 24. Test of stability of the small bright structure: (**a**) equi-intensity lines of input light; (**b**) motion of the switching front; (**c**) light intensity reflected from the sample

A more explicit test of the existence of such independent localized structures was performed by injection of spatially narrow, temporarily short light pulses into the illuminated area [30]. Figure 25a shows a collection of bright spots resulting from illumination of the area. The narrow pulse was first directed at the spot marked "a". As can be seen in Fig. 25b, this switches the bright spot "a" to dark. All other spots remain unchanged.

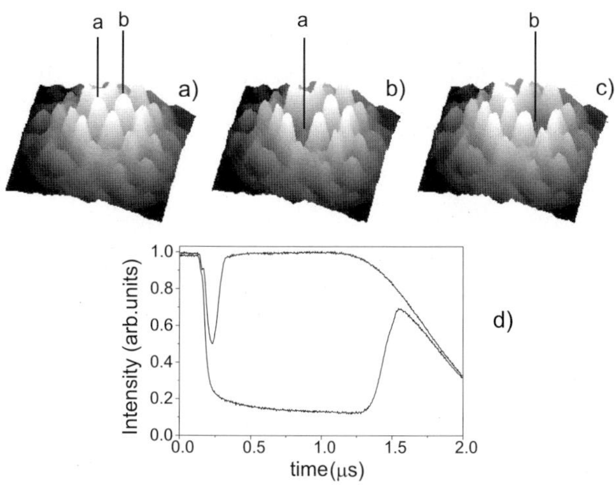

Fig. 25. Switching of individual bright-spots of a bright-spot cluster (see text). The nonlinearity is defocusing and dispersive

The second pulse was directed at the bright spot "b", which was also switched to dark, all other spots remaining unchanged, as seen in Fig. 25c. As the Figs. 25a,c are time-average-pictures that do not directly indicate the switching, Fig. 25d gives the intensity at the center of a switched spot as a function of time. The upper trace corresponds to a pulse energy that is not sufficient for switching, and no permanent switch of the bright spot results. With sufficient energy of the pulse, however, permanent switching occurs, i.e. the intensity remains small throughout the illumination (until the reduction of the background intensity near the end of the illumination returns the resonator to a monostable condition). Thus, in this case, individual bright spots are found which can be switched independently of the rest of the system. This makes the bright spots observed very "soliton-like" [30].

It was not posible to experiment with single bright spots in order to show their soliton nature, under the largely dispersive conditions used, because the collection of bright spots appears largely as a consequence of linear filtering of the high-finesse, high-Fresnel-number resonator. For details, see [30].

In order to suppress this linear ("noise induced") structure, we chose to work at lower resonator finesse, i.e. closer to the band edge or the exciton line, where, moreover, the dissipative solitons predicted in [16] are expected to be more likely. Figure 26 shows observations (the pictures were taken as snapshots of 50 ns duration) under these conditions.

Figure 26a shows a switched area (the resonator field is high in the dark area because observation was performed in reflection) surrounded by a switching front. For small intensities such a switched area collapses into the structure shown in Fig. 26b, which shows all the characteristic features of a bright soliton (dark due to observation in reflection), particularly the spatial oscillations around it [31].

We have recently been able to switch this structure on and off with a narrow pulse, similarly to what was done to observe Fig. 25d.

Interestingly, at higher illumination intensity, "dark" solitons (bright in reflection) appear. Figure 26c shows such a soliton, embedded in the up-

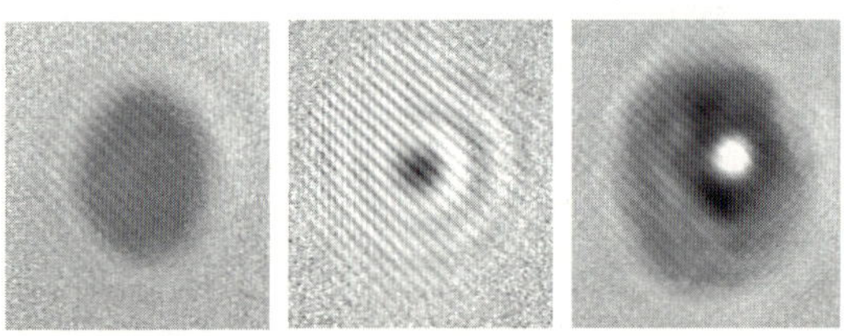

a) b) c)

Fig. 26. Switched structures: reflectivity (reflected light/incident light) of the sample. (**a**) Switched domain (limited by a contour of Maxwellian intensity); (**b**) bright soliton (dark spot in reflection); (**c**) dark soliton (bright spot in reflection) switched area as expected [31]. Such dark solitons were predicted in [32]. As predicted there, we have found that these "dark" solitons are less stable than the bright ones. They appear to move, and we find that for smaller intensities they tend to pulse in a regular fashion, somewhat similarly to what was predicted in [32].

A hint of the nonlinear nature of these structures comes from the brightness of the light reflected from the structure. Quantitative intensity measurement [31] shows that the intensity of the light reflected at the center of the structure is almost twice as high as the illumination intensity. This means a reflectivity higher than one. This has to be interpreted as meaning that the structure collects light from its surroundings and emits it at its center.

Figure 27, finally, shows that more than one soliton can exist, even under our conditions, which are limited by finite laser power and spatial nonuniformity of the illuminating field [31]. With these solitary structures in semiconductor microresonators , information processing and storage should be possible.

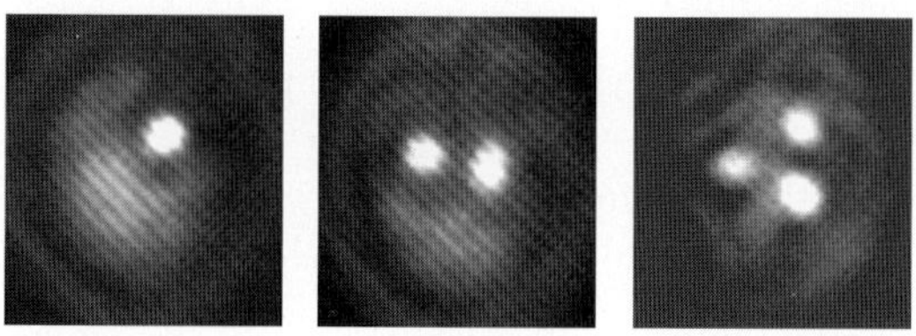

a) b) c)

Fig. 27. Dark solitons observed in the switched area (intensity)

Acknowledgments

This work was supported by ESPRIT projects PASS and PIANOS. The growth of semiconductor Fabry–Perots resonators by I. Sagnes is gratefully acknowledged.

References

1. V.E. Zakharov, A.B. Shabat, Sov. Phys. JETP **34**, 62 (1972)
2. P.V. Manuyshev, A. Villeneuve, G.I. Stegeman, J.S. Aitchison, Electron. Lett. **30**, 726 (1994)
3. P.L. Kelley, Phys. Rev. Lett. **15**, 1005 (1965)
4. J.M. Marburger, Theor. Prog. Quantum Electron. **4**, 35 (1975)
5. M. Shih, P. Leach, M. Segev, M. Garret, G. Salamo, G. Valley, Opt. Lett. **21**, 324 (1996); W.E. Torruellas, B. Lawrence, G.I. Stegeman, Electron Lett. **32**, 2092 (1996)

6. W.J. Firth, A. Lord, A.J. Scroggie, Phys. Scr. **T67**, 12 (1996)
7. M. Tlidi, P. Mandel, R. Lefever, Phys. Rev. Lett. **73**, 64 (1994); W.J. Firth, G.K. Harkness, Asian J. Phys. **7**, 665 (1998)
8. M. Kreuzer, H. Gottschling, T. Tschudi, R. Neubecker, Mol. Cryst. Liq. Cryst. **207**, 219 (1991)
9. B. Fischer, O. Werner, M. Horowitz, Appl. Phys. Lett. **58**, 2729 (1991)
10. N.N. Rozanov, Opt. Spectrosc. **70**, 1342 (1991).
11. E.A. Ostrovskaya, Y. Kivshar, D.E. Skryabin, W.J. Firth, Phys. Rev. Lett. **83**, 296 (1999)
12. N.N. Rosanov, *Progress in Optics*, Vol. 35, ed. by E. Wolf, (Elsevier Science, Amsterdam, 1996)
13. V.Yu. Bazhenov, V.B. Taranenko, V.M. Vasnetsov, Proc. SPIE **1840**, 183 (1992);
14. M. Saffman, D. Montgomery, D.Z. Anderson, Opt. Lett. **19**, 518 (1994)
15. K. Staliunas, Phys. Rev. A **48**, 1573 (1993)
16. W.J. Firth, A. Scroggie, Phys. Rev. Lett. **76**, 1623 (1996)
17. M.P. Petrov, S.I. Stepanov, A.V. Khomenko, *Photorefractive Crystals in Coherent Systems*, (Springer, Berlin, Heidelberg, 1991)
18. K. Staliunas, V.B. Taranenko, G. Sleky, R. Viselga, C.O. Weiss, Phys. Rev. A **57**, 599 (1998)
19. P.V. Jacobsen, J.V. Moloney, A.C. Newell, R. Judik, Phys. Rev. A **45**, 129 (1992)
20. K. Staliunas, G. Slekys, C.O. Weiss, Phys. Rev. Lett. **79**, 2658 (1997)
21. L. Haward, N. Kopell, Stud. Appl. Math. **56**, 95 (1977); Q. Ouyang, H.L. Swinney, Chaos **1**, 4 (1991).
22. S.A. Akmanov, J.E. Dyahov, A.S. Chirkin, *Introduction to Statistical Radio Physics and Optics*, (Nauka, Moscow, 1981)
23. K. Staliunas, V.J. Sanchez-Morcillo, Opt. Commun. **139**, 306 (1997)
24. V.B. Taranenko, K. Staliunas, C.O. Weiss, Phys. Rev. Lett. **81**, 2236 (1998)
25. S. Trillo, M. Haeltermann, A. Sheppard, Opt. Lett. **22**, 970 (1997)
26. M. Vaupel, C.O. Weiss, Phys. Rev. A **51**, 4078 (1995)
27. L. Spinelli, G. Tissoni, M. Brambilla, F. Prati, L.A. Lugiato, Phys. Rev. A **58** 2542 (1998); D. Michaelis, U. Peschel, F. Lederer, Phys. Rev. A **56**, R3366 (1997)
28. B.G. Sfez, J.L. Oudar, J.C. Michel, R. Kuszelewicz, R. Azoulay, Appl. Phys. Lett. **57**, 1849 (1990); I. Abram, S. Iung, R. Kuszelewicz, G. Le Roux, C. Licoppe, J.L. Oudar, Appl. Phys. Lett. **65**, 2516 (1994)
29. W.J. Firth, A.J. Scroggie, Europhys. Lett. **26**, 521 (1994)
30. V.B. Taranenko, R. Kuszelewicz, E. Ganne, C.O. Weiss, Phys. Rev. A **61**, 063818 (2000)
31. V.B. Taranenko, R. Kuszelewicz, E. Ganne, C.O. Weiss, Appl. Phys. B **72**, 377
32. D. Michaelis, U. Peschel, F. Lederer, Opt. Lett. **23**, 1814 (1998)

Nonlinear Magneto-Optic Solitons

Allan D. Boardman and Ming Xie

Summary. Fundamental ideas concerning the use of magneto-optic materials to control spatial solitons are introduced. Following a short review of the availability of ferrite materials and their modifications, within the current technology, some fascinating magneto-optic configurations are introduced. This is followed by a derivation of the envelope equations for the TE and TM modes. It is then shown that an elegant degree of control of spatial solitons can be achieved through the use of current-strip electrodes, which behave like potential functions. This is substantiated with a simple Lagrangian analysis, and several electrode configurations are analyzed to demonstrate an impressive degree of contol over soliton behavior. Since these systems are very easy to create experimentally, it is possible, in principle, to devise any degree of sophistication. The approach therefore contains a lot of promise for future chip-level all-optical processing.

1 Introduction to Magneto-Optical Materials

Magneto-optics was once described as the stepchild of integrated optics [1]. Some of this impression originated from a desire simply to "insert" magneto-optics into known designs rather than to address and control the fascinating complexity of the materials. A major task, for example, is to understand the manner in which an external magnetic field, applied to a waveguide containing a third-order optically nonlinear $\left(\chi^{(3)}\right)$ material and magneto-optical elements, can control bright solitons [2–4]. This typical magneto-optical phenomenon is the motivation for this chapter.

Magneto-optic materials have been brought to a high state of readiness by the magnetism community [5,6], which maintains a very strong interest in magneto-optic recording media and optical nonreciprocal devices. These include periodic structures and ultrathin films, with the general aim being to exploit the controllable magnetic properties now available [7]. For waveguides, low propagation loss must be achieved, however; and some candidates for this property are Ca-doped films and CdIG films. In the 350–850 nm range, Co/Pt–SiO–Al on glass has similar exploitable properties, as do TbFeCo films at 633 nm. The classic materials, on the other hand, are rare-earth iron garnets (YIG). In fact, epitaxially grown YIG films are common, with the added possibility that YIG/semiconductor structures could be used to achieve amplification during propagation. Fe/GaAs film structures can also be used in this way. In any case, the loss associated with propagation in YIG films can be dramatically lowered with bismuth substitution, so loss should not be a problem for spatial-soliton applications which operate on a scale of the order of the diffraction length, typically a few millimeters. The substantial points then are that YIG, expitaxially grown onto single-crystal substrates, has an excellent optical quality, is transparent in the 1.1 μm to 1.6 μm range, and has a magnetization that is saturable by small, easily generated magnetic fields, using current-strip electrodes. Using the latter to control spatial solitons is an attractive way forward and will be the main topic of this chapter.

The orientation of the magnetization M relative to the propagation direction of a wave [7–9] in a slab of magneto-optic material has been widely studied in three traditional forms: polar, longitudinal, and transverse. The dielectric tensors associated with these configurations can be conveniently written as follows [3,4]:

$$\text{polar:} \qquad \varepsilon = \begin{pmatrix} n^2 & 0 & -\mathrm{i}Qn^2 \\ 0 & n^2 & 0 \\ \mathrm{i}Qn^2 & 0 & n^2 \end{pmatrix} , \tag{1}$$

$$\text{longitudinal:} \qquad \varepsilon = \begin{pmatrix} n^2 & -\mathrm{i}Qn^2 & 0 \\ \mathrm{i}Qn^2 & n^2 & 0 \\ 0 & 0 & n^2 \end{pmatrix} , \tag{2}$$

$$\text{transverse:} \qquad \varepsilon = \begin{pmatrix} n^2 & 0 & 0 \\ 0 & n^2 & -\mathrm{i}Qn^2 \\ 0 & \mathrm{i}Qn^2 & n^2 \end{pmatrix} , \tag{3}$$

where n is the refractive index and Q is called the magneto-optical parameter. Typically, for YIG at 1.152 μm, $n^2 = 4.963$, and $Qn^2 = 3.4 \times 10^{-4}$ [7].

2 Envelope Equations for Longitudinal and Transverse Configurations

The coordinates of the propagation and a simple planar structure are defined in Fig. 1; z is the propagation direction, x is a transverse direction lying in the plane of the guiding structure, and y is the normal coordinate direction. All the waves have a fast time variation $\exp(\mathrm{i}\omega t)$, where ω is the usual angular frequency and t is the time. The wave equation is, therefore, ignoring losses,

$$\nabla(\nabla \cdot \boldsymbol{E}) - \nabla^2 \boldsymbol{E} = \frac{\omega^2}{c^2} \left(\varepsilon \boldsymbol{E} + \boldsymbol{P}_{\mathrm{M}} + \boldsymbol{P}_{\mathrm{NL}} \right) , \tag{4}$$

where c is the velocity of light in vacuo, ε is the dielectric-constant distribution for the given structure that exists in the absence of any magnetic effects, and $\boldsymbol{P}_{\mathrm{M}}$ and $\boldsymbol{P}_{\mathrm{NL}}$ represent the polarizations introduced by the magneto-optic and nonlinear effects, respectively. The TE and TM waves are defined

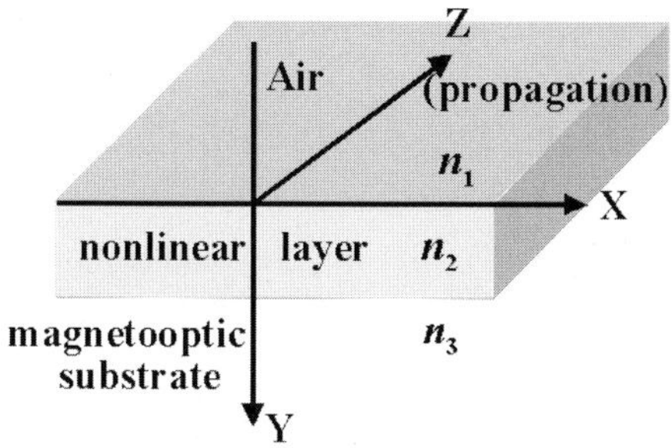

Fig. 1. Sketch of a planar (slab) waveguide structure unbounded in the $\pm x$ directions

here as follows:

$$\text{TE: } \boldsymbol{E}_1 = A_1(x,z)\xi_x(y)\exp\left(-\mathrm{i}\frac{\omega}{c}\beta_1 z\right)\exp(\mathrm{i}\omega t)\hat{\boldsymbol{x}}\ , \tag{5}$$

$$\text{TM: } \boldsymbol{E}_2 = A_2(x,z)\left[\xi_y(y)\hat{\boldsymbol{y}} + \xi_z(y)\hat{\boldsymbol{z}}\right]\exp\left(-\mathrm{i}\frac{\omega}{c}\beta_2 z\right)\exp(\mathrm{i}\omega t)\ . \tag{6}$$

Here $\hat{\boldsymbol{x}}$, $\hat{\boldsymbol{y}}$, $\hat{\boldsymbol{z}}$ are unit vectors, β_1, β_2 are dimensionless wavenumbers (effective refractive indices), and $\xi(y)$ accounts for the modal field shape. $A_{1,2}(x,\ z)$ are amplitudes that slowly vary as the propagation proceeds along z, and contain a transverse x dependence as well.

The part of the material polarization that is due to the magneto-optic properties is

$$\boldsymbol{P}_\mathrm{M} = \begin{pmatrix} 0 & \mathrm{i}\varepsilon_{xy} & \mathrm{i}\varepsilon_{yz} \\ -\mathrm{i}\varepsilon_{xy} & 0 & \mathrm{i}\varepsilon_{yz} \\ -\mathrm{i}\varepsilon_{xz} & -\mathrm{i}\varepsilon_{yz} & 0 \end{pmatrix} \begin{pmatrix} A_1\xi_x\exp\left(-\mathrm{i}\frac{\omega}{c}\beta_1 z\right) \\ A_2\xi_y\exp\left(-\mathrm{i}\frac{\omega}{c}\beta_2 z\right) \\ A_2\xi_z\exp\left(-\mathrm{i}\frac{\omega}{c}\beta_2 z\right) \end{pmatrix} \exp(\mathrm{i}\omega t)\ . \tag{7}$$

Hence, use of the results

$$\begin{aligned} &\int \boldsymbol{E}_1^* \cdot \nabla(\nabla\cdot\boldsymbol{E})\,\mathrm{d}y - \int \boldsymbol{E}_1^*\cdot\nabla^2\boldsymbol{E}\,\mathrm{d}y - \frac{\omega^2}{c^2}\int \varepsilon\boldsymbol{E}_1^*\cdot\boldsymbol{E}\,\mathrm{d}y \\ &= -\frac{\omega^2}{c^2}|A_1|^2\int\varepsilon|\xi_x|^2\,\mathrm{d}y - 2\mathrm{i}\frac{\omega}{c}\beta_1 A_1^*\frac{\partial A_1}{\partial z}\int|\xi_z|^2\,\mathrm{d}y - |A_1|^2\int\varepsilon|\xi_x|^2\,\mathrm{d}y \\ &+ A_1^*\frac{\partial^2 A_1}{\partial x^2}\int|\xi_x|^2\,\mathrm{d}y\ , \end{aligned} \tag{8}$$

$$\int \boldsymbol{E}_1^*\cdot\boldsymbol{P}_\mathrm{M}\,\mathrm{d}y = A_1^*A_2\int\xi_x^*\left(\mathrm{i}\varepsilon_{xy}\xi_y + \mathrm{i}\varepsilon_{xz}\xi_z\right)\exp\left(-\frac{\omega}{c}\gamma z\right)\mathrm{d}y\ , \tag{9}$$

where $\gamma = \beta_2 - \beta_1$, leads to

$$2\mathrm{i}\frac{c}{\omega}\frac{\partial A_1}{\partial z} = \mathrm{i}A_2\left(\overline{\varepsilon}_{xy} + \mathrm{i}\overline{\varepsilon}_{xz}\right)\exp\left(-\mathrm{i}\frac{\omega}{c}\gamma z\right) + \frac{c^2}{\omega^2\beta_1}\frac{\partial^2 A_1}{\partial x^2}\ , \tag{10}$$

and the averages over the waveguide structure are

$$\overline{\varepsilon}_{xy} = \frac{\int \varepsilon_{xy}\xi_y\xi_x^* \, dy}{\beta_1 \int |\xi_x|^2 \, dy}, \qquad \overline{\varepsilon} = -\frac{\int \varepsilon_{xz}(i\xi_z)\xi_x^* \, dy}{\beta_1 \int |\xi_x|^2 \, dy} . \tag{11}$$

Note that ξ_z is imaginary, so that $\mathrm{i}\xi_z$ is real. In a similar fashion, the equation for A_2 emerges as

$$2\mathrm{i}\frac{c}{\omega}\frac{\partial A_2}{\partial z} = \frac{c^2}{\omega^2\beta_2}\frac{\partial^2 A_2}{\partial x^2} - \mathrm{i}A_1\left(\overline{\varepsilon}_{xy} - \mathrm{i}\overline{\varepsilon}_{xz}\right)\exp\left(\mathrm{i}\frac{\omega}{c}\gamma z\right) + 2\frac{\omega^2}{c^2}\overline{\varepsilon}_{yx}A_2 \,, \tag{12}$$

where

$$\overline{\varepsilon}_{yz} = \frac{c}{\omega}\frac{\int \varepsilon_{yz}\xi_y \dfrac{\partial \xi_y}{\partial y} \, dy}{\beta_2^2 \int \left(|\xi_y|^2 + |\xi_z|^2\right) dy} . \tag{13}$$

Both equations take into account diffraction through the $\partial^2/\partial x^2$ terms and account for magneto-optical effects through off-diagonal dielectric-tensor elements, suitably averaged over the waveguide structure. This averaging offers, immediately, a chance to optimize the magneto-optic influence upon spatial-soliton propagation.

It is interesting that even though the nonlinearity is weak in real materials, it has competition from the magneto-optic influence. A complication is that there is considerable opportunity to make the role of nonlinearity a very complex issue. Given that the purpose of this chapter is to highlight the elegant electromagnetic issues arising from the introduction of magneto-optics, however, only the simplest physically acceptable nonlinearity will be introduced here.

This nonlinearity is the Kerr third-order type, and the polarization $\boldsymbol{P}_{\mathrm{NL}}$ can be expressed as $P_x\hat{\boldsymbol{x}} + P_y\hat{\boldsymbol{y}} + P_z\hat{\boldsymbol{z}}$. For example, the x component for an isotropic medium is

$$P_x = \frac{3}{4}\left[2\chi_{xxyy}\left(|E_x|^2 + |E_y|^2 + |E_z|^2\right)E_x + \chi_{xyyx}\left(E_x^2 + E_y^2 + E_z^2\right)E_x^*\right] , \tag{14}$$

in which ε_0, the permittivity of free space, has been absorbed into the susceptibility components χ_{xxyy}, χ_{xyyx}. The definitions of the TE and TM parts of the electric-field vectors requires the evaluation of

$$\boldsymbol{E}_1^* \cdot \boldsymbol{P}_{\mathrm{NL}} = E_x^* P_x \,, \qquad \boldsymbol{E}_2^* \cdot \boldsymbol{P}_{\mathrm{NL}} = E_y^* P_y + E_z^* P_z \,. \tag{15}$$

After these have been integrated over y, we obtain the following definitions of the nonlinear coefficients for self-phase modulation for the TE wave,

$$\overline{\chi}_{\mathrm{e}} = \frac{3}{4}\frac{\int \left(2\chi_{xxyy} + \chi_{xyyx}\right)|\xi_x|^4 \, dy}{\beta_1 \int |\xi_x|^2 \, dy} ; \tag{16}$$

for self-phase modulation for the TM wave,

$$\overline{\chi}_{\mathrm{m}} = \frac{3}{4}\left(\frac{\int 2\chi_{xxyy}\left(|\xi_y|^2+|\xi_z|^2\right)^2 \mathrm{d}y + \int \chi_{xyyx}\left(\xi_y^2+\xi_z^2\right)\left(\xi_y^{*2}+\xi_z^{*2}\right)\mathrm{d}y}{\beta_2\int\left(|\xi_y|^2+|\xi_z|^2\right)\mathrm{d}y}\right);\tag{17}$$

for cross-phase modulation between TE and TM waves,

$$\overline{\chi}_{\mathrm{c}} = \frac{3}{2}\frac{\int \chi_{xxyy}\left(|\xi_y|^2+|\xi_z|^2\right)|\xi_x|^2\,\mathrm{d}y}{\beta_2\int\left(|\xi_y|^2+|\xi_z|^2\right)\mathrm{d}y};\tag{18}$$

and the four-wave mixing coefficient, responsible for energy exchange between the polarizations,

$$\chi_{\mathrm{F}} = \frac{3}{4}\frac{\int \chi_{xyyx}\left(\chi_y^{*2}+\xi_z^{*2}\right)\chi_x^2\,\mathrm{d}y}{\beta_2\int\left(|\xi_y|^2+|\xi_z|^2\right)\mathrm{d}y}.\tag{19}$$

These lead to the coupled equations

$$\begin{aligned}2\mathrm{i}\frac{c}{\omega}\frac{\partial A_1}{\partial z} &= \frac{c^2}{\omega^2\beta_1}\frac{\partial^2 A_1}{\partial x^2} + \mathrm{i}A_2\left(\overline{\varepsilon}_{xy}+\mathrm{i}\overline{\varepsilon}_{xz}\right)\exp\left(-\mathrm{i}\frac{\omega}{c}\gamma z\right)\\ &\quad + \overline{\chi}_{\mathrm{e}}A_1|A_1|^2 + \overline{\chi}_{\mathrm{c}}A_1|A_2|^2 + \overline{\chi}_{\mathrm{F}}A_1^*A_2^2\exp\left(-2\mathrm{i}\frac{\omega}{c}\gamma z\right),\end{aligned}\tag{20}$$

$$\begin{aligned}2\mathrm{i}\frac{c}{\omega}\frac{\partial A_2}{\partial z} &= \frac{c^2}{\omega^2\beta_2}\frac{\partial^2 A_2}{\partial x^2} - \mathrm{i}A_1\left(\overline{\varepsilon}_{xy}-\mathrm{i}\overline{\varepsilon}_{xz}\right)\exp\left(\mathrm{i}\frac{\omega}{c}\gamma z\right) + \frac{2\omega^2}{c^2}\overline{\varepsilon}_{yz}A_2\\ &\quad + \overline{\chi}_{\mathrm{m}}A_2|A_2|^2 + \overline{\chi}_{\mathrm{c}}A_2|A_1|^2 + \overline{\chi}_{\mathrm{F}}A_2^*A_1^2\exp\left(2\mathrm{i}\frac{\omega}{c}\gamma z\right).\end{aligned}\tag{21}$$

These equations have been derived for an isotropic material, which has only two independent third-order susceptibility tensor components χ_{xxyy} and χ_{xyyx}. This means that $\overline{\chi}_{\mathrm{m}}$, $\overline{\chi}_{\mathrm{e}}$, $\overline{\chi}_{\mathrm{c}}$, and $\overline{\chi}_{\mathrm{F}}$ are not independent of each other.

3 Transverse Case: Potential Analysis

For the waveguide shown in Fig. 1, beam propagation occurs along the z axis and is perpendicular to a magnetization created by an applied field pointing along the x axis. The waveguide structure is planar and consists of a nonlinear layer of thickness d. It has an upper, air boundary and is sitting upon a thick (mathematically infinite) magnetized substrate. Magneto-optically, this is the transverse case, as defined in the introduction.

The appropriate magneto-optic material tensor couples the E_z and E_y components of the electric field. Furthermore, since the E_x components travel independently, only the TM wave needs to be considered. Because this is the case, the envelope equation for the transverse case, found by setting $A_1 = 0$, $\bar{\varepsilon}_{xy} = 0$, $\bar{\varepsilon}_{xz} = 0$, is

$$2\mathrm{i}\frac{c}{\omega}\frac{\partial A_2}{\partial z} = \frac{c^2}{\omega^2\beta_2}\frac{\partial^2 A_2}{\partial x^2} + 2\bar{\varepsilon}_{yz}A_2 + \overline{\chi}_m A_2|A_2|^2 \,. \tag{22}$$

This equation has spatial-soliton solutions because of the nonlinear term, but a constant $\bar{\varepsilon}_{yz}$ does nothing except cause the soliton to experience an additional constant phase shift. This is easy to see without any complicated analysis because a simple transformation will permit the $2\bar{\varepsilon}_{yz}A_2$ and the $2\mathrm{i}(c/\omega)(\partial A_2/\partial z)$ terms to be combined. Secondly, $\bar{\varepsilon}_{yz}$ contains the gradient $\partial\xi_y/\partial y$ in its numerator, and $\partial\xi_y/\partial z \propto \xi_z$. Hence, a significant refractive-index step is needed at the boundary to achieve a reasonable value of the longitudinal field component E_z.

Equation (22) can be transformed into a more compact form. First, the amplitude can be redefined in terms of a function ψ, i.e.

$$A_2^* = \frac{c}{\omega D_0}\frac{\sqrt{2}}{\sqrt{\beta_2\overline{\chi}_\mathrm{m}}}\psi \,. \tag{23}$$

Also, the magneto-optic parameter

$$\nu = 2\beta_2\frac{\omega^2}{c^2}D_0^2\bar{\varepsilon}_{yz} \tag{24}$$

can be introduced, where D_0 is the half-maximum beam width. Finally, the transformations

$$x \to D_0 x' \,, \qquad z \to L_\mathrm{D} z' \,, \tag{25}$$

can be effected, where $L_\mathrm{D} = 2\beta_2(\omega/c)D_0^2$ is the diffraction length, i.e. the distance over which the beam will double in size owing to diffraction. After all of this processing, and when the dashes are dropped as being redundant in the final analysis, the basic TM equation becomes an equation of the nonlinear Schrödinger type

$$\mathrm{i}\frac{\partial\psi}{\partial z} + \frac{\partial^2\psi}{\partial x^2} + \nu\psi + 2|\psi|^2\psi = 0 \,. \tag{26}$$

Typically, $\overline{\chi}_\mathrm{m} = 5.5 \times 10^{-11}\ \mathrm{m^2\,V^{-2}}$, $D_0 \sim 10\ \mu\mathrm{m}$, $\beta_2 = 2.67$, $d = 0.3\ \mu\mathrm{m}$, $n_1 = 1$, $n_2 = 3.32$, $n_3 = 2.3$, $L_\mathrm{D} = 2.16$ mm, and $\nu = 1.6$. This data corresponds to an AlGaAs film on a simple magnetic-garnet substrate, or a more complex magnetic garnet such as $Y_{2.5}Ce_{0.5}Fe_5O_{12}$.

The Lagrangian density obtained from (26) is

$$L = \frac{\mathrm{i}}{2}\left(\psi^*\frac{\partial\psi}{\partial z} - \frac{\partial\psi^*}{\partial z}\psi\right) - \left|\frac{\partial\psi}{\partial z}\right|^2 + |\psi|^4 + \nu|\psi|^2 \,, \tag{27}$$

and a suitable trial function is

$$\psi = \eta \operatorname{sech}\left[\eta(x - x_0)\right] \exp\left(\mathrm{i}\frac{\xi}{2}(x - x_0) + i\phi\right) \tag{28}$$

where x_0 is the location of the center of the localized state (a soliton), η is the amplitude, ξ is the inclination of the spatial-soliton beam to the propagation direction, and ϕ is a phase factor. Following Whitham [10] the most useful quantity is the average Lagrangian

$$\begin{aligned} \boldsymbol{L} &= \int_{-\infty}^{\infty} L\, dx \\ &= \eta\left(\xi\frac{\partial x_0}{\partial z} - 2\frac{\partial \phi}{\partial z}\right) + \frac{2}{3}\eta^3 - \frac{\xi^2}{2}\eta + \eta^2 \int \nu(x) \operatorname{sech}^2\left[\eta(x - x_0)\right]\, dx\ , \end{aligned} \tag{29}$$

where, for the first time, the possibility that the magneto-optic parameter ν must be a function of x, the transverse coordinate, if it is to have any control properties, is admitted.

The application of the Euler–Lagrange equations

$$\frac{\partial L}{\partial f} = \frac{\mathrm{d}}{\mathrm{d}z}\frac{\partial L}{\partial(\partial f/\partial z)} - \frac{\partial L}{\partial f} = 0\ , \tag{30}$$

where $f = \eta$, x_0, ξ, ϕ, then leads to the equations

$$\frac{\partial x_0}{\partial z} = \xi, \quad \frac{\partial \eta}{\partial z} = 0, \quad \frac{\partial \xi}{\partial z} = \eta\frac{2}{\partial x_0} \int \nu(x) \operatorname{sech}^2\left[\eta(x - x_0)\right]\, dx\ , \tag{31}$$

$$\begin{aligned} \frac{\partial \phi}{\partial z} &= \eta^2 + \frac{\xi^2}{4} + \eta \int \nu(x) \operatorname{sech}^2\left[\eta(x - x_0)\right]\, dx\ , \\ &\quad + \eta^2 \int (x - x_0)\nu(x) \operatorname{sech}^2\left[\eta(x - x_0)\right] \tanh\left[\eta(x - x_0)\right]\, dx\ . \end{aligned} \tag{32}$$

Equation (30) shows that the soliton amplitude is constant and that (32) can be used to calculate the phase change, once η, ξ, x_0 are known. If $\nu =$ constant, for example, then the phase equation is simply

$$\frac{\partial \phi}{\partial z} = \eta^2 + \frac{\xi^2}{4} + \nu\ . \tag{33}$$

Equation (33) demonstrates that, as expected, $\nu =$ constant contributes only an additional phase shift.

If in (32) ν can be kept under the integral sign by making ν some function $\nu(x)$, then (30) yields

$$\frac{\mathrm{d}^2 x_0}{\mathrm{d}z^2} = \frac{\partial}{\partial x_0} \int \nu(x) \operatorname{sech}^2(x - x_0)\, x\, \mathrm{d}x\ , \tag{34}$$

where, because η = constant, we can arbitrarily set $\eta = 1$. Multiplying both sides of (34) by $\mathrm{d}x_0/\mathrm{d}z$ results in the first integral of (34), which is

$$\frac{1}{2}\left(\frac{\mathrm{d}x_0}{\mathrm{d}z}\right)^2 - \int \nu(x)\,\mathrm{sech}^2(x-x_0)\,\mathrm{d}x = \text{constant}\,. \tag{35}$$

It is very clear from this equation that an effective potential energy controls the soliton behavior. This potential function is

$$U(x_0) = -\int \nu(x)\,\mathrm{sech}^2(x-x_0)\,\mathrm{d}x\,. \tag{36}$$

All that is necessary now is to devise some physical situations to create the function $\nu(x)$, and hence the potential barrier/well that is represented by $U(x_0)$.

4 Transverse Case: Control of Solitons by an Electrode Structure

A planar guiding structure, unmodified by electrode control, is shown in Fig. 1. The magneto-optic material happens to be the substrate, but this is not a vital point, because what is going to be investigated is the placement of an electrode structure on the upper surface of the guide (the interface between the nonlinear material and air) to create a magnetic field that drops rapidly to zero in the ($\pm x$) directions. In this way, a magnetic field exists that creates a magnetization that is either saturated or unsaturated over a limited region of the magnetic substrate.

A magneto-optical material experiences *saturation* of its magnetization at a particular value of the magnetic field. This means that increasing the magnetic field beyond a value of about 300 Oe, in many cases, will not increase the magnetization ($\boldsymbol{M}$) any further. Since $\nu(x)$ is proportional to the magnetization, this means that $\nu(x)$ has a clear maximum value associated with the saturation phenomenon. The magnetic-field ($\boldsymbol{H}$) distribution created by an electrode structure can be readily calculated from Maxwell's equations, but its relationship to $\nu(x)$ is not so straightforward. This relationship is complicated by the absence of an analytical formula connecting $\boldsymbol{M}$ to $\boldsymbol{H}$. Nevertheless, a simple hyperbolic-tangent function provides a model that is very close to the observed magnetization versus magnetic-field behavior, for many magnetic materials. Accordingly, the model $\nu = A\tanh(BH/H_\mathrm{s})$ is assumed here, where H is the magnetic field and A and B are just empirical constants, selected so that tanh $\to 1$ in the region of high magnetic field. H_s is simply the saturation magnetic field, thus making $\nu(x)$ reach a saturation value $\nu(x) = A$. For smaller H, the tanh function falls away rapidly, thus making $\nu(x) \to 0$ as $x \to \pm\infty$.

Electrode structures have been used routinely in the past to create linear couplers [11], which consist of silver electrodes 20 μm wide, carrying as little as

1 mA of electric current. For simplicity, it is assumed here that the magnetic field that they produce can be approximated by that produced by an infinitely long thin wire, or, indeed, a set of such wires. Several electrode structures are shown in Fig. 2, and the coordinate systems for two cases are shown in Fig. 3. For a single wire, the modulus of the magnetic field at the boundary is

$$H = \frac{I}{2\pi\sqrt{x^2 + d^2}} \,. \tag{37}$$

Assuming that $I = 60$ mA and $d = 0.3$ μm, the maximum field directly under the wire is $H(0) = I/2\pi d = 400$ Oe. Suppose that the magnetic substrate becomes saturated at 300 Oe $= 23.87 \times 10^3$ A m^{-1}; then the x position at which the magnetic field is enough to saturate the magnetic material is $x = 0.27$ μm. For $\nu = A\,\tanh(BH/H_s)$, taking $A = 1.6$ as an example, Fig. 4 shows plots of H, ν, and $\nu\,\sin\gamma$, which are relevant to this single-wire case; $\nu\,\sin\gamma$ is used because it is the field parallel to the x axis which participates in saturating the magnetization of the magnetic substrate. In the calculations reported here the current (wire) strip is assumed to be infinitely thin but, of course, it has width in practice. It should be expected, therefore, that $\nu(x)$

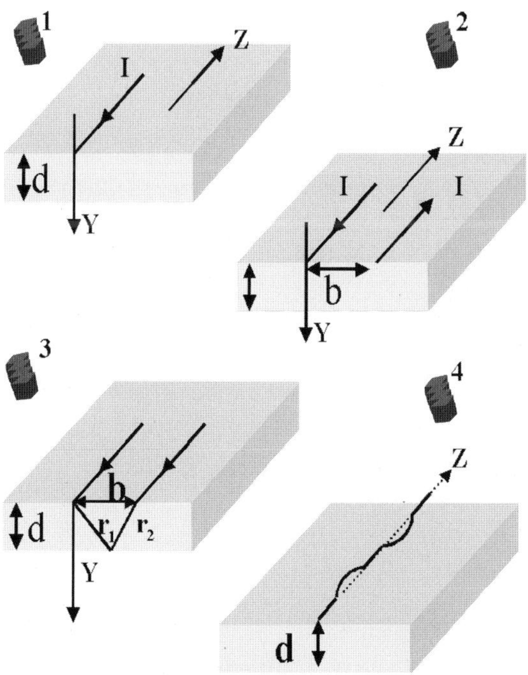

Fig. 2. Some current-strip-line configurations. In practice the current would flow along the top surface in thin metallic strips. In sketches 1–4 the current flows in wires in the directions shown

is somewhat wider than the theoretical result in Fig. 4. This is not actually a very significant effect, so it will be ignored from now on. Note that the potential U is opposite in sign to ν. This was established by a variational analysis. Note also that as the beam width D_0 changes, so do ν and the power P. In fact, as D_0 increases, P decreases, and vice versa. In practice, what will happen is that high-power solitons will cross the finite-$\nu(x)$ region but that smaller-power beams will breakup.

If two wires are used, the possibility of antiparallel currents is introduced. In the geometry shown in Fig. 3,

$$|\boldsymbol{H}_1| = \frac{I}{2\pi r_1}\ , \ |\boldsymbol{H}_2| = \frac{I}{2\pi r_2}\ , \ r_1 = \sqrt{d^2 + x^2}\ , \ r^2 = \sqrt{d^2 + (x-b)^2}\ ,$$

$$\cos\gamma_1 = x/r_1\ , \ \sin\gamma_1 = d/r_1\ , \ \cos\gamma_2 = (x-b)/r_2\ , \ \sin\gamma_2 = d/r_2\ ,$$

$$\boldsymbol{H} = \boldsymbol{H}_1 + \boldsymbol{H}_2\ , \quad H = \frac{I}{2\pi r_1 r_2}\sqrt{r_1^2 + r_2^2 + 2[x(x-b) + d^2]}\ ,$$

$$\sin\theta_1 = \frac{bd}{r_1\sqrt{r_1^2 + r_2^2 + 2[x(x-b) + d^2]}}\ ,$$

$$\cos\theta_1 = \frac{r_1^2 + x(x-b) + d^2}{\sqrt{r_1 r_1^2 + r_2^2 + 2[x(x-b) + d^2]}}\ ,$$

$$\cos(\theta_1 + t) = \frac{d}{r_1 r_2\sqrt{r_1^2 + r_2^2 + 2[x(x-b) + d^2]}}\left[r_1^2 + (x-b)^2 + d^2\right].$$

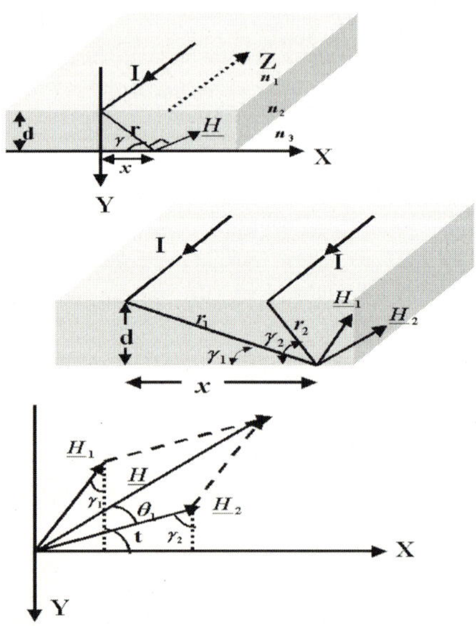

Fig. 3. Coordinate systems and vector magnetic fields due to a single wire and a pair of wires

It is the component of the field $H\cos(\theta_1 + t)$ that is responsible for saturating the magnetic medium. Figure 4 shows the magneto-optical parameters for a single current strip. A positive value is shown (negative potential) but either $\nu > 0$ or $\nu < 0$ can be achieved since $\nu \propto I$. The current can easily be switched between the $\pm z$ directions. This is a *nonreciprocal* effect; it could be produced photonically and is therefore very fast. The convention adopted here is that solitons propagate along the positive z axis and that a current flowing along the negative z direction gives a magnetic field pointing along the positive x axis. This in turn gives a positive ν and hence a negative potential.

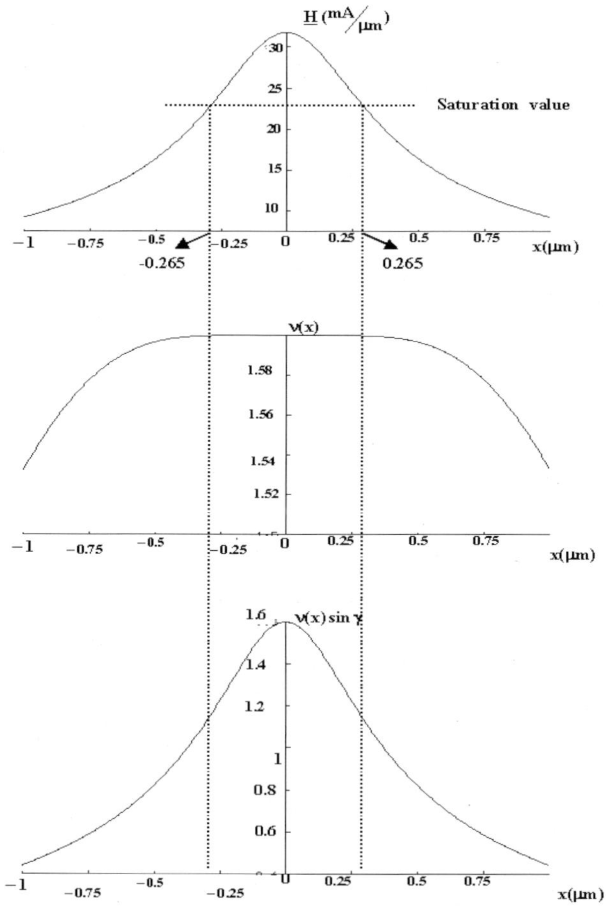

Fig. 4. Distribution of magnetic field and magneto-optic parameters for a single current strip (wire)

The results show that use of a beam width of 6 μm and $\nu = \pm 0.6$ permits transmission for $\nu = 0.6$ but creates a reflection for $\nu = -0.6$. However, if the beam diameter is increased to $D_0 = 10.0$ μm for $\nu = 1.6$, this leads to a breakup of the beam, thus indicating that the power loss creates an instability.

Figure 5 shows some results for two current strips, with currents in the same direction and $D_0 = 10$ μm. A variety of soliton behaviors is obtained. For example, a positive potential results in reflection occurring when $I =$ 100 mA, but transmission occurs at $I = 50$ mA. Also, if the soliton is located initially between the wires, then trapping occurs for a positive potential but escape occurs for a negative potential.

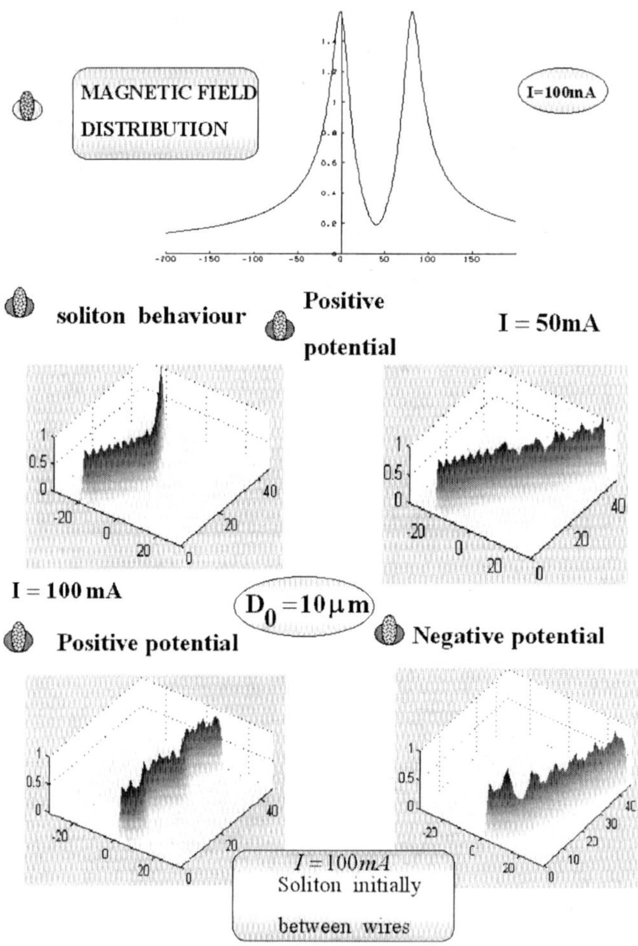

Fig. 5. Soliton dynamics controlled by two current strips supporting currents flowing in the same direction

For reversed current strips, in which the currents are flowing in opposite directions, the soliton dynamics are also interesting. For $I_1 = 100\,\text{mA}$ and $I_2 = 200\,\text{mA}$, $\nu(x)$, shown in Fig. 6, has a complicated shape. Note that the (x, z) scales of the system are also shown and that the transverse extent is measured on a micron scale while the propagation length is on a millimeter scale. The dynamics for $I_1 = 100$ mA, $I_2 = 200$ mA show that a soliton encountering this structure bounces off it, whereas for $I_1 = 100$ mA, $I_2 = 140$ mA there is some energy splitting. If one current is switched off, $\nu(x)$ becomes a simple, regular function. In principle, there is no limit to the

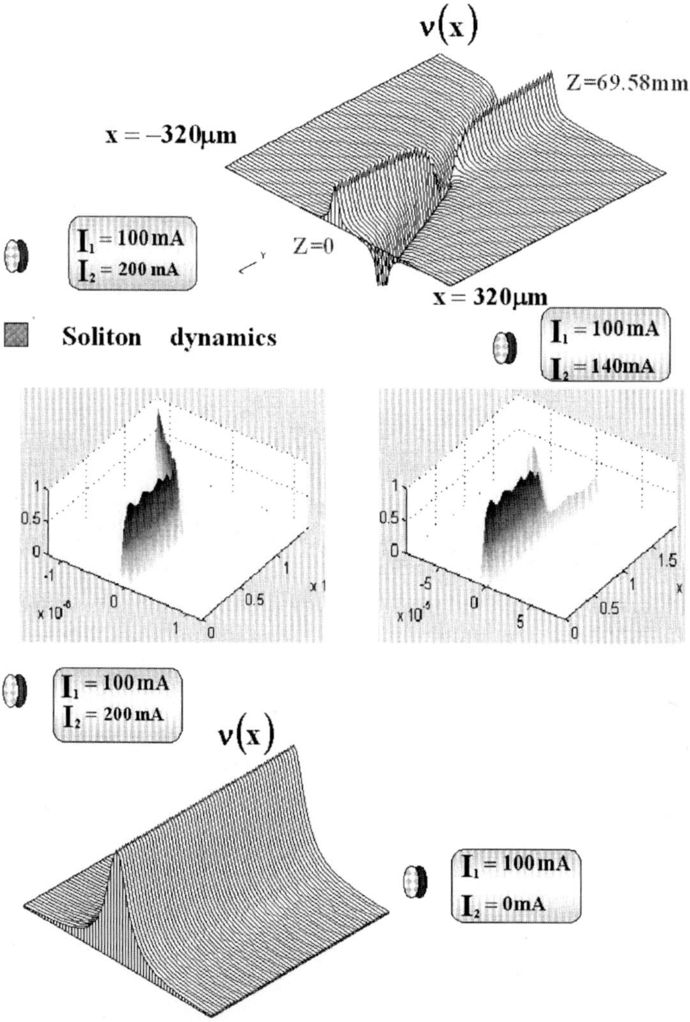

Fig. 6. Results obtained with reversed current strips

well/barrier structures that can be created, and the fabrication processes are well within the capabilities of current technology. Even curved strips can be used, as Fig. 2 shows rather well. If b is the distance of a point on the current path from the z axis then b can, for example, be chosen arbitrarily as

$$b = \begin{cases} 0\,, & z < NT/8 \\ 40\cos\left(2\pi\dfrac{(z-NT/8)}{NT/4}\right) - 1\,, & \dfrac{NT}{8} \le z \le \dfrac{3}{8}NT \\ 0\,, & \dfrac{3}{8}NT \le z \le \dfrac{5}{8}NT \\ -40\cos\left(2\pi\dfrac{(z-NT/8)}{NT/4}\right) + 1\,, & \dfrac{5}{8}NT \le z \le \dfrac{7}{8}NT \\ 0\,, & z \ge \dfrac{7}{8}NT \end{cases}\,, \tag{38}$$

where NT is arbitrary. Some results obtained with a current strip following this path are shown in Fig. 7, which, once again, illustrate a strong degree of spatial-soliton control.

5 Conclusions

This chapter was designed to introduce the basic idea of including magneto-optic layers, claddings, or substrates into the standard, planar waveguide format. The latter is attractive for integrated optics because there the aim is to produce a "chip-level" format that will participate in, and control, all-optical processing operations in the future. A very attractive way forward is to use spatial-soliton beams in which the diffraction length is the operative length scale. The reason is that diffraction operates over only the order of millimeters and so fits the "chip" format very well. Once solitons are created, then controlling their dynamics becomes an important issue, and magneto-optics is put forward here as a very attractive option. It has been shown here that building a planar structure upon a magneto-optic substrate leads to a number of possibilities, ranging from using pure TM waves to using coupled TE–TM waves. To illustrate the control aspects in an elegant way, TM waves were selected, within what is called the transverse configuration. This configuration uses a transversely varying magneto-optic parameter, created by deploying a number of electrode structures. These electrode structures will, in practice, be narrow strips of metal, which can be created in any desired format. Even the simplest of them gives an impressive degree of control over the soliton dynamics. A number of interesting cases were presented to illustrate the capability of this type of structure; the effect can be expressed in terms of potential barriers, or wells. Added, or buried, electrode structures may well become a feature of the all-optical chip technology of the future. This will give a real possibility of manipulating the solitons in any way that is desired. The availability of magneto-optic materials is assured by the global technology that is driving linear magneto-optics. Experimental work in this area cannot be very far away.

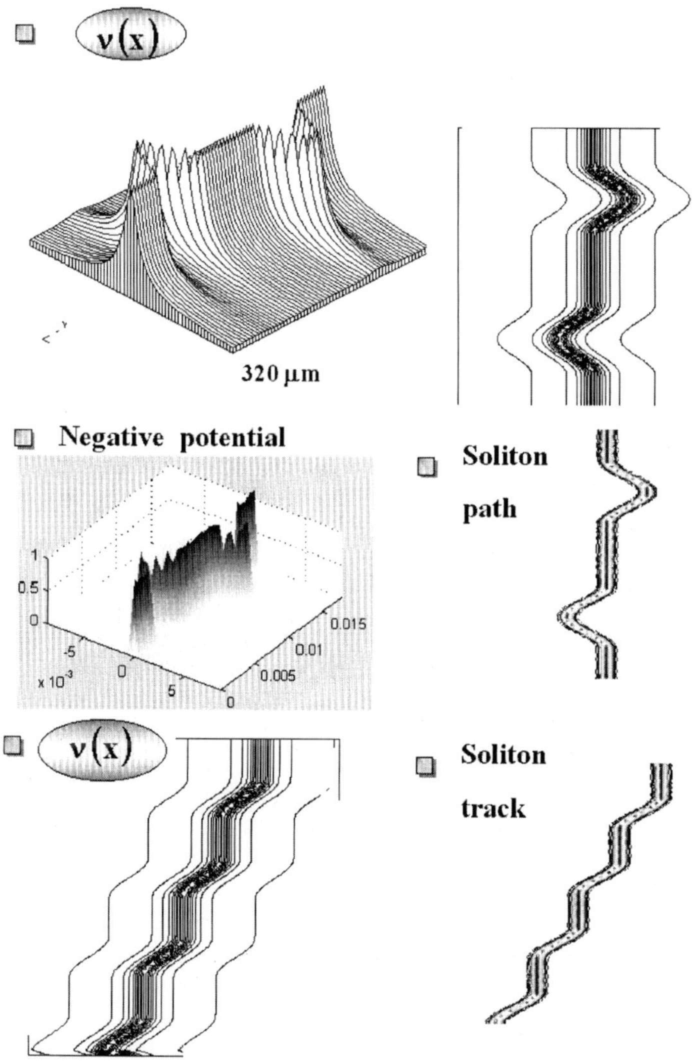

Fig. 7. Results obtained with curved single current-strip

References

1. D.S. Stancill, IEEE J. Quantum Electron. **27**, 61 (1991).
2. A.D. Boardman, K. Xie, Phys. Rev. Lett., **76**, 4591 (1995).
3. A.D. Boardman, K. Xie, Phys. Rev. E **55**, 1899 (1997).
4. A.D. Boardman, K. Xie, J. Opt. Soc. Am. B **15**, 3102 (1997).
5. U. Pustogowa, W. Hubner, K.H. Benneman, Phys. Rev. B **49**, 10031 (1994).
6. M. Wallenhorst, M. Niemoller, H. Dotsch, P. Hertel, R. Gerhardt B. Gather, J. Appl. Phys. **77**, 2902 (1995).
7. A.K. Zvezdin, V.A. Kotov, *Modern Magnetooptics and Magnetooptical Materials* (Institute of Physics, Bristol, 1997).
8. A.M. Prokhorov, G.A. Smolenskii, A.N. Ageev, Sov. Phys. Usp. **27**, 339 (1984).
9. A.V. Sokolov, *Optical Properties of Metals* (Blackie, London, 1967).
10. G.B. Whitham, *Linear and Nonlinear Waves* (Wiley, New York, 1974).
11. T. Mizumoto, S. Mashimo, T. Ida, Y. Naito, IEEE Trans. Magnetooptics **29**, 3417 (1993).

Coupled Propagation of Light and Matter Waves: Solitons and Transverse Instabilities

Mark Saffman and Dmitry V. Skryabin

Summary. We study the coupled propagation dynamics of ultracold atoms and quasiresonant coherent light fields in the framework of a mean field description. For large atom-field detuning we obtain two coupled equations for the evolution of the atomic ground state and the radiation field. Modulational instabilities and solitary solutions of these equations are found.

1 Introduction

In 1962 Askar'yan introduced the notion of self-focusing of radiation in a nonlinear medium [1]. The physical situation he considered was that of the gradient force due to spatial variations in the average value of the radiation intensity, pushing electrons or atoms into or out of the center of the beam. As he noted in the final paragraph of his paper "... actions of intense radiation in the medium can be so strong that a gradient is produced in the properties of the medium in the beam and outside of it, which results in the waveguide propagation of the beam...". The waveguiding envisioned in [1] was due to local changes in the density of the medium. Although spatial solitons have been studied in the intervening years in a great variety of media, it is uncertain whether the mechanism of Askar'yan has in fact been observed. Most experimental studies of self-focusing have exploited electronic or orientational nonlinearities that do not rely upon a density change. In media with thermal nonlinearity the density does change owing to the presence of the beam, although it is the intensity, not the gradient of the intensity, that is responsible for the nonlinearity. As thermal media are generally self-defocusing , only dark optical have been observed in this case [2]. Photorefractive media which have been the subject of much work in recent years, perhaps come closest to supporting the type of waveguiding envisaged by Askar'yan. There the induced nonlinearity is caused by a modification of the trapped charge density that is driven by the gradient of the optical intensity [3].

In the years since 1962, tremendous advances have been made in techniques for the optical manipulation of neutral atoms [4,5]. The compression and rarefaction described in [1] have been observed, using a thermal beam of neutral atoms copropagating with a focused laser beam [6]. Furthermore, it is now a routine matter to obtain optical cooling of atoms to subrecoil temperatures, where the atomic momentum is even less than the momentum due to a single photon. The wave and diffractive aspects of atomic propagation have been observed unambiguously [7], and sources of spatially coherent atomic waves, or atom lasers, are being actively developed [8–10].

These developments in atomic and optical physics make it possible today to study the nonlinear propagation of radiation and matter in a unified fashion. Although experimental observations remain a topic for the (not too distant) future, several authors have already studied the nonlinear propagation of light and matter theoretically. Most of these recent studies of the interaction of cold atoms with light have been done in the approximation in which the spatial dependence of the optical field is assumed to be unaffected by the atomic dynamics. This approximation has been used successfully to describe trapping and cooling of atoms by optical fields. The basic model describing this process in the case of a near resonant optical field is a set of coupled Schrödinger equations for the atomic wavefunctions of the ground and excited states, where the interaction with the optical field is described in the dipole approximation [11,12]. If the atomic medium is sufficiently dense, then multiple scattering of photons by atoms becomes an important process and dipole–dipole corrections have to be taken into account [13–15]. The latter corrections result in nonlinear potential terms in the Schrödinger equation, which have been shown to lead to the formation of atomic solitons in a traveling-wave optical field [14] or to the formation of gap atomic solitons in a standing-wave field [13]. Another source of atomic nonlinearity is collisions, which in the zero-temperature approximation also lead to the appearance of a cubic nonlinearity in the Schrödinger equation, transforming it into the so-called Gross–Pitaevskii (GP) equation . The GP equation has been widely used to describe the recent wave of experimental results on Bose–Einstein condensation (BEC) in alkali vapors [16].

Although a rigorous theoretical approach to describing the joint dynamics of optical and atomic fields in the quantum regime has been developed by several groups of authors [12,13,15,17–20], there have been only a few investigations of the coupled dynamics. In [21] both the longitudinal and the transverse dynamics of the atomic beam were accounted for, but transverse variations of the optical field were neglected. In this approximation, the effects of mutual trapping of the atomic and optical fields do not appear. A first calculation of mutual trapping of an optical beam and a hot atomic gas was presented in [22]. In another work [23], diffraction of both optical and atomic fields was taken into account, and filamentation of the coupled fields was studied. However, the model used in [23] was not derived from first principles, and therefore a more rigorous treatment is desirable. Below, we use a quantum model that describes the mutual spaciotemporal dynamics of an optical field and an ultracold Bose gas of two-level atoms [20]. To study solutions of these equations, we use the mean-field approximation, i.e. we replace all operators by their expectation values. We develop a self-consistent description of the optical and atomic nonlinearities that goes beyond the adiabatic procedure used previously to eliminate the wavefunction of the excited state. As a result of this procedure, we arrive at coupled mean-field equations of the nonlinear Schrödinger (NLS) type describing interaction of paraxial optical and atomic beams. We study the stability of the continuous-wave solutions and coupled solitonic solutions of the resulting system.

The chapter is organized as follows. In Sect. 2 we derive the coupled propagation equations using a two-level atomic model. We point out the analogy with parametric three-wave mixing in the low-density limit. Adiabatic elimination of the excited state then leads to coupled equations for the optical field and the atomic ground state, which we use as a basis for studies of modulational instability and solitary solutions. In Sect. 3 we study the linear instability of plane wave solutions, which leads to convective instabilities of the coupled fields. In Sect. 4 we study nonlinear solitonic solutions of the coupled equations. In Sect. 5 we consider the observability of modulational instability and opto-atomic solitons in relation to the present experimental limits on atom laser sources.

2 Governing Equations

The mean values of the components of the matter-field wavefunction $\hat{\Phi} = \hat{\Phi}_{\mathrm{g}} \mathrm{e}^{-\mathrm{i}E_{\mathrm{g}}t/\hbar} + \hat{\Phi}_{\mathrm{e}} \mathrm{e}^{-\mathrm{i}E_{\mathrm{e}}t/\hbar}$ of a Bose gas of two-level atoms, evolving in the classical optical field $(A\mathrm{e}^{-\mathrm{i}\omega_{\mathrm{L}}t} + A^{*}\mathrm{e}^{\imath\omega_{\mathrm{L}}t})/2$, obey the following system of coupled Schrödinger equations [12,13,20]:

$$\mathrm{i}\hbar\partial_t \Phi_{\mathrm{g}} = -\frac{\hbar^2}{2m}\nabla^2\Phi_{\mathrm{g}} - \frac{1}{2}\mu A^{*}\Phi_{\mathrm{e}}\mathrm{e}^{\mathrm{i}\Delta t} + \frac{4\pi\hbar^2 a_{\mathrm{gg}}}{m}|\Phi_{\mathrm{g}}|^2\Phi_{\mathrm{g}} \,, \tag{1a}$$

$$\mathrm{i}\hbar\partial_t \Phi_{\mathrm{e}} = -\frac{\hbar^2}{2m}\nabla^2\Phi_{\mathrm{e}} - \frac{1}{2}\mu A\Phi_{\mathrm{g}}\mathrm{e}^{-\mathrm{i}\Delta t} - \mathrm{i}\hbar\frac{\gamma}{2}\Phi_{\mathrm{e}} \,, \tag{1b}$$

where the indices g and e denote the ground and excited states, μ is the matrix element of the atomic dipole moment, m is the mass of an atom, $\Delta = \omega_{\mathrm{L}} - \omega_{\mathrm{a}}$, $\hbar\omega_{\mathrm{a}} = E_{\mathrm{e}} - E_{\mathrm{g}}$, γ is the spontaneous decay rate of the excited atoms, a_{gg} is the scattering length corresponding to interactions of the ground-state atoms, and $\nabla^2 = \partial_x^2 + \partial_y^2 + \partial_z^2$. The wavefunctions are normalized so that $\int \mathrm{d}^3x\,(|\Phi_{\mathrm{g}}|^2 + |\Phi_{\mathrm{e}}|^2) = N$, where N is the number of atoms. A term corresponding to population of the ground state due to spontaneous decay of the excited state has been neglected in (1a), as it will be negligible in the limit of large detuning considered below. The latter limit also allows us to eliminate Φ_{e} adiabatically. As we shall show below, our adiabatic method allows comparison of the atom–optical nonlinearities due to dipole, dipole–dipole and collisional interactions and due to the Kerr effect. By taking the collisional term in the form of a cubic nonlinearity of the atomic wavefunction, we have made an implicit assumption that the atoms in the ground state are at zero temperature. The equation for the slowly varying amplitude of the optical field is

$$\frac{2\mathrm{i}\omega_{\mathrm{L}} n^2}{c^2}\partial_t A = -\boldsymbol{\nabla}^2 A - \frac{\omega_{\mathrm{L}}^2 n^2}{c^2} A - \frac{\omega_{\mathrm{L}}^2}{c^2\varepsilon_0} P\,, \tag{2}$$

where c is the velocity of light in vacuum and n is the index of refraction of the medium. The polarization due to dipole interaction with the atoms is

$$P = \mu \Phi_{\mathrm{g}}^* \Phi_{\mathrm{e}} \mathrm{e}^{\mathrm{i}\Delta t}. \tag{3}$$

Equations (1–3) can also be used to describe optical manipulation of cold atoms above the critical temperature for condensation [11], but the effect of atomic collisions in this regime can be disregarded: $a_{\mathrm{gg}} = 0$. In this regime the wavefunction Φ is that of a single atom, and the atomic polarization should be multiplied by the density of atoms.

In dense media the optical field A should include corrections due to the induced polarization. A self-consistent quantum treatment of these local-field corrections can be found in [19,20]. The results of those treatments can be recovered by formal substitution of the classical expression [24]

$$A \to A + \frac{P}{3\varepsilon_0}$$

into the equations for the atomic wavefunctions.

In the following, we shall assume monochromatic, and paraxial optical and atomic fields, and therefore make the substitutions

$$\begin{aligned} &A(x,y,z,t) \to A(x,y,z)\mathrm{e}^{\mathrm{i}k_{\mathrm{L}}z}\,, \\ &\Phi_{\mathrm{g}}(x,y,z,t) \to \Phi_{\mathrm{g}}(x,y,z)\mathrm{e}^{\mathrm{i}(k_{\mathrm{a}}z-\omega t)}\,, \\ &\Phi_{\mathrm{e}}(x,y,z,t) \to \Phi_{\mathrm{e}}(x,y,z)\mathrm{e}^{\mathrm{i}((k_{\mathrm{L}}+k_{\mathrm{a}})z-(\omega+\Delta)t)}\,, \end{aligned} \tag{4}$$

where $k_{\mathrm{L}} = \omega_{\mathrm{L}} n/c$ is the optical wavenumber, $k_{\mathrm{a}} = m v_{\mathrm{a}}/\hbar$ is the atomic wavenumber, m is the atomic mass, v_{a} is the velocity of the atomic beam, and $\hbar\omega = m v_{\mathrm{a}}^2/2$. The detuning is now given by $\Delta = \omega_{\mathrm{L}} - \omega_{\mathrm{a}} - k_{\mathrm{L}} v_{\mathrm{a}}$, which accounts for the Doppler shift due to the mean atomic motion. With these substitutions, (1–3) take the form

$$\begin{aligned} &(\mathrm{i}\sigma\partial_\zeta + \partial_\xi^2 + \partial_\eta^2)\Phi_{\mathrm{g}} \\ &\quad = -\sqrt{|\delta|}\Omega^*\Phi_{\mathrm{e}} + \frac{m w_{\mathrm{L}}^2}{\hbar^2}\left(\frac{4\pi\hbar^2 a_{\mathrm{gg}}}{m}|\Phi_{\mathrm{g}}|^2 - \frac{\mu^2}{6\varepsilon_0}|\Phi_{\mathrm{e}}|^2\right)\Phi_{\mathrm{g}}\,, \end{aligned} \tag{5a}$$

$$\begin{aligned} &(\mathrm{i}\sigma\partial_\zeta + \partial_\xi^2 + \partial_\eta^2)\Phi_{\mathrm{e}} \\ &\quad = -\sqrt{|\delta|}\Omega\Phi_{\mathrm{g}} - \frac{m w_{\mathrm{L}}^2}{\hbar^2}\frac{\mu^2}{6\varepsilon_0}|\Phi_{\mathrm{g}}|^2\Phi_{\mathrm{e}} - \delta\left(1 + \mathrm{i}\frac{m w_{\mathrm{L}}^2}{\hbar}\frac{\gamma}{2\delta}\right)\Phi_{\mathrm{e}}\,, \end{aligned} \tag{5b}$$

$$(\mathrm{i}\partial_\zeta + \partial_\xi^2 + \partial_\eta^2)\Omega = -\frac{k_{\mathrm{L}}^2 w_{\mathrm{L}}^4 m \mu^2}{4\varepsilon_0\hbar^2\sqrt{|\delta|}}\Phi_{\mathrm{g}}^*\Phi_{\mathrm{e}}\,, \tag{5c}$$

Table 1. Parameters of several atomic species for which Bose–Einstein condensation has been observed; $a_0 = 5.3 \times 10^{-11}$ m is the Bohr radius

Parameter	Units	^{7}Li	^{23}Na	^{87}Rb
$m \times 10^{26}$	kg	1.2	3.9	15
λ_{cooling}	μm	0.671	0.589	0.780
$\gamma/2\pi$	MHz	5.83	9.71	6.0
$\omega_{\text{R}}/2\pi$	kHz	62.8	24.8	3.74
I_{sat}	W/m^2	25.3	61.9	16.5
$\mu \times 10^{29}$	C m	2.0	2.1	2.5
a_{gg}	a_0	−27.6	260	110
$4\pi\hbar^2 a_{\text{gg}}/m \times 10^{49}$	J m^3	−0.17	0.50	0.056
$\mu^2/6\varepsilon_0 \times 10^{49}$	J m^3	74	83	119

where

$$\delta = k_{\text{L}}^2 w_{\text{L}}^2 \left(\frac{\Delta}{2\omega_{\text{R}}} - \sigma \right) , \tag{6}$$

$\omega_{\text{R}} = \hbar k_{\text{L}}^2/(2m)$ is the single-photon recoil frequency, $\sigma = k_{\text{a}}/k_{\text{L}}$, $\zeta = z/(k_{\text{L}} w_{\text{L}}^2)$, $(\xi, \eta) = \sqrt{2}(x, y)/w_{\text{L}}$, and w_{L} is a characteristic transverse size of the beams.

$$\Omega = \frac{\mu A}{2\hbar} \frac{k_{\text{L}} w_{\text{L}}^2}{v_{\text{a}}} \frac{\sigma}{\sqrt{|\delta|}}$$

is the Rabi frequency multiplied by the time required for an atom to travel one diffraction length of the optical beam and by the dimensionless constant $\sigma/\sqrt{|\delta|}$.

Equation (5) can be simplified slightly by considering typical parameter values for the atomic species used most widely in Bose–Einstein condensation experiments. Table 1 lists some intrinsic and derived values for Li, Na , and Rb atoms. For sufficiently large detunings, we can neglect the dissipative correction to δ on the right-hand side of (5b), which is due to spontaneous emission. To see this, we note that for $\sigma \sim 1$, $\delta \simeq w_{\text{L}}^2 m \Delta/\hbar$, and the magnitude of the imaginary correction in the last term of (5b) becomes $|\gamma/2\Delta|$. We can therefore neglect spontaneous decay for reasonably large detunings, say $\Delta \geq 10\gamma$.

With this simplification, and assuming a low-density limit where the cubic terms in (5a) and (5b) are small, we see that (5) is equivalent to the equations of parametric three-wave mixing, where Φ_{g} and Ω are the signal and idler waves, and Φ_{e} the sum frequency. It should be noted, however, that the detuning parameter δ for optical–atomic parametric mixing will typically be exceptionally large. For a fixed value of σ, we can choose ω_{L} to make the term

Table 2. Nonlinear coefficients in (8a) and (8b) for an average atomic density $\bar{N} = 10^{20}$ m^{-3}, intensity $I = 0.1 I_{\rm sat}$, $\Delta/\gamma = 100$, $\sigma = 1$, and $w_{\rm L} = 2 \times 10^{-6}$ m.

Atom	$\beta_{\rm dd}$	$\beta_{\rm col}$	$1/\lvert\delta\rvert$	$\lvert\Omega\rvert^2$	$\lvert\psi\rvert^2$
^{7}Li	1.9×10^{-3}	-7.3	6.2×10^{-7}	2.0	1.0
^{23}Na	1.5×10^{-3}	78	1.1×10^{-7}	11	0.88
^{87}Rb	2.6×10^{-3}	25	4.8×10^{-8}	26	1.2

in parentheses in (6) equal to zero. Let us assume that $\omega_{\rm L}$ can be fixed with an accuracy corresponding to the recoil frequency; then the residual detuning is $\delta_{\rm R} = k_{\rm L}^2 w_{\rm L}^2/2$. Even for a narrow beam with a width of the order of several microns, we find $\delta_{\rm R} \sim 30$. For more realistic detunings with $\Delta \geq 10\gamma$, we find $\delta \gtrsim 10^5$; see Table 2.

Such large values of δ make the full three-wave system unsuitable for numerical modeling. We shall therefore proceed by adiabatically eliminating the excited atomic state. The extremely large value of δ justifies this approach. It should also be noted that in [19,20] the collisional term in (5a) was neglected in favor of the dipole–dipole interaction as physical estimates showed that, even for a saturation parameter of only 0.01, the dipole–dipole nonlinearity was dominant. Although the dipole–dipole coefficient is much larger than the collisional coefficient, as can be seen from the last lines of Table 1, the dipole–dipole term is only dominant when the detuning is very small. For sufficiently large detuning, the excited atomic state is depleted and the collisional nonlinearity always dominates; see Fig. 1. We therefore retain both the dipole–dipole and the collisional terms. We consider the dipole–dipole terms and the left-hand side in (5b) as a correction, and after two steps of the adiabatic procedure we find

$$\Phi_{\rm e} = -\frac{s\Omega\Phi_{\rm g}}{\sqrt{|\delta|}}\left[1 - \frac{1}{|\delta|}\left(\frac{s w_{\rm L}^2 m\mu^2}{6\varepsilon_0\hbar^2}|\Phi_{\rm g}|^2 + |\Omega|^2\right) + O\left(\frac{1}{|\delta|^2}\right)\right], \tag{7}$$

where $s = \mathrm{sgn}(\delta)$. Here we have also neglected the nonlinear, diffraction term which is proportional to $(\sigma - 1)$. The coefficient on the right-hand side of (7) has the dimensions of density and can be used to scale the ground-state wavefunction

$$\psi = \Phi_{\rm g}\frac{k_{\rm L} w_{\rm L}^2 \mu}{2\hbar}\sqrt{m/(\varepsilon_0|\delta|)}\,,$$

which gives

$$(\mathrm{i}\sigma\partial_\zeta + \partial_\xi^2 + \partial_\eta^2)\psi = \left[s\left(1 - \frac{|\Omega|^2}{|\delta|}\right)|\Omega|^2 - 2\beta_{\rm dd}|\Omega|^2|\psi|^2 + \beta_{\rm col}|\psi|^2\right]\psi\,, \tag{8a}$$

$$(\mathrm{i}\partial_\zeta + \partial_\xi^2 + \partial_\eta^2)\Omega = \left[s\left(1 - \frac{|\Omega|^2}{|\delta|}\right)|\psi|^2 - \beta_{\rm dd}|\psi|^4\right]\Omega\,. \tag{8b}$$

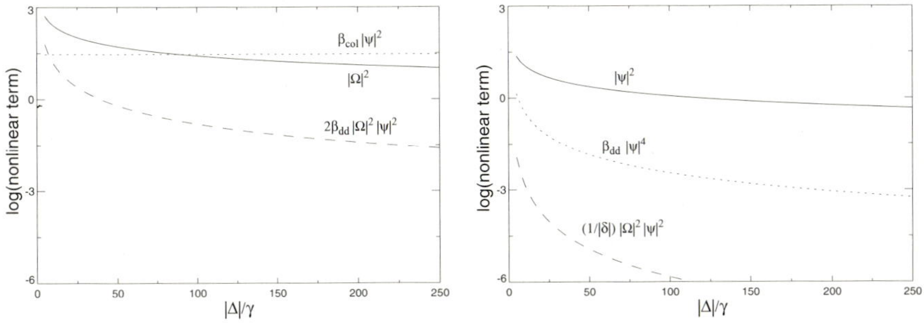

Fig. 1. The Variation of the nonlinear terms in (8a) (*left*) and (8b) (*right*) with detuning. The physical parameters were taken from Table 1 for ^{87}Rb, with atomic density $\bar{N} = 10^{20}\ \mathrm{m}^{-3}$, optical intensity $I = 0.1 I_{\mathrm{sat}}$, $\sigma = 1$, and $w_{\mathrm{L}} = 2 \times 10^{-6}$ m

Here,

$$\beta_{\mathrm{dd}} = \frac{2}{3k_{\mathrm{L}}^2 w_{\mathrm{L}}^2}$$

and

$$\beta_{\mathrm{col}} = \frac{16\pi\varepsilon_0\hbar^2 a_{\mathrm{gg}}|\delta|}{k_{\mathrm{L}}^2 w_{\mathrm{L}}^2 m\mu^2}$$

characterize the strengths of the dipole–dipole and collision interactions relative to the dipole interaction. Characteristic values of the nonlinear coeficients and field strengths are shown in Table 2.

The nonlinear terms in (8) correspond to distinct physical processes. In (8a), for the atomic dynamics, $s|\Omega|^2\psi$ gives light-induced focusing or defocusing due to the dipole interaction; the next term, proportional to $|\Omega|^4$, is a higher-order correction; $-2\beta_{\mathrm{dd}}|\Omega|^2|\psi|^2\psi$ is a focusing nonlinearity (independent of the sign of δ) due to dipole–dipole coupling; and $\beta_{\mathrm{col}}|\psi|^2\psi$ is a focusing or defocusing collisional term, depending on the sign of the scattering length a_{gg}. In (8b), for the optical dynamics, $s|\psi|^2\Omega$ gives a focusing or defocusing effect that is linear in the local atomic density; $-(s/|\delta|)|\Omega|^2|\psi|^2\Omega$ is a focusing or defocusing Kerr nonlinearity that is cubic in the optical field, and $-\beta_{\mathrm{dd}}|\psi|^4\Omega$ is a dipole–dipole correction that gives higher-order focusing. Note that the method used in [19,20] for adiabatic elimination of the excited state failed to provide a result that accounted explicitly for the Kerr nonlinearity of the far-detuned two-level medium.

The nonlinear terms scale with different powers of $|\delta|$, as is seen in Fig. 1.[1] Considering the atomic equation first, the collisional term is independent of δ and so will always dominate for sufficiently large detuning. The dipole term

[1] Figure 1 is plotted in terms of the physical detuning Δ. For the parameter regime shown in the figure and $\sigma = 1$, Δ and δ are proportional to each other.

decays as $|\delta|^{-1}$ and the dipole–dipole term falls off as $|\delta|^{-2}$. For sufficiently high densities, the dipole and dipole–dipole terms may be equal at some finite value of $|\delta|$. The nonlinear terms in the optical equation (8b) scale as $|\delta|^{-1}$, $|\delta|^{-2}$, and $|\delta|^{-3}$ for the density-dependent term, the dipole–dipole correction, and the Kerr nonlinearity, respectively. We see that the Kerr nonlinearity is much weaker than any other term for all values of $|\delta|$ within the regime of applicability of the adiabatic elimination.

Equations (8) can be used to describe a number of different physical situations. Assuming $|\psi| =$ constant, (8b) reduces to the NLS equation used to describe propagation of paraxial beams in media with a Kerr nonlinearity. Assuming $|\Omega| =$ constant and $\beta_{\mathrm{dd}} = \beta_{\mathrm{col}} = 0$, (8a) reduces to an equation describing trapping and scattering of atoms by a far-detuned optical field. Assuming $|\Omega| =$ constant and $\beta_{\mathrm{dd}} \neq 0$, we arrive at so-called "nonlinear atom optics" [14]. And, finally, for $\Omega = 0$ and $\beta_{\mathrm{col}} \neq 0$, we recover the paraxial version of the stationary GP equation describing Bose–Einstein condensates.

Equations (8) conserve the normalized power of the optical beam and the number of atoms:

$$\int \mathrm{d}\xi\,\mathrm{d}\eta\,|\Omega|^2 = \frac{k_{\mathrm{L}}^2 w_{\mathrm{L}}^2 \sigma^2 \mu^2}{\varepsilon_0 \hbar^2 c |\delta| v_{\mathrm{a}}^2}\,P\,, \tag{9}$$

$$\int \mathrm{d}\xi\,\mathrm{d}\eta\,|\psi|^2 = \frac{k_{\mathrm{L}}^2 w_{\mathrm{L}}^2 m \mu^2}{2\varepsilon_0 \hbar^2 |\delta| v_{\mathrm{a}}}\,F\,, \tag{10}$$

where P is the optical-beam power in watts and F is the atomic-beam flux in atoms per second. The average atomic density is given by $\bar{N} = F/(v_{\mathrm{a}}A)$, where A is the cross-sectional area of the beam.

3 Modulational Instability

Before proceeding with analysis of the set (8), we recognize that the Kerr nonlinearity can be safely neglected for the physical parameters of interest (see Table 2). In the remainder of this chapter, we shall therefore use the equations

$$(\mathrm{i}\sigma\partial_\zeta + \partial_\xi^2 + \partial_\eta^2)\psi = \left(s|\Omega|^2 - 2\beta_{\mathrm{dd}}|\Omega|^2|\psi|^2 + \beta_{\mathrm{col}}|\psi|^2\right)\psi\,, \tag{11a}$$

$$(\mathrm{i}\partial_\zeta + \partial_\xi^2 + \partial_\eta^2)\Omega = \left(s|\psi|^2 - \beta_{\mathrm{dd}}|\psi|^4\right)\Omega\,. \tag{11b}$$

The simplest solutions of (11) are plane waves, which can be sought in the form $\psi = \psi_0 \mathrm{e}^{\mathrm{i}\kappa_\psi \zeta}$, $\Omega = \Omega_0 \mathrm{e}^{\mathrm{i}\kappa_\Omega \zeta}$, where ψ_0, κ_ψ, Ω_0, and κ_Ω are real constants. To study the stability of these solutions, we perturb them by the weak periodic modulations

$$\begin{aligned}\psi &= \{\psi_0 + [U_\psi(\zeta) + \mathrm{i}W_\psi(\zeta)]\cos \boldsymbol{k}\cdot\boldsymbol{r}\}\,\mathrm{e}^{\mathrm{i}\kappa_\psi\zeta}\,,\\ \Omega &= \{\Omega_0 + [U_\Omega(\zeta) + \mathrm{i}W_\Omega(\zeta)]\cos \boldsymbol{k}\cdot\boldsymbol{r}\}\,\mathrm{e}^{\mathrm{i}\kappa_\Omega\zeta}\,,\end{aligned} \tag{12}$$

where $\boldsymbol{k}$ is the wavevector of a Fourier component of the perturbation, and $\boldsymbol{k}\cdot\boldsymbol{r} = k_\xi\xi + k_\eta\eta$. Substituting (12) into (11) and separating real and imaginary parts of the linearized equations, we arrive at the linear system

$$\partial_\zeta \varepsilon = \hat{L}(|\boldsymbol{k}|^2)\varepsilon \ , \ \varepsilon = (U_\psi, W_\psi, U_\Omega, W_\Omega)^{\mathrm{T}} \ , \tag{13}$$

where $\hat{L}$ is a 4×4 real matrix. The presence of any eigenvalue $\lambda(|\boldsymbol{k}|^2)$ of $\hat{L}$ (such that $\hat{L}\boldsymbol{e} = \lambda\boldsymbol{e}$) with a positive real part implies instability. We have found that the plane wave solutions are modulationally unstable for a wide range of experimentally relevant parameters. Details of this study will be presented elsewhere. Here we shall restrict ourselves to showing some results of a numerical simulation of (11) with realistic initial conditions in the form of spatially bounded atomic and optical beams. If the area of the beams is large enough, then modulational instability develops in a fashion qualitatively similar to the case of spatially unbounded plane wave solutions.

For $\beta_{\mathrm{col}} < 0$ the condensate is modulationally unstable even without an optical field. It breaks into a set of filaments, which collapse with increasing ζ. This collapse is due to the critical character of the cubic nonlinearity in propagation with two transverse dimensions [25]. However, it is clear that our model can correctly predict only the initial stage of filamentation, not the collapse itself. This is due to the fact that the local atomic density inside a filament becomes so high that three-body and higher-order collisions become important, leading to the necessity of taking into account quintic and higher nonlinearities of the ground-state wavefunctions.

The effect of the optical field exhibits itself in the most striking way for $\beta_{\mathrm{col}} > 0$, when in its absence the condensate is modulationally stable. In Figs. 2–4 we show numerically calculated dynamics of the coupled atomic

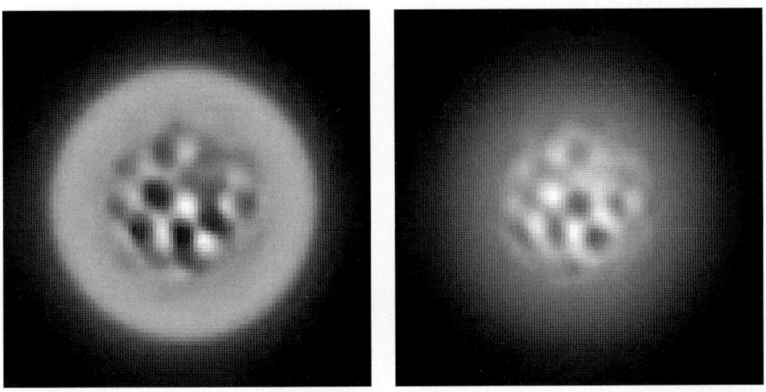

Fig. 2. Interleaved filamentation pattern arising as a result of modulational instability of the coupled atomic (*left*) and optical (*right*) beams at $\zeta = 0.5$, for $s = 1$. The parameters for the numerical simulation were $\psi(\zeta = 0) = 4\mathrm{e}^{-(\xi^2+\eta^2)/6}$, $\Omega(\zeta = 0) = 6\mathrm{e}^{-(\xi^2+\eta^2)/6}$, $\beta_{\mathrm{dd}} = 1.6 \times 10^{-4}$, and $\beta_{\mathrm{col}} = 3.5$. The size of the computational window was 25, and the initial noise level at $\zeta = 0$ was 1%

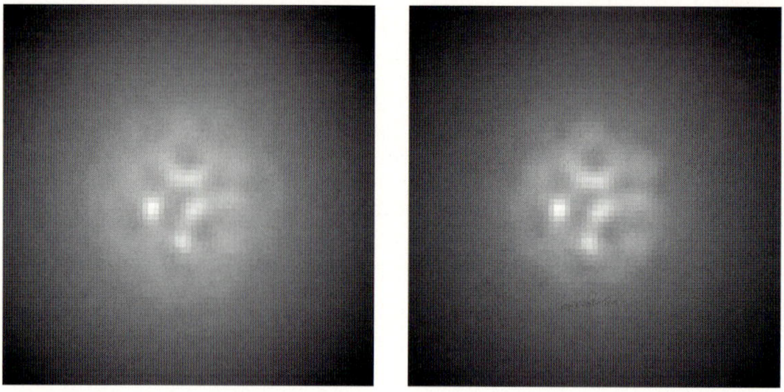

Fig. 3. Pattern of coincident filaments arising as a result of modulational instability of the coupled atomic (*left*) and optical (*right*) beams at $\zeta = 0.5$, for $s = -1$. Initial conditions for numerical simulation: $\psi(\zeta = 0) = 4\mathrm{e}^{-(\xi^2+\eta^2)/8}$, $\Omega(\zeta = 0) = 6\mathrm{e}^{-(\xi^2+\eta^2)/8}$. The size of the computational window was 25 units. Initial noise 1%

and optical beams. The values of β_{dd} and β_{col} given in the caption of Fig. 2 were also used in the simulations shown in Figs. 3–6. We have found that when the optical intensity exceeds some critical value, both beams develop filamentary structures; see Figs. 2 and 3. For unbounded plane wave solutions, the threshold effect is absent, i.e. coupled waves are always unstable.

The character of the filamentary pattern depends dramatically on the sign of detuning, s. If the dipole interaction induces a defocusing nonlinearity and $s = 1$, then the maxima of the atomic field in the filamentary pattern coincide with the minima of the optical field and vice versa; see Fig. 2. If, however, $s = -1$, i.e. the dipole nonlinearity is focusing, then the maxima of the atomic and optical fields coincide, as do as their minima; see Fig. 3.

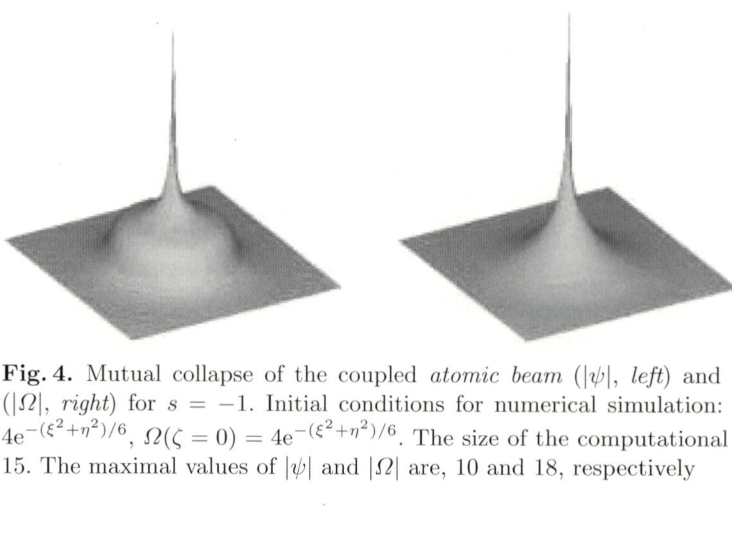

Fig. 4. Mutual collapse of the coupled *atomic beam* ($|\psi|$, *left*) and *optical beam* ($|\Omega|$, *right*) for $s = -1$. Initial conditions for numerical simulation: $\psi(\zeta = 0) = 4\mathrm{e}^{-(\xi^2+\eta^2)/6}$, $\Omega(\zeta = 0) = 4\mathrm{e}^{-(\xi^2+\eta^2)/6}$. The size of the computational window was 15. The maximal values of $|\psi|$ and $|\Omega|$ are, 10 and 18, respectively

This effect is similar to the modulational instability of coupled orthogonally polarized optical waves [26]. The major differences between the case under consideration and the purely optical case are, first, that the focusing optical Kerr nonlinearity corresponds to the defocusing dipole nonlinearity and, second, that in the optical case, modulational instability is significantly affected by nonlinear self-action, while here nonlinear cross-coupling is the strongly dominating process. For $s = -1$ the filaments collapse upon propagation, because dipole and dipole–dipole interactions lead to strong focusing effects which cannot be outweighed by the defocusing due to collisions ($\beta_{\rm col} > 0$) and the optical Kerr nonlinearity. Decreasing the transverse size of the atomic and optical beams leads to the collapse of the beams as whole without formation of multiple filaments, as shown in Fig. 4. Collapse can be avoided providing non-locality of the dipole–dipole and two-body interactions are taken into account.

The difference between the filamentary patterns appearing for $s = 1$ and $s = -1$ can be interpreted on the basis of the symmetries of the system; see [26] for more details. Modulational instability can be understood in terms of continuation of the neutral eigenvectors (i.e. those with zero eigenvalues) of $\hat{L}(|\boldsymbol{k}|^2 = 0)$ into the region of nonzero $|\boldsymbol{k}|$. The neutral eigenvectors can be generated by applying the infinitesimal phase transformations

$$(\psi, \Omega) \to (\psi e^{i\alpha}, \Omega e^{i\alpha}), \ (\psi, \Omega) \to (\psi e^{i\phi}, \Omega e^{-i\phi}) \,. \tag{14}$$

Infinitesimal variations of α and ϕ generate eigenvectors $\boldsymbol{e}_\alpha = (0, \psi_0, 0, \Omega_0)^{\rm T}$ and $\boldsymbol{e}_\phi = (0, \psi_0, 0, -\Omega_0)^{\rm T}$. It is clear that growth of perturbations along the direction in the phase space determined by the continuation of $\boldsymbol{e}_\phi$ generates a pattern with interleaving minima and maxima of the two fields, and this scenario is realized for $s = 1$. Conversely, continuation of $\boldsymbol{e}_\alpha$ becomes unstable for $s = -1$, and a filamentary pattern with coincident peaks appears.

4 Solitons and Vortices

The coupled equations (11) have a variety of different types of solitary solutions. This topic is still at the very beginning of its study, and here we shall merely enumerate some possible solutions and present numerical results verifying the existence of two of them.

The simplest diffraction-free solitary solutions can be sought in the form

$$\psi = \psi_0(\varrho) e^{i\kappa_\psi \zeta + i m_\psi \theta} \,, \quad \Omega = \Omega_0(\varrho) e^{i\kappa_\Omega \zeta + i m_\Omega \theta} \,, \tag{15}$$

where $\varrho = |\xi + i\eta|$ is the distance from the origin and $\theta = \arg(\xi + i\eta)$ is the polar angle. The amplitudes $\psi_0(\varrho)$, $\Omega_0(\varrho)$ now obey the stationary versions of (11), with $i\partial_\zeta$ replaced by $-\kappa_\psi$ and $-\kappa_\Omega$, respectively; m_ψ and m_Ω are integer numbers indicating the orders of the phase singularities at the beam center. To fully specify solutions of the ordinary differential equations for

Fig. 5. Atomic vortex (*left*) guiding a bright optical soliton (*right*); $s = 1$. The size of the computational window was 15. The maximal values of $|\psi|$ and $|\Omega|$ are, 1.45 and 2.4, respectively

$\psi_0(\varrho)$ and $\Omega_0(\varrho)$, one has to supplement them by boundary conditions. The boundary conditions at the origin are determined by the values of $m_{\psi,\Omega}$. If m is different from zero, then the corresponding amplitude has to be zero at the origin to provide continuity of the solution. If m is zero, then the corresponding amplitude has a zero derivative at the origin. As $\varrho \to \infty$, the amplitudes ψ_0 and Ω_0 should tend either to zero or to one of the plane wave solutions.

To gain intuitive insight into the problem of the existence of coupled light–matter solitons, it is useful to use the analogy between solitary waves and linear waveguides. For $\Omega = 0$ and $\beta_{\text{col}} < 0$, two-body collisions lead to the possibility of the existence of purely atomic bright solitons and of ring-like solutions with a nested phase dislocation (vortex) [27]. Atomic vortices located on an infinite background [28] exist for $\Omega = 0$ and $\beta_{\text{col}} > 0$. The density distribution created by these solutions can be considered as an effective potential for the optical field. In the case of the vortex on an infinite background, the potential is attractive for $s = 1$, and in the case of solitons on a zero background, it is attractive for $s = -1$. Thus, in these two cases, the atomic solitons and vortices can guide bright optical beams. An example of such a configuration is shown in Fig. 5. Preliminary numerical simulations indicate that the structure shown in Fig. 5 is dynamically stable.

It is clear that bright–bright solitons that exist for $\beta_{\text{col}} < 0$ and $s = -1$ are unstable because the focusing collisional nonlinearity is enhanced by the dipole and dipole–dipole interactions, and the small defocusing Kerr term cannot counterbalance it. However, bright–bright solitons can also be found for $\beta_{\text{col}} > 0$ and $s = -1$; see Fig. 6. In this case the growth rate of the instability is significantly weakened because the collisional nonlinearity is defocusing.

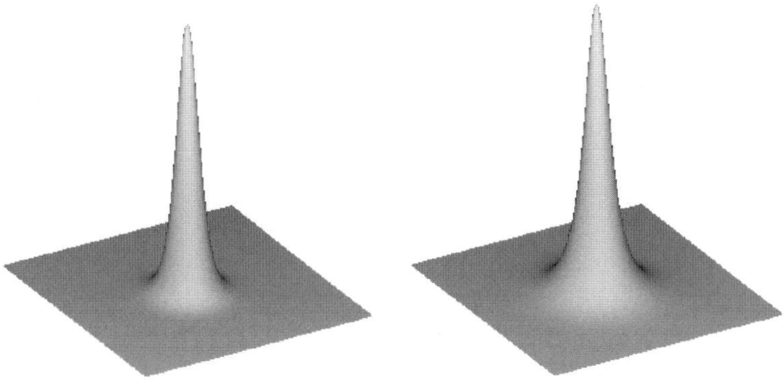

Fig. 6. Coupled bright atomic (*left*) and optical (*right*) solitons; $s = -1$, the other parameters are those for Rb. The size of the computational window was 15. The maximal values of $|\psi|$ and $|\Omega|$ are, 5.5 and 2.6, respectively

5 Discussion

The results presented in the preceding sections demonstrate that a rich variety of spatial structures may appear in the coupled propagation of optical and atomic waves. Although the filamentation instability is deleterious to the transmission of well-collimated beams with smooth wavefronts, spatial instabilities are potentially useful for splitting and combining optical and atomic waves. As a first step towards applications, it is necessary to estimate physical parameters for which the structures shown in Figs. 2–6 could be observed. A detailed discussion will be given elsewhere. Our aim here is to indicate characteristic physical values for these phenomena.

The physical propagation lengths corresponding to the calculated filamentation patterns can be estimated as follows. Assuming Rb atoms, the nonlinear coefficients used in Fig. 2 correspond to $w_{\mathrm{L}} = 8$ μm and $\Delta/\gamma = 14$. The physical box size in Fig. 2 is then 140 μm on each side, and the physical propagation length is $z \simeq 0.26$ mm. The intensity at the center of the Gaussian is $I_{\mathrm{peak}}/I_{\mathrm{sat}} \simeq 1.4 \times 10^{-2}$, and the peak density is $N_{\mathrm{peak}} \simeq 1.4 \times 10^{20}\ \mathrm{m}^{-3}$. The transverse scale of the filamentation is of order 8 μm. These values indicate that observation of filamentation patterns should be possible with current atom laser sources.

Let us estimate also the spatial size of the coupled atomic and optical solitons shown in Fig. 6. Using again parameters for Rb, $w_{\mathrm{L}} = 19$ μm, and $\Delta/\gamma = 14$, we obtain a physical box size of 85 μm on each side, $I_{\mathrm{peak}}/I_{\mathrm{sat}} \simeq 2.6 \times 10^{-3}$, and $N_{\mathrm{peak}} \simeq 2.6 \times 10^{20}\ \mathrm{m}^{-3}$. Under these conditions the solitons have a transverse diameter of order 15 μm. The corresponding linear diffraction length is of order 2 mm; so unambiguous observation of solitary propagation would require an interaction length of at least half a centimeter. This would require a continuous atom laser source lasting about one second, for a beam propagation velocity equal to the recoil velocity. The atom-laser source described in [9] provided pulses of up to 100 ms duration. We can conclude from the above estimates that observation of solitary propagation is not terribly far from the limits set by current atom laser experiments.

In conclusion, the equations derived and studied here provide a basis for investigating the coupled spatial dynamics of coherent optical and atomic waves. Depending on the physical parameters in force, in particular the amount of detuning between the optical field and the atomic transition, different nonlinear effects dominate. For small detunings and low atomic densities, the dipole atom–field coupling dominates, while for large detunings and high densities, the two-body collisional nonlinearity is dominant. Local-field effects give a dipole–dipole nonlinearity, which appears as a correction to the above terms. Modulational instability is observed for wide beams. The nonlinear stage of the modulational instability itself or localized initial conditions lead to a rich variety of possible solitary structures. There is no doubt that much remains to be discovered in the course of exploring the possibilities inherent in the coupled dynamics of optical and atomic fields.

References

1. G.A. Askar'yan, Effects of the gradient of a strong electromagnetic beam on electrons and atoms, Zh. Éksp. Teor. Fiz. **42**, 1567–1570 (1962) [Sov. Phys. JETP **15**, 1088–1090 (1962)].
2. G.A. Swartzlander, Jr., D.R. Andersen, J.J. Regan, H. Yin, A.E. Kaplan, Spatial dark-soliton stripes and grids in self-defocusing materials, Phys. Rev. Lett. **66**, 1583–1586 (1991); erratum, Phys. Rev. Lett. **66**, 3321 (1991).
3. M. Segev, G.C. Valley, B. Crosignani, P. Di Porto, A. Yariv, Steady-state spatial screening solitons in photorefractive media with external applied field, Phys. Rev. Lett. **73**, 3211–3214 (1994); A.A. Zozulya, D.Z. Anderson, Propagation of an optical beam in a photorefractive medium in the presence of a photogalvanic nonlinearity or an externally applied electric field, Phys. Rev. A **51**, 1520–1531 (1995).
4. A.P. Kazantsev, G.I. Surdutovich, V.P. Yakovlev, *Mechanical Action of Light on Atoms*, (World Scientific, Singapore, 1990).
5. C.S. Adams, M. Sigel, J. Mlynek, Atom optics, Phys. Rep. **240**, 143–210 (1994).
6. J.E. Bjorkholm, R.R. Freeman, A. Ashkin, D.B. Pearson, Observation of focusing of neutral atoms by the dipole forces of resonance-radiation pressure, Phys. Rev. Lett. **41**, 1361–1364 (1978).
7. O. Carnal, J. Mlynek, Young's double-slit experiment with atoms: a simple atom interferometer, Phys. Rev. Lett. **66**, 2689–2692 (1991).
8. M.-O. Mewes, M.R. Andrews, D.M. Kurn, D.S. Durfee, C.G. Townsend, W. Ketterle, Output coupler for Bose–Einstein condensed atoms Phys. Rev. Lett. **78**, 582–585 (1997); M.R. Andrews, C.G. Townsend, H.-J. Miesner, D.S. Durfee, D.M. Kurn, W. Ketterle, Observation of interference between two Bose condensates, Science **275**, 637-641 (1997).
9. I. Bloch, T.W. Hänsch, T. Esslinger, Atom laser with a cw output coupler, Phys. Rev. Lett. **82**, 3008–3011 (1999).

10. E.W. Hagley, L. Deng, M. Kozuma, J. Wen, K. Helmerson, S.L. Rolston, W.D. Phillips, A well-collimated quasi-continuous atom laser, Science **283**, 1706–1709 (1999).
11. A.F. Bernhardt, B.W. Shore, Coherent atomic deflection by resonant standing waves, Phys. Rev. A **23**, 1290–1301 (1981); L. Mandel, E. Wolf, *Optical Coherence and Quantum Optics*, (Cambridge University Press, Cambridge, 1995) Chap. 15.
12. J. Javanainen, Spectrum of light scattered from a degenerate Bose gas, Phys. Rev. Lett. **75**, 1927–1930 (1995).
13. G. Lenz, P. Meystre, E.M. Wright, Nonlinear atom optics, Phys. Rev. Lett. **71**, 3271–3274 (1993); G. Lenz, P. Meystre, E.M. Wright, Nonlinear atom optics: general formalism and atomic solitons, Phys. Rev. A **50**, 1681–1691 (1994).
14. W. Zhang, D.F. Walls, B.C. Saunders, Atomic soliton in a traveling wave laser beam, Phys. Rev. Lett. **72**, 60–63 (1994).
15. W. Zhang, D.F. Walls, Quantum field theory of interaction of ultracold atoms with a light wave: Bragg scattering in nonlinear atom optics, Phys. Rev. A **49**, 3799–3813 (1994).
16. F. Dalfovo, S. Giorgini, L.P. Pitaevskii, S. Stringari, Theory of Bose–Einstein condensation in trapped gases, Rev. Mod. Phys. **71**, 463–512 (1999).
17. Y. Castin, K. Mølmer, Maxwell–Bloch equations: A unified view of nonlinear optics and nonlinear atom optics, Phys. Rev. A **51**, R3426–R3428 (1995).
18. O. Morice, Y. Castin, J. Dalibard, Refractive index of a dilute Bose gas, Phys. Rev. A **51**, 3896–3901 (1995).
19. H. Wallis, Mean field for a dilute Bose gas in the light field of an evanescent wave, Phys. Rev. A **56**, 2060–2067 (1997).
20. K.V. Krutitsky, F. Burgbacher, J. Audretsch, Local-field approach to the interaction of an ultracold dense Bose gas with a light field, Phys. Rev. A **59**, 1517–1527 (1999).
21. W. Zhang, B.C. Sanders, W. Tan, Self-trapping and self-focusing of a coherent atomic beam, Phys. Rev. A **56**, 1433–1437 (1997).
22. Yu.L. Klimontovich, S.N. Luzgin, Possibility of combined self-focusing of atomic and light beams, Pis'ma Zh. Eksp. Teor. Fiz. **30**, 645–647 (1979) [JETP Lett. **30**, 610–612 (1979)].
23. M. Saffman, Self-induced dipole force and filamentation instability of a matter wave, Phys. Rev. Lett. **81**, 65–68 (1998).
24. C. Kittel, *Introduction to Solid State Physics*, 7th ed., (Wiley, New York, 1996), Chap. 13.
25. E.A. Kuznetsov, A.M. Rubenchik, V.E. Zakharov, Soliton stability in plasmas and hydrodynamics, Phys. Rep. **142**, 103–165 (1986).
26. D.V. Skryabin, W.J. Firth, Modulational instability of bright solitary waves in incoherently coupled nonlinear Schrödinger equations, Phys. Rev. E **60**, 1019–1029 (1999); erratum **60**, 6242 (1999).
27. D.V. Skryabin, W.J. Firth, Dynamics of self-trapped beams with phase dislocation in saturable Kerr and quadratic nonlinear media, Phys. Rev. E **58**, 3916–3930 (1998).
28. L.P. Pitaevskii, Vortex lines in an imperfect Bose gas, Zh. Éksp. Teor. Fiz. **40**, 646–651 (1961). [Sov. Phys. JETP **13**, 451–454 (1961)].

Index

Springer Series in
OPTICAL SCIENCES

New editions of volumes prior to volume 60

1 **Solid-State Laser Engineering**
By W. Koechner, 5th revised and updated ed. 1999, 472 figs., 55 tabs., XII, 746 pages

14 **Laser Crystals**
Their Physics and Properties
By A. A. Kaminskii, 2nd ed. 1990, 89 figs., 56 tabs., XVI, 456 pages

15 **X-Ray Spectroscopy**
An Introduction
By B. K. Agarwal, 2nd ed. 1991, 239 figs., XV, 419 pages

36 **Transmission Electron Microscopy**
Physics of Image Formation and Microanalysis
By L. Reimer, 4th ed. 1997, 273 figs. XVI, 584 pages

45 **Scanning Electron Microscopy**
Physics of Image Formation and Microanalysis
By L. Reimer, 2nd completely revised and updated ed. 1998,
260 figs., XIV, 527 pages

Published titles since volume 60

60 **Holographic Interferometry in Experimental Mechanics**
By Yu. I. Ostrovsky, V. P. Shchepinov, V. V. Yakovlev, 1991, 167 figs., IX, 248 pages

61 **Millimetre and Submillimetre Wavelength Lasers**
A Handbook of cw Measurements
By N. G. Douglas, 1989, 15 figs., IX, 278 pages

62 **Photoacoustic and Photothermal Phenomena II**
Proceedings of the 6th International Topical Meeting, Baltimore, Maryland,
July 31 - August 3, 1989
By J. C. Murphy, J. W. Maclachlan Spicer, L. C. Aamodt, B. S. H. Royce (Eds.),
1990, 389 figs., 23 tabs., XXI, 545 pages

63 **Electron Energy Loss Spectrometers**
The Technology of High Performance
By H. Ibach, 1991, 103 figs., VIII, 178 pages

64 **Handbook of Nonlinear Optical Crystals**
By V. G. Dmitriev, G. G. Gurzadyan, D. N. Nikogosyan,
3rd revised ed. 1999, 39 figs., XVIII, 413 pages

65 **High-Power Dye Lasers**
By F. J. Duarte (Ed.), 1991, 93 figs., XIII, 252 pages

66 **Silver-Halide Recording Materials**
for Holography and Their Processing
By H. I. Bjelkhagen, 2nd ed. 1995, 64 figs., XX, 440 pages

67 **X-Ray Microscopy III**
Proceedings of the Third International Conference, London, September 3-7, 1990
By A. G. Michette, G. R. Morrison, C. J. Buckley (Eds.), 1992, 359 figs., XVI, 491 pages

68 **Holographic Interferometry**
Principles and Methods
By P. K. Rastogi (Ed.), 1994, 178 figs., 3 in color, XIII, 328 pages

69 **Photoacoustic and Photothermal Phenomena III**
Proceedings of the 7th International Topical Meeting, Doorwerth, The Netherlands,
August 26-30, 1991
By D. Bicanic (Ed.), 1992, 501 figs., XXVIII, 731 pages

Springer Series in
OPTICAL SCIENCES

70 **Electron Holography**
By A. Tonomura, 2nd, enlarged ed. 1999, 127 figs., XII, 162 pages

71 **Energy-Filtering Transmission Electron Microscopy**
By L. Reimer (Ed.), 1995, 199 figs., XIV, 424 pages

72 **Nonlinear Optical Effects and Materials**
By P. Günter (Ed.), 2000, 174 figs., 43 tabs., XIV, 540 pages

73 **Evanescent Waves**
From Newtonian Optics to Atomic Optics
By F. de Fornel, 2001, 277 figs., XVIII, 268 pages

74 **International Trends in Optics and Photonics**
ICO IV
By T. Asakura (Ed.), 1999, 190 figs., 14 tabs., XX, 426 pages

75 **Advanced Optical Imaging Theory**
By M. Gu, 2000, 93 figs., XII, 214 pages

76 **Holographic Data Storage**
By H.J. Coufal, D. Psaltis, G.T. Sincerbox (Eds.), 2000
228 figs., 64 in color, 12 tabs., XXVI, 486 pages

77 **Solid-State Lasers for Materials Processing**
Fundamental Relations and Technical Realizations
By R. Iffländer, 2001, 230 figs., 73 tabs., XVIII, 350 pages

78 **Holography**
The First 50 Years
By J.-M. Fournier (Ed.), 2001, 266 figs., XII, 460 pages

79 **Mathematical Methods of Quantum Optics**
By R.R. Puri, 2001, 13 figs., XIV, 285 pages

80 **Optical Properties of Photonic Crystals**
By K. Sakoda, 2001, 95 figs., 28 tabs., XII, 223 pages

81 **Photonic Analog-to-Digital Conversion**
By B.L. Shoop, 2001, 259 figs., 11 tabs., XIV, 330 pages

82 **Spatial Solitons**
By S. Trillo, W.E. Torruellas (Eds), 2001, 194 figs., 7 tabs., XX, 454 pages

83 **Nonimaging Fresnel Lenses**
Design and Performance of Solar Concentrators
By R. Leutz, A. Suzuki, 2001, 139 figs., 44 tabs., XII, 272 pages

84 **Nano-Optics**
By S. Kawata, M. Ohtsu, M. Irie (Eds.), 2002, 258 figs., 2 tabs., XIII, 313 pages

Printing (Computer to Film): Saladruck Berlin
Binding: Stürtz AG, Würzburg